A Geography of Ohio

Lake Erie
Williams
Fulton
Lucas
Ottawa
Wood
Henry
Defiance
Sandusky
Erie
Lorain
Cuyahoga
Lake
Geauga
Ashtabula
Trumbull
Portage
Summit
Medina
Huron
Seneca
Hancock
Putnam
Paulding
Van Wert
Allen
Hardin
Wyandot
Crawford
Richland
Ashland
Wayne
Stark
Mahoning
Columbiana
Mercer
Auglaize
Marion
Morrow
Holmes
Carroll
Tuscarawas
Knox
Logan
Union
Shelby
Jefferson
Coshocton
Delaware
Darke
Harrison
Champaign
Licking
Miami
Franklin
Muskingum
Guernsey
Belmont
Clark
Madison
Preble
Montgomery
Fairfield
Perry
Noble
Monroe
Greene
Pickaway
Morgan
Fayette
Hocking
Washington
Butler
Warren
Clinton
Ross
Athens
Vinton
Hamilton
Highland
Clermont
Pike
Meigs
Jackson
Brown
Adams
Gallia
Scioto
Lawrence
OHIO
RIVER
OHIO
100 miles
The Laboratory for Cartographic and Spatial Analysis, The University of Akron, 1992

A Geography of Ohio

Edited by
Leonard Peacefull

The Kent State University Press
Kent, Ohio, and London, England

Library of Congress Catalog Card Number 95-40344
ISBN 0-87338-525-X
Manufactured in the United States of America

05 04 03 02 01 00 99 98 97 96 5 4 3 2 1

Revised edition, originally published as *The Changing Heartland: A Geography of Ohio,* © 1990 by Ginn Press. Printed with permission.

Library of Congress Cataloging-in-Publication Data

A geography of Ohio / edited by Leonard Peacefull.
p. cm.
"Revised edition, originally published as The changing heartland : a geography of Ohio, c1990 by Ginn Press"—T.p. verso.
Includes bibliographical references and index.
ISBN 0-87338-525-X (alk. paper)
1. Ohio—Geography. I. Peacefull, Leonard. II. Changing heartland.
F491.8G46 1996
917.71—dc20 95-40344
CIP

British Library Cataloging-in-Publication data are available.

Contents

Introduction

Leonard Peacefull

As we near the end of the twentieth century it is time to reflect upon all that has happened over the past century and how events have contrived to produce a new geography of Ohio. In this time the state has undergone a vast transformation. Once the preeminent manufacturing heart of the United States, the engine room that drove the nation's economy, it no longer is the center of manufacturing. Formerly proud industrial cities now have little or no manufacturing and the current economic base is geared toward service industries. Ohio has become the home office of multinational companies that continue to manufacture outside of Ohio and the United States. These changes have affected the people and impacted the land; consequently they have altered the geography of the state. Change has come about most rapidly in the last twenty-five years. The social upheavals of the 1960s and the several oil crises of the 1970s that eventually led to the economic recessions of the 1980s are events that contributed to the demise of industry in many towns throughout the state. These events had great effect on the people of Ohio. For many, their lifestyles changed. Some were in different jobs and a great number moved location—out of the city and into the rural areas. Most importantly peoples' attitudes changed: they had more leisure time and were able to travel around the state, and they became aware of the environment and began to use the state's resources in new and more conservational ways. Consequently, there is a new and different geography of Ohio resulting from the metamorphosis taking place in the closing decades of the twentieth century.

This book will address the new geography in a framework that is uniquely Ohio. The themes pursued are divided into four sections, or landscapes: physical, cultural, urban, and economic. Each chapter is written by an expert in that area. Within some of the chapters are short, complementary essays, or vignettes, on special topics. The first section sets the framework for our geographic treatise, providing a foundation for the later chapters. Following chapter 1's discussion of the shape of the land and its geologic history, chapter 2 describes the varying climates of Ohio. This very useful chapter takes an in-depth look at tornadoes and their effect on the state. Next, because Ohio is very much an agricultural state, with its soils playing an important part in the economy, chapter 3 takes the reader on a tour of the various soil types in Ohio, briefly illustrating their formation and use. The physical landscape section ends with a description of the state's natural resources, which have been largely responsible for Ohio's industrial development. A knowledge of the shape of the land, the climate that affects it, and the resources available impacts on the decisions that humans make about the activities they undertake on the land that is Ohio.

With the knowledge gained in the first section the reader can appreciate the developments of human activity so richly described, first in chapter 5, on the early aboriginal peoples, and then in chapter 6, on the cultural influences of the European occupation. One feature distinct to each of the various groups that came to Ohio is their individual architecture, illustrated in a short vignette at the end of the latter chapter. Early industrial and economic development is covered in chapter 7, and in chapter 8, the last chapter of this section, the present population, including Ohio's many and varied ethnic patterns, is discussed.

The third section of the book deals with the urban landscape. Chapter 9 looks at large cities, taking first a general overview of current urban growth in Ohio before concluding with seven vignettes on the larger cities. Each vignette examines the changes that have taken place or are taking place in our state today and describes how a particular city has coped with these changes, especially the demise of the manufacturing base that for many years formed the economic cornerstone of many of these cities. Next, chapter 10 describes the development of the small town in Ohio—their spatial patterns and the factors that have influenced their growth. Of particular note is a section within this chapter dealing with the origin and distribution of Ohio place-names.

The economic section of the book begins with an examination of energy production and consumption. Chapter 11 concentrates on the production and distribution of electrical energy, so important to the state's economy that few activities could survive without it. The following chapter takes a thoughtful look at current agricultural activity, examining first the essential ingredients for agriculture, then the changes that have occurred in the industry, and finally regional variations in agricultural practices within the state. Chapter 13, dealing with manufacturing, closely scrutinizes recent changes in the manufacturing sector. To illustrate current practices in manufacturing, this chapter includes detailed examinations of three industries: automobiles, soft drinks, and steel. Recognizing that we could not move around the state or sell our manufactured goods without a transportation infrastructure, chapter 14 gives an account of the various modes of transport available to Ohioans—roads, railroads, airways, and, included for good measure, bicycles. The penultimate chapter chronicles one of the most rapidly growing industries in the state: recreation and tourism. Here the motivation for people to recreate and the various forms that recreation can take are considered in some detail.

This work concludes with a description of one particular area within the state: the southeast. Encapsulated here are examples of the changes affecting the rest of the state—a declining industrial base, an increase in leisure activities, and in-migration of city workers looking for rural domeciles—that together are creating a new economic base for the future. The southeast is awakening to the twenty-first century, just as is the rest of the state. In that new century there will be new challenges, new lifestyles, and different ways of earning a living in this national heartland. A new geography of the state is emerging, one that is reflected in chapters of this book.

CHAPTER ONE

The Land

Leonard Peacefull

The Ohio landscape contains a great variety of features that combine to create a fascinating and diverse geography. From the undulating hills of the southeast to the distinctly flat lake plain of the northwest, the topography constantly changes. A traveler crossing the state will not be confronted with the same aspect for long. These changes result from a geological history that has seen many modifications in the shape and position of the land.

The Landform Regions

Ohio has a distinct advantage over its neighboring states in that it lies across two of the main physical regions of North America: the Appalachian system in the East, and the Central Lowlands drained by the Mississippi. The eastern side of the state lies firmly on the Appalachian Plateau, where Paleozoic formations retain almost the same undisturbed bedding as they had before the upthrust (Paterson 1984). The western side of the state is part of the Central Lowlands—Mississippi–Great Lakes section, not flat by any means but possessing less of a roller-coaster appearance than the east. When compared to the undulations in other states, however, Ohio, with its elevation varying little more than 700 feet from the plateau to the lakes, is relatively flat. This flatness is evident in western Ohio, but the eastern half of the state presents a distinctly different terrain. These contrasts are the result of the controlling influences of the underlying rocks and the effects of the last ice age. Besides the distinct contrast between the northwest and the eastern parts of Ohio there are other variations to be found in the different landform regions within the state (Fig. 1.1). Of the many physiographic regions of the eastern United States identified by Fenneman, three are located in Ohio. These are the Central Lowland Province, the Appalachian Plateau Province, and the Interior Low Plateau Province (Fenneman 1938).

The western half of Ohio lies within the Central Lowland Province, a province that has been subjected to intense glacial activity for much of the last half-million years. Here the landscape was fashioned by ice sheets that scoured and scraped, irreversibly changing the surface before they melted, leaving behind the material they had been carrying. Two distinct subregions can be recognized within this province: the *Till Plains,* formed by glacial deposition; and the *Lake Plain,* formed by glacial meltwater.

The eastern half of the state is part of the Appalachian Plateau Province. Although this region presents a fairly uniform plateau surface deeply dissected by rivers, the northern and western section also experienced glaciation, which further altered surface features. Thus this region can be subdivided into the *unglaciated plateau* and the *glaciated plateau.*

The last of the three physiographic regions Fenneman identified in Ohio is the Interior Low Plateau, also called the Lexington Plain. Only a small fragment of this region, which extends deep into Kentucky, has been isolated in Adams County, on Ohio's southern border.

The Till Plains

The Till Plains draw their name from glacial material, or till, deposited by great ice sheets which covered the region during the Recent Ice Age. Although referred to as a plain, this region contains the highest point in the state: Campbell Hill in Logan County. Measuring

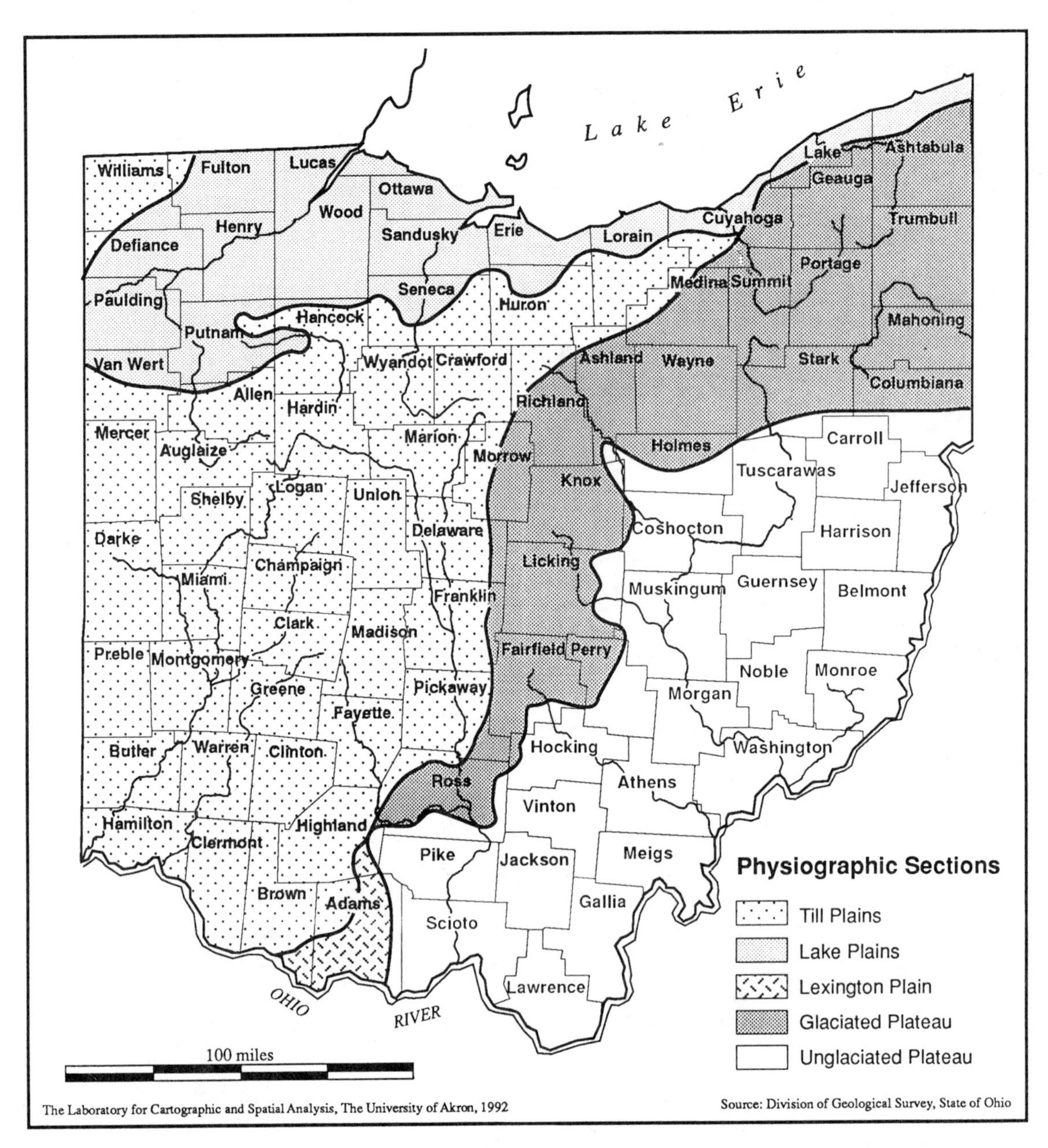

Fig. 1.1. Ohio's Physiographic Sections

1,550 feet above sea level, this peak is on the Bellefontaine Outlier, a section of bedrock curved upward in the shape of a dome and raised still further by material the glaciers left behind.

One distinctive feature of the Till Plains is the series of *glacial moraines* that cross from east to west. These moraines consist of material deposited when the warming climate melted the ice sheets. As the climate changed, the ice that covered this part of the state receded, not in a continuous movement but in a series of phases. Where the ice began to melt the forward movement was abated while there was still a continual flow to the rear of the ice front. In some instances these ice fronts remained stationary for several hundred years, allowing millions of tons of rock debris to pile up. The moraines this debris created grew to heights exceeding 100 feet and widths of several miles. Those formed where the ice front stopped are called *terminal moraines;* those created as the ice receded in stages are called *recessional moraines.* The most northern of the recessional moraines is the Defiance moraine, an escarpment that sweeps across the region dominating the flat plain that surrounds it. Towards the southern end of the region the deposited material comes, not directly from ice sheets, but from outwash streams that flowed from melting ice fronts. These initially fast-flowing streams carved wide valleys that were later filled with sand and gravel from the glacier.

The furthest advance of the last ice sheet that covered the state, the Wisconsin, is marked by the Hartwell and the Cuba terminal moraines. South of these moraines, the Till Plains are covered by material deposited in two previous ice advances: the Illinoian and the Kansan. Consequently the surface has suffered the processes of weathering over a longer period. Nutrients from the surface layers of soil have leached downward where a hardpan has formed about 9 feet below the surface. This hardpan is an impervious layer that restricts drainage, keeping upper soil layers wet for much of the year.

The glaciers that moved back and forth across the landscape have affected drainage patterns. Advancing ice blocked many streams, creating lakes that in turn found fresh outlets and cut new watercourses. The Ohio River is a product of such an occurrence during the Kansan ice advance. The proto–Ohio River, known as the Teays River, drained a large portion of the eastern United States lying west of the Appalachians. Finding its path blocked by the Kansan ice sheet, the Teays evolved into the Ohio River by cutting a new course. Other rivers also found their old courses blocked and formed new ones. Another example, the Little Miami River near Clifton in Greene County, cut a spectacular gorge, 70 feet deep.

The Lake Plain

Stretching along the entire south shore of Lake Erie and spreading into the Maumee Valley is a flat plain that was once the bottom of a larger lake. This lake has been known by several different names, each representing a different stage in its growth and recession. Composed of glacial meltwaters that were obstructed by glacial debris to the south and higher ground to the east, the surface of this lake reached a height of 790 feet above sea level, more than 220 feet higher than the present level of Lake Erie. Not only was the lake probably much deeper than the present lake, it extended over a greater area, reaching as far as present-day Fort Wayne, Indiana. With the changes in the ice advances and retreats the lake's surface level altered.

The changing lake levels created many *lake terraces;* these can be easily distinguished as one traverses the Lake Plain, from either the Till Plains or the Glaciated Plateau, to the shores of the modern lake. These fluctuating lake levels also left their mark in the form of sand ridges, dunes, and sandbars, all of which are now covered by oak woods. These *beach ridge* formations, with their well-drained soil and higher elevation, were ideal sites for early settlements and routeways.

The lacustrine sediments that accumulated on the bottom of the lake produced highly fertile organic soils and some of the best farmland in the state. However, the high clay content of this soil contributes to its poor drainage. Early settlers called the area, lying adjacent to the Maumee Valley, the Great Black Swamp. Today's farmers of the Lake Plain have invested in thousands of miles of drain tile to keep the area dry and productivity high.

The Unglaciated Plateau

The state's southeastern corner, comprising about a fourth of its total area, is the only section not ravaged by Pleistocene glaciation. Though rugged in appearance, with deep river valleys and seemingly high hilltops, the elevation of these hills rarely exceeds 1,400 feet. In fact, there is little variation in the general elevation of 1,200 feet. This surface is thought by some to be the remnant of an old erosion surface called the Harrisburg Peneplain.

A *peneplain* is formed from mountains worn down by subaerial erosion into a low, gently undulating surface uniformly level to the eye and deeply entrenched by river valleys. Introduced by W. M. Davis in 1889 (Thornbury 1969), the concept of peneplains has been hotly debated, with some scientists claiming that peneplains are completely misunderstood and falsely recognized. This debate is outlined by Thornbury (1969), who suggests the name *former erosion surfaces.*

Surrounding the Harrisburg Peneplain, another former erosion surface approximately 950 feet above sea level can be recognized. Both surfaces have been cut into heavily by the drainage system of the region so that some of the valley floors are 300 feet below the surface of the plateau. Located in the center of the plateau, the Flushing Escarpment acts as a watershed between the streams that flow directly into the Ohio and Muskingum Rivers.

The rugged terrain and poor soils of this region are not conducive to a flourishing population. Agriculture is confined to the valley floors or the windswept plateau. But the area is blessed in other ways, possessing as it does extensive deposits of coal, oil, and gas that boost its economy.

The Glaciated Plateau

Northwest of the unglaciated plateau lies a region that has experienced extensive glaciation. Here, the plateau surface has been modified by the glacial debris that filled in valleys and buried watercourses, and the surrounding hills have been worn down by the ice that passed over them, changing the course of streams as it went. Consequently the old drainage pattern has been considerably altered. The Cuyahoga River, for example, which once flowed south to join the proto–Ohio River, now has a horseshoe bend, just before it reaches the city of Akron, that takes it north to join Lake Erie. The Grand River also pursues an odd course, seeming a misfit in its wide, glacially enlarged valley. And most interesting is the Salt Creek in Hocking County that, upon finding its path blocked by the Illinoian ice, backed up and eventually spilled over the watershed at its headward end, cutting a steep-sided gorge down which the ponded water escaped. As a consequence of this drainage reversal, the tributaries upstream from the gorge join Salt Creek in an unusual way. Instead of flowing in the same direction as the main channel, they flow in the opposite direction, joining the stream at an acute angle.

The topography of the glaciated plateau is littered with evidence of the Ice Age: low drumlin swarms in Wayne County, and kame terraces and kettle depressions in Portage County. The variation in soils as their corresponding drift deposits change makes for an interesting and varied landscape. Some of the soils allow productive agriculture, while others are so barren that they are best left in a natural state, supporting woodland and scrubland.

Of all the soils in northern Ohio, the most interesting are the *mucks:* highly organic soils formed by the filling of kettle depressions. A *kettle depression* develops when a remnant of a retreating ice sheet is stranded, much like a landlocked iceberg, creating a depression that fills with water as the ice melts. The resulting *kettle lake,* being fairly shallow, is occupied by various forms of aquatic vegetation. When this vegetation dies it sinks to the mud at the bottom of the lake, adding layers with the passing years. Over time the lake becomes choked with this mixture of vegetation and mud, and the water disappears, gradually drying the mud and leaving the highly organic muck soil. The Hartville mucks in Stark County are highly productive examples of this process, enabling farmers to glean two or three crops per year. However, due to their highly combustible nature, they must be constantly irrigated in the dry summer weather. Figure 1.2 shows a kame and kettle landscape in northern Stark County. The kettle depressions in the photo are nearly filled by rotting vegetation, and eroded material from the surrounding land. Figure 1.3 shows the dark muck soils of the Hartville mucks. They lie somewhat lower than the road because the dry topsoil has been eroded by summer winds.

Industry has taken advantage of the glacial deposits. Gravel is extracted for roadways, and sand, for construction. Today many fine outcrops of glacial drift have disappeared under the advance of the bulldozer and the backhoe.

The Lexington Plain

The Lexington Plain, or Bluegrass region, is a small portion of the Kentucky landscape that has been isolated in Ohio. The Appalachian Plateau rises distinctly in the east, while in the west the boundary is indistinct within the Till Plains. Occupying most of Adams County and small portions of Highland and Brown Counties is a landscape of flat-topped hills controlled by the Silurian and Ordovician rocks that dip gently eastward. Limestone and dolomite are the controlling factors, standing out above the easily eroded shales in the form of steep-sided cliffs and flat-topped hills. The surface is pitted with sinkholes caused by the solution of the underlying limestone and the resulting collapse of the surface above. Though the most common sinkhole is funnel shaped (Thornbury 1969), topographically sinkholes may appear as mere depressions, or they may extend deep underground. As solution continues and loosened joint blocks fall, these holes may enlarge into shafts and connect with vaulted chambers (Holmes 1969).

Fig. 1.2. Kame and Kettle Landscape in Northern Stark County
(Photo by Leonard Peacefull)

Fig. 1.3. Swamp Road, Stark County, Crossing the Hartville Mucks
(Photo by Leonard Peacefull)

Table 1.1

The Major Periods in the Geologic Timetable

ERA	PERIOD	AGE Beginning of period Million years before present
Cenozoic	Quaternary	2
	Tertiary	65
Mesozoic	Cretaceous	135
	Jurassic	190
	Triassic	225
Palaeozoic	Permian	280
	Pennsylvanian	320
	Mississippian	345
	Devonian	400
	Silurian	440
	Ordovician	500
	Cambrian	570
Pre-Cambrian		4600+

Streams may change course and disappear down these shafts, leaving dry valleys downstream. Agriculture is limited in this region, with cultivation more common on the gentler slopes and hilltops, and the valley floors and steep hillsides left covered by woods.

Geological History

The geological history of Ohio extends back into the earliest times of the geological record. The bedrock consists of limestone, shale, sandstone, and conglomerate—materials that formed as sedimentary deposits in or near an ocean that covered the area during the Paleozoic era, 280 to 570 million years before the modern period. The progress of Ohio's geological history is shown in Table 1.1.

Some sands and muds settled out of the waters of shallow seas, forming deposits when the seawater evaporated leaving great salt flats. Other deposits formed from the residue of flood plains and river deltas, or the black muck that accumulated in swampy forests. Each rock type reflects the environment under which its sediments formed. Climate, depth and turbidity of water, and runoff from the surrounding land all contribute to the formation of the different types of sedimentary rocks. Thus as environmental conditions changed, so too did rock types, with the oldest rocks at the bottom of the rock succession and the youngest at the top. During the millions of years that followed, internal movements have raised these rocks into a dome shape. This dome has been gradually worn down, peeling away one layer from another, until at the center of the dome the oldest rocks—those from the Cambrian, Ordovician, and Silurian periods—come to the surface (Figures 1.4 and 1.5)

Each period in Ohio's geological history has a different story. This is due to the influence of *plate tectonics* (movements of the earth's crust) that shifted the surface that today covers Ohio to and from other locations on the earth. This surface was therefore subjected to different climates, and its rocks eroded at different rates. These would have been coupled with the surface runoff in the form of streams and rivers transporting the eroded material to the sea, where the depth of the sea and the motion within it influenced the type of rocks that formed. (For a discussion on these topics refer to Press and Siever 1982.)

The Pre-Cambrian Basement

The sedimentary rocks of the Paleozoic era rest on much older and entirely different rocks. These rocks—granites and metamorphics—are akin to those found in Canada. Granite is a coarse-grained rock with many crystalline structures. It was formed deep below the surface as hot, molten magma that slowly cooled. The slower the cooling, the larger the crystal lattice in the rock. Metamorphic rocks were subjected to so much heat and pressure—a pressure cooker environment—that their chemical structure changed (Press and Siever 1982). These rocks still lie far below the surface and offer little to the landscape or economy.

The Lower Paleozoic Eras

In the late Cambrian and Early Ordovician times, a sea spread over the area now known as the Midwest, and Ohio lay in a shallow sea that teemed with life. Trilobites, primitive sponges, jellyfish and brachiopods flourished in these waters. When they died, their skeletons and shells sank to the sea floor, piling up over the years and solidifying into the rock material of

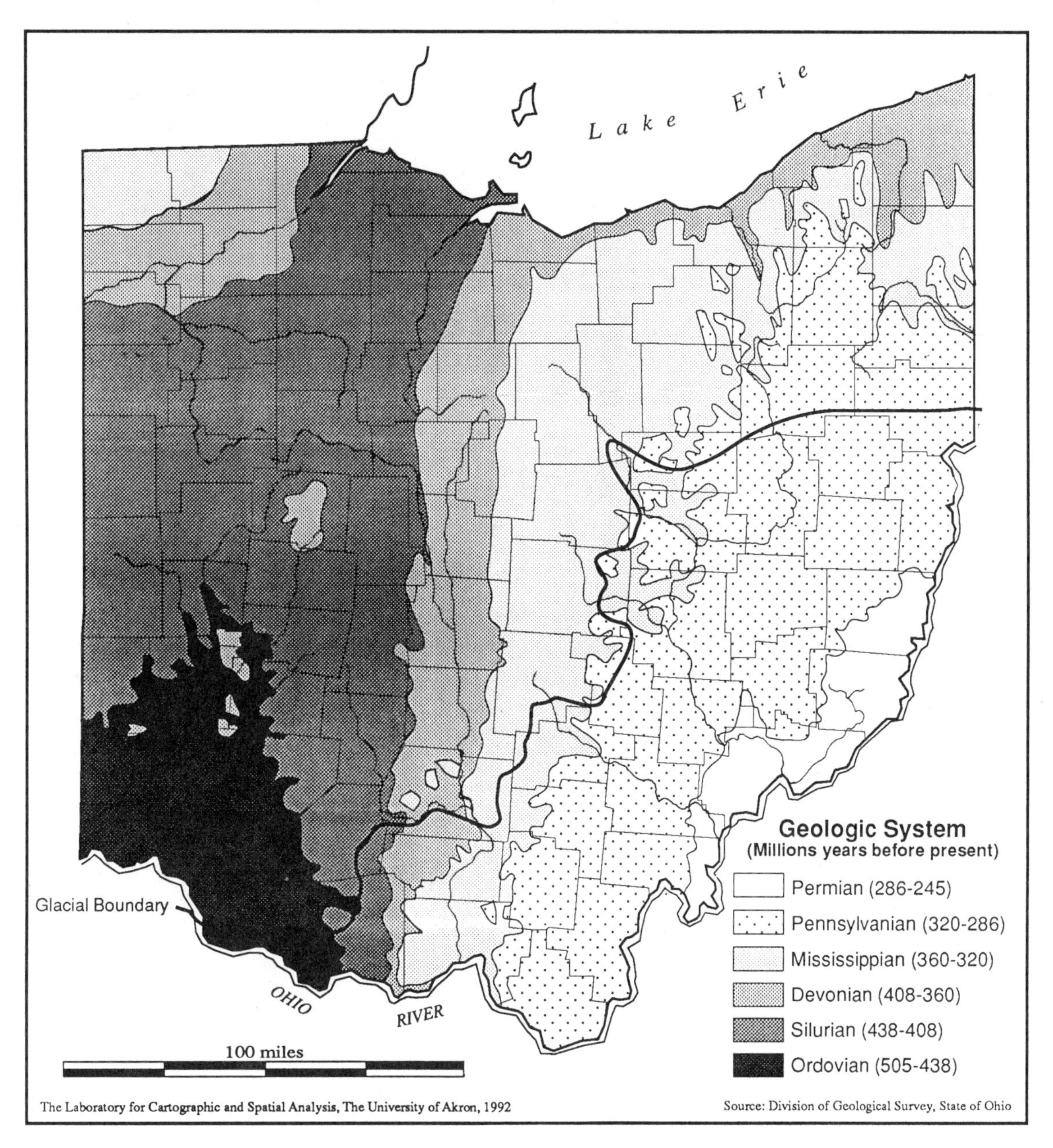

Fig. 1.4. Ohio's Geologic System

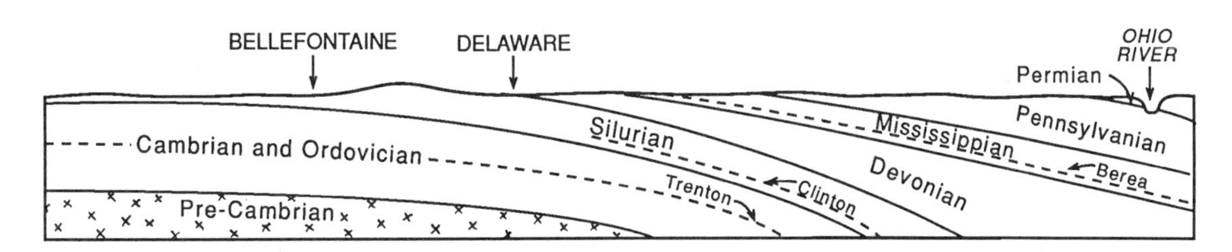

Fig. 1.5. Geological Cross Section of Ohio

limestone and dolomite that today are full of these fossilized remains.

The youngest of these rocks comprise the Ordovician outcrop in southwest Ohio. This outcrop, the Cincinnati Anticline, forms the crest of the low dome mentioned earlier; it is economically important because of its oil deposits, formed from the microscopic life that once thrived in the sea.

The Silurian Seas

Toward the end of the Ordovician and the beginning of the Silurian, Ohio was dry land, but in the middle Silurian the seas returned to cover all but the southeast corner of the state. This sea was warm, much like the Caribbean, and abounding with life. The rocks of this period, limestones and dolomites, form the great limestone belt in western Ohio. The state's Silurian limestones have a high concentration of magnesium carbonate and are a very useful mineral resource. They form the basis of construction material, ammonium, glass, agricultural lime, roadstone, and railroad ballast.

Near the end of the Silurian period the climate changed to a hot arid environment. The seas became shallow and eventually dried up. As the water evaporated, two minerals, gypsum and halite, formed in huge beds on the exposed sea floor. Both minerals are commercially important. Gypsum, the primary raw material in plaster and wallboards, is mined at Gypsum in Ottawa County; halite, commonly known as rock salt, is mined at several locations in northwestern Ohio.

The Devonian: A Time of Fish

During the early part of the Devonian period much of Ohio and adjacent portions of Indiana and Kentucky formed a low island surrounded by warm, shallow water. As time passed, the sea rose so that by the middle Devonian virtually the entire island lay beneath the clear sea. Here, coral reefs supported a variety of life. The shells, corals, and mud that collected on the sea floor formed a light gray, highly fossiliferous limestone known as the Columbus Limestone. A distinctive feature of the Columbus Limestone is the presence of fossil fish. These fish fossils are some of the earliest and can be up to 25 feet long. These fish did not have bony skeletons like those of today; their fossil record consists of scales and teeth. The Columbus is a pure limestone, making it a valuable industrial mineral. It is used in the manufacture of cement, concrete, and building stone, and most importantly as a flux in the smelting of iron ore. The rock is extensively mined, from Columbus to Marblehead.

Toward the end of the Devonian the accumulation of material in the sea changed. This was brought about by mountain building to the east, an event that caused streams to flow faster and erode quicker, enabling them to carry a far greater load down into the sea. Consequently the sea became muddier, producing the Ohio Shale, a rock containing few fossils that easily breaks into thin plates and flakes. One notable feature is the presence of material similar to coal that developed from the remnants of plant life washed from land to sea and carbonized. The shale also contains iron concretions, about the size of baseballs, formed from mud that collected around small foreign bodies such as sand grains or plant remnants. The shale itself is of little commercial value, but it contains pockets of natural gas that can be exploited.

The Mississippian Period

The Mississippian period occurred approximately 20–45 million years ago, when rivers cascaded from the mountains down into seas that covered northern and eastern Ohio. The seas were muddier than in Devonian times; hence corals were unable to survive, but shelled creatures such as brachiopods and gastropods flourished. The land was well covered with tall trees and ferns, and animal life consisted mostly of amphibians.

A dark shale, similar to that formed in the Devonian, developed during the Mississippian. Known as the Bedford Shale, this rock is easily eroded but in some places stands out as cliff formations. These cliffs are formed where the shale is overlain by the more resistant Berea Sandstone, a fine-grained sandstone composed of the fine quartz crystals that were deposited in a shallow sea. It is a valuable commercial rock, used in building-stone, grindstone, and other industrial purposes.

Much of the dramatic scenery found in several of the state's parks is composed of a coarse sandstone interspersed with pebbles. Called Black Hand Sandstone, its formation began when northward-flowing streams deposited their material in delta areas. This material was later covered by mud, and it solidified into rock, producing the high cliffs, bluffs, and gorges that today attract many visitors.

The Pennsylvanian: The Age of Coal

At the end of the Mississippian, a shallow sea covered most of the state, including the deltas that produced the Black Hand Sandstone. The climate was warm, probably tropical like that of Central America today. On land, vegetation thrived, while around the coast a swamp forest played a major role in producing the rocks of this period. This swamp forest extended over much of what is now Ohio, eastern Pennsylvania, West Virginia, Kentucky, and Indiana. The dead and decaying material of the forest—leaves, branches, and tree trunks—accumulated in the swamp. Periodically the sea rose, burying the decaying matter under a layer of sand and clay. The clay effectively sealed the plant matter from the decaying effects of oxygen and in many cases preserved the entombed material. Over the years, more trees died and the sea encroached onto the land, covering successive layers in mud. The weight of all these sediments over millions of years compressed the deposits into coal.

The coal deposits were not the first rocks formed during the Pennsylvania period. After the Mississippi sea had receded, a large delta formed in eastern Ohio. Into this delta drained the rivers that were eroding the mountains to the north and east, bringing with them small, pea-sized pebbles that later cemented into a rock called the Sharon Conglomerate. Another rock formed around this time is the Vanport Flint, a siliceous limestone that breaks into sharp-edged fragments. The Indian tribes who occupied Ohio prior to European settlement used this flint to make arrowheads, spears, scrapers, and others tools. Although this mineral was at one time important (it is the state mineral), it does not have the commercial value of another important Pennsylvanian deposit: coal.

Most of eastern Ohio's surface rocks—primarily limestones, shales, siltstones, sandstones, and clay, all interspersed with coal deposits—were laid down toward the end of the Pennsylvanian period. This succession of rocks, collectively called the *coal measures,* ranges up to 2,000 feet thick. Because the coal deposits are successively covered by other rocks, it is easy to speculate that the sea covering the area when the coal measures formed fluctuated dramatically. Sometimes it was calm, deep, and warm, teeming with life; sometimes it was shallow and turbid; still other times it almost disappeared, leaving the coal swamp forest. The coal deposits in Ohio are all bituminous, or soft, coal, which burns well but gives off a good deal of sulfur. This is because the coal contains iron pyrite (FeS_2), a mineral that forms where there is an absence of oxygen, just as coal forms. Not only is pyrite commercially useless, but when bituminous coal is burned, the pyrite emits sulfurous pollutants into the atmosphere. These pollutants mix with the weak solution of hydrochloric acid (H_2CO_3) in rainwater and with carbon dioxide (CO_2) in the atmosphere to produce acid rain, a major problem in eastern North America.

By the end of the Paleozoic era, the waters had receded leaving Ohio permanently above sea level. This is believed to be so because no sediments have been found that were laid down in a sea after the Permian period. For the next 223 million years, Ohio was subjected to all kinds of weathering and erosion. Many of these erosion surfaces are evident today in eastern Ohio. By the end of the Tertiary period, just over 2 million years ago, Ohio was probably rolling hill country drained by well-developed rivers. Continental movement had brought the land into roughly its present position, and unlike the tropical climates of the Pennsylvanian, temperatures were growing distinctly colder.

Glaciation in Ohio

Approximately 700,000 years ago, 1.5 million years into a geological epoch known as the Pleistocene, Earth's climate gradually cooled and ice caps formed over the polar regions. Why this happened is not fully understood. Geologists disagree among themselves; so do climatologists. One suggestion is there are periodic changes in the rate the sun burns its nuclear fuel (Press and Siever 1982). Another relates to the earth's procession around the sun. As the earth revolves around the sun its orbit is liable to slight variations; additionally, the earth wobbles on its axis, a phenomenon known as the *Chandler Wobble* (Chinery 1971). These two variations were correlated with Earth's glacial periods by Milankovich in the 1920s and 1930s. Climatically the evidence points to a period of rapid evaporation from the oceans just before the onslaught of the Ice Age, and increased precipitation that in the polar regions fell as snow. This snow continued for a considerable length of time, building up huge deposits that gradually turned to ice and glaciers.

A glacier is a mass of ice and snow that shows evidence of either present or former movement (Thornbury 1969). Two distinct types of glacier have been identified. An *ice cap* covers a large area, forming a

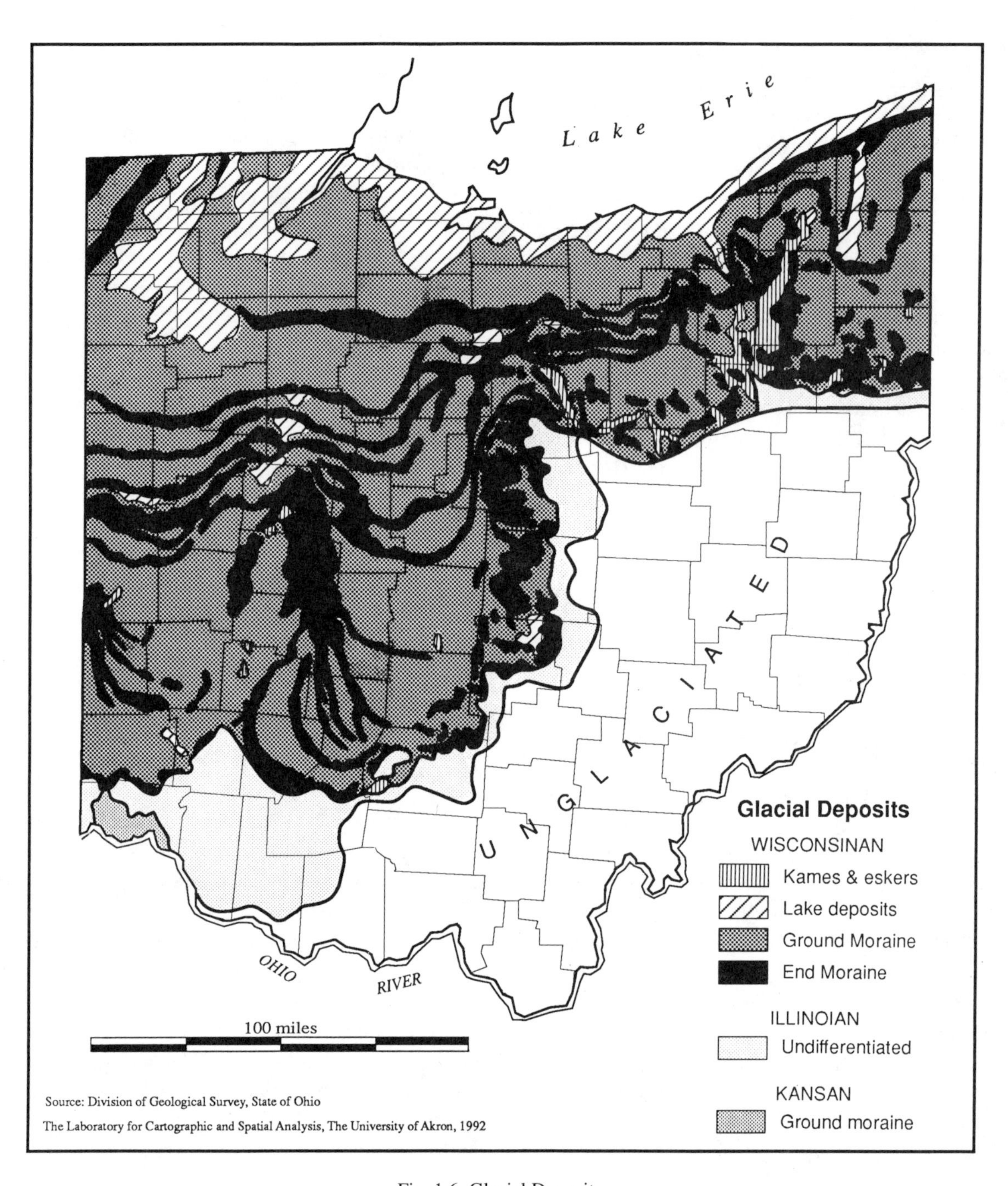

Fig. 1.6. Glacial Deposits

continuous sheet that moves in different directions, while an *ice stream* occupies a smaller, more marked course, such as a former river valley. Ice streams are more often associated today with the term *glacier* than ice caps. In North America the ice cap that developed as a result of the climate changes and rotational variations was centered around the Hudson Bay area of Canada. From here it spread out in two great lobes, the Labrador and the Keewatin, respectively east and west of the ice cap's center.

It must not be thought that from the beginning of the Ice Age, 1.8 million years ago, to its end, 8,000–15,000 years ago, there was one continuous ice sheet covering North America. Abundant evidence shows that there were periods within the Ice Age when the climate improved and at times was even milder

Table 1.2
Glacial Periods of the Last Ice Age

GLACIAL STAGE	INTERGLACIAL PERIOD
Wisconsin	
	Sangamon
Illinoian	
	Yarmouth
Kansan	
	Aflonian
Nebraskan	

than at the present (Miller 1961). During these periods of warming, called *interglacials,* the ice retreated, sometimes as far as its center of origin. Four ice advances are recognized in North America, each named for the state in which there is significant evidence of its presence (Table 1.2). It is obvious that with each successive ice advance, evidence of previous glacial stages was obliterated. Only where a later ice advance failed to penetrate can evidence of previous stages be seen. In Ohio the majority of the landforms resulting from glaciation can be attributed to the last ice advance, the Wisconsin. However, some evidence does remain of the more extensive Illinoian ice sheet, and of fragments of the Kansan ice sheet (Figure 1.6).

The Wisconsin Ice Advance

About 40,000 years ago the last of the major ice advances began. It took 6,000 years for the ice to reach Ohio and another 12,000 years for the ice to reach the central part of the state. Traveling at speeds averaging 160–220 feet per year, the ice split into four main lobes as it moved southward. The two lobes to the west extended roughly to the Scioto and Miami Valleys, while two smaller lobes moved along the Grand River and Killbuck Creek Valley in the east. A speed of 160–220 feet per year does not seem very fast but these are averages; there may have been times when the ice was moving in excess of 600 feet per year, and other times when it remained stationary for weeks on end. At the center of the sheet the ice must have been over two miles thick, judging by the present ice caps in Antarctica and Greenland, which both exceed 10,000 feet in thickness. As the ice spread over the landscape, it thinned out: around the north coast of the state it was approximately 1 mile thick, and at the advancing front, 50–200 feet thick.

The ice's advance across the state led to considerable erosion. Filled with debris accumulated from its movement across the Canadian Shield, the glacier abraded the landscape, tearing off its surface. Near the advancing front, the spread of the glacier slowed, and the glacier dumped its load in mounds called *moraines.* Thus the present surface of the glaciated parts of the state are a consequence of either glacial erosion or glacial deposition.

Glacial Erosion. Glacial erosion helped shape the landscape, from the glacial grooves on Kelly's Island to the broad expanse of Grand River Valley. These and other landforms were the direct result of the glaciers that moved over the landscape, plucking at the surface, removing rocks and debris that then became a constituent part of the glacier, thereby adding their abrasive qualities to the moving ice. This loosened material acted like a giant piece of sandpaper. Some rocks were smoothed, while others were left with small grooves or striations. When a glacier fills a former river valley, it plucks and scratches at the valley's sides and floor, giving that valley a broad, steep-sided appearance. These U-shaped valleys are left occupied by streams too small to have carved out such broad features.

Glacial Deposition. There are two types of landforms created by glacial deposits. *Till* or *drift deposits* are deposited directly by the glacier and consist of heterogeneous mixtures of rock fragments of all shapes and sizes. The other type, *indirect deposits,* result when meltwater runs off the leading edge of the ice sheet. These indirect deposits are usually sorted and stratified as layers of clays, silts, sands, and gravels.

The depositional process takes place along the margins and underneath the glacier. When the rate at which the ice and snow melts from the glacier, a process known as *ablation,* equals the rate at which the glacier itself is moving, then the ice front becomes stationary. The rock debris continues to pile up, and when this stagnant ice front melts, the debris is exposed as an elongated mound of material called a *moraine.* The moraine at the line of the glacier's furthest advance is called a *terminal moraine.* As the ice retreats, it does so in stages, and where it pauses other moraines are laid down. These *recessional mor-*

aines can be seen across the state as long, narrow, low ridges. The Wabash moraine lies between Celina, Kenton, Bucyrus, and Akron. The Defiance moraine, the northernmost moraine shown on the map of glacial deposits (Figure 1.6), sweeps south around Cleveland before turning west as a definite ridge on the till plain in Seneca and Hancock Counties.

Besides moraines, another form of direct glacial deposition, called *drumlins,* can be seen on the landscape. Drumlins are low hills varying between a few feet to over one hundred feet high and shaped like half an egg. They are composed of till but may exhibit some stratified material; it is quite usual for larger examples to have a bedrock core. Drumlins always occur in groups parallel to one another and aligned with the flow of the glacier. This grouping is called a *swarm* and has given rise to the term "basket of eggs topography," because of its shape.

As noted earlier, indirect deposition is the result of meltwater flowing from an ice front. The outwash material usually consists of sandy gravel and can produce such features as kames, eskers, and outwash valleys, for example the Great Sandy Valley in Carroll County.

Kames are isolated or clustered mounds of poorly stratified rock material deposited by meltwater at the edge of a stagnant ice sheet (Buchanan 1974). Kames may be accompanied by depressions, called *kettles,* that occur when an isolated block of ice, lodged in the depression between two kames, melts and forms a lake. The numerous kames in Portage, Stark, and Summit Counties arose from the material deposited between the ice lobes that protruded southwards. There are also some isolated cases of kame and kettle topography in Logan and Green Counties as well as some particularly interesting kames and kettles along State Route 43 between Kent and Hartville. If the streams that formed the kame continued for a longer period of time, then the deposit formed would stretch backward as the ice retreated (Holmes 1969). Such a feature is called an esker and is composed of sandy gravel material. Eskers are rare in Ohio, though it may be possible to discern one among the trees at Mogadore Reservoir in Portage County.

When the climate finally changed and the ice receded from the state for the last time, the landscape was left permanently altered. In some places the glaciers left rich, fertile soils, while in others the soil is so poor that very little can be done with or on it. In the northwest a much larger Lake Erie covered parts of the state with a rich lacustrine deposit that led to the fertile, if somewhat heavy, soil of the Great Black Swamp. Even the unglaciated region of the southeast was altered. During the Ice Age periglacial conditions existed similar to those of the tundra regions of northern Canada. Mosses, lichens, and low plants produced soils of a different nature than those which existed before the ice came. To say that this area escaped the effects of the Ice Age is not true; all areas of the state have experienced change during the past 750,000 years.

The landscape is a product of change, with the underlying bedrock giving shape to the land and erosional forces continually at work on the surface. Up until 10,000 years ago, ice sculpted the landscape; and when the ice left, rivers and surface drainage cut new courses, making more changes. Yet in the last 200 years, humans have become the chief moulders of the land, and probably the most severe erosional force in the state. Open-cast mines tear the surface, roads cut through hillsides, and urban developments create whole new landscapes. It is on this surface that the present-day geography of Ohio is taking shape.

REFERENCES

Buchanan, R. O. 1974. *An Illustrated Dictionary of Geography.* London: Heinemann.

Chinnery, M. 1971. *Understanding the Earth.* Sussex: Artemis Press.

Fenneman, N. 1938. *Physiography of Eastern United States.* New York: McGraw-Hill.

Holmes, D. L. 1969. *Elements of Physical Geology.* London: Nelson.

Miller, A. 1961. *Climatology.* 9th ed. London: Methuen.

Paterson, J. H. 1984. *North America: A Geography of Canada and the United States.* 7th ed. New York: Oxford University Press.

Press, F., and R. Siever. 1982. 3d ed. *Earth.* San Francisco: W. H. Freeman.

Thornbury, W. 1969. 2d ed. *Principles of Geomorphology.* New York: Wiley.

CHAPTER TWO

Climate and Weather

Thomas W. Schmidlin

Ohio's Climatic Setting

The climate of a region is determined by its elevation, latitude, prevailing air currents, and proximity to oceans or large lakes. Ohio's location—at low elevations in the middle latitudes directly south of the Great Lakes—gives it a climate with four distinct seasons, large seasonal temperature ranges, frequent precipitation, and a changeability of weather typical of the middle latitudes. While prevailing winds are from the southwest, one or two weather disturbances affect Ohio each week, bringing changes in wind direction and some precipitation. These weather disturbances, such as fronts or low-pressure areas, may bring warm subtropical air from the south or cold arctic air from the north.

Disturbances generally arrive along one of three major storm tracks. Those from the northern plains or Canadian Prairies are sometimes called Alberta Clippers. They are fast-moving storms from the midcontinent that normally bring light precipitation. A more significant storm track comes from the southern Plains, often tapping into moisture from the Gulf of Mexico and bringing significant rain or snow to Ohio. A third storm track originates along the Gulf Coast and follows the Appalachians or Atlantic Coast northward, sometimes causing heavy rain or snow, usually in eastern Ohio.

Local Effects on Weather and Climate

In addition to large-scale weather disturbances—such as highs, lows, fronts, and air masses—Ohio has certain local features that affect the climate in certain areas. These include elevation, Lake Erie, and cities.

Elevation. Ohio's elevation ranges from 440 feet along the Ohio River at Cincinnati to 575 feet along Lake Erie to over 1,200 feet in some parts of the central, northeast, and southeast portions of the state. This is a comparatively small elevation range but some of the effects elevation causes are evident in Ohio's climate. Precipitation and cloud cover increase as air is lifted over the higher elevations, causing hilltops and ridges to receive a few more inches of rain and snow each year and more days of fog and drizzle than the valleys and lower elevations. Temperatures are generally cooler at higher elevation, but this difference is only 1–2 degrees. A more noticeable difference occurs in hilly terrain on clear, calm nights when cool air, which is heavier than warm air, flows down the slopes into the valley bottoms. These valleys become "frost pockets" and may be several degrees colder than the nearby ridges. This explains why most of the coldest temperatures recorded in Ohio have occurred in the valleys of the hilly southeast.

Lake Erie. Lake Erie covers 10,000 square miles and forms most of Ohio's northern border. The lake has a large impact on the climate of Ohio's "north coast." Lake waters change temperature more slowly than land, delaying the change of seasons along the shore. Because these waters warm gradually in spring, land within a few miles of the shore remains cooler than the rest of Ohio during April, May, and June. After reaching a temperature of 75°F–80°F in August, Lake Erie cools slowly during autumn and early winter, tempering the cold waves of October, November, and December and pushing back the first autumn freeze 2–4 weeks. This allows the commercial production of peaches, grapes, and other tender crops in North

Coast communities. Lake Erie also adds moisture to the air during autumn and winter. Evaporation is greatest during these seasons because the lake waters are much warmer than the air. The added moisture results in frequent cloudiness across northeast Ohio during late autumn and winter. The moisture evaporated from Lake Erie by cold air masses also gives "lake-effect" rain and snow to northeast Ohio during the cold season.

City Climates. All settlements of Ohio, from the smallest rural community to the largest city, create a local urban climate that differs from the adjacent natural landscape (Schmidlin 1989a; Kochar and Schmidlin 1990). The buildings, pavement, vehicles, and sparseness of vegetation cause local "heat islands." A heat island is most evident in fair weather, when the temperature difference between a city and the nearby rural areas may be 1°F–2°F in small towns and 10°F or more in large cities. In cities they put heat stress on residents and contribute to higher summer air-conditioning costs and lower winter heating costs in urban buildings.

Temperature

The Seasonal Cycle

The seasonal cycle of temperature in Ohio is controlled by changes in the amount of solar radiation received on the surface and movement of cold and warm air masses across eastern North America. An average seasonal temperature cycle for Columbus is shown in Figure 2.1. This temperature pattern is a result of the seasonal cycle of solar radiation with a low sun and shorter days in winter and a high sun and longer days in summer. In any particular year, the seasonal temperature cycle is not a smooth pattern of warming and cooling but a series of warm and cool spells controlled by air masses. Winter arrives, for example, not as a gradual cooling trend through November and December but as a succession of cold spells, each colder than the last, with a few days of warmer weather scattered among them. The temperatures of Ohio's four seasons are described in the next sections.

Winter

Winter begins, in the astronomical sense, about December 22, but wintry weather arrives in Ohio at least a month sooner. It is common to consider December, January, and February to be Ohio's winter months, but wintry weather may begin in November, and it usually continues into March. Because days are short, with only 9 to 10 hours of sunlight, and skies are often cloudy, temperatures are more affected by changes in air masses than by the sun. Airflow patterns in the upper troposphere, 20,000–35,000 feet above the earth's surface, determine whether warm air masses from the south or cold air masses from the north will flow into a region. In a typical winter these patterns shift every few days, bringing several days of cold weather, followed by a period of mild weather, followed by another cold spell, and so on. If the airflow patterns in the upper troposphere become "blocked" and do not change much for a week or two, then an extended period of very cold weather or unusually mild weather will result. For example, during the winter of 1976–77 the airflow in the upper troposphere came from the northwest and for many weeks blew in arctic air masses, making that winter Ohio's coldest of the century. A similar pattern made the December of 1989 Ohio's coldest on record, with temperatures 10°F below average. On the other hand, during December 1982 an airflow pattern from the south produced very mild temperatures, 10°F above average.

Ohio's winter temperatures commonly fluctuate across the freezing point. Typical daytime highs are in the 30s to low 40s, and overnight lows are in the teens to low 20s. (Figures 2.2 and 2.3). A cold spell will bring 5–10 days with temperatures continuously below freezing and minimum temperatures below 0°F. A mild spell of winter weather brings temperatures of 50°F or warmer. These sharp changes in winter temperature cannot be accurately forecast more than 5–7 days in advance.

Very cold mornings with temperatures of 0°F or below occur on fewer than five days a year along Lake Erie's shore and in the Ohio River Valley (Figure 2.4). The remainder of the state expects 5–15 days of subzero weather. The coldest temperature experienced in southern Ohio and along the Lake Erie shoreline usually falls in the 0°F to –10°F range. Lower temperatures, ranging –5°F to –15°F, occur in the rest of the state (Edgell 1992). The coldest temperature of the winter often arrives 1–2 days after a snowstorm, as strong northwest winds pull arctic air southward over the snow-covered ground. Though most likely during January or February, this can occur anytime from late-November to March. The coldest temperature recorded in Ohio came on February 10,

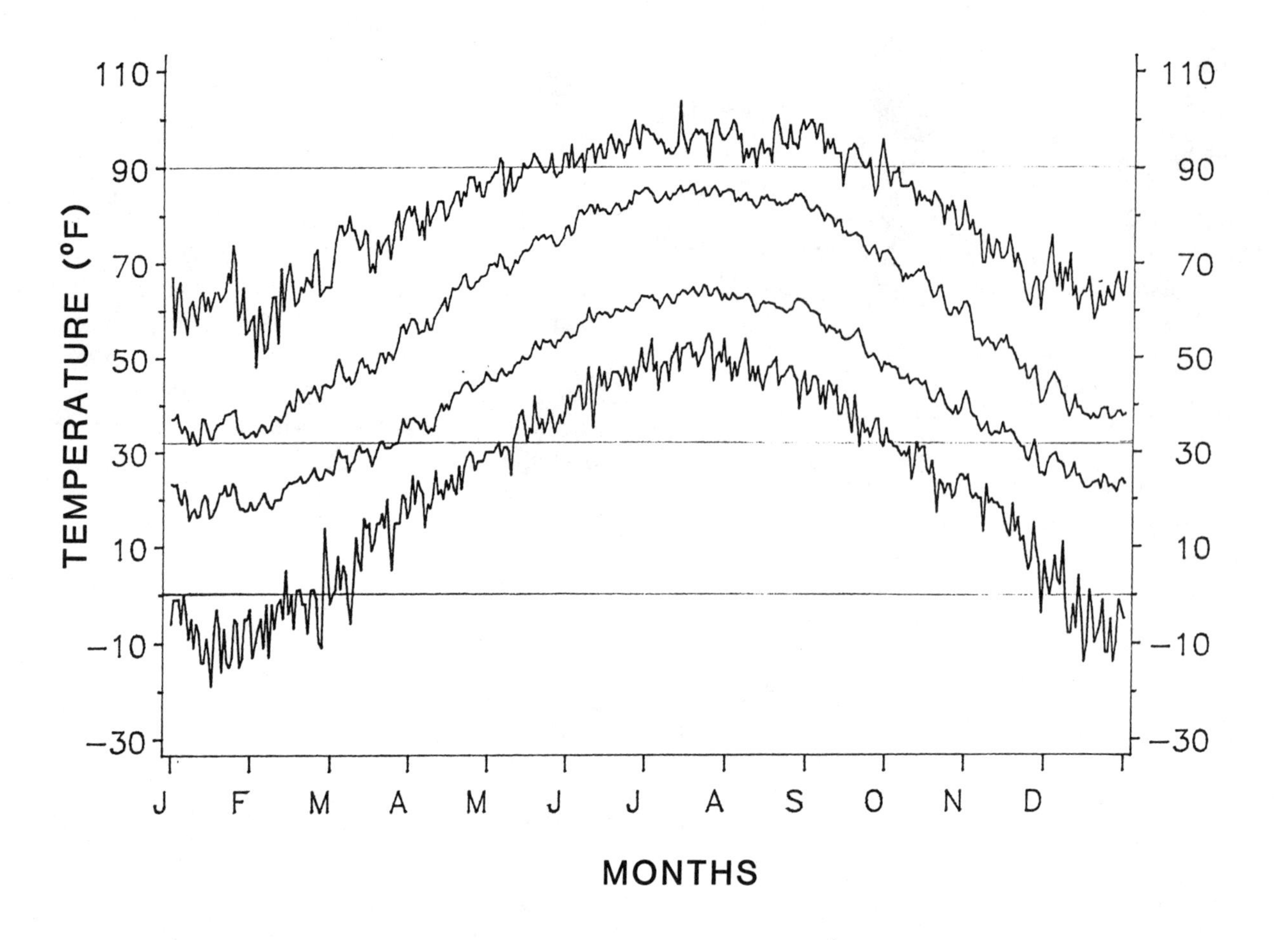

Fig. 2.1. The seasonal cycle of daily temperatures at Columbus. From the top, lines are record maximum, average maximum, average minimum, and record minimum.

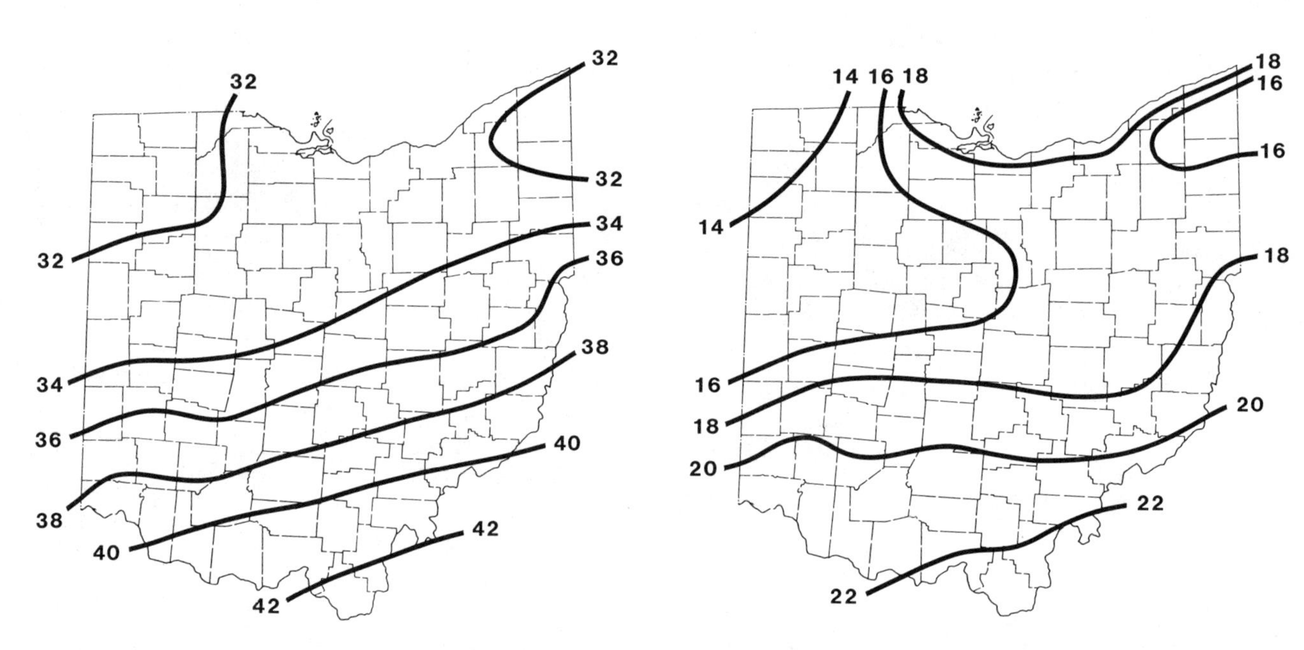

Fig. 2.2. Average daily high temperature in January.

Fig. 2.3. Average daily low temperature in January.

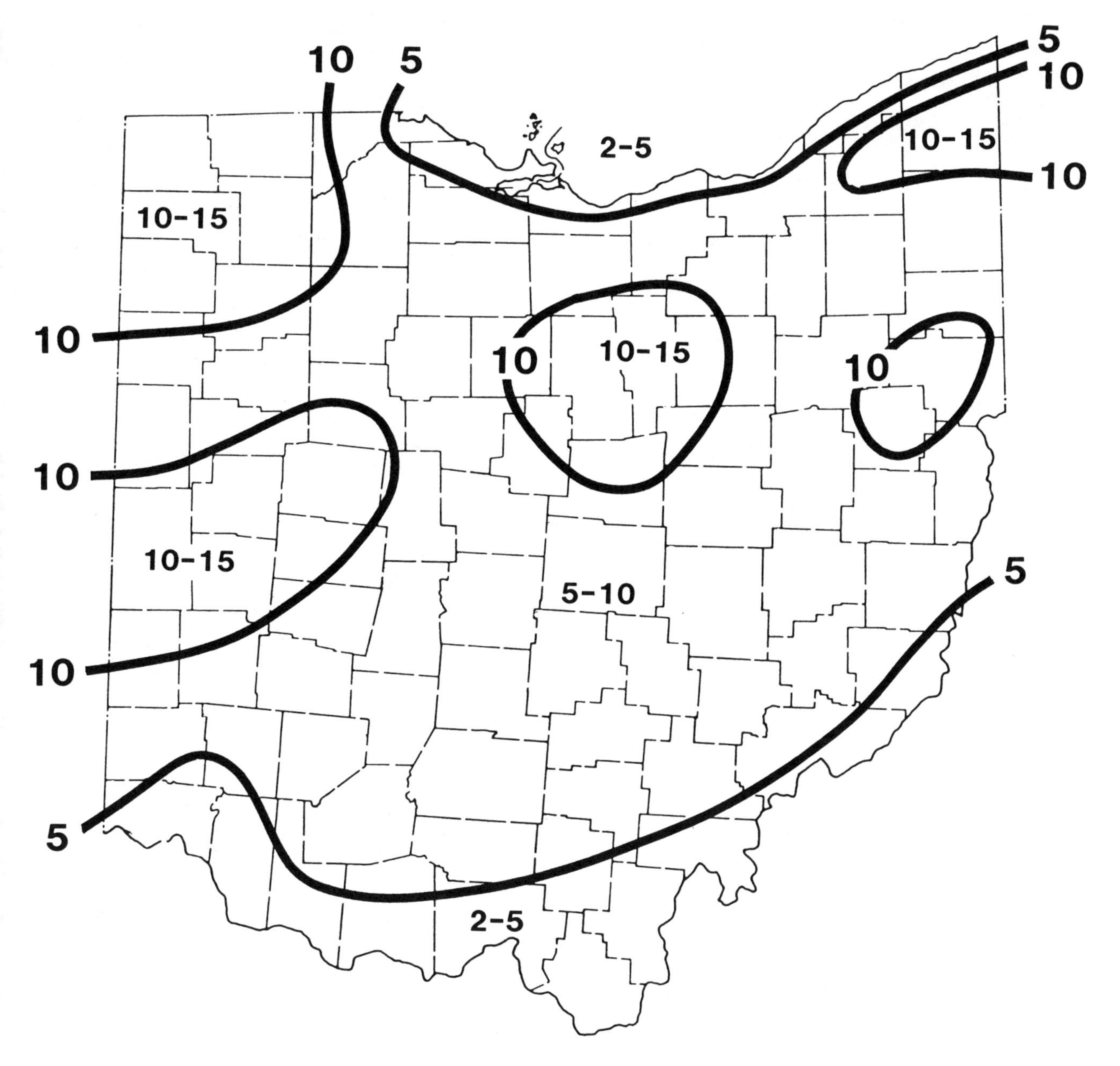

Fig. 2.4. Average annual number of days with a temperature 0°F or colder.

1899, when the mercury plunged to −39°F at Milligan in Perry County (Figure 2.5). Several major cities, including Cleveland, Columbus, Dayton, Akron, and Cincinnati, reached temperatures of −20°F or colder during January 1994.

Spring

Spring is a time of transition from winter to summer, a time when the upper airflow pattern and the jet stream shift northward and temperatures warm. Spring's arrival is sporadic, however, and winter does not give up easily. The first warm days with temperatures in the 60s and 70s come by late March, but snow and freezing weather usually return before reliably warmer weather arrives in late April and May (Figures 2.6 and 2.7). As spring advances, warm temperatures last longer and cold spells dwindle. The last freezing temperature experienced along the Lake Erie shoreline and in far southern Ohio occurs during late

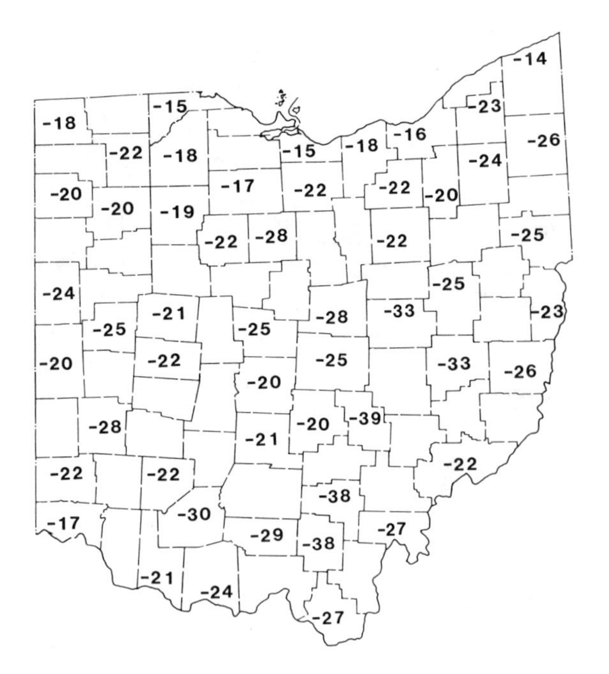

Fig. 2.5. Low temperatures on February 10, 1899.

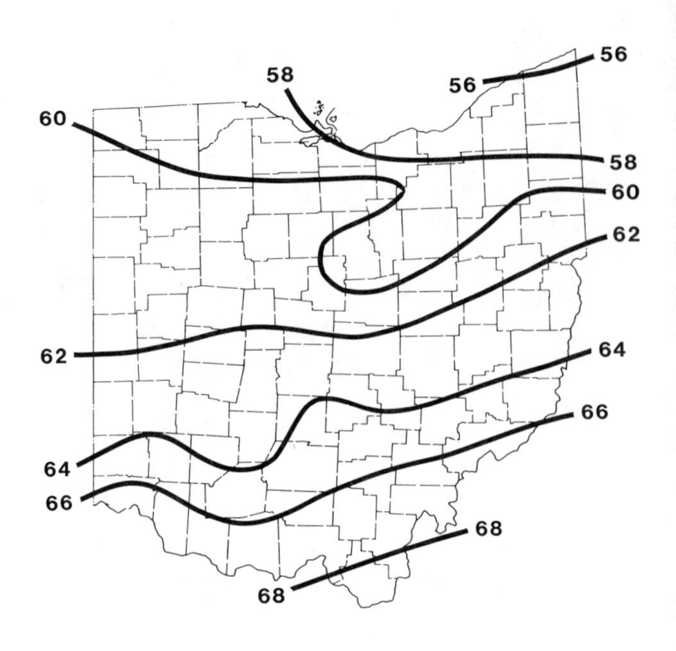

Fig. 2.6. Average daily high temperature in April.

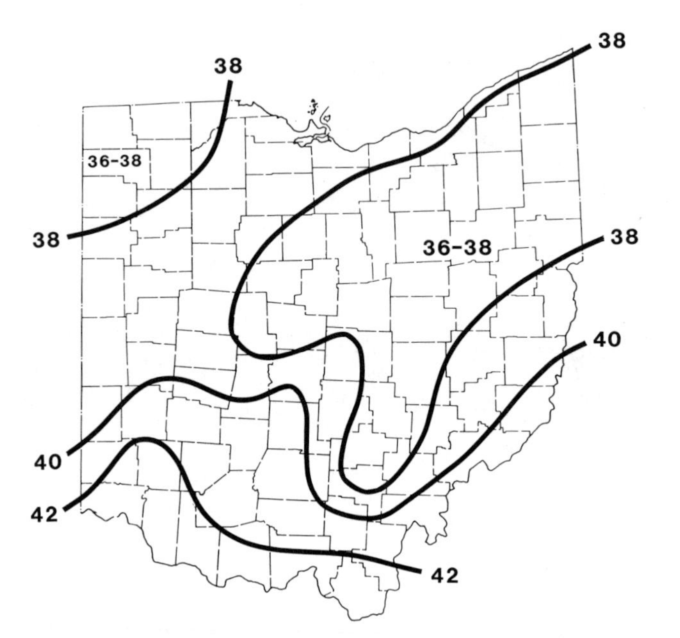

Fig 2.7. Average daily low temperature in April.

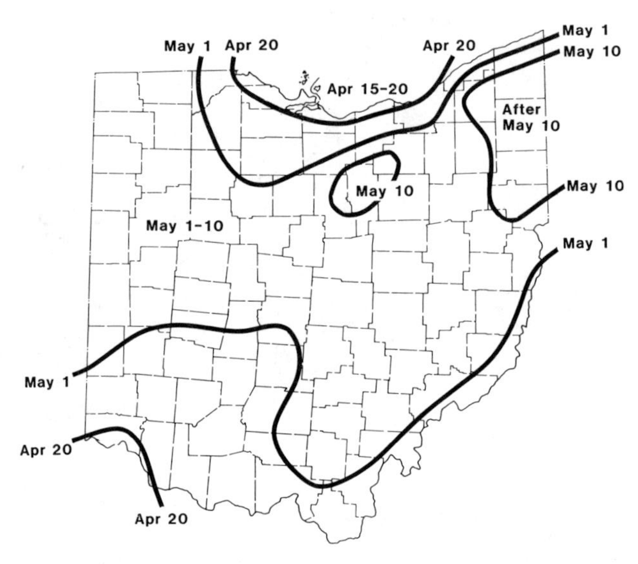

Fig. 2.8. Average date of the last spring freeze (32°F).

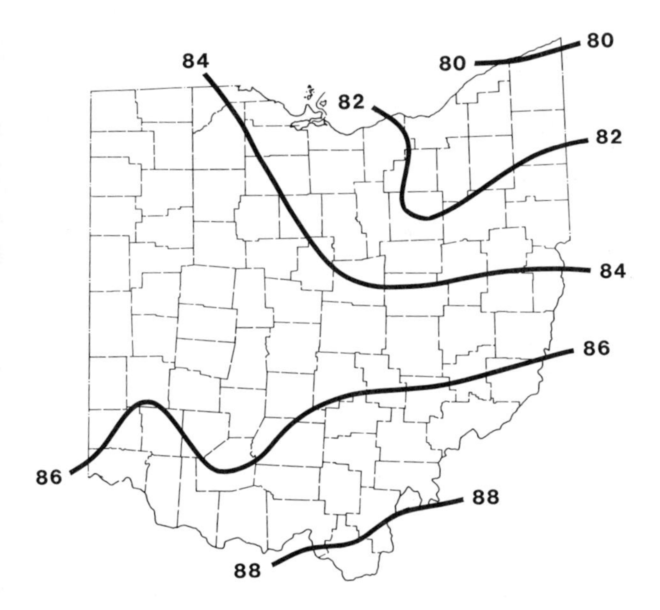

Fig. 2.9. Average daily high temperature in July.

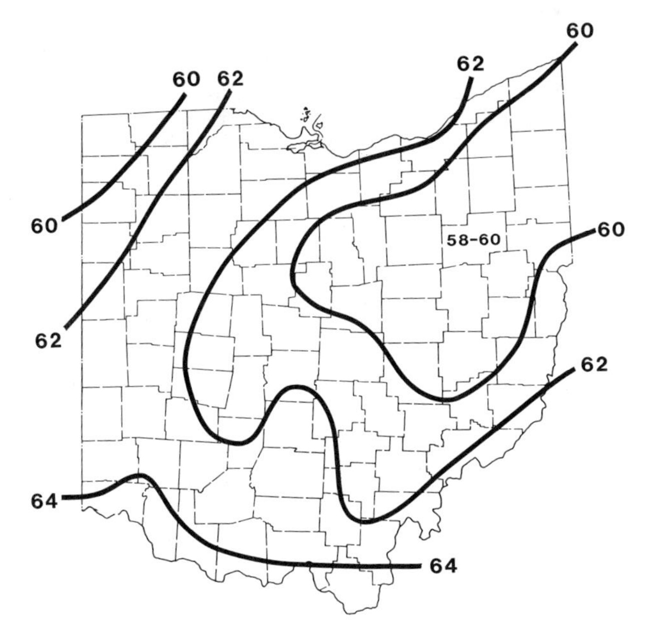

Fig. 2.10. Average daily low temperature in July.

April (Figure 2.8). Much of central and northwest Ohio is subjected to a freeze during early May, and the higher valleys of the north-central and northeast regions may expect one in mid May. On rare occasions, freezing temperatures have been recorded in those regions as late as mid-June.

Summer

Long days with 14–15 hours of daylight, sunny skies, and south winds bring warm temperatures by late May. Summer temperatures are commonly in the 80s during the afternoon and the 60s overnight (Figures 2.9 and 2.10). Temperatures are not only warmer but also less variable than during other seasons. The major storm track retreats northward to Hudson Bay during summer, so weather disturbances crossing Ohio are weaker and less frequent than in winter. Cold air is trapped in the Arctic, so only relatively small changes in temperature occur across Ohio during the summer months.

Crisp and cool summer weather with days in the 70s and nights in the 50s arrives on north winds with high pressure from Canada. As the high pressure moves east, south winds bring tropical air northward and hot weather returns. Very hot weather with temperatures over 90°F comes to Ohio when a high-pressure system remains over the southeastern United States for several days. High humidity with overnight temperatures in the 70s make these days uncomfortable. These conditions are most common in southern Ohio but usually last only a few days before they are interrupted by cooler air from the north (Figure 2.11). The hottest official temperature recorded in Ohio was 113°F, near Gallipolis on July 21, 1934 (Figure 2.12). The cool waters in the center of Lake Erie reduce the number of hot days in far northeast portions of the state.

Autumn

Autumn begins during September, when days grow shorter and air masses from the north are cooler. This is a pleasant season, with abundant sunshine in September and October and fewer wet days than during summer or winter. Temperatures cool so that by October, afternoon highs are in the 60s and overnight lows fall to the 40s (Figures 2.13 and 2.14).

Autumn's first freezing temperatures end the growing season, killing tender vegetation such as corn, beans, and many annual flowers. This first freeze is expected during early October across much of central and northern Ohio, inland from Lake Erie (Figure 2.15). Along the lake, its warm waters delay the first autumn freeze until late October, just as in warmer southern Ohio, the Ohio River similarly affects the land near its shores. Freezing temperatures have occurred as early as mid-September in most inland areas. These early freezes are most common in the

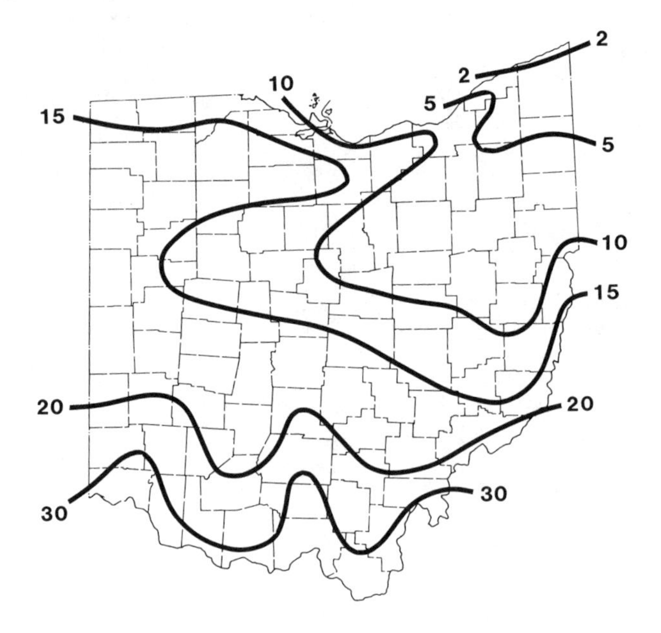

Fig. 2.11. Average annual number of days with a temperature 90°F or warmer.

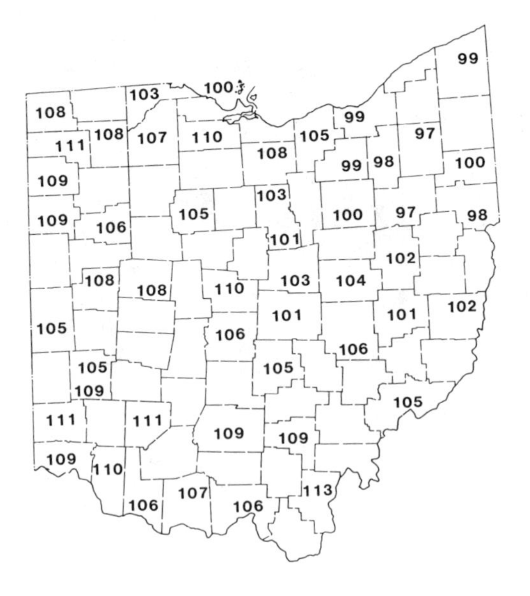

Fig. 2.12. High temperatures on July 21, 1934.

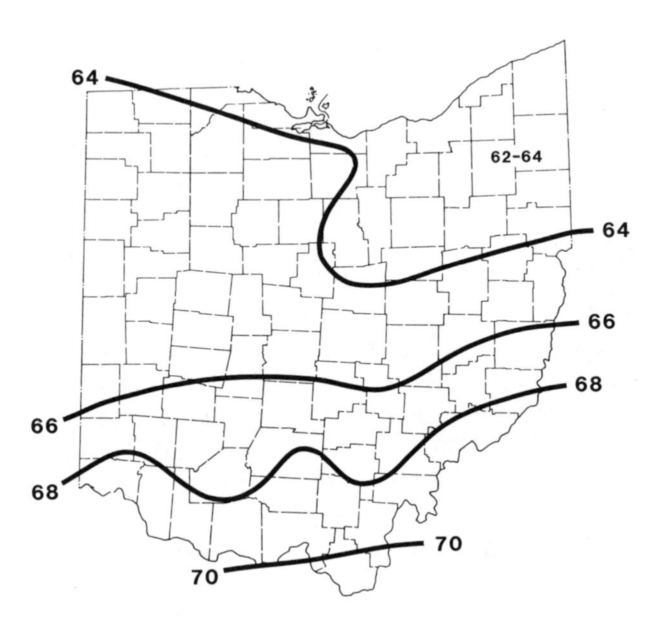

Fig. 2.13. Average daily high temperature in October.

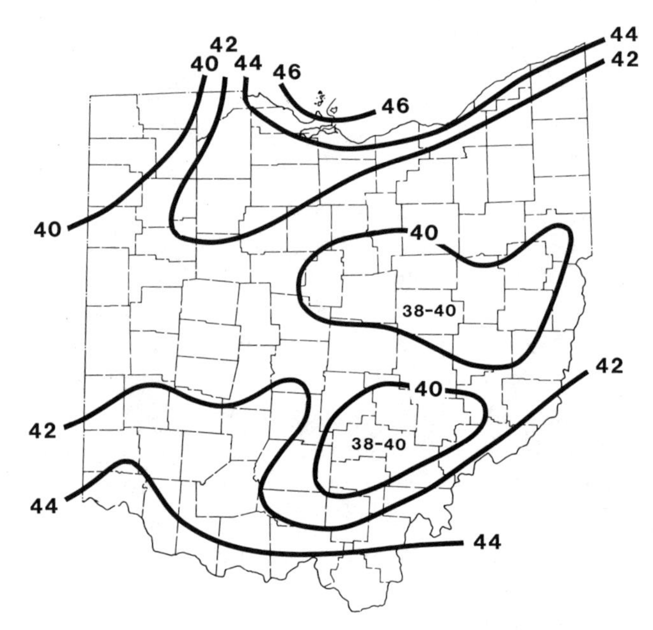

Fig. 2.14. Average daily low temperature in October.

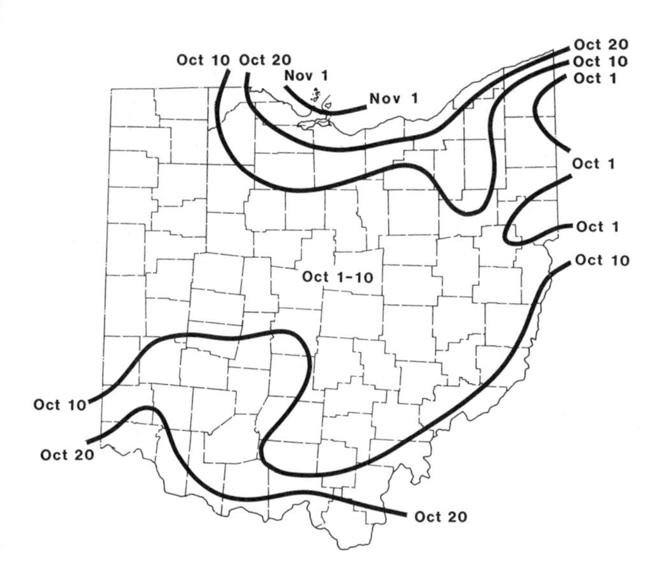

Fig. 2.15. Average date of the first autumn freeze (32°F).

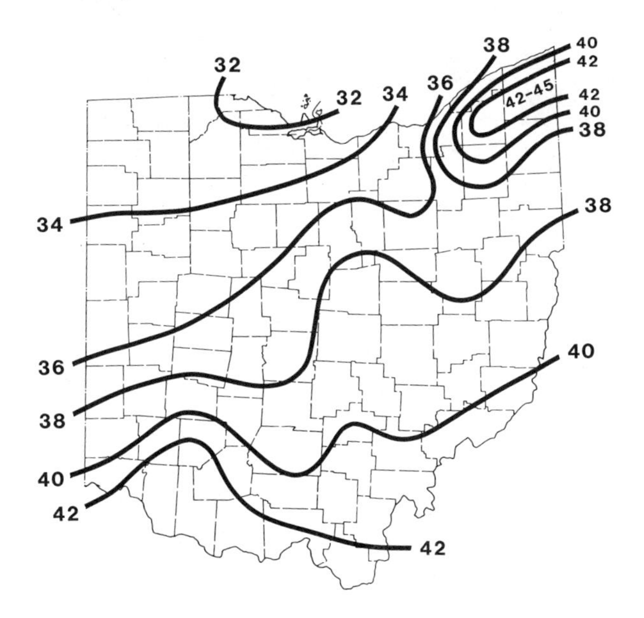

Fig. 2.16. Average annual precipitation in inches.

high valleys of eastern Ohio. The first snowflakes may fall during October, but significant snowfalls are rare before November. The character of autumn changes in November, becoming cloudier with colder air. Freezing temperatures are common by mid-November, and the first significant snows arrive during this month. The transition from autumn to winter is completed by December.

Precipitation

Rainfall

Precipitation is normally abundant and frequent in Ohio. The low-pressure weather disturbances that originate in the western or southern United States often move toward the Great Lakes–Ohio Valley region. These weather disturbances and their associated fronts affect Ohio once or twice a week during the cold season and about once a week in summer. They bring clouds, rain or snow, shifting winds, and temperature changes. Thus some precipitation can be expected every few days.

Ohio's average annual precipitation (rain plus the melted equivalent of snow) ranges from about 32 inches at the western end of Lake Erie to over 42 inches at the highest elevations of northeast and southern Ohio (Figure 2.16). Southern Ohio receives more precipitation than northern Ohio because of its proximity to the Gulf of Mexico moisture source. An exception is the moist, hilly, northeastern portion of the state, where lake-effect rain and snow are common in late autumn and winter. The wettest regions of Ohio do not necessarily have the most number of wet days. Although southern Ohio receives more moisture than the north, the north, especially the northeast, has more days with rain or snow.

Ohio does not have a distinct dry season. The summer months may seem drier than winter or spring because there are fewer rainy days, but in fact summer showers and thunderstorms tend to bring more precipitation than disturbances in other seasons. The greatest average monthly rainfall, about 4 inches, occurs during June or July, while the lowest monthly average, about 2.4 inches, occurs during February. Dry weather hits Ohio when a high–pressure system stagnates over the region, diverting weather disturbances and bringing clear skies and dry air. These are most common during summer and early autumn, and may last 1–2 weeks. During the early summer of 1988, this weather pattern brought severe drought to Ohio, with very little rain for 8 weeks. Droughts of this magnitude are rare, occurring about five times in a century.

Ohio also suffers its share of excessive rainfall. Summer thunderstorms may drop several inches in a few hours, triggering flash floods of streams. Though such floods develop in a matter of minutes, and often with little warning, they are usually localized and cause only minor damage, flooded roads and farm fields being the most common result. However, flash floods in the hilly terrain of eastern and southern Ohio can be more severe, washing away cars and homes. On July 4, 1969, severe thunderstorms dropped over 10 inches of rain in a few hours across north-central Ohio, causing floods that drowned twenty-four people and closed hundreds of roads, including the Ohio Turnpike. Another flash flood, on June 14, 1990, killed twenty-six people, this time in Belmont County. Floods can also occur during the winter and spring when a general rain of 2–4 inches falls onto frozen or snow-covered ground. These floods may affect big areas of the state and cause a gradual rise in larger rivers, such as the Ohio, Scioto, Great Miami, and Maumee. In March 1913 heavy rains of this type led to extensive flooding, causing over 300 deaths statewide.

Ohio normally experiences thunderstorms on 35–45 days per year, most commonly between April and September and most frequently in the south. These storms may be accompanied by lightning, which can strike the ground, a tree or other natural feature, or a building or tower. Though it generally seeks out the tallest structure, lightning can strike nearly anywhere. Contrary to popular wisdom, lightning does strike the same place twice; in fact, tall buildings and towers may be struck hundreds of times over their lifetimes. Such structures can be partially protected with lightning rods that carry the dangerous current safely into the ground.

Freezing rain occurs several times each winter. This type of rain is especially treacherous for drivers and pedestrians because it falls as liquid from warmer clouds and then freezes on cold roads and sidewalks. Most instances of freezing rain, or *ice storms,* last only a few hours and cause little damage, excepting that caused by the slippery surfaces. A serious ice storm will bring several hours of freezing rain that coats surfaces, including tree limbs and wires, with an inch or more of ice, disrupting traffic and causing hundreds of trees and power lines to collapse. Fortunately widespread severe ice storms are rare in Ohio.

Snowfall

Ohio falls within the transitional snow zone of North America. Winter snow is common but temperatures rarely stay below freezing for more than a week. Snow does not accumulate to great depths as it does in the colder northern climates (Edgell 1988). The snow season begins in mid-November in northern Ohio and in early December in the south. Snowfall is heaviest during January and February and commonly extends into mid-March. Snowfall is rare by mid-April, but light snows have occurred as late as May.

There is a wide range in the average snowfall across the state (Figure 2.17). While most areas receive 20–35 inches each winter, lake-effect snow downwind of Lake Erie contributes to heavier snowfall in north-central and northeast Ohio. Average snowfall increases to over 80 inches in the "snowbelt" east of Cleveland (Miller and Weaver 1971). Higher elevations and snow squalls from Lake Erie give Chardon, located in the core of the snowbelt in northern Geauga County, an average snowfall of 105 inches (Schmidlin, 1989b).

The heaviest single snowfall of the winter usually measures 4–8 inches, except in the northeast snowbelt where a 12–16-inch snowfall is expected each winter. In most areas, the biggest snowstorms on record fall in the 20–24-inch range, but these average only 1 per century in any particular location. One of Ohio's deepest snowfalls—30–40 inches in the extreme eastern counties—occurred over several days during the Thanksgiving holidays in 1950 (Figure 2.18).

The deepest snow cover is usually 6 inches, except in the snowbelt of northeast Ohio, where an accumulation of 12–18 inches occurs in most winters (Schmidlin 1989b). Though the snow is usually deepest in January or February, the largest buildup can occur anytime from November to April. A winter of unusually cold temperatures or a series of large snowstorms can result in deeper-than-average snow. Most of Ohio has experienced snow depths of 20 inches or more. These are rare events, occurring about every 30–50 years. In the snowbelt, depths of 20 inches are more common, and undrifted snow has reached 36 inches after major lake-effect snowstorms in Geauga, Lake, and Ashtabula Counties (Schmidlin 1989b).

The probability of having a snow cover increases during early December, as temperatures fall toward winter values. During mid-January the daily probability of having a snow cover of 1 inch or more is over 80 percent in the snowbelt, 50–70 percent elsewhere in the north, and 35–50 percent in central and southern Ohio (Figure 2.19). Thus, though Ohioans must be prepared for snow for about 5 months each winter, those living outside the snowbelt cannot depend on a mid-winter snow cover for sledding or other snow sports.

The blizzard is a particularly dangerous form of snowstorm. In addition to bringing deep snow, a bliz-

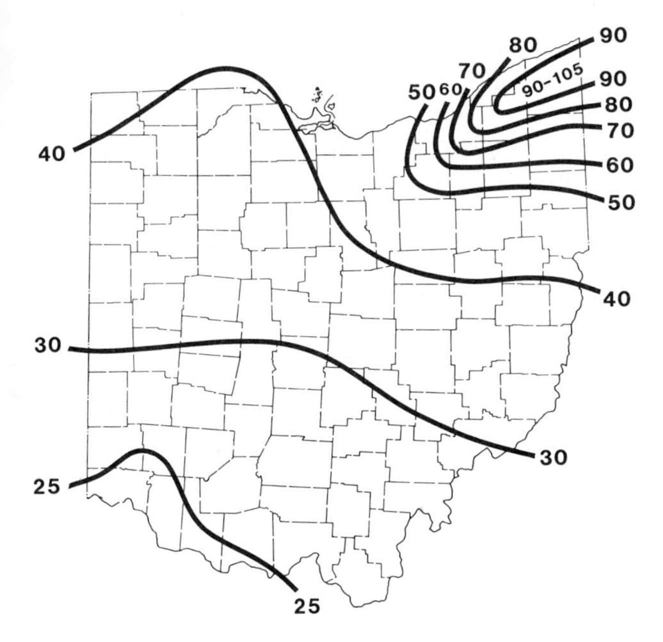

Fig. 2.17. Average annual snowfall in inches.

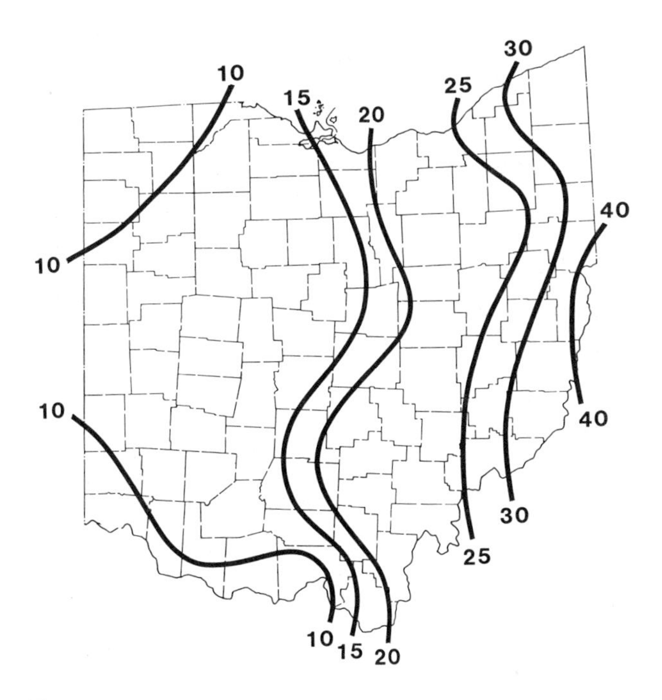

Fig. 2.18. Snowfall during the storm of 24–28 November 1950. A total of forty-four inches was measured at Steubenville.

zard sustains winds over 35 miles per hour and temperatures below 20°F for several hours. This combination of snow, wind, and cold makes most travel impossible and can be deadly to anyone caught outdoors. Blizzard conditions are rare in Ohio, usually occurring only in small areas of the snowbelt and lasting only a few hours. Statewide blizzards have struck twice this century: on January 12, 1918, and on January 26, 1978. The 1978 blizzard immobilized the state for several days and caused the deaths of thirty-six people.

Winds

It is rare for winds to blow from the same direction for more than a few days. Though the most common direction, or *prevailing wind,* is from the southwest, weather disturbances can bring winds from any direction in any season. In fact it is not uncommon for wind direction to change several times in a day. During summer, warm winds from the south or southwest prevail, with occasional cool breezes from the north or east. During winter, cold winds from the north or northwest are more common, yet mild south winds may blow, even in the cold season.

Wind speeds are usually light, averaging 5–10 miles per hour, with the strongest occurring during the afternoon and lightest at night. This daily pattern is most evident in summer, when nights are fairly calm. Winds are lighter in cities because of the interference of tall buildings. Buildings may also create "wind tunnels," small areas where pedestrians have trouble walking on windy days. Winds are, on average, 50 percent stronger during the cold season, November–April, than during summer; however, strong winds may occur in any season. Winds over 50 miles per hour, which damage vegetation and buildings, may occur with spring and summer thunderstorms or with the passage of a strong, low-pressure storm system through the region during the cold seasons. Most communities have experienced severe thunderstorms accompanied by winds over 75 miles per hour. Fortunately rare, these winds can cause widespread damage to trees, communication wires, and buildings. Tornadoes, as discussed later, typically have wind speeds near 100 miles per hour, but Ohio's strongest tornadoes have winds over 200 miles per hour.

Sunshine

Ohio's position in the northern middle latitudes means that the sun is above the horizon approximately 9½ hours in December and January and 14½ hours in June and July. In addition to this variability in day length, there are large seasonal differences in cloud cover that affect the amount of time the sun actually shines on

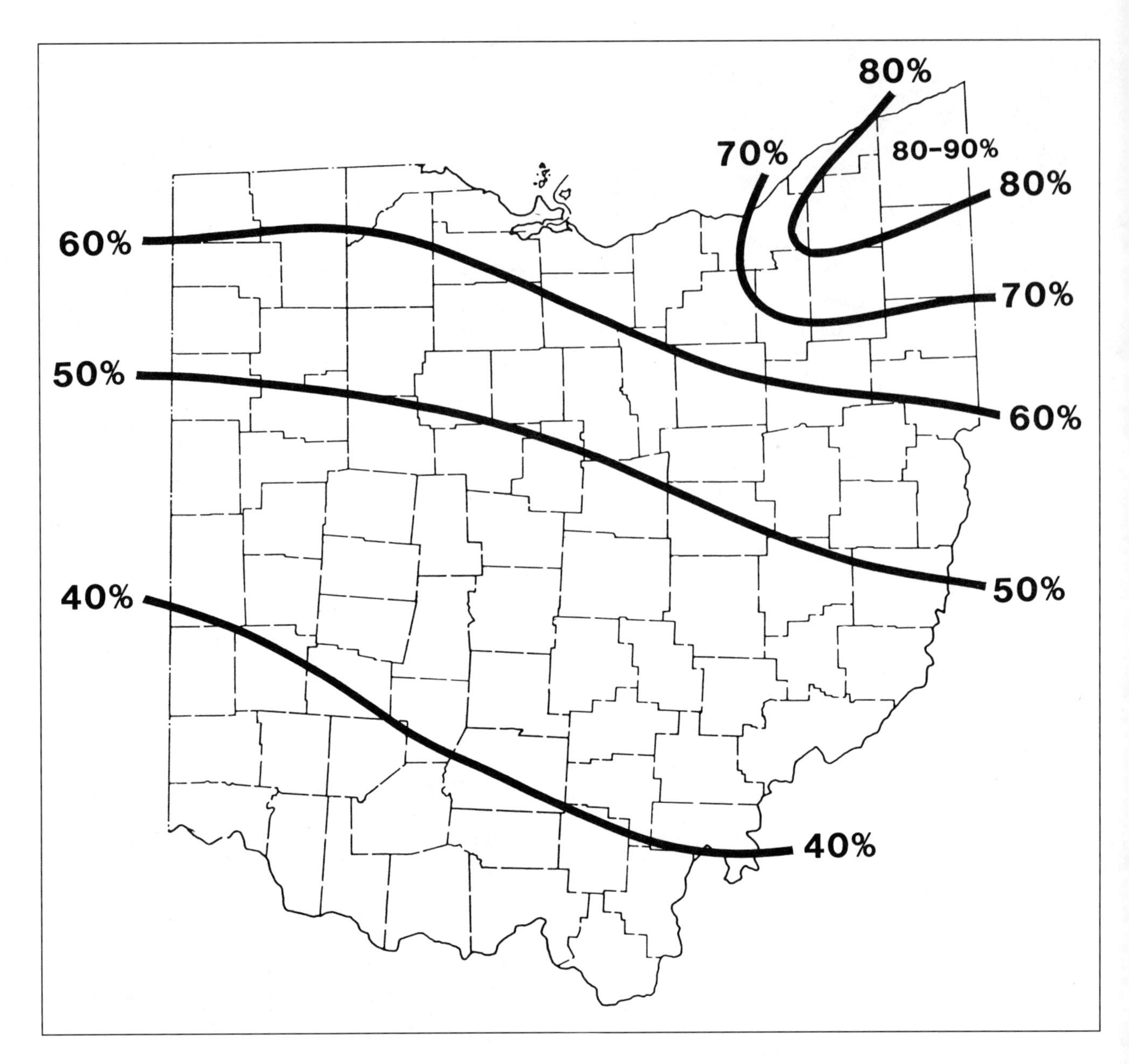

Fig. 2.19. Probability of a snow cover of one inch or more during mid-January.

the state. Summer and early autumn are relatively sunny periods, with storm systems weaker than in winter and high pressure giving many clear or partly cloudy days. During these seasons, the sun shines through about 60 percent of daylight hours.

As November unfolds and winter approaches, storm systems grow stronger and more frequent, cloudiness increases, and sunshine diminishes. During the cold season the sun shines 25–40 percent of daylight hours, with most winter months bringing only 4–5 clear days. The cloudiest portion of Ohio during autumn and winter is the northeast, where lake-effect clouds and snow are the rule. But during summer, the North Coast cities of Cleveland and Toledo are Ohio's sunniest locations.

Tornadoes

Ohio is at the northeastern edge of North America's "Tornado Alley" that is centered in the Great Plains. As Figure 2.20 shows, tornadoes, which are also called *twisters* or *cyclones*, are most common in the western,

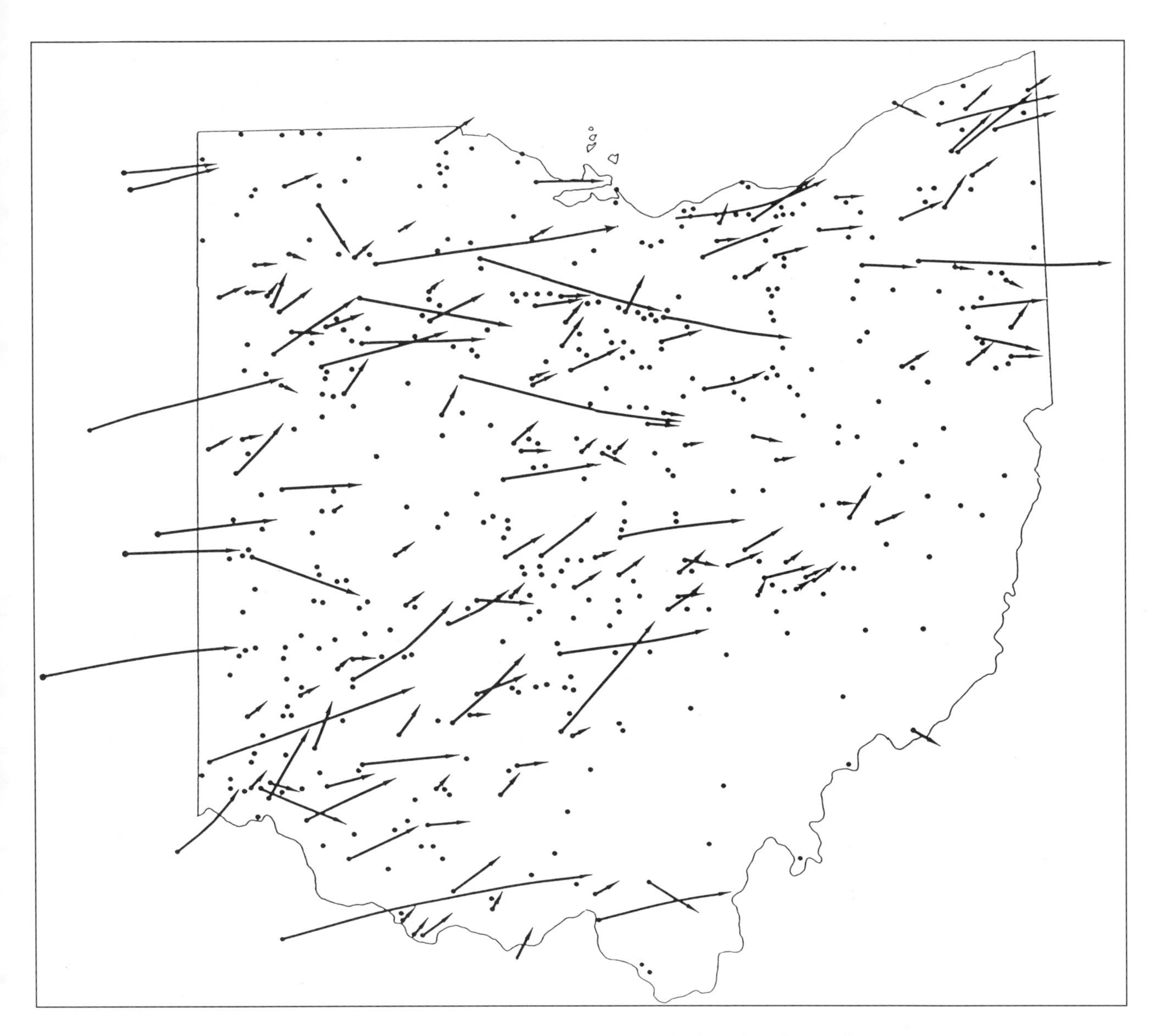

Fig. 2.20. Tornado paths during 1950–1988. Paths less than three miles are shown as a dot. Some tornadoes were not on the ground continuously along their entire path.

central, and northern portions of the state. Since 1950, when the government began keeping detailed records, Ohio has averaged fourteen tornadoes annually (Schmidlin 1988). None were reported during the drought year of 1988, but sixty-one touched down in Ohio during 1992.

The typical Ohio tornado is 100 yards wide, is on the ground for about one-half mile, and has wind speeds near 100 miles per hour (Schmidlin 1988); it causes only minor damage, rarely killing or injuring anyone. About 3 percent of the tornadoes that touch the state are large, violent storms. One-half-mile wide, grounded 20 miles or more, and containing winds estimated at over 200 miles per hour, these storms average just once in ten years but leave a lasting impression, bringing about most of the state's tornado-related injuries and fatalities as they cause extensive damage to farms, homes, and communities in their paths. Tornadoes killed 158 and injured approximately 4,200 people in Ohio during the years 1950–1994.

The deadliest tornado in Ohio history struck Sandusky and Lorain on Saturday, June 28, 1924. Much of downtown Lorain was destroyed and about eighty

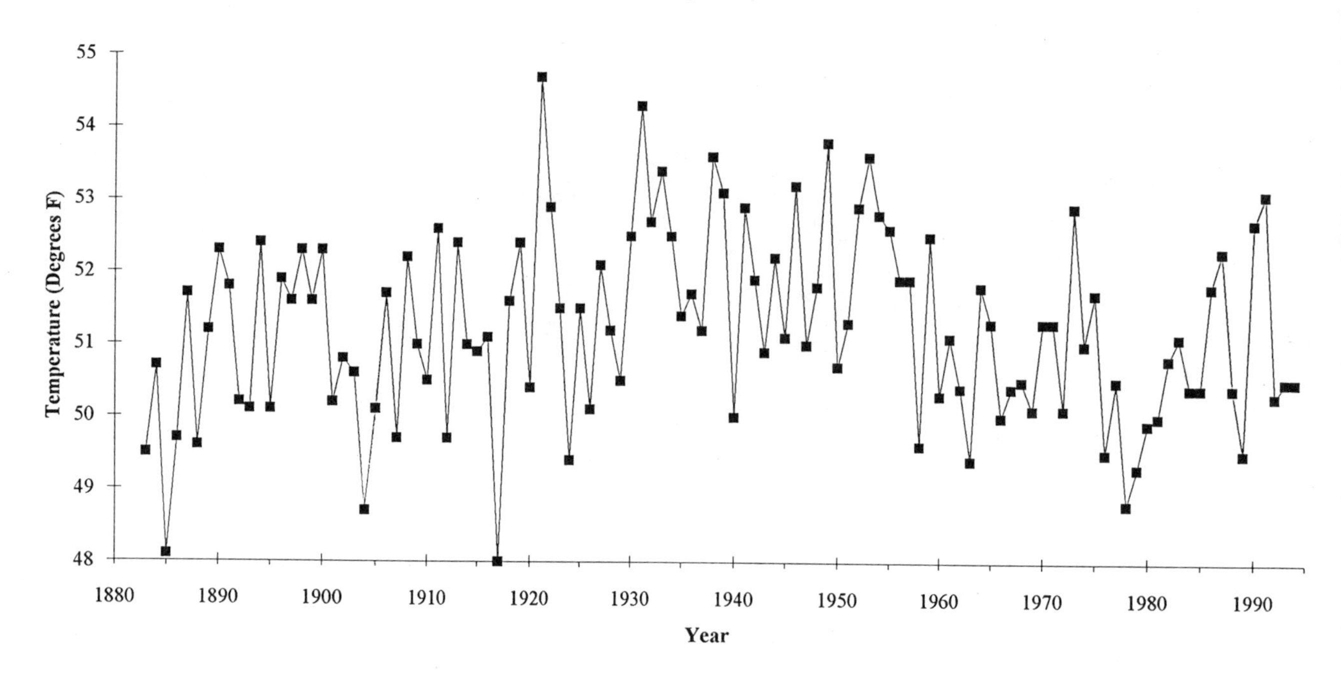

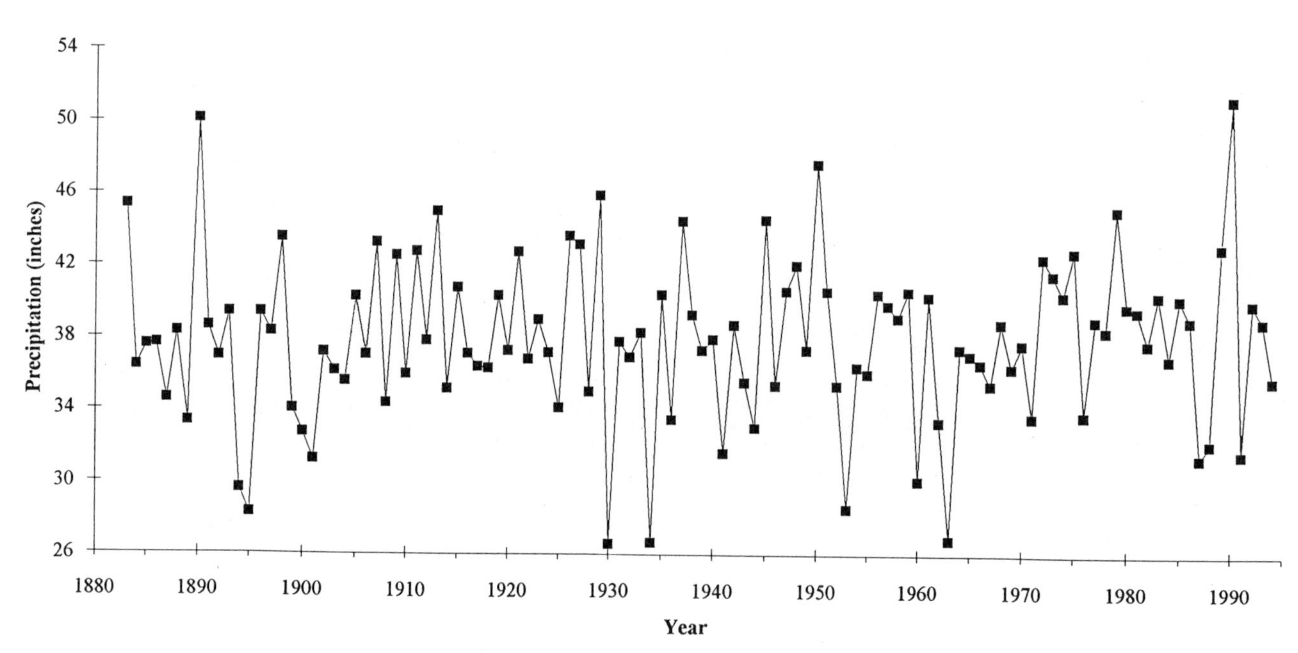

Fig. 2.21. Statewide average annual temperature (top) and precipitation (bottom) during 1883–1994.

persons were killed. Several large tornadoes swept across western and northern Ohio on the evening of April 11, 1965. These "Palm Sunday" storms killed fifty-five persons, with the greatest losses occurring in Allen, Lorain, and Lucas Counties. The deadliest single tornado in recent history killed thirty-two when it devastated much of Xenia on April 3, 1974. On May 31, 1985, several large tornadoes touched down in the northeast, killing eleven in Trumbull County and many more across the border, in Pennsylvania.

Climate Change

Earth's climate is always changing; in fact several ice ages have occurred over the past million years. During the last Ice Age, which ended about 15,000 years ago, much of Ohio was covered for several thousand years by hundreds of feet of ice. Warmer climates, such as those that exist now, prevail between ice ages, when most glacial ice is confined to the polar regions. Another ice age is likely, but not for thousands of years. Changes in climate are generally too slow to be noticed within one person's lifetime. Variability in weather patterns from year to year—such as unusually cold or mild winters, or wet or dry spells—are normal climatic variations, not to be interpreted as changes in climate. Figure 2.21 shows 110 years of statewide averages in temperature and precipitation, revealing some recent trends in temperature (Karl et al. 1983). The years 1880–1890 and 1900–1910 were cold in Ohio. During the period 1930–1955, temperatures warmed, in most years notably exceeding the 110-year average. This was followed by substantial cooling around 1960, with the 1960s, 1970s, and early 1980s much cooler than the mid-twentieth century (Rogers and Yersavich 1988). Precipitation, on the other hand, while varying from year to year, did not change in any identifiable way. The dry years of 1930, 1934, and 1963 are obvious, and the 1960s tended to be dry and the 1970s wet, but there were no clear trends.

There has been much concern and speculation recently about potential climate change caused by carbon dioxide and other "greenhouse gases" that industrial society puts into the atmosphere. Some scientists speculate that these pollutants will bring on rapid greenhouse warming of Earth by the middle of the twenty-first century, while others argue that the Industrial Revolution began over 100 years ago, and there is little evidence of any global warming so far. Our understanding of the atmosphere is so crude, opponents of the theory add, that we cannot accurately predict next week's weather, let alone the weather of the next century. About all that is known for certain is that climate will change in the future, as it has in the past, and humans may have an impact on those changes. Large changes in climate can cause shifts in vegetation and agriculture, affect ocean currents and sea level, and create hazards or opportunities for society that did not exist before.

REFERENCES

Edgell, D. J. 1988. An Analysis of the Snow Water Equivalent Measurement in Ohio with Applications for Snow Melt Climatology. Master's thesis, Kent State University, Kent, Ohio.

———.1992. The Climatology of the Extreme Minimum Winter Temperature in Ohio. Ph.D. dissertation, Kent State University, Kent, Ohio.

Karl, T. R., L. K. Metcalf, M. L. Nicodemus, and R. G. Quayle. 1983. *Statewide Average Climatic History, Ohio 1883–1982*. Historical Climatology Series 6-1. Asheville, N. C.: National Climatic Data Center.

Kochar, N. and T. W. Schmidlin. 1990. Heat Island of the Akron-Canton Airport. *Geographical Bulletin* 32: 46–55.

Miller, M. E., and C. R. Weaver, 1971. *Snow in Ohio*. Research Bulletin 1044. Wooster, Ohio: Ohio Agricultural Research and Development Center.

Rogers, J. C., and A. Yersavich. 1988. Daily Air Temperature Variability Associated with Climatic Variability at Columbus, Ohio. *Physical Geography* 9:120–38.

Schmidlin, T. W. 1988. Ohio Tornado Climatology, 1950–85. *Preprints of the 15th Conference on Severe Local Storms*. Boston, Mass.: American Meteorological Society.

———.1989a. The Urban Heat Island at Toledo, Ohio. *The Ohio Journal of Science* 89:38–41.

———.1989b. Climatic Summary of Snowfall and Snow Depth in the Ohio Snowbelt at Chardon. *The Ohio Journal of Science* 89:101–8.

CHAPTER THREE

Ohio Soils

Thomas L. Nash and Timothy D. Gerber

Soil is a natural body consisting of layers, or *horizons*, of mineral and/or organic constituents of variable thickness, which differ from the geologic parent material in their morphological, physical, chemical, and mineralogic properties. *Pedology,* the term given to the study of soils as naturally occurring phenomena, takes into account their composition, distribution, and method of formation. Though the science of pedology is relatively new, over the past fifty years a large body of work has appeared that covers various aspects of Ohio's soils. In part, this is because the information on the physical environment that is required to produce such texts began accumulating fifty years ago from significant numbers of purely pedological investigations throughout Ohio and the rest of the world. Soil scientists worldwide disagree about the nomenclature and classification of soils, with the effect that information on modern pedologic classification of soil materials has changed in this period. The terminology used herein is that of the Soil Taxonomy produced by the USDA. This chapter will present a modest level of information about the fundamental concepts in pedology, and then examine the characteristic soils of Ohio, their relationship to each other, and their utilization. For a more detailed description of individual soil types the reader is referred to the individual county soil survey report published by the U.S. Department of Agriculture, Natural Resources Conservation Service.

Fundamental Soil Characteristics

Agriculturalists and gardeners usually regard soil as the extreme upper few inches of the earth's crust that are either cultivated or permeated by land plant roots. This idea is limited, allowing neither a full appreciation of the wide differences between soils nor a full comprehension of their potential and problems. The soil scientist recognizes not only the surface layer but the many others underneath, considering the relationship between the soil and its physical environment, especially such factors as natural drainage and vegetation. The first step toward understanding soils is to dig a hole into the surface of the ground, which will expose a layered pattern usually characterized by different colors. Each layer is called a *soil horizon,* and the set of layers in a single hole is called a *soil profile.* Though a soil profile is normally considered to be a two-dimensional cross section through the upper portion of the earth (Figure 3.1), soil actually extends laterally in all directions, forming a three-dimensional continuum. The horizon sequence of any one soil profile is unique. Over distance, one or more horizons make a slow, lateral change into other horizons with markedly different properties. The distance through which specific horizons change is seldom the same; one or more may change completely over a given distance while others continue unaltered and with a greater spatial distribution.

Numcrous studies have indicated that many soil properties are related to both the gradient of the slope and the position of the soil on a slope. The term *catena* is used to describe a group of soils on a hillslope that vary laterally but retain a distinct similarity for a variety of geomorphic and pedologic reasons. These natural conditions are also called *toposequences* (Figure 3.2). Although most soil catenas are depicted as two-dimensional cross sections, a

SOIL PROFILE CHARACTERISTICS

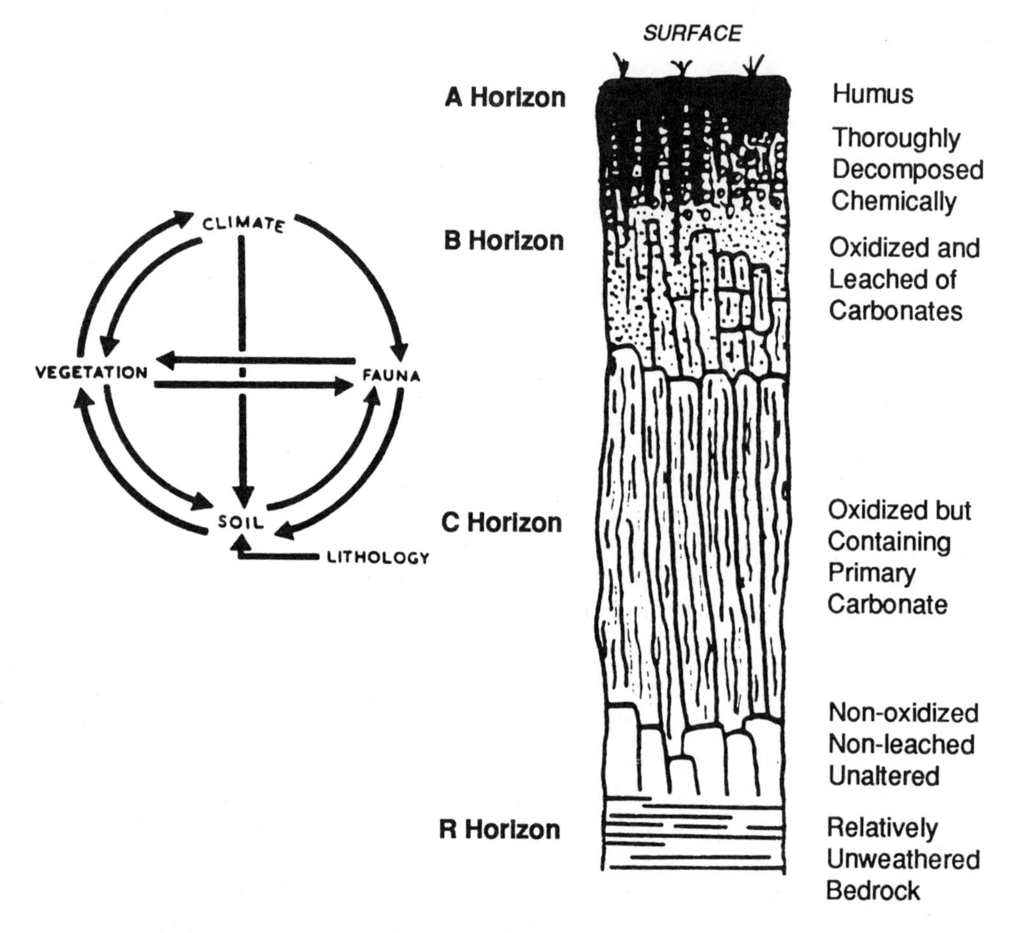

A Horizon - zone from which materials are dissolved, leached, washed downward; first, leaching of CaCO3; second, clay formation and removal; third, oxidation and downward movement of Fe and Al; is zone of maximum biological activity; black, dark brown, dark gray. (Topsoil)

B Horizon - zone of deposition of leached and washed materials from above; accumulation of clay minerals, of Fe and Al; yellowish brown, brown, grayish brown. (Subsoil)

C Horizon - zone of parent material, upper part is oxidized and farther down is not; C Horizon sometimes absent. (Substratum)

R Horizon - unweathered bedrock.

Fig. 3.1. Soil Profile Characteristics

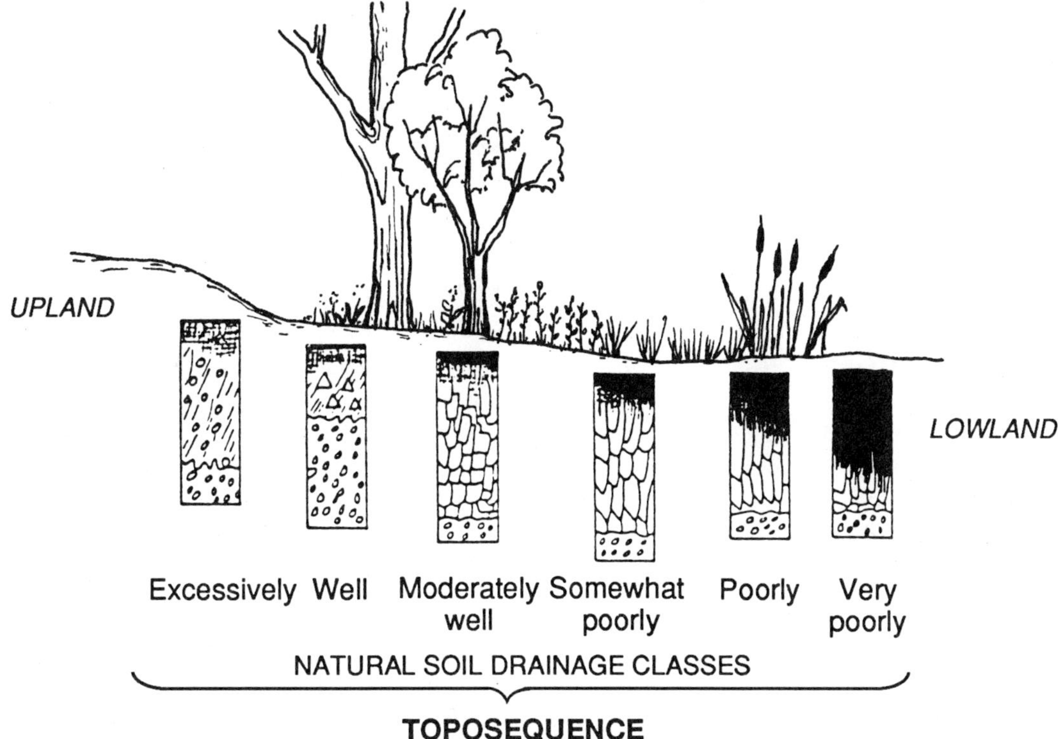

Fig. 3.2. Natural Soil Drainage

three-dimensional concept, such as a small watershed, would be more accurate. Soil catenas are better understood in terms of lateral flow of water from the higher, upland parts of the watershed to the lower terraces and bottomlands. It is common in Ohio to find markedly different soils juxtaposed in rolling topography where convex slopes are at a slightly higher elevation than adjacent concave slopes or closed depressions. Soils on convex slopes are usually well or moderately well drained, while those in depressions are somewhat poorly to very poorly drained, with indications of varying degrees of wetness exhibited in the profile. In broad, flat, upland positions soil drainage is more closely associated with the texture of the soils. Coarser textured soils are generally more permeable than finer textured soils, and these differences are reflected in the soil profile. It should be understood that even coarse-textured soil materials can be poorly drained if found in depressions.

Factors of Soil Formation

Since the early studies of Dokuchaev in Russia and Hilgard in the United States, soil scientists have identified five main factors that define the soil system (FitzPatrick 1983): climate, organisms, topography, geologic parent material, and time. These factors and their processes provide insight into the distribution of soils in Ohio.

Climate-Soil Relationships

The climate is considered by many scientists to be the most important factor in determining the properties of soils. Moisture and temperature are the two aspects most significant in controlling soil properties. Moisture is important because water in any form is involved in manifold physical, chemical, and biochemical processes that function within the soil, and the amount of moisture received by the soil surface influences the weathering and leaching conditions deep into the soil profile. Temperature influences the rate of chemical and biochemical processes.

Organism-Soil Relationships

Within a relatively constant regional climate, vegetation and fauna will vary from site to site, and thus soil properties can vary accordingly. Some changes in soil characteristics are quite subtle and can be observed only as variations in soil chemistry, such as pH or exchangeable cations attached to particles of clay and organic matter. The biotic factor in soil formation is difficult to assess because of the dependence of both organisms and soil on climate and the interaction of soil and vegetation. In many parts of Ohio, such effects could be interpreted as microclimatological influences associated with vegetation differences.

Topography-Soil Relationships

Topography, and in particular local relief, controls much of the distribution of soils in Ohio, to such an extent that soils of markedly contrasting morphologies and characteristics can merge laterally with one another and yet be in equilibrium with local environmental conditions. Soil characteristics and soil moisture conditions vary laterally with topography. Orientation of slope, for instance, affects the microclimate and hence the soil; steepness of slope affects soil properties through the rates of surface water runoff and erosion; rolling terrain causes wide variation because lower areas are likely to receive accumulations of water runoff and sediment derived from surrounding higher areas; and low areas can be affected by the high water table, which significantly influences soil profiles.

An excellent example of the influence of slope orientation on soil is the study of Finney, Holowaychuk, and Heddleson in southern Ohio (1962). Several NW-SE trending valleys were studied, all of which possessed parent material of mostly sandstone colluvium. The microclimate varies quite significantly with orientation, with SW-facing slopes, for example, illustrating higher temperatures on the leaf-litter, as well as greater annual fluctuations. Soil moisture values follow the temperature differences, with soils on NE-facing slopes generally more moist than those on SW-facing slopes. Vegetation correlates with moisture and temperature trends. A mixed oak association is dominant on the SW-facing slopes, whereas a mixed mesophytic plant association is found on NE-facing slopes. These combined microclimatic and vegetation differences produce soil differences. The NE-facing slopes are characterized by a thicker "A" horizon.

Influence of Parent Material on Weathering and Soil Formation

Geologic parent materials influence soil properties to varying degrees. Minerals which make up rocks and glacial materials vary in their resistance to weathering—some weather quite rapidly, others more slowly. Texture influences the rate and depth of leaching, and this relates to many soil properties. Grain size in rocks and texture of till affect the rate of weathering, with coarser materials weathering more rapidly than fine materials. Parent materials exert control on the clay minerals that form, but once weathering release has taken place, the micro- and macro-environments within the soil determine whether certain ions or other constituents are selectively removed or remain behind. This ultimately determines what kind of clay forms within the soil profile.

Soil Development with Time

All rock and mineral weathering brought about through the actions of climate, organisms, and topography is time dependent. The time needed to produce various combinations of weathering and soil features varies, but generally those soil properties associated with the buildup of organic matter develop rapidly, while those associated with the weathering of minerals develop slowly. For example, granitic clasts in outwash in Ohio do show progressively greater alteration with age, but even in Illinoian-age outwash their interiors are not strongly weathered (Lessig 1961). Many of the properties of soils require a long time to form.

In general, most soils in the state exhibit an *argillic* horizon, which is a part of the "B" horizon that has been enriched with clay carried downward from the upper part of the soil profile by percolating water. Globally such soils are associated with cool and humid to hot and semiarid regions and with mostly middle to late Pleistocene geomorphic surfaces that have been stable for at least several thousand years. Most soils in Ohio are classified as Alfisols, with the better drained ones further classified as Hapludalfs and the poorer drained ones further classified as Epiaqualfs or Endoaqualfs.

While an interesting study in its own right, pedogenic processes cannot be covered in detail in this book. However, it is important to relate the application of soil information to the overall understanding of Ohio's present-day geography.

Characteristic Soils of Ohio

Since the 1960s there has been an outpouring of publications dealing with the characterization, genesis, classification, and geography of soils throughout the United States. The completion of the *FAO Soil Map of the World* and the publication of *Soil Taxonomy* are excellent examples of such works (Foth and Schafer 1980). Although these publications and thousands of individual county soil survey reports have greatly increased our knowledge and understanding of soils, they have addressed themselves primarily to a professional audience and present much detail about a limited geographic area. This section of the chapter synthesizes and organizes some of this knowledge to make it readily available to students, teachers, and others interested in the land of Ohio.

The soils of Ohio formed through the effects of natural forces, including living organisms, on rock and mineral material exposed over long periods of time on the earth's surface. Since these factors did not interact uniformly throughout the state, some 450 different soils formed and are currently recognized in Ohio. These soils are grouped into eight major soil regions based on the type of rock and mineral materials and the length of time that this material was exposed on the surface. Figures 3.3a and 3.3b describe each of the major soil regions. Figure 3.4 illustrates the soil profile characteristics for significant regional soils.

Northeastern Ohio Soils

E. Larry Milliron

Land use in northeastern Ohio is highly diversified. While dairy operations, grain farming, and truck crops are common, one also finds Christmas tree plantations, orchards, and nurseries. The region's high population makes land use and management highly competitive. About 30 percent of the land is forested.

The most important soil management concerns for agriculture are erosion in sloping areas and seasonal wetness in depressions and broad flats. Although there is less cropland in this region than in western and northwestern Ohio, a high proportion of the soils are still designated as "prime farmland."

The Wisconsin glacier deposited till that constitutes the unconsolidated mantle overlying the sandstone and shale bedrock that dominates this region. Because of the scarcity of limestone bedrock in the area, the till has a relatively low lime content, which gradually increases from east to west. Clay content in the till generally increases from south to north.

The most common soils of northeastern Ohio are Canfield and Mahoning soils, which formed in glacial till under coniferous vegetation and later deciduous forest conditions. While argillic horizons are identified in both soils, Canfield soils are classified as Fragiudalfs instead of Hapludalfs because of a distinctive genetic subsoil horizon called a *fragipan*. Mahoning soils are Epiaqualfs. The two soils are associated with different kinds of till, so they are not commonly adjacent to each other. However, both soils have a silt loam surface horizon with moderate organic matter content. The pH in both soils increases with depth in the subsoil, generally from strongly or very strongly acid to neutral, and the substrata have carbonates.

Canfield soils have a fragipan—which is dense and brittle and severely restricts water movement—in the lower part of the subsoil below about 1½–2½ feet. Though they are moderately well drained, permeability is slow in the fragipan, even though the clay content there is only about 25–32 percent. They have low available water capacity because roots are restricted to the zone above the fragipan. Slopes commonly range 2–12 percent.

Mahoning soils have more clay in the subsoil, generally 35–45 percent, which causes the permeability to be slow or very slow. They are somewhat poorly drained and have moderate available water capacity because roots are not restricted to the upper subsoil. Gray colors and carbonates are present at shallower depths than in the Canfield soils. Slopes range 0–6 percent.

Northwestern Ohio Soils

Timothy D. Gerber

The soils of the lake plain in northwestern Ohio fall in the easternmost corner of the American Corn Belt and have high natural fertility. Nearly all farmland here is cropland, with only about 7 percent forested. Seasonal

OHIO'S SOIL REGIONS

SOILS FORMED IN:

High Lime Glacial Lake Sediments and Glacial Till
Major soils: Hoytville, Latty, Nappanee, Paulding, Toledo
Occupy about 12% of the state.

Very deep, very poorly drained and somewhat poorly drained soils on broad flats with some rises. Formed in glacial lake sediments and glacial till high in lime. Natural fertility is high. Due to seasonal wetness, high clay content, slow or very slow permeability and flat topography; artificial drainage is needed for optimum crop production. The soils are used primarily for corn, soybeans, wheat and hay. Tomatoes and sugar beets are important specialty crops.

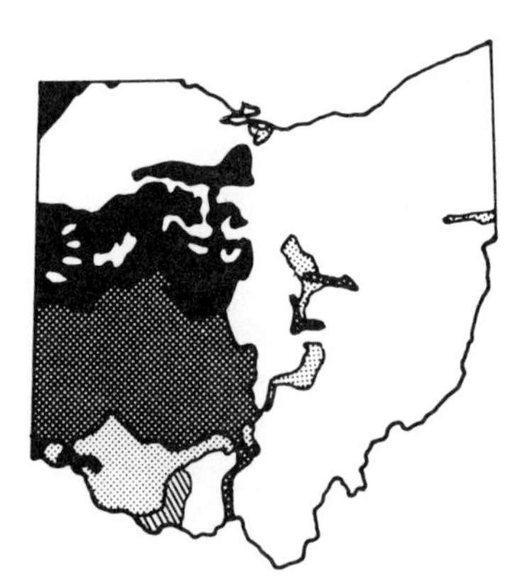

SOILS FORMED IN:

High Lime Glacial Drift of Wisconsin Age
Major soils:
Subregion A (to the north)- Blount, Glynwood and Pewamo

Subregion B (to the south)- Brookston, Crosby, Eldean, Fincastle and Miamian
Occupy about 29% of the state.

Very deep, well drained to very poorly drained and nearly level to sloping. Formed in glacial drift derived mostly from limestone and dolomite. In **Subregion A** formed in moderately fine textured glacial till while in **Subregion B** formed in medium textured glacial till and in glacial outwash. Natural fertility is moderate to high. Due to the slope, slow or moderate permeability and seasonal wetness; erosion control and artificial drainage are needed for optimum crop production. The soils are used primarily for corn, soybeans, wheat and hay grown under a mixed livestock and cash grain system of farming. Areas adjacent to metropolitan centers of the region are rapidly being developed for nonfarm uses.

Glacial Drift of Illinoian Age
Major soils: Avonburg, Clermont, Rossmoyne and Homewood
Occupy about 7% of the state.

Very deep, poorly drained to well drained on broad nearly level and gently sloping flats interspersed with strongly sloping valleys. Formed in glacial drift derived mainly from limestone, sandstone or shale overlain in many areas by loess. More deeply leached, lower in natural fertility, more acidic and much older than soils formed in high lime lake sediments and glacial drift. Due to seasonal wetness, slow or very slow permeability and gently sloping to strongly sloping topography; artificial drainage and erosion control are needed for optimum crop production. Cash grain and general farming are the major types of agriculture. Tobacco is an important specialty crop.

Limestone and Shale
Major soils: Bratton, Eden and Jessup
Occupy about 1% of the state.

Moderately deep to very deep, moderately well drained and well drained soils on gently sloping and sloping ridges, as well as steep hillsides bordering narrow valleys. Formed mainly in limestone or interstratified limy shale and limestone. Natural fertility is high or medium. Erosion control is needed for optimum crop production. Most of the region is used for pasture and woodland. Tobacco is an important specialty crop on the less sloping soils.

Source: Ohio Department of Natural Resources

Fig. 3.3a. Ohio's Soil Regions

OHIO'S SOIL REGIONS

continued

SOILS FORMED IN:

Low Lime Glacial Lake Sediments, Glacial Till and Beach Ridge Deposits

Major soils: Canadice, Conneaut and Oshtemo

Occupy about 2% of the state.

Very deep, poorly drained to well drained soils on nearly level flats broken by parallel, undulating beach ridges. Formed in glacial lake sediments, glacial till and beach ridge deposits derived mainly from sandstone and/or shale. Low in lime content. Generally low or medium in natural fertility. Due to seasonal wetness, slow or very slow permeability and flat topography of the Canadice and Conneaut soils; artificial drainage is needed for optimum crop production. Droughtiness and an erosion hazard limit crop production on Oshtemo soils. Much of the region is urbanized and used for housing, industry, transportation and related nonfarm uses.

Low and Medium Lime Glacial Drift of Wisconsin Age

Major soils: Bennington, Canfield, Mahoning, Platea, Venango

Occupy about 22% of the state.

Very deep, somewhat poorly drained and moderately well drained, nearly level and gently sloping soils on broad undulating areas broken in places by a few higher hills, as well as sloping to steep breaks bordering stream valleys. Formed in glacial drift of Wisconsin age derived mainly from sandstone and shale with lesser amounts of limestone. Natural fertility is medium. Due to seasonal wetness, slow or very slow permeability and slope; artificial drainage and erosion control are needed for optimum crop production. General farming and dairying are the principal types of agriculture. Areas adjacent to large metropolitan centers are no longer farmed and are waiting nonfarm development or are already developed.

Sandstone, Siltstone and Shale

Major soils: Coshocton, Gilpin, Rarden, Shelocta, Westmoreland and Upshur

Occupy about 27% of the state.

Moderately deep to very deep, well drained and moderately well drained soils on gently sloping and sloping ridgetops, as well as moderately steep and very steep hillsides. Formed mainly in acidic sandstone, siltstone and shale. Some formed in limestone or in limey shale. Natural fertility is medium or low and soils are generally acidic. Erosion control needed for optimum crop production. Because much of the region is too steep or stony for cultivation, a relatively large proportion of the region is forested or pastured. General farming is practiced on many of the less sloping areas. Strip mining is common in some areas and contributes sediment and toxic acid to some streams.

Source: Ohio Department of Natural Resources

Fig. 3.3b. Ohio's Soil Regions

wetness is a severe problem, and both surface and tile drainage are commonly used to increase crop yields by lowering the water table, particularly in April and May. The soils are severely limited for use as highway foundations, building sites, or septic tank absorption fields because of the seasonally high water table, slow runoff, restricted permeability, and low bearing strength.

The Wisconsin glacier remained stationary for a time near the north border of Ohio and impounded water that created a huge glacial lake. Even some of the previously deposited till was reworked by the water, which produced lacustrine sediments. This created a landscape with soils developed from fine-textured calcareous till and lake deposits. The most common soils of this region are Hoytville and Nappanee soils, both Epiaqualfs that developed under swamp forest vegetation—primarily elm, ash, and soft maple trees. They formed in late Wisconsinan glacial

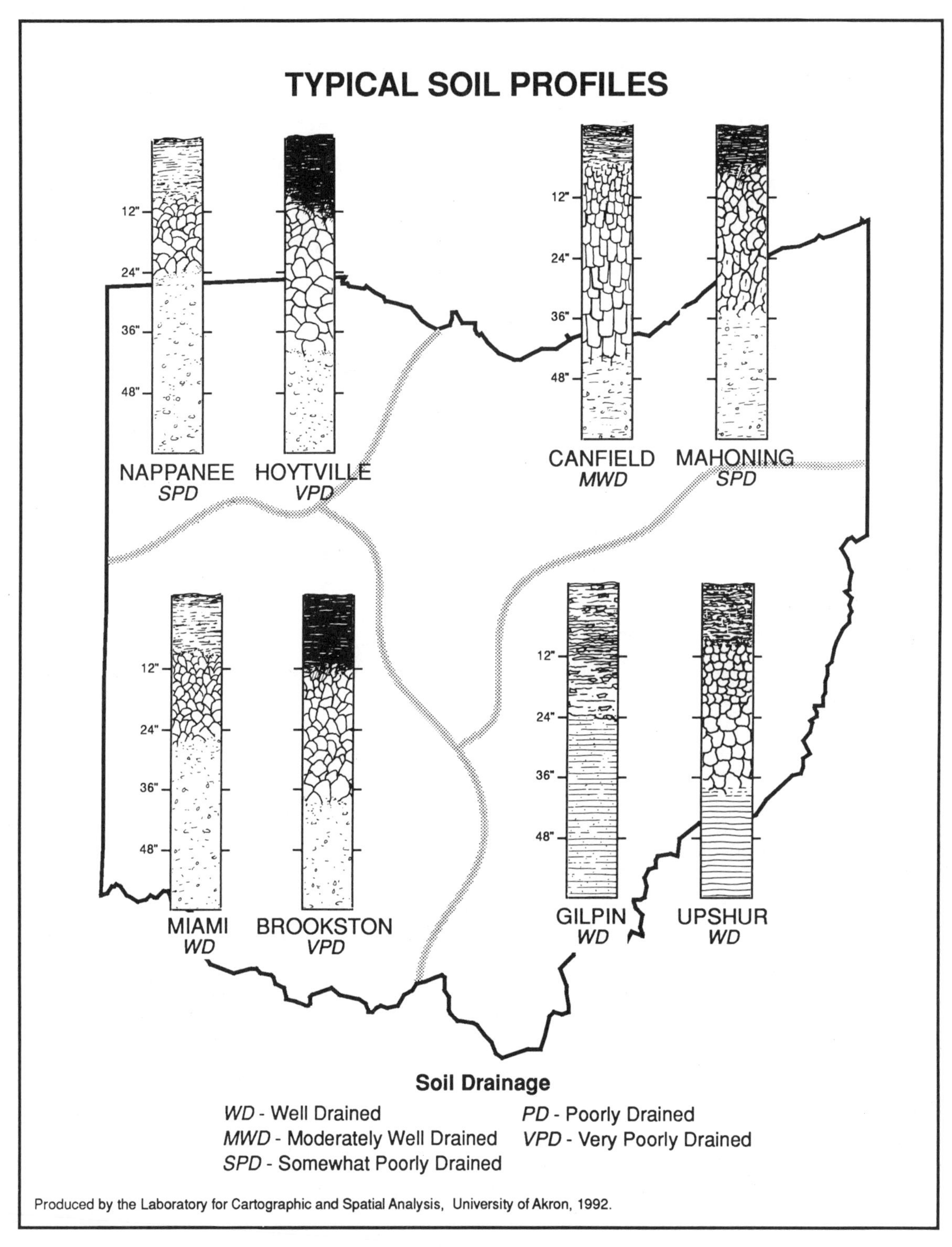

Fig. 3.4. Typical Soil Profiles

till that was clay-enriched from readvances of the ice across clay-filled glacial lake beds. Both soils have slow permeability and a moderate available water capacity because of the relatively high clay content—commonly 35–50 percent—in the subsoil and substratum. The subsoil is generally slightly acid to mildly alkaline (pH 6.1–7.8), with pH gradually increasing with soil depth.

Hoytville soils, which are found in depressions and broad flats with less than 2 percent slope, are very poorly drained. The surface horizon is commonly clay or silty clay loam, with moderate to high organic matter content. Subsoil colors are mainly shades of gray with brown mottles. Carbonates are leached from the upper 3–4 1/2 feet.

Nappanee soils, which appear on low knolls and on side slopes along drainageways with less than 6 percent slope, are somewhat poorly drained. The surface horizon is commonly silty clay loam or silt loam, with moderate organic matter content. Subsoil colors are mainly shades of brown with grayish mottles and coatings on the surfaces of the natural aggregates. Carbonates are leached from the upper 1 1/2–3 1/2 feet.

Western Ohio Soils
K. Ed Miller

The soils of western Ohio also comprise part of the easternmost section of the American Corn Belt and thus are used dominantly for cropland, with only about 7 percent forested. They have moderate to high natural fertility.

Seasonal wetness is a major concern in this region, but only in depressions and broad flats. Since western Ohio also includes more sloping areas, erosion creates problems for agriculture; whereas slope, restricted permeability, and low strength are limitations for urban development. Two lobes of the Wisconsin glacier, which are separated by a massive bedrock high point near Bellefontaine in Logan County, created alternating sequences of ground moraines and end moraines in this region. The variation in landscapes that resulted has played a major part in the origin and development of the soils. Since the bedrock across which the ice moved was dominantly limestone, carbonaceous materials make up about 1/4–1/2 of the glacial drift.

The most common soils of this region are Miamian and Brookston soils, which formed in loamy glacial drift and in a thin mantle of loess in places. Trees were the dominant vegetation during formation, with swamp forests and wetland prairies common on the extensive flats. Though argillic horizons are identified in both soils, Brookston soils are classified as Argiaquolls instead of Endoaqualfs because of their thick, dark surface horizon. Miamian soils are Hapludalfs. Both soils have moderately slow permeability because of the relatively high clay content in the subsoil. Although the substratum or "C" horizon is generally dense, it has only about 20–27 percent clay.

Miamian soils are well-drained soils found on knolls, side slopes, and ridgetops. Slopes range from nearly level to very steep. The surface horizon is commonly silt loam with moderate organic content, unless the soil has been eroded enough that subsoil material has been mixed by plowing. Miamian soils have a moderate available water capacity, and subsoil colors are mainly shades of brown. The pH in the subsoil increases with soil depth from as low as 4.5 to as high as 7.8. Carbonates are leached from the upper 1 1/2–3 1/2 feet.

Brookston soils are very poorly drained soils; they occur in shallow, closed depressions and on broad flats with less than 2 percent slope. The surface horizon is commonly silty clay loam with moderate or high organic matter content. Brookston soils have a high available water capacity, and subsoil colors are mainly shades of gray and brown in a mottled pattern. The subsoil is generally neutral or mildly alkaline, and carbonates are leached from the upper 2 1/2–5 feet.

Southeast Ohio Soils
Charles E. Redmond

Southeast Ohio is characterized by rugged topography: hills with long, steep side slopes and narrow, sloping ridgetops are separated by narrow valleys that carry swiftly flowing streams. About 50 percent of the land is forested. Crops are largely confined to ridgetops and valley bottoms, and controlling erosion is the main concern in the management of cropland and pastures. Using equipment on the steep slopes and conserving moisture are other concerns. The ridgetops have the best potential for building sites, although shallowness to bedrock and seasonal wetness are limitations in some areas. Slippage is a severe limitation in the placement of buildings and roads on the lower part of many slopes.

This part of the state has not been subjected to the leveling and mixing effects of the continental glaciers. Consequently, the soils formed from a variety of Mississippian-, Pennsylvanian-, and Permian-age bedrocks. The different types of rock are interbedded, resulting in a very mixed pattern of soil parent materials. This pattern is further complicated by the movement of weathered rock material downslope.

Because of the mixed pattern of parent materials, soil complexes or mixtures of soils are common. Gilpin and

FORESTS OF OHIO
(at the time of the earliest land surveys)

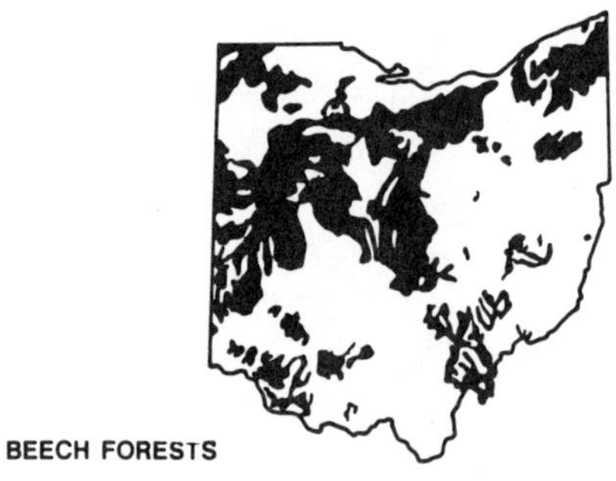

BEECH FORESTS

These forests were characterized by a large fraction of beech, sugar maple, red oak, white ash, and white oak, with scattered individuals of basswood, shagbark hickory, black cherry, and more rarely, cucumbertree. The most familiar types were beech-sugar maple and "wet beech" on poorly drained flatlands. In the dissected Allegheny Plateau, tuliptree, red maple, and/or sugar maple were associated with beech, generally in the valleys, forming the beech-maple-tuliptree subtype.

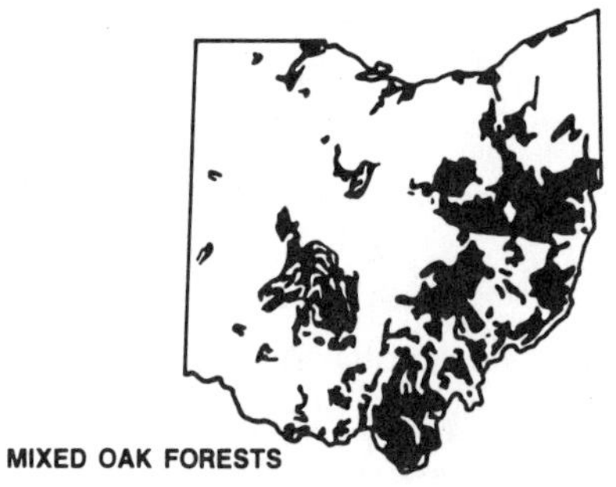

MIXED OAK FORESTS

These forests included a wide variety of primary forest types, of which the most widespread were the white oak-black oak-hickory type and the white oak type. A white oak-black oak-chestnut type occurred in the low-lime glaciated plateau, chiefly on hilltops, and extended down the south-facing slopes. Covering ridge tops of the unglaciated Allegheny Plateau were forests of white oak-black oak and chestnut oak-chestnut, with sour gum, flowering dogwood, sassafras, Virginia pine, pitch pine, and/or shortleaf yellow pine locally.

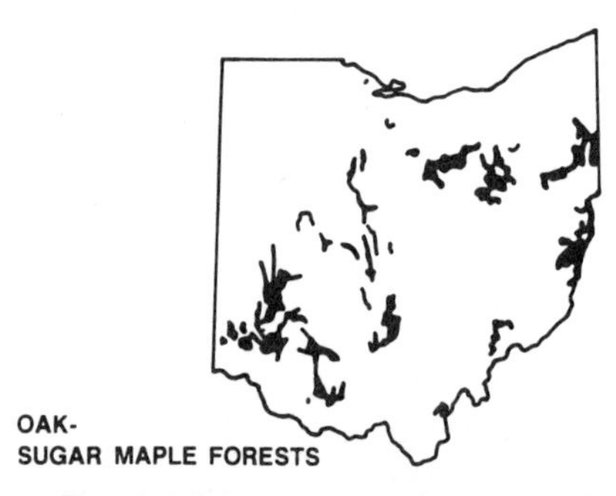

**OAK-
SUGAR MAPLE FORESTS**

These included xero-mesophytic forests usually lacking beech, chestnut, red maple, and tuliptree. Dominants included white oak, red oak, black walnut, black maple, sugar maple, white ash, red elm, basswood, bitternut, and shagbark hickory. Of indicator value today in the areas formerly occupied by these forests are Ohio buckeye, northern hackberry, honey locust, and blue ash. Local components often included black cherry, Kentucky coffee-tree, chinquapin oak, redbud, and eastern red-cedar.

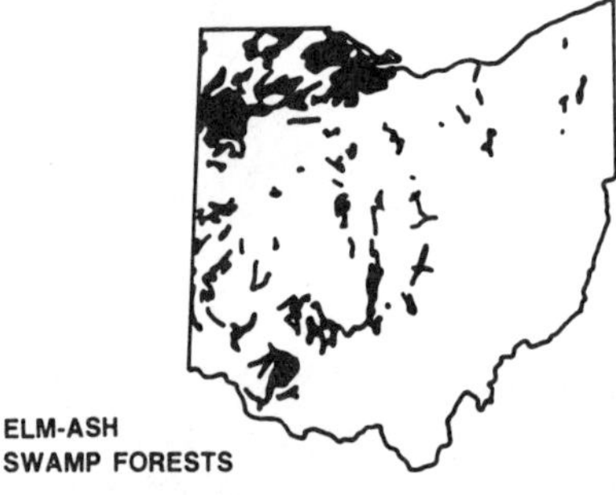

**ELM-ASH
SWAMP FORESTS**

These forests were consistent in having among the dominant trees of the canopy white elm, black ash, and/or white ash, silver maple, and/or red maple. Extremely wet phases contained cottonwood and/or sycamore. Better-drained phases or transitions are bur oak-big shellbark hickory and red oak-basswood. These "swamp oak-hickory" communities were enriched locally with swamp white oak, pin oak, white oak, black walnut, and tuliptree.

MIXED MESOPHYTIC FORESTS

Mixed mesophytic forests had no single species comprising a very large fraction of the dominants. Segregates included oak-chestnut-tuliptree, oak-hickory-tuliptree, white oak-beech-maple, and hemlock-beech-chestnut-red oak. North-facing slopes of the Portage Escarpment in Lake and Ashtabula Counties were covered with the latter forest type. The mixed mesophytic forests of southwestern Ohio were generally different from those of eastern Ohio in that the former contained a large fraction of white basswood, and tuliptree.

Source: Ohio Department of Natural Resources

Fig. 3.5. Forests of Ohio

WETLANDS OF OHIO
(at the time of the earliest land surveys)

PRAIRIE GRASSLANDS

Most of these grass-dominated communities were located on wet lands and were dominated by tall grasses such as the giant reed grass, slough grass, bluejoint, and/or big bluestem. Drier prairies and borders were dominated by big bluestem, little bluestem, switch grass, and/or Indian grass.

OAK SAVANNAS (OAK OPENINGS)

Black oak and white oak were abundant in open thin groves or scattered clumps or individuals. Associated species included trembling aspen, red maple, pin oak, sassafras, prairie willow, big bluestem, little bluestem, panic grass, Indian grass, sandbur grass, blueberry, aromatic wintergreen, wild strawberry, and chokeberry. Wet depressions supported shrubby cornel, buttonbush, and willow thickets.

MARSHES AND FENS

Marshes differed from wet prairies in that grasses composed only a small fraction of the vegetation. Where water stood through the summer to a depth of one to three feet, tall emergent aquatics grew, including bulrushes, giant reed grass, wild rice, cattail, bur-reed, wapato, pickerelweed, and rose-mallow. More frequent but less extensive were the sedge meadow communities.

Fens are habitats supplied with cold water, artesian in origin, containing dissolved bicarbonates of calcium and magnesium. Shrubby cinquefoil, swamp birch, and hoary willow appear to be restricted to fens. Characteristic forbs are grass-of-Parnassus, fringed gentian, Canadian burnet, Kalm's lobelia, Riddell's goldenrod, and Ohio goldenrod.

SPHAGNUM PEAT BOGS

These occurred in undrained areas, e.g. kettle holes in moraines, and contained remnants of the boreal evergreen forest including tamarack, poison sumac, leatherleaf, cotton sedge, round-leaved sundew, northern pitcher-plant, and *Sphagnum* species.

BOTTOMLAND HARDWOOD FORESTS

These forests occupied older valleys and terraces of major streams as well as recent alluvium. They included vegetation types of variable composition: beech-white oak, beech-maple, beech-elm-ash-yellow buckeye, elm-sycamore-river birch-red maple, and sweet gum-river birch.

Source: Ohio Department of Natural Resources

Fig. 3.6. Wetlands of Ohio

Upshur are two common soils of this region, and in many areas they are present in a complex pattern of alternating bands around the hillsides. Both soils developed under mostly oak forest vegetation and have argillic horizons, but though Upshur soils are Hapludalfs, Gilpin soils are classified as Hapludults because of their acidity. Gilpin and Upshur soils are both well drained and are present on ridgetops and side slopes with up to 70 percent slope. Gilpin soils are only 20–40 inches deep to bedrock, commonly olive brown siltstone that is rarely hard enough to interfere with excavations to 5 feet. They have moderate permeability and a low available water capacity. The surface horizon is commonly silt loam with low to moderate organic matter content. The subsoil is mainly yellowish brown in color and strongly or very strongly acid (pH 4.5–5.5).

Upshur soils are deeper than 40 inches to soft shale bedrock. They have slow permeability and a moderate available water capacity. Although some areas have a silt loam surface horizon with moderate organic matter content, many eroded areas have much more clay and less organic matter on the surface. The subsoil is mainly reddish brown clay. The pH in the subsoil ranges widely, from 4.5 to 7.8, with higher pH conditions—a result of carbonates in the bedrock—occurring in some places.

Natural Vegetation

The original natural vegetation of Ohio has been almost completely altered for agricultural and settlement purposes. Whereas in 1800 primeval forests covered more than 25 million acres, in 1940 only some 3.7 million acres remained intact. Today this area is continually reduced, as suburban land uses spread over the countryside. Figures 3.5 and 3.6 portray the distribution and character of natural vegetation in the state at the time of settlement.

The Black Swamp (based on Noble 1975)

Perhaps the most striking example of vegetation change has been in northwestern Ohio, where today there is little remaining evidence of the tangle of swamp forest, the Black Swamp, which originally stretched from Paulding and Van Wert Counties to the marshy Lake Erie shore in Lucas and Ottawa Counties. This great barrier to early east-west movement was slowly drained, cleared, settled, and turned to cultivation between 1860 and 1885. The extent of the Black Swamp approximates quite closely that of glacial Lake Maumee and occupies an area almost the size of Connecticut. Within this virtually level and naturally poorly drained area grew a dense forest dominated by elm and ash, but containing also oak, birch, cottonwood, and poplar. Elsewhere in the state, especially within the areas of glaciation, other patches of swamp forest developed, but were nowhere of large extent.

The Black Swamp—"a region of either standing water or so wet as to ooze water when walked upon" (Kaatz 1955) covered by a thick forest vegetation made more impenetrable by windfalls and scattered tangles of trees blown down during tornadoes or other violent windstorms—was a most formidable barrier to early movement. Accounts of the Battle of Fallen Timbers (1794) give vivid descriptions of one area obstructed by windfallen trees. The early missionary Zeisberger conveys some idea of the difficulty of travel by noting that a journey from Sandusky to the Maumee River, a distance of between 30 and 35 miles, in the dry season (October 1791) took two and a half days by horseback (Bliss 1885, 30). As a result of the Black Swamp barrier, westward movement was diverted northward via Lake Erie or by land across southern Ontario from New York to Michigan or south of the swamps along the Ohio River. The significance of this diversion is shown by the fact that between 1830 and 1890 the population of the six southeasternmost counties of Michigan exceeded that of the nine northwesternmost counties of Ohio (Kaatz 1955). During this period, settlement within the Black Swamp was limited at first to the better drained river levees. Only after the completion of drainage operations around 1885 was the potential for settlement in northwestern Ohio realized.

REFERENCES

Birkeland, P. W. 1984. *Soils and Geomorphology.* New York: Oxford University Press.

Bliss, E. 1885. *Diary of David Zeisberger, a Moravian Missionary among the Indians of Ohio.* Cincinnati: R. Clark Clarke and Co.

Finney, H. R., N. Holowaychuk, and M. R. Heddleson. 1962. The Influence of Microclimate on the Morphology of Certain Soils of the Allegheny Plateau of Ohio. *Soil Science Society of America Proceedings* 26:28–92.

FitzPatrick, E. A. 1983. *Soils: Their Formation, Classification, and Distribution.* London: Longman.

Foth, H. D., and J. W. Schafer. 1980. *Soil Geography and Land Use.* New York: Wiley.

Kaatz, M. R. 1955. The Black Swamp: A Study in Historical Geography. *Annals, Association of American Geographers* 45:1–35.

Lessig, H. D. 1961. The Soils Developed on Wisconsin and Illinoian Age Glacial Outwash Terraces along Little Beaver Creek and Adjoining Upper Ohio Valley. *Ohio Journal of Science* 6:286–94.

Noble, A. G., and A. J. Korsok. 1975. *Ohio: An American Heartland.* Columbus: Ohio Department of Natural Resources.

Peattie, R. 1923. *Geography of Ohio.* Columbus: Geological Survey of Ohio.

Soil Conservation Service. 1967. *Soils: Distribution, Principal Kinds of Soil; Orders, Suborder and Great Group.* Sheet No. 86. In National Atlas, U.S. Geological Survey. Washington, D.C.

Wilding, L. P., N. E. Smeck, and G. F. Hall, eds. 1983. *Pedogenesis and Soil Taxonomy.* New York: Elsevier.

Editor's Note: The authors would like to thank the following three soil scientists, who contributed the regional vignettes to this chapter. K. Ed Miller, Soil Survey Coordinator, Ohio Department of Natural Resources, Division of Soil and Water Conservation. He has coauthored soil survey manuscripts for Marion, Hardin, and Clark Counties.

E. Larry Milliron, Area Resource Soil Scientist, Natural Resources Conservation Service, USDA. He received his bachelor's degree from Ohio State University. His experience is in soil survey and soil interpretations. His works include *Soil Survey of Cedar County, Nebraska,* USDA, 1987.

Charles E. Redmond, Area Resource Soil Scientist, U.S. Natural Resources Conservation Service. He received his Ph.D. from Michigan State University and has coauthored soil survey publications for Erie, Richland, Ashland, and Knox Counties in Ohio.

CHAPTER FOUR

Ohio's Mineral Resources

Leonard Peacefull

The industrial activity which brought Ohio economic prominence in the twentieth century was based on the state's mineral wealth. At one time or another in Ohio's history, it has led the nation as a producer of coal, oil, natural gas, salt, sandstone, limestone, common clay, and iron ore. Mining has long been a tradition in Ohio, dating from the time when the aboriginal peoples quarried the Vanport Flint to make arrowheads, spear points, and knives. In 1994, only Fulton, out of Ohio's eighty-eight counties, was without mining activity, although some of the operations in other counties were limited.

The nature of the underlying geology creates the potential for modern mining activities. Mineral resources can be divided into two groups: *industrial minerals* and *energy minerals* (Table 4.1). Historically the energy mineral coal has been the dominant mineral (Prosser 1988); however in recent years, two industrial minerals—limestone, and sand and gravel—have individually surpassed coal in tonnage produced. Nonetheless coal remains the most valuable mineral, accounting for half of the state's mining income (Figure 4.1). Although iron ore once played an important part in the mining economy of the state, no metal ores are mined in Ohio today.

Industrial minerals—limestone, sand and gravel, sandstone, salt, and some minor minerals—are used as a raw material in other industries, while energy minerals—chiefly coal, natural gas, and oil—are used to produce the electrical energy that powers industrial equipment and domestic appliances, as fuel in automobiles as well as home heating and cooking devices, and in industrial applications. In terms of income or value of production, the energy mineral production in the state is the most important. Whereas industrial minerals are evenly distributed across the state, energy minerals are, with one or two exceptions, confined to the east.

Methods of Production

Mineral resources are extracted in a number of ways: surface mining, also called *opencast* or *open pit* mining; underground mining, through either drift or deep shaft mines; pumping, from wells either shallow to the surface or deep in the ground (used to extract oil, gas, and brine); or dredging (used to remove sands and gravels from Lake Erie and the river beds).

Surface Mining

The solid minerals are mined by either an underground method or the opencast method. Surface mining is the most advantageous approach from a mining point of view; however it does cause more environmental problems. In surface mining the topsoil and the overlying rocks, or *overburden,* are removed to reveal the workable coal seam. In small mines excavators do the job, but large enterprises use drag lines (Figure 4.2). The largest is "Big Muskie" operated by the Central Ohio Coal Company at their Muskingum mine in Muskingum and Noble Counties.

Several environmental problems arise during and after a mine's operation. First, the stripping of topsoil and overburden increases the potential for soil erosion and disrupts surface drainage patterns. Second, mine waste is washed away by the newly created drainage patterns, carrying pollutants to local waters as well as to faraway streams and lakes. Third, when mined land is left barren and unreclaimed, it becomes a

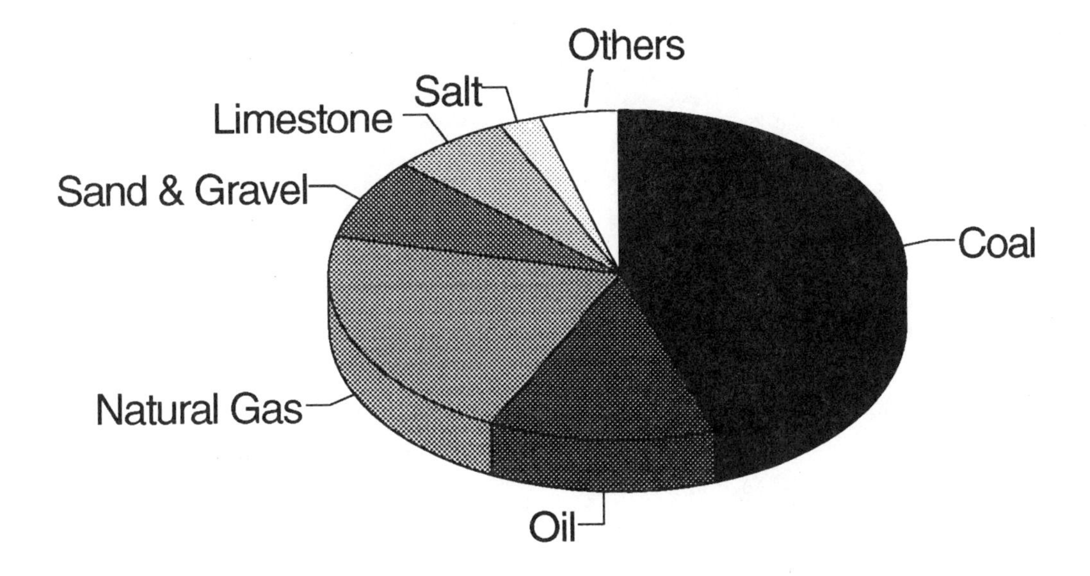

Fig 4.1 Resource Production - by Value - 1991

Source ODNR Division of Geological Survey

Fig. 4.1. Resource Production by Value, 1991

Table 4.1
Mineral Production 1991

	PRODUCTION TONS*	VALUE DOLLARS
Coal	29,213,746	816,705,900
Sand & Gravel	42,793,746	144,384,916
Limestone	53,944,380	118,611,303
Salt	2,172,911	42,106,635
Sandstone & Conglomerate	1,862,155	25,317,526
Clay	2,572,179	7,628,331
Shale	1,930,713	3,137,442
Gypsum	199,857	1,898,642
Peat	18,595	112,323
Natural Gas (1)	144,373,377	341,850,552
Oil (2)	9,158,332	176,264,798
Total Value		$1,678,018,368

Sources: Weisgarber 1992 and McCormac 1993
* Except Natural Gas and Oil
(1) Production in Million Cubic Feet
(2) Production in Barrels

Fig. 4.2a. Dragline, East Ohio Limestone (Photo by Leonard Peacefull)

Fig. 4.2b. Central Ohio Coal Company's "Big Muskie" (Photo by Leonard Peacefull)

Fig. 4.3. Reclaimed Land in Holmes County (Photo by Leonard Peacefull)

derelict scar on the landscape. Legislation was enacted in the 1970s to compel mine owners to return land to its original contour (Figure 4.3). Although this legislation has gone a long way toward remedying the problem, unfortunately mines left unreclaimed prior to the 1970s remain as open scars on the landscape because their operator has gone out of business and can no longer be traced and forced to reclaim the land.

Underground Mining

Salt and coal are the only minerals mined underground. Two primary methods are used: the *longwall method* and the *continuous miner process*. The longwall method uses two rotating cutting drums to strip the mineral off in large blocks. One drum cuts the upper part of the seam and the other, the lower part. The machine is attached to a conveyor that carries the coal away from the mining face. Up to 90 percent of the coal can be extracted using this method. About 70 percent of the coal produced in Ohio is mined in this fashion.

In the continuous miner process a machine with cutting teeth mounted at its front end cuts into the coal seam, then passes the cut material back to a conveyor that discharges it into a wagon on a small railroad, which in turn shuttles the material to the main conveyor system. Unlike the longwall method, which works an area measuring hundreds of feet by hundreds of feet, the continuous miner is used in sections only 15–20 feet wide. Pillars containing a large proportion of coal are left to support the roof and are mined out only when the mining operation has reached its end. The continuous miner has a capacity of up to 15 tons per minute, but because considerable time is needed to secure the roof, move the machine to another location, or wait for the tram to return with empty wagons, continuous miners work at about 30 percent efficiency. Mines using continuous miners produce 30 percent of Ohio's underground coal.

Wells

The extracting of nonsolid minerals—oil, natural gas, and brine—requires that wells be dug that can pump

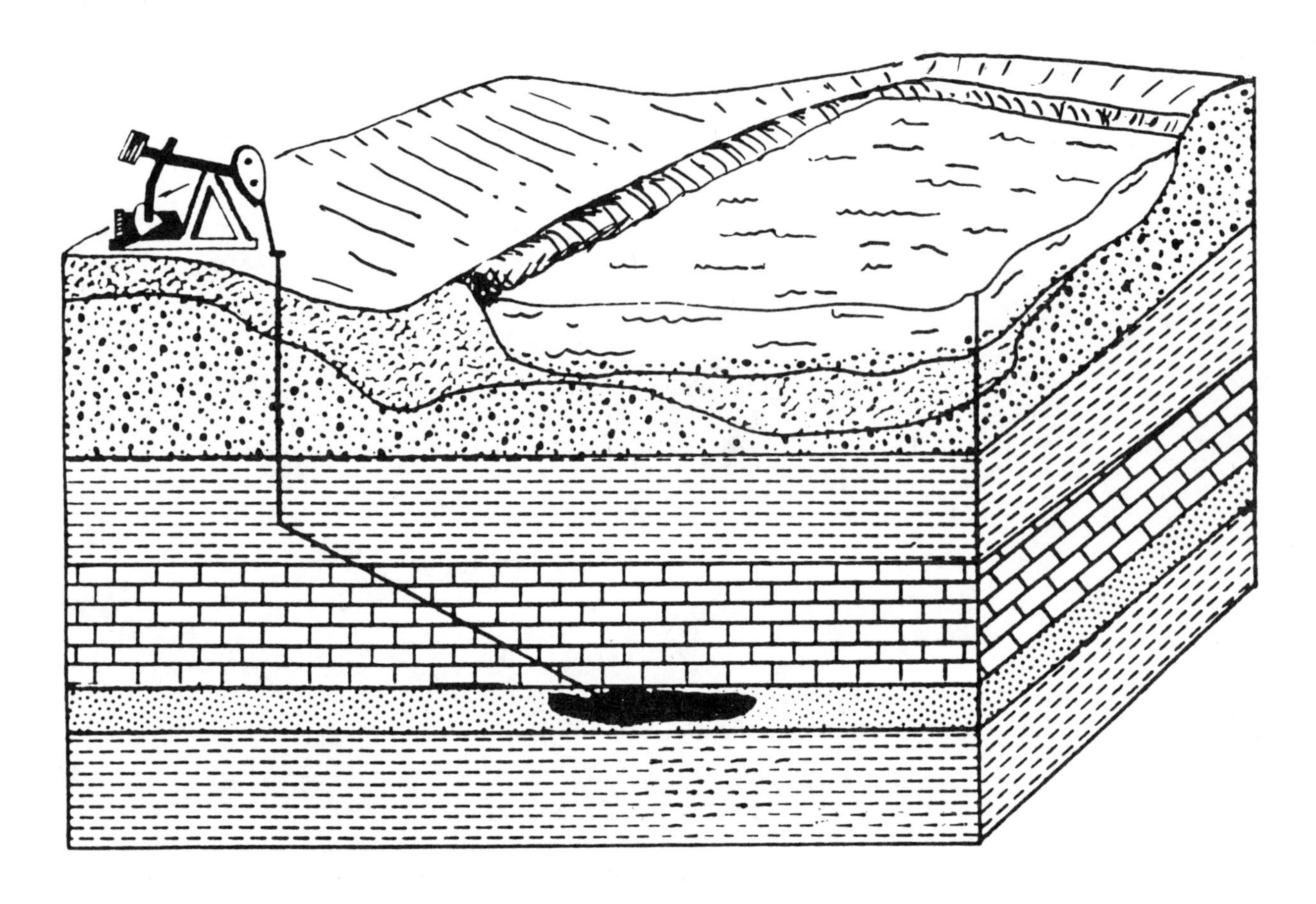

Fig. 4.4. Diagram of Directional Drilling (after McCormac 1993)

the mineral from its natural storehouse. All of the new wells drilled in the state today are for either oil or natural gas. Two different tools are used to drill these wells: the rotary drill or the cable tool rig. The rotary drill is favored because it can reach greater depths than the cable tool rig. Cable tool wells reach down 199–4,623 feet, with an average depth of 1,800 feet, while rotary-drilled wells average 4,465 feet (McCormac 1993).

Directional Drilling. This is used to target areas where vertical drills cannot reach. Various causes could necessitate this. For example, the target area may be covered by water (Figure 4.4) or under an environmentally sensitive area such as a wetland, or it may be located where zoning regulations do not permit drilling. In such cases a well is drilled vertically to a predesignated depth, then deviated at an angle designed to encounter the producing formation (McCormac 1993).

Well Classification. Wells are classified as either *development wells* or *exploratory wells.* A development well is one completed in a known oil- or gas-bearing formation. An exploratory well is one drilled into a formation that is known to contain oil or gas, or that is more than one mile from the nearest well completed in an oil- or gas-bearing formation. Exploratory wells are either *outpost wells* or *wildcats.* An outpost well is placed and drilled so as to extend the production area of a partially developed pool. Outposts are usually 1–2 miles from the main well. A wildcat is either drilled into a formation believed to contain either oil or gas, or is located over 2 miles from the nearest productive area. A wildcat may be drilled in a productive area but into a stratum either above (*shallow pool*) or below (*deep pool*) the current productive layer. A *new-pool* wildcat is drilled into a layer between two existing productive pools. An example of this is a proposed Devonian Shale well in an area where a Mississippian Berea well and a Silurian Clinton well are in production.

Dredging

Dredging for sand and gravel in Lake Erie is done by specially converted freight carriers, which all use the same method. A long steel and rubber pipe, called a *drag*, with a boxlike free end about 12 inches or more in diameter, called a *hood*, is lowered to the bed of the lake. As the hood is dragged along the bottom, a grid on its front breaks up the compacted sand and prevents undesirable debris from getting into the drag. The loose material is sucked up the pipe by huge centrifugal suction pumps as a slurry of sand and water. These powerful pumps can bring sand to the surface from depths of up to 90 feet. After passing through the pump, the slurry moves down the trough that runs the length of the cargo bin of the dredger. This trough has evenly spaced openings that are sized to sort the material and drop it into the hold. Due to their lower specific gravity, undesirable sand, silt clay, and shell particles are carried along in suspension and washed overboard. Commercial dredging is currently permitted only in authorized sections of Lake Erie; they are Maumee Bay, offshore from Lorain-Vermillion (two sites), and offshore from Fairport Harbor. All areas are located several miles into the lake to eliminate the possibility of damage to the shoreline (Liebenthal 1990).

Coal

Coal is the most important mineral mined in Ohio. For over 100 years it formed the basis of Ohio's industrial growth and strength. The smokestack industries of the northern cities relied heavily on the state's coal to heat production plants, fuel steel-producing furnaces, and generate electricity. In other parts of the state industry was dependent on coal to run its machinery. With the decline of the heavy industries of the north and a downturn in manufacturing statewide, it would be logical to conclude that coal has lost dominance and is no longer significant in the state's mineral production. This is not entirely so. While it is true that coal can no longer claim to be the leader in tonnage produced, in terms of value, its importance has risen. During the years 1969–1987, coal's contribution to the total value of the state's resource production increased from one-third to one-half. An obvious reason for this increase is the considerable rise in the price of coal at the mine. In 1969 the average price of coal at the mine was $4.11 per ton; by 1987 it had risen to $30.14 per ton, declining in 1991 to an average $25.70 per ton. These prices have in some areas added to farmers' incomes. Farmers who allow small mining operators on their property are entitled to royalties paid by the mine operator. Considering that one acre-foot of coal will produce 1,800 tons, it is quite easy to realize that a modest income can be gained from the development of a small acreage.

During the past 30 years, coal's industrial importance has declined, primarily because of the move to more environmentally friendly sources of energy production. The problem with Ohio's coal is that its high sulfur content produces pollutants, and these are recognized as a contributing cause of acid rain. Consequently industrial users are seeking alternative sources of electricity, such as coal with a lower sulfur content from other parts of the country.

Ohio's coal-bearing rocks are of Pennsylvanian age (see chapter one) and are confined to the eastern and southern portions of the state. Figure 4.5 shows the distribution of coal production in 1995. Belmont County is the leading producer, and four of its adjacent counties are among the top six producers. Meigs County, which in 1969 only produced 0.02 percent of the state total, now ranks second in production (Figure 4.5). The Belmont coalfield in the center of Ohio's coal-mining region has long been the leading producer. This field spans across Belmont, Harrison, Jefferson, and Monroe Counties, all near the Ohio River, and is based on the Pittsburgh No. 8 coal seam. A new field, the Meigs coalfield in the southern part of the coal-mining region, is based on the Clinton No. 4 seam. Production in this field goes on in Meigs and adjacent Vinton and Jackson Counties. This is an underground mining operation that uses the longwall method. It is interesting to compare further the production data for 1990 with that of 1969 (Table 4.2 and Figure 4.6). In 1969 Belmont and Harrison produced nearly 50 percent of Ohio's total output of 51.1 million tons (Korsok 1975). Today these counties produce less than 31 percent of the total output of 28.6 million tons.

There are 192 active mines in operation in the state producing salable coal. Two operations in Columbiana

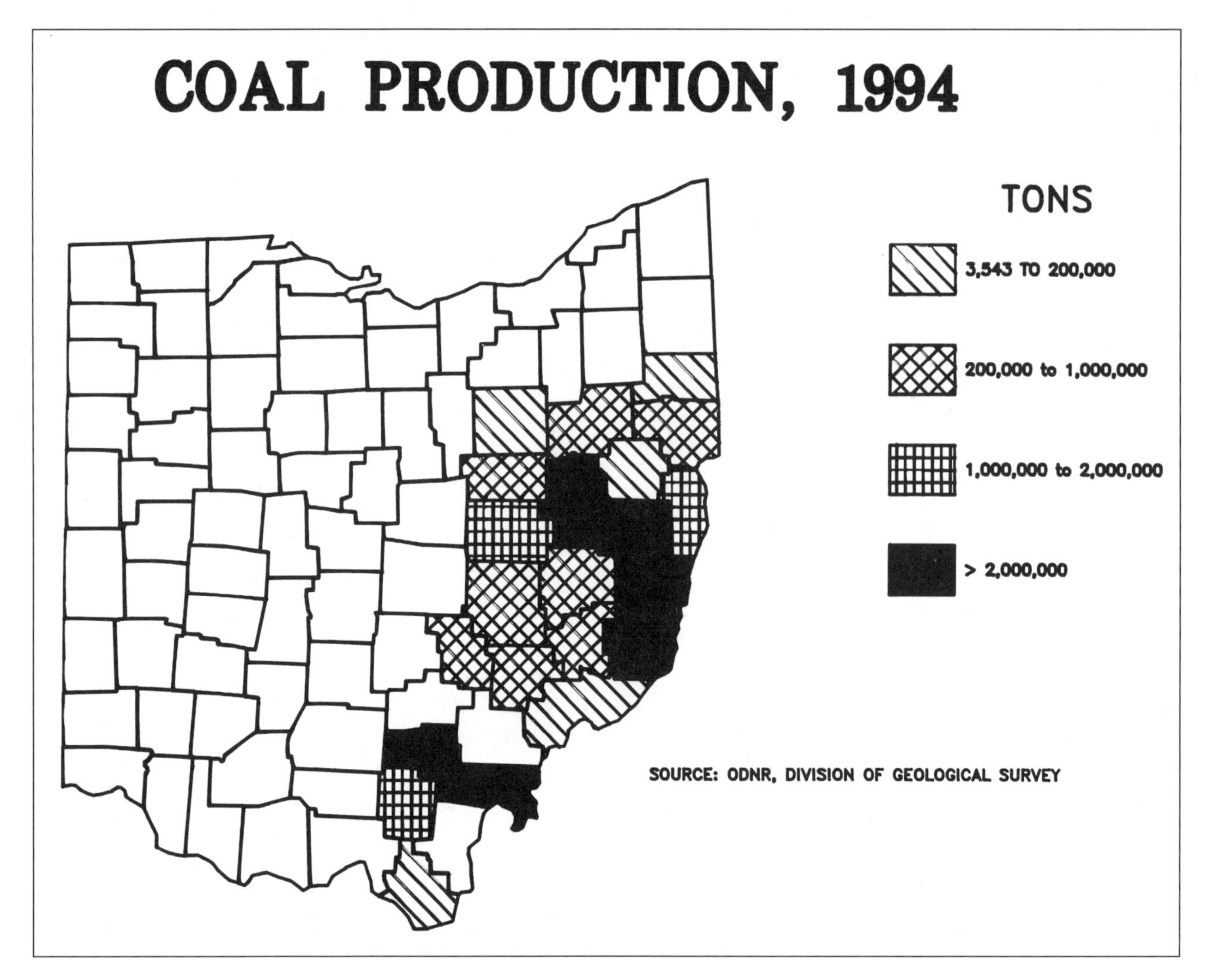

Fig. 4.5. Coal Production, 1994

County produce coal from both underground and surface mines. When a mine crosses a county line, therefore operating in two counties, confusing anomalies can arise because the mine operators must report production by county. For example, Vinton County has twelve surface mines and production data from an underground mine that is not counted as a Vinton mine because it extends from Meigs County. The production total for Wayne County comes entirely from a surface mine that extends from neighboring Holmes County. Similarly there is only one surface mine in Morgan County that extends from neighboring Noble County. If in a particular year one of the mines that extends into an adjacent county does not work over the county line and therefore does not produce in that county, then the data would show a void and not be mapped, suggesting erroneously that this county was devoid of any coal resources.

The Industrial Minerals

Ohio ranks thirteenth in the nation in the value of its industrial mineral production (Prosser 1988). The mineral resources included in the category are limestone, sand and gravel, salt, shale, sandstone, clay, gympsum, and peat. In 1994, one or more of these minerals were extracted from eighty-six of the eighty-eight counties. There was limited nonfuel production in Defiance and Jefferson Counties. Figure 4.8 shows the production and income figures for the

Table 4.2
Coal Production 1990

COUNTY	PRODUCTION TONS	VALUE AT MINE ($ 000s)	% OF STATE TOTAL
Athens	130,259	3,408	0.40
Belmont	5,080,847	131,485	15.49
Carroll	1,013,297	23,517	3.09
Columbiana	1,060,926	19,887	3.23
Coshocton	2,008,597	59,146	6.12
Guernsey	156,140	4,085	0.48
Harrison	2,042,798	55,010	6.23
Hocking	96,850	1,622	0.30
Holmes	459,129	7,850	1.40
Jackson	1,200,683	26,033	3.66
Jefferson	2.081,033	54,025	6.34
Lawrence	316,025	6,923	0.96
Mahoning	190,622	5,706	0.58
Meigs	4,391,761	173,976	13.39
Monroe	2,325,659	79,242	7.09
Muskingum	367,030	13,297	1.12
Noble	3,857,332	134,107	11.76
Perry	1,394,668	38,492	4.25
Stark	170,635	2,802	0.52
Tuscarawas	1,903,228	36,434	5.80
Vinton	2,458,461	75,717	7.49
Washington	64,281	1,056	0.20
Wayne	34,022	340	0.10
Total	32,804,283	954,160	100

Source: ODNR Division of Geological Survey

five major industrial minerals. The distribution of industrial mineral production is shown in Figure 4.7. Although limestone ranks first, sand and gravel is the most widespread commodity, produced in sixty-six counties and Lake Erie. At the other end of the scale, gypsum is the least widespread, produced at only one site—in Ottawa County.

Limestone

As long ago as 1838, the state geologist W. W. Mather commented, "Limestone is the most valuable building material among rocks of Ohio" (Collins 1987). In those days limestone was used not only for buildings, hearthstones, and windowsills but also for whitewashing and plastering. Even then, Mather was not underestimating the value of Ohio limestone. Today Ohio ranks second in the nation in limestone production, accounting for 12 percent of the total output (U.S. Bureau of Mines). Figure 4.9 shows the limestone producing counties, and Table 4.3 shows production data for 1994. Erie County is the state's leading producer, followed by Franklin, Ottawa, Wyandot, and Delaware, each producing over 3 million tons annually. They account for 14.2 percent of the state's production. Sales of limestone netted over $276 million in 1994, at an average price of $4.42 per ton (Weisgarber 1995).

The majority of the state's limestone production is located in the northwestern and central counties, with a smaller amount produced in the southeast. If one compares this with chapter 1's geological map of the

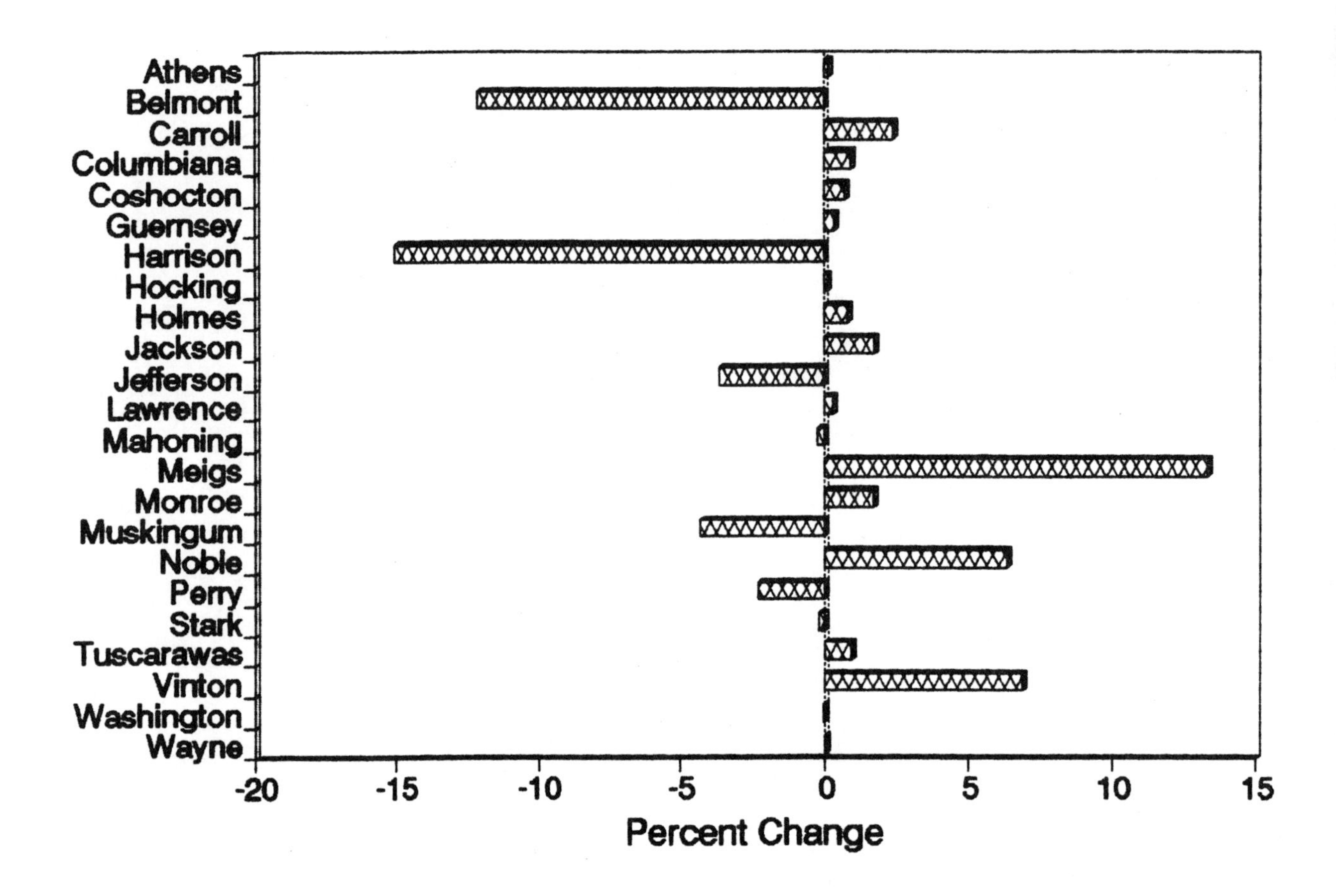

Fig. 4.6. Changes in Coal Production from 1969 to 1990

state (Figure 1.2), it can be seen that production is associated with three different ages and types of limestone. In the northwest the Silurian Dolomite, a hard, almost pure magnesium carbonate, is mined. The dolomite is used to manufacture heat-resistant materials, smelt iron ore, and create agricultural lime. In the central counties of Delaware, Franklin, Erie, and Crawford, the Devonian period Columbus Limestone is quarried. The Columbus Limestone is used in cement, is crushed for concrete, and is cut for use as building stone; it is also important as a flux in the smelting process of steel manufacture. Finally, in the southern and eastern counties the limestone is of Pennsylvanian age and associated with the carbonaceous deposits of the coal measures; in fact in some instances both limestone and coal are mined at the same site. Most of this production is used in road construction, with some going to cement manufacture.

Limestone is used more extensively than any of the other minerals produced in the state (Figure 4.10), and of all its uses, construction is the most absorbing and the most extensive. Most limestone producing counties supply the construction industry with material for making and resurfacing roads and for building. While the majority of Sandusky County's production is burnt as lime, other uses of limestone are for riprap—a guard against bank erosion on lakes and rivers—and in railroad ballast.

Sand and Gravel

Sand and gravel production began when the need for aggregate, used to build highways and the ever burgeoning cities, rose dramatically at the beginning of the century. Even today, sand and gravel production is associated with the fortunes of the construction industry, and production may fluctuate along with the state of the economy. Ohio's output of industrial sand ranks fourth in the nation, albeit a modest 5 percent of the country's total. Extraction of these minerals is the

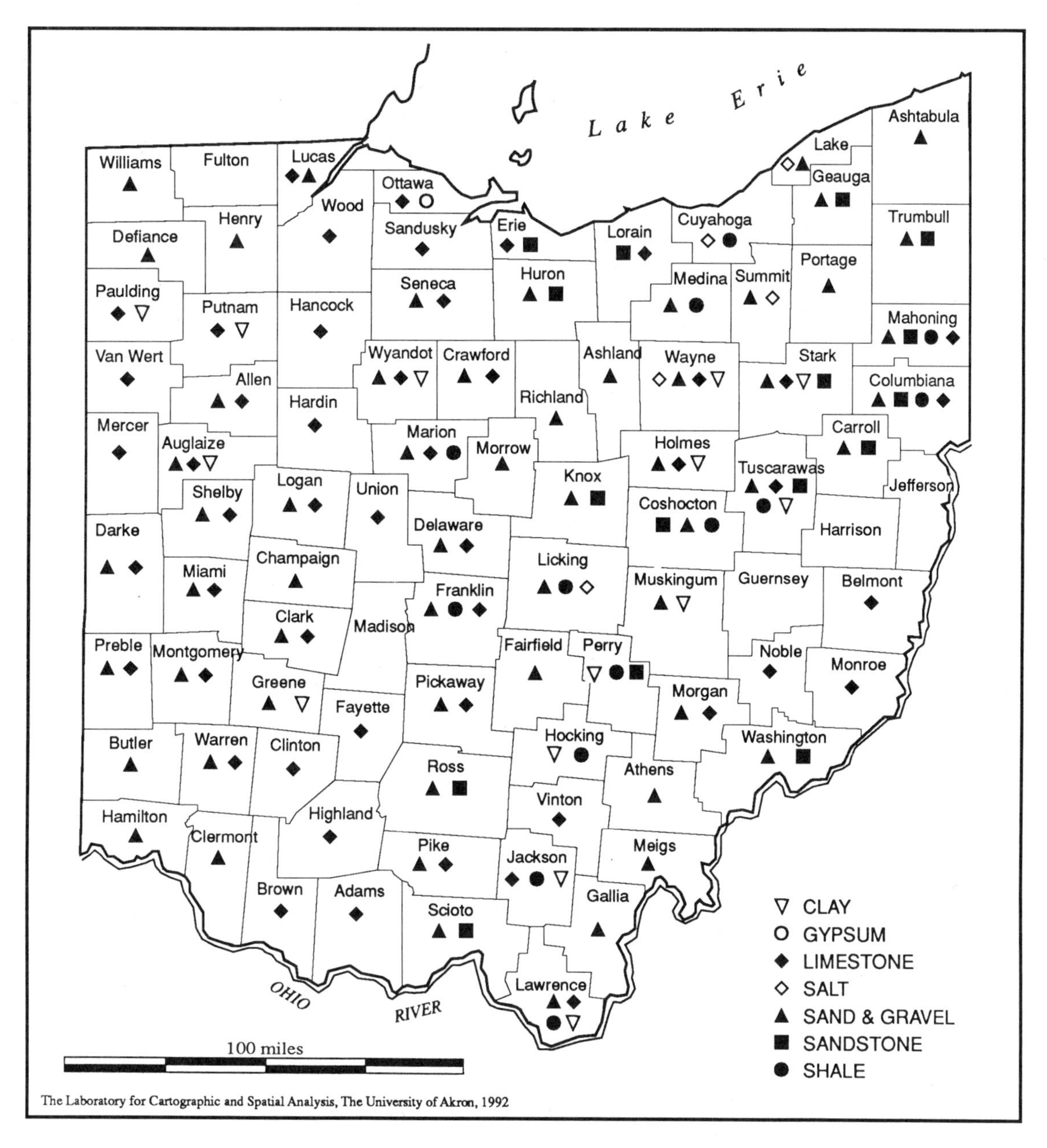

Fig. 4.7. Distribution Map, Industrial Minerals

state's most extensive mining activity. In 1994, 66 counties were engaged in the process, with the leading producers being Butler, at 5,552,042 tons; Hamilton, at 5,220,397 tons; Franklin, at 4,459,933 tons; and Portage, at 4,387,187 tons. Much of the production takes place along the Little Miami River south of Dayton (Crowell, Weisgarber, and Scheerens 1991). There are also 165,000 tons dredged from lake Erie.

A major proportion of sand and gravel production occurs in the glaciated region. This material was either dumped by the glacial lobes as they swept across the state or was left behind in moraines when warmer climate forced the glaciers to recede. In the unglaciated region, sand and gravel was deposited as outwash sands by runoff streams that flowed from the melting ice lobe. Thus, extraction today takes place in modern river valleys or postglacial outwash valleys. The Muskingum and Miami River Valleys are two locations where deposits of this type are being worked. The ridges dredged in Lake Erie are actually

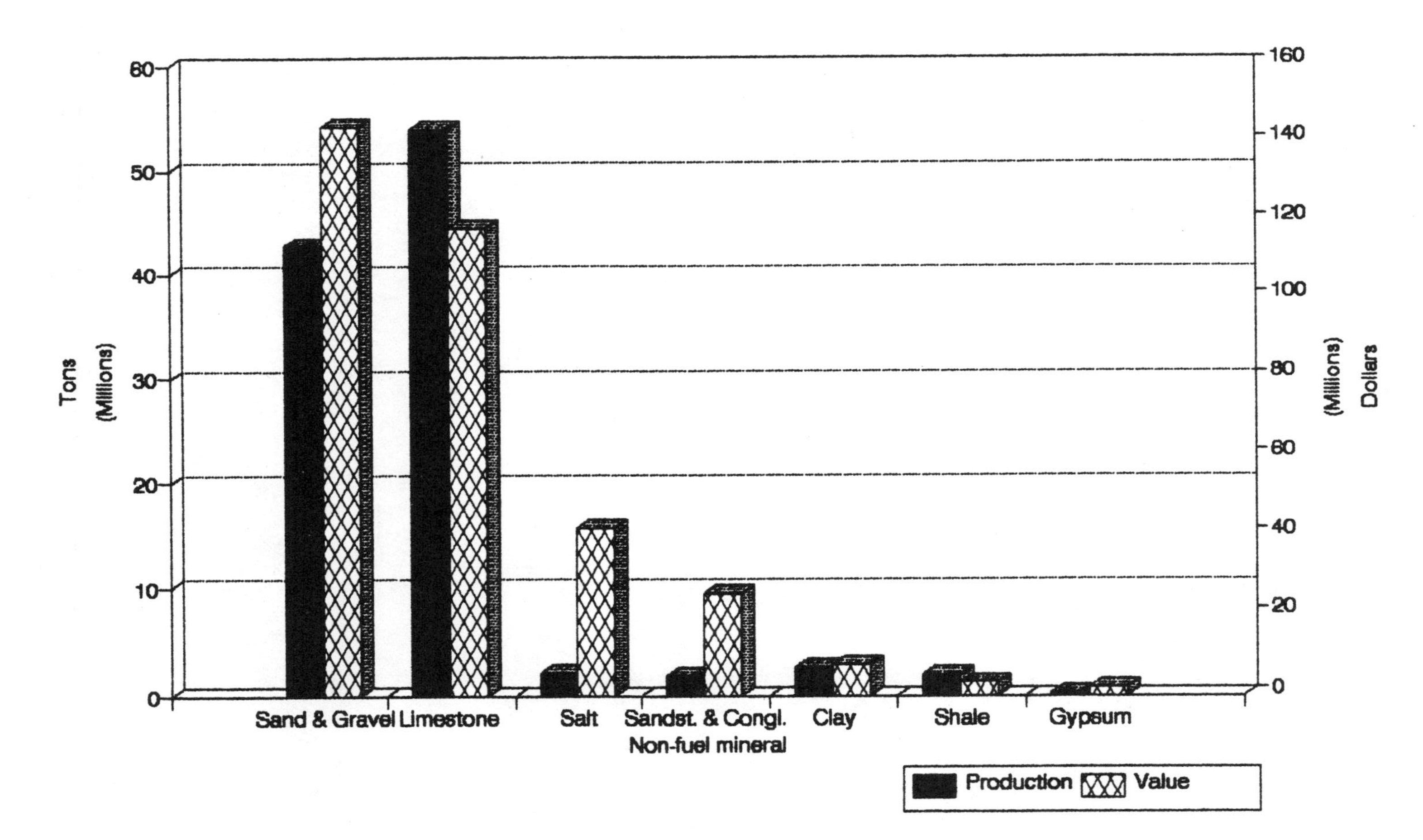

Fig. 4.8. Value of Non-fuel Minerals, 1991

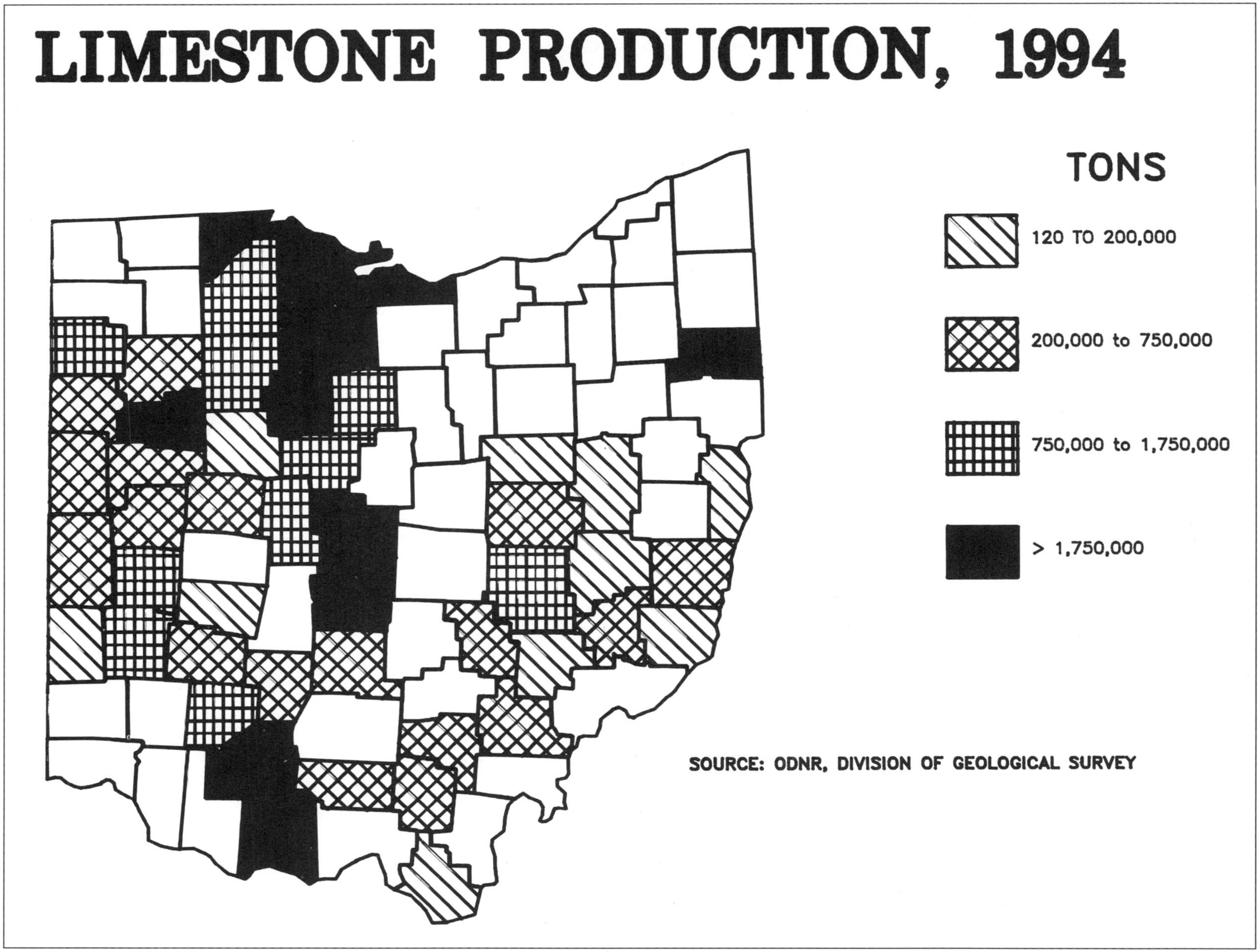

Fig. 4.9. Limestone Production, 1994

Table 4.3
Limestone Production 1990

COUNTY	PRODUCTION TONS	COUNTY	PRODUCTION TONS	COUNTY	PRODUCTION TONS
Adams	1,859,651	Hardin	188,454	Pickaway	217,037
Allen	1,317,436	Highland	929,220	Pike	268,357
Athens	339,078	Holmes	87,240	Preble	82,876
Auglaize	585,442	Jackson	783,061	Putnam	202,258
Belmont	135,932	Lawrence	75,962	Sandusky	3,206,527
Brown	4,716	Logan	678,217	Seneca	1,506,470
Carroll	4,084	Lucas	2,482,425	Shelby	556,861
Clark	121,142	Mahoning	1,978,616	Stark	117,896
Clinton	715,698	Marion	840,420	Tuscarawas	148,889
Crawford	1,271,069	Mercer	427,181	Union	661,575
Darke	400,707	Miami	682,404	Van Wert	459,340
Delaware	2,908,589	Monroe	72,350	Vinton	513,642
Erie	5,0313,762	Montgomery	1,085,643	Warren	41,921
Fayette	544,084	Morgan	17,680	Wayne	206,750
Franklin	4,311,825	Muskingum	2,051,688	Wood	1,275,483
Greene	947,330	Noble	176,507	Wyandot	3,728,851
Guernsey	95,079	Ottawa	3,341,155		
Hancock	1,249,756	Paulding	1,452,167	Total	52,388,503

Source: ODNR Division of Geological Survey

submerged glacial moraines deposited by receding ice lobes. Each moraine can contain different types of material; thus a dredging operation may visit several of the sites before it can collect a required load.

Sand and gravel are used much like limestone; in fact the two are often combined to create an efficient product, used mainly in road construction and resurfacing, in building, and in making Portland Cement, concrete, and bituminous concrete. A small proportion is used in filtration plants. Sand from Lake Erie is primarily used in masonry cement, in concrete, and for beach replenishment (Liebenthal 1990).

Salt

Salt, or *halite,* is an evaporite that formed in the Silurian period. It is extracted in two ways: mining; or pumping, a process in which water is pumped into wells drilled in the salt beds. This process, called *brining,* dissolves the salt and then pumps it back to the surface as brine. The existence of salt licks in various parts of the state was known to Indians long before the arrival of any settlers. Salt was a precious commodity on the frontier in the late eighteenth and early nineteenth centuries; consequently salt was the first mineral extracted by Ohioan industry. By the time of the Civil War, Ohio was the nation's leading producer. In 1889 rock salt was discovered near Cleveland, a discovery that resulted in the development of a large-scale salt-mining operation in northeast Ohio (Collins 1987). Today salt is extracted at only five sites in Ohio. Two of these are underground mines situated adjacent to Lake Erie in Lake and Cuyahoga Counties, while brining takes place at Barberton in Summit County, Rittman in Wayne County, and Newark in Licking County. Ohio's output of 4.3 million tons ranks the state fourth in salt production in the country.

In the early days salt was used in the dairy industry, in ceramic glazing, in ice cream production, and of course as table salt. The brining operators have attracted chemical industries to their neighborhoods, while the two rock salt mines produce salt for ice control on the highways.

Shale

In terms of tonnage produced, shale ranks fifth among Ohio's industrial minerals. However, the shale that is produced for common clay products ranks first in the

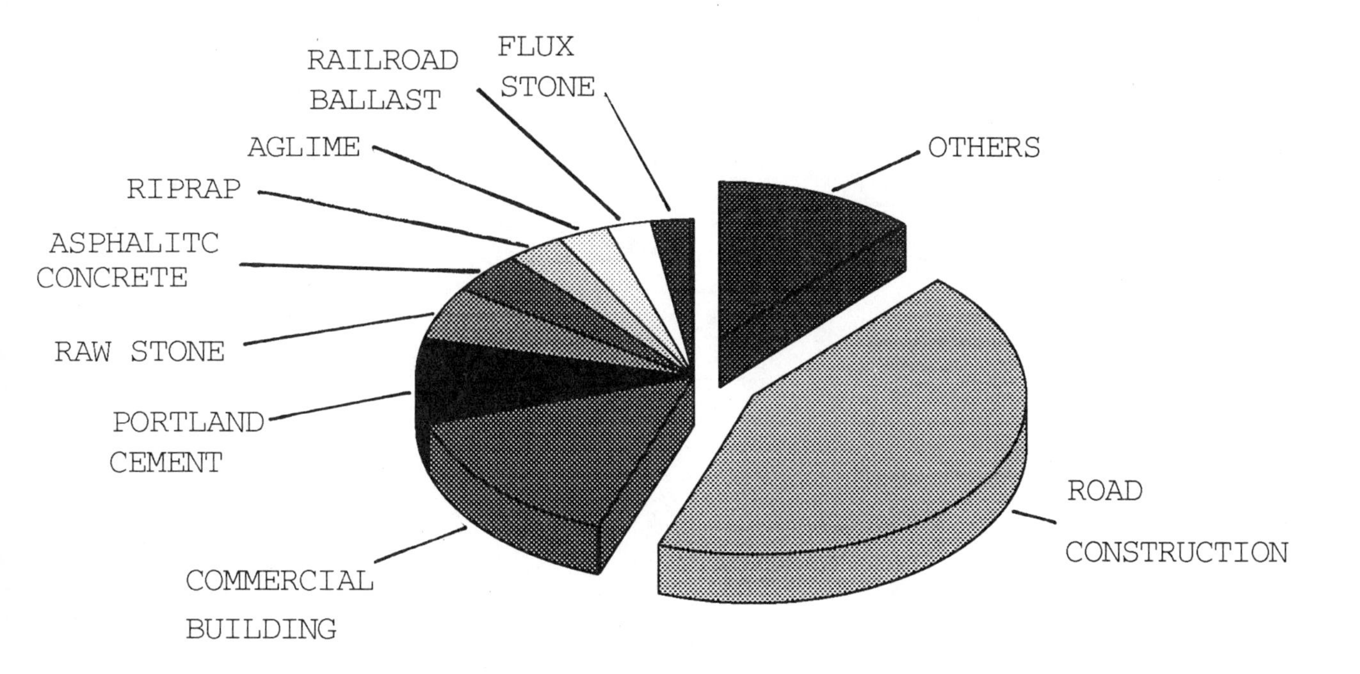

Fig. 4.10. The Uses of Limestone

nation, contributing to the development of pottery and brick-making industries. Most shale is mined in counties east of the Scioto-Olentangy Rivers, on the Mississippi and Pennsylvanian Shales. The leading county is Stark, producing 63 percent of the state's total. Tuscarawas County ranks second with 11 percent, all of which is used as a light-weight aggregate, and Marion County ranks third with 14 percent, all of which is common clay.

In some places along its western edge, the Mississippi Shale, known as the Bedford Shale, resembles a red mudstone. Found primarily in Jefferson Township in Franklin County, this material is mined for brickmaking. Brick was first produced in Marietta in the late eighteenth century. Though most of the early brick works exploited alluvial and glacial clays (Collins 1987), as more settlers poured into the state, shale's place in the industry grew. Its other uses include vitrified products—such as bathroom fittings—roof tile, paving bricks, terra cotta, and sewer and drain tile.

Sandstone and Conglomerate

Like so many of Ohio's minerals, sandstone has played an important role in the development of the nation. By 1819 Ohio was the leading sandstone producer, with the stone shipped out first by canal and later by railroad.

Today sandstone and conglomerate are mined in seventeen counties, all east of the Scioto Valley. Though mostly of Pennsylvanian age, a few outcrops of Mississippian age are quarried, mined either as dimension stone—which is cut out in blocks to be used in building or as grindstones—or as crushed stone—which is used as foundry stone, glass sand, riprap, aggregate, filler gravel, and in polishing and grinding processes as well as various other uses, including pool tabletops. Most of the rock cut for dimension stone is from a series called Massillon Sandstone. One area where this rock occurs spreads across northwestern Coshocton County, southwestern Holmes County, and eastern Knox County. The

Fig. 4.11. Belden Brick Works at Sugar Creek (Photo by Leonard Peacefull)

Sharon Conglomerate (see chapter one) is quarried in Geauga and Mahoning Counties on the Ross-Pike county border.

In Lorain, Erie, and Huron Counties, between Amherst and Wakeman, Berea Sandstone is quarried. This is a fine-grained sandstone, useful as building stone, that has been used for buildings in various parts of the country and is currently exported to Canada, Europe, and Australia (Collins 1987). These quarries in northern Ohio are noted to be among the deepest and largest in the world. Another occurrence of the Berea, but one that is no longer mined, is Deep Lock Quarry in Peninsula, now part of the Cuyahoga Valley National Recreation Area. Here it is possible to see the grindstones that were once shipped out via the Ohio and Erie Canal.

Clay, Gypsum, and Peat

The other industrial minerals quarried in the state are clay, gypsum, and peat. Clay is produced in thirty counties; the combined production of these counties, when added to the common clay products of the shale, makes Ohio the nation's leading producer. Most clay production comes from a semicircle from Lawrence up through Portage County. Tuscarawas, Stark, Coshocton and Paulding Counties accounted for 66 percent of the 1994 total. Much of the clay produced is mined along with coal. Such material is called *fire clay* because it is capable of resisting great heat. The term fire clay was once used to describe all clay associated with coal seams but now refers specifically to those clays with refracting properties. It is chiefly used to line furnaces, where it is subjected to intense heat. Non–fire clay products are used in the landfill industry and in cement manufacture.

Gypsum, another evaporite mined in the state, was discovered in the early nineteenth century; its mining production began in 1822. The mineral occurs at only one site, Gypsum in Ottawa County, where it is manufactured into plaster and wallboard at a nearby plant.

Peat, which is not commonly considered a mineral, is produced at sites in Portage, Champaign, and Williams Counties. In 1994 combined production was 8,446 tons, a decline from the 1990 total of 38,690 tons. Well known to gardeners as a mulch, it is also used for worm culture and as a legume inoculate (Weisgarber 1995).

Oil and Natural Gas

Although there are now over 64,000 active wells in Ohio, the state's oil industry can be traced back to humble beginnings in Noble County when, in 1814, two men prospecting for salt discovered oil. The magnitude of their discovery was not appreciated at the time because there was no commercial market for the product. This may have been the first discovery of oil in the United States; it preceded Colonel E. L. Drake's Titusville Well in Pennsylvania by some forty-four years. Ohio's first commercial oil and gas well became productive in 1860, at Macksburg in Washington County. Today this county still leads the way in Ohio prospecting. After 1860 a true commercial market was established, leading to further development of the oil industry. By the end of the last century, Ohio had become the leading oil and gas producer in the nation.

Anyone traveling through the eastern counties in the early 1980s could not fail to notice the increasing amount of drilling taking place. New wells seemed to be sprouting up in cornfields all over. Well-head pumps that had been idle for years sprang to life, nodding up and down as they brought the precious resource to the surface. A rebirth of Ohio's oil and natural gas industry was taking shape, triggered by the rise in the world crude petroleum price to $35 per barrel. In 1981, 5,585 drilling operations were reported, a record for one year in terms of the number of drilling operations (Sneeringer 1984). The previous record was set in 1909 at 3,591. The old record was surpassed again in 1982 when 4,743 drilling operations were reported. However, all records were shattered the following years when 6,260 drilling operations were reported. A new boom in the state's oil fortunes was underway. Since then, world oil prices have declined and production in Ohio has followed suit.

In 1983 Ohio's oil producers set production records, producing almost 15 million barrels, a value of $421.3 million. Unfortunately, with the dramatic fall in world oil prices in the mid-1980s, oil production declined to 8.7 million barrels in 1994, a market value of $136.2 million. On the other hand natural gas has fared somewhat better. Production rose to 186,480,000 thousand cubic feet (MCF) in 1984 but has declined to 130,855,248 MCF in 1994 (Table 4.1 and Figure 4.12).

Oil and Gas Formation. Petroleum and natural gas are hydrocarbons formed from the decayed remains of organisms that lived in the shallow seas that have come and gone throughout geological history. The era of the rock in which these energy minerals are found today does not always signify the time period in which the organisms lived. This chemically altered matter has, in many cases, moved through and between the pore spaces of rock formations until it became trapped in what are today economically viable concentrations called *pools.* These pools are not underground lakes of oil or natural gas, as the term may imply, but simply areas where hydrocarbons saturate the pore spaces of porous rock. Usually there is a layer of impervious rock that halts the progress of the hydrocarbons, causing a buildup. There are many varieties of trap: often oil is trapped at the top of an anticlinal structure; or it may be caught in a fault-induced trap; or, in some instances, a salt dome acts as trap. Whatever the case, it is rare that an oil-bearing strata can be recognized from the surface. In fact the pool may be hundreds of feet underground. The discovery of a rock pool is a very scientific and painstaking exercise. Maps of the formation and the corresponding underground structure are produced from seismic survey techniques in which boreholes are drilled to identify the actual rocks and their characteristics before a decision is made to drill. Even then the act of drilling does not always guarantee a productive well; in many instances the prospectors may find nothing and classify the well as dry.

The Distribution of Oil and Gas. The major oil and natural gas fields found in Ohio at the end of the twentieth century are confined to the eastern half of the state. A line from western Cuyahoga County to the center of Jackson County would delineate the axis of the main oil field. East of this line are large patches—in Washington and Stark Counties—and a more extensive tract that extends from Muskingum County through Morgan County to Athens and Meigs Coun-

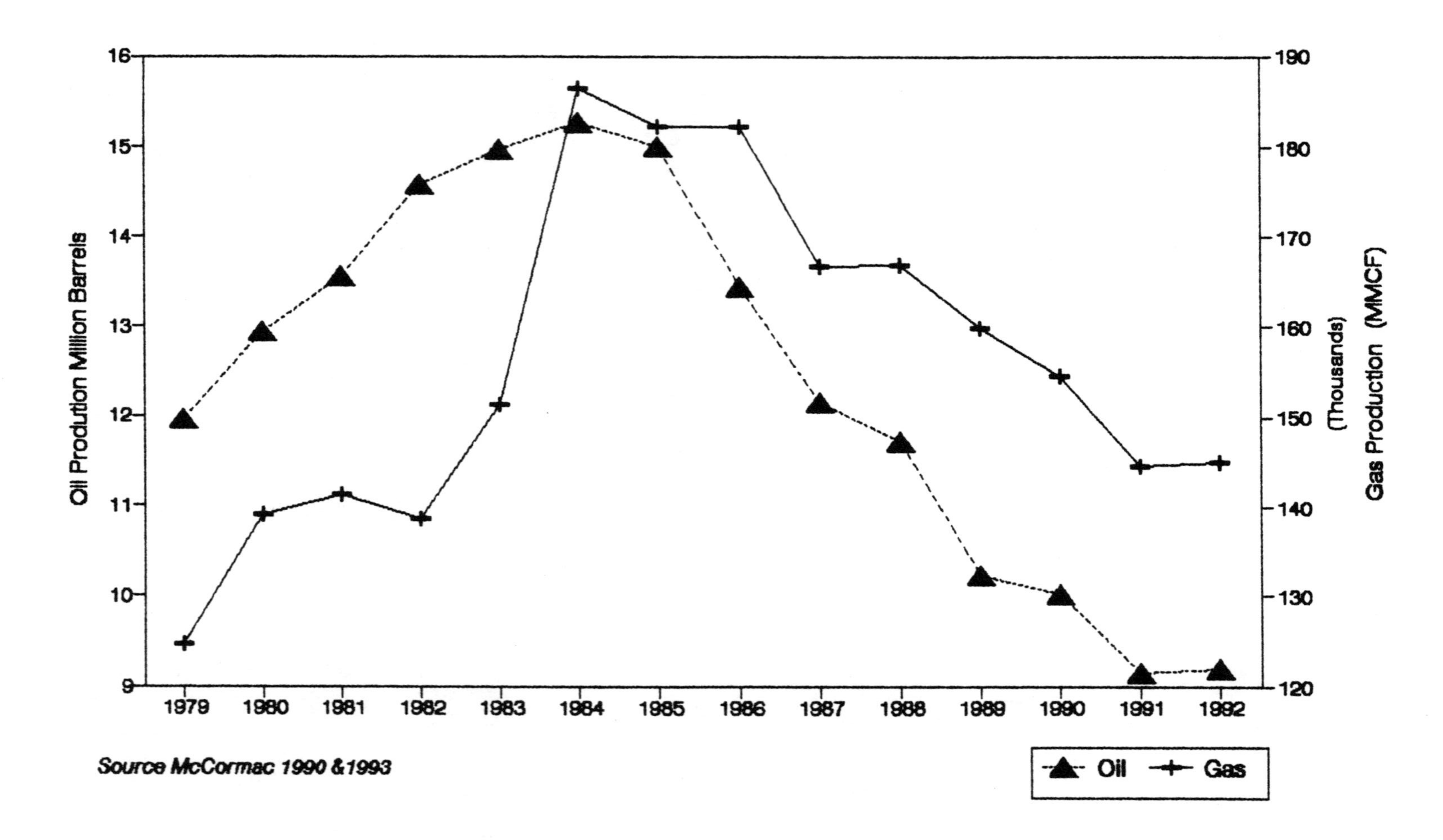

Fig. 4.12. Oil and Natural Gas Production, 1979–1992

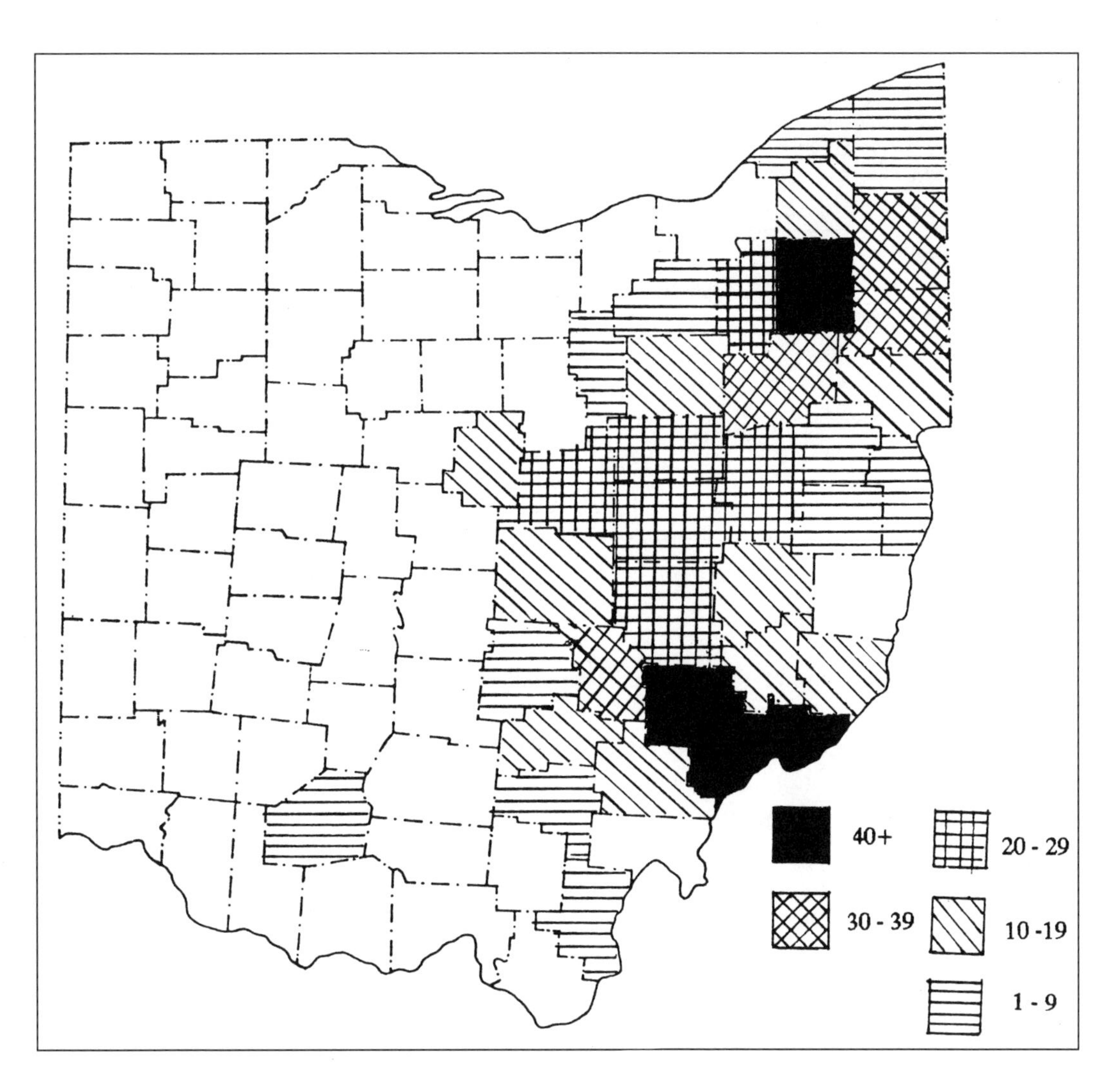

Fig. 4.13. New Well Production by County

Table 4.4
Wells Drilled by Producing Zones, 1992

PRODUCING ZONE	SHALLOW DEPTH IN FEET	DEEP DEPTH IN FEET	NUMBER OF WELLS	PERCENT PRODUCTIVE
Pennsylvanian	199	1,326	28	60.71
Berea (L. Miss.)	380	2,087	82	82.93
Ohio Shale (Devon.)	272	3,986	43	95.35
Newburg (Silurian)	1,962	2,020	1	0
Clinton (Silurian)	2,097	6,428	402	98.51
Rose Run (L. Ordovician)	3,634	7,702	112	58.04
Trempealeau (Cambrian)	2,420	7,166	62	33.87
Kerbell (Cambrian)	2,375	7,950	16	6.25
Pre-Cambrian	3,148	5,930	2	0

Source: McCormac 1993

ties. As Table 4.4 shows, the oil and natural gas from these areas can come from relatively shallow wells in Pennsylvanian-age sandstone or deeper wells in Cambrian-age dolomite. The Clinton Sandstone of Silurian age is the most productive and attracts the largest number of new wells.

Two areas located west of the main fields deserve mention. In northwest Ohio a fairly extensive field curves away from Sandusky Bay toward the Indiana line at Mercer County. This field was active during the late-nineteenth century, especially the Ordovician Trenton formation. Today it is largely inactive, with little new drilling taking place. The other site, in Morrow County, concentrates on the Knox Dolomite of Cambrian-Ordovician age. The exploration and mining of this field caused a minor boom in the 1960s. Today this field remains active with twenty-three new wells drilled in 1992.

Drilling Activity. Drilling in the state is very diverse as to the depth of the well and nature of the oil-bearing strata. The average well is approximately 4,100 feet deep, with the deepest going down to 10,200 feet and the shallowest around 90 feet. Each of the geological periods represented in Ohio's rocks, except the Permian, contain oil-bearing formations. The most actively productive, with 98 percent of its wells producing, is the Clinton Sandstone formation. This formation still attracts the most new well activity. The production for each formation in 1992 is given in Table 4.4. In 1994, 660 wells were drilled of which 77 percent were productive. The illustration in Figure 4.13 shows new oil and gas production by county. Washington County, which had the most new wells drilled in 1992, also had the most diverse drilling activity. Wells were drilled to seven different geological zones, ranging from the Pennsylvanian Wolf Creek Sandstone to the Silurian Clinton Sandstone (McCormac 1993).

The Ordovician Rose Run Sandstone formation below the Clinton Sandstone has attracted a lot of exploratory attention in recent years. One reason is that if the strata turns out to be dry, there is a potential that the Clinton Sandstone will be productive. Consequently nonproducing wells are plugged back to the Clinton Sandstone. This does not always happen, as McCormac (1993) notes. Since 1987, only 72 percent of these dry wells have been plugged back to produce from the Clinton Sandstone. Of 925 wells drilled into the Rose Run, 75 percent have been drilled since 1982; most have been drilled since 1987. Muskingum County has seen the most recent activity in the development of this formation, although Coshocton and Holmes Counties lead with the most wells drilled since 1987.

In 1992, drilling fell below 1,000 new wells for the year for the first time since 1938. This may not reflect a trend but a consequence of the general economic climate at the beginning of the 1990s. But although the boom years of oil and gas production may be past, considerable interest remains in Ohio's oil and gas industry, with the market value of combined production exceeding $453 million in 1994. And in spite of there being no significant new reservoir discoveries since 1968, applications for new well permits are continually being submitted.

REFERENCES

Collins, H. 1987. *A Historical Sketch of the Mineral Industries of Ohio.* Information Circular 54. Columbus, Ohio: Ohio Department of Natural Resources, Division of Geological Survey.

Crowell, D., S. Weisgarber, and C. L. Scheerens. 1990. *Mineral Industries Map of Ohio.* Columbus, Ohio: Ohio Department of Natural Resources, Division of Geological Survey.

Korsok, A. 1975. The Extractive Industries of Ohio. In *Ohio: An American Heartland,* by A. G. Noble and A. J. Korsok. Bulletin 65. Columbus, Ohio: Ohio Department of Natural Resources, Division of Geological Survey.

Liebenthal, D. 1990. The Lake Erie Sand and Gravel Industry in Ohio. In *1988 Report on Ohio Mineral Industries,* ed. S. Lopez. Columbus, Ohio: Ohio Department of Natural Resources, Division of Geological Survey.

McCormac, M. 1990. *1989 Ohio Oil and Gas Developments.* Columbus, Ohio: Ohio Department of Natural Resources, Division of Oil and Gas.

———.1993. *1992 Ohio Oil and Gas Developments.* Columbus, Ohio: Ohio Department of Natural Resources, Division of Oil and Gas.

Prosser, L. J. 1988. *Ohio Mineral Yearbook.* Washington, D.C.: U.S. Department of the Interior, Bureau of Mines.

Sneeringer, M. 1984. *1983 Report on Ohio Mineral Industries.* Columbus, Ohio: Ohio Department of Natural Resources, Division of Geological Survey.

Weisgarber, S. 1991. *1990 Report on Ohio Mineral Industries.* Columbus, Ohio: Ohio Department of Natural Resources, Division of Geological Survey.

———.1995. *Draft of 1994 Report on Ohio Mineral Industries.* Columbus, Ohio: Ohio Department of Natural Resources, Division of Geological Survey.

CHAPTER FIVE

Aboriginal Cultures and Landscapes

Jeffrey J. Gordon

The Earliest Aboriginal Migrations

Although Ohio has long been the home of many native peoples, they are not indigenous to either the state or the Western Hemisphere. Scattered and controversial remains of human occupation, some dating back 20,000–40,000 years, have been found in various locations in the hemisphere. Dates of 13,000 years ago have been established, with 14,000–17,000 years ago possible, for the earliest occupation of parts of Ohio. American Indians are the only aboriginal Ohioans, as they were the first to successfully inhabit this area.

In view of the total human time span on Earth, people were late arrivals in the Western Hemisphere. The first peopling of the hemisphere including Ohio coincided with the late stage of the last Ice Age. During the most recent ice advance, the Wisconsin, climate was colder and wetter and continental glaciation extended into the middle latitudes. A markedly different physical geography from that which exists today permitted early human migration by land into the Western Hemisphere. Pursuing game animals, these Ice Age peoples traveled from Siberia in northeastern Asia across the Bering Land Bridge into Alaska, then spread throughout the North American continent (including Ohio) and eventually into South America. Also known as Beringia, the Bering Land Bridge was a large, 1,300-mile-wide plain which emerged intermittently to replace the Bering Strait. This land bridge arose as sea levels dropped—the covering water temporarily encased as ice and snow—exposing shallow areas of the seas as dry land. Migration across this barrier-free landscape from the Eastern to the Western Hemisphere became possible for long intervals; the interhemispheric change in location was most likely unnoticed by these early travelers, who encountered a similar environment on both ends of Beringia. These first immigrants into the Western Hemisphere probably followed the Mackenzie River Valley, which offered an ice-free corridor through the ice sheet southward across Canada into the interior of the United States and so to Ohio. The presence of game in great numbers and humans in small numbers must have constituted a paradise for hunters. These people traveled in small groups, such as extended families, in many small-scale migrations spreading over millennia, rather than in just one or several large-scale migrations.

Climate eventually warmed again to close the Ice Age; glacial melting definitively overtook glacial accumulation, resulting in the steady retreat of the continental ice sheets to their present polar latitude locations in Greenland and Antarctica. Sea levels correspondingly rose, submerging Beringia and reestablishing the Bering Strait, thereby terminating the relatively easy land migration from Asia to North America. Millennia later, early Inuit and Aleuts, the only other aboriginal Americans, would make this same migration, only by boat. Although interhemispheric migration ceased with the recurrence of the Bering Strait, an event which unfortunately submerged much evidence of these first immigrants, intrahemispheric migration within the Americas (including Ohio) continued, even up to and during the historic period of European exploration and colonization of the continental interiors. Geographic mobility was typical, often leading to sequent occupance, as one group replaced another in a given area. It was more common for hiatuses to occur between human occupations than for a group to remain in a given locale over a long period with its culture evolving in situ. Like the Eastern Hemisphere, the Western Hemisphere also had internal cultural dynamics, as peoples interacted through space while changing their cultures over time. The supposed state of equilibrium achieved by the American aborigines and altered by the coming of the Europeans never in fact existed.

Aboriginal Culture Stages in Ohio

The terminology used to discuss the Western Hemisphere prior to the coming of the Europeans such as *pre-Columbian* (before Columbus's 1492 landfall) or *prehistoric* (peoples without written records) is largely artificial because it was a later creation of the white settlers, never used by the aborigines. Although European explorers, for example, took credit for "discovering" this hemisphere, terming it the "New World" or the "Americas" and labeling its inhabitants "Indians," such nomenclature reflected both their ethnocentrism and early incorrect geographical knowledge. Archaeologists also have created artificial constructs. Lacking a more complete understanding, they divide aboriginal occupation chronologically in different ways, then, as better data become available, refine their categories. At present, classification of aboriginal Ohio is divided into three major culture stages: Paleo-Indian, Archaic, and Woodland. Each stage seems to have gradually evolved into the next, meanwhile experiencing fluctuating periods of marked cultural development and decline that correlated with the impact of changing climate and cultural innovation on subsistence patterns. These broad culture stages are further subdivided by archaeologists into temporal periods called *horizons* and geographical regions called *complexes*.

Paleo-Indian

Paleo-Indians of the Late Pleistocene were the earliest immigrants in Ohio for whom there is sufficient artifactual evidence. They entered and thinly occupied Ohio from the south and west, advancing as the glacial ice retreated northward. However, the presence of earlier peoples existing in smaller numbers and possessing generalized cultures and simpler technologies cannot be ruled out. Sketchy archaeological findings of some crude choppers and chopping tools suggest that a pre-projectile society predated the Paleo-Indians with whom these artifacts were not associated.

Paleo-Indians are also referred to as the Early Hunters or Big Game Hunters because their culture revolved around the hunting of now extinct Pleistocene megafauna, especially the mammoth and Pleistocene bison in the Early and Late Paleo-Indian horizons respectively. Other Pleistocene species in Ohio include the mastodon, musk-ox, giant beaver, ground sloth, saber-tooth cat, and caribou. These huge mammals, having thick, protective coats of hair, were ice age adaptations that lived in periglacial environments. These rich tundra biomes arose along the edges of glaciers where meltwater runoff created lush vegetation—grasses, mosses, sedges, and lichens—with park-tundra of coniferous spruce woodlands beyond. Herds of large herbivores were attracted to these favorable ecological zones possessing sufficient browse. Paleo-Indians followed, apparently dependent on the herds for nearly total sustenance in the form of meat for food; hides and furs for clothing and shelter; sinew for cordage; and bones for tools, weapons, utensils, ornaments, and shelter.

Evidence that Paleo-Indians coexisted with and hunted ice age animals has been uncovered from kill sites in which such animal bones were found in direct association with projectile points. The animal skeletons also revealed incised markings indicating butchering activity, and the absence of tailbones suggests that the hides were removed, because the tail usually remains with the hide in the skinning process. Bones were also fire blackened and marrowbones broken, indicating cooking activity. These hunters left carved renderings of mammoths in nearby Pennsylvania and New Jersey. Paleo-Indians were motivated, full-time hunters of megafauna because a single kill could supply an extended family or band for many days or even weeks. Recent evidence, for example, of an 11,000-year-old Paleo-Indian meat-cache site uncovered from southern Michigan revealed large chunks of butchered mastodon tied up and tethered to the bottom of what was then a glacial pond, employing a refrigeration technique that preserved the meat for later consumption.

Paleo-Indians could successfully kill large prey because they possessed a highly specialized hunting technology. Their stone tools included a variety of flaked knives, gravers, choppers, scrapers, and perforators. Their major weapon was the spear, powerfully propelled by the use of an atlatl or spear-thrower. Long predating bows and arrows, these wooden spears had detachable bone foreshafts with sharp-edged flint or bone projectile points. Each hunter probably carried several point-tipped foreshafts. The spear may have had a cord tied to it, allowing the hunter to quickly and safely retrieve, rearm, and hurl it again at a missed or wounded target. This terrestrial technology parallels the harpoon used for marine-mammal hunting by the Inuit and Northwest Coast Indians in later times. The flint projectile points were 2–6 inches long, carefully flaked, quite thin to maximize penetration, and commonly given a central lengthwise groove—called a *flute*—where attached to the foreshaft. Unique to Paleo-Indian culture, these specialized fluted points represented the most sophisticated hunting technology ever used by Ohio's aborigines. Excavation of similar artifacts in Siberia revealed that the Paleo-Indian tool kit and culture did not originate in situ here, where no precursors have been found, but were developed prior to their emigration from the Eastern Hemisphere. A recently uncovered Paleo-Indian antler spear point found in a sinkhole in a cave near Findlay in Hancock County and dated to 11,000 years ago is among the oldest artifacts yet found in Ohio. Paleo-Indians probably hunted in small groups to coordinate a strategy, such as stampeding their prey into sites from which escape was difficult. Driven into bogs, swamps, lakes, streams, and pits, and stampeded off bluffs, the megafauna could be more easily approached and killed, especially if other people were already stationed in place to block the animals' escape and to help kill them.

The big game hunting livelihood of the Paleo-Indians necessitated a nomadic existence; there is no evidence of permanent habitation. A remarkable Paleo-Indian site recently discovered in Sharon Township in Medina County dating to over 12,000 years ago, however, has yielded evidence of wooden supports and other wooden structural remains. Little is known of other subsistence activities they may have engaged in, but there is evidence of gathering and fishing. Kill sites and campsites reflecting temporary habitation, including ephemeral cave and rock shelter sites that provided protection from the elements, have been found; quarry sites at flint deposits, such as those along the Walhonding River in Coshocton County, also have been uncovered. Paleo-Indians favored higher ground locations—ridges, stream-cut terraces, or hills, for example—for their temporary habitation, most likely because better drainage and relative dryness offered greater comfort, and elevation offered a vantage point from which to spy out distant prey. Their material culture was necessarily severely limited in quantity and to easily portable possessions according to what they and their dogs could carry, perhaps aided by the travois utilized by later peoples. A nomadic lifestyle precluded innovation or adoption of more advanced cultural forms such as ceramics, horticulture, domestic animals (except dogs), and social units larger than extended families, clans, or bands (such as tribes which arise much later). Despite its great geographical range and temporal span, the Paleo-Indian stage remains the least known because of its great antiquity, scanty human populations, lack of permanent habitation, and paucity of material culture.

The retreat of the Wisconsin glaciation and continuing ecological changes ended the hitherto successful Paleo-Indian culture. Climate became warmer and drier, dooming both the periglacial-adapted megafauna and the Paleo-Indian culture dependent on it.

Archaic

With the end of the Ice Age, its periglacial environment, and the resultant loss of megafauna about 10,000 years ago, the raison d'être of the specialized Big Game Hunters no longer existed. Although the Late Paleo-Indian people did not die out, their culture did. About 8,000–10,000 years ago the Archaic stage began to replace the Paleo-Indian stage throughout the hemisphere, including Ohio, reflecting a correlation between climatic and cultural change. Archaic peoples settled Ohio in a northward progression following the ecological succession of the coniferous forests by the encroaching deciduous forests, which were better adapted to the warmer post–Ice Age climate. This sequence occurred because deciduous forests possess a greater carrying capacity and a more exploitable biomass of game animals and plant foods than do coniferous forests. Ohio became markedly different environmentally in the postglacial era, and the Archaic peoples took advantage of the favorably changing ecological conditions.

All American Indians are descended from the Paleo-Indians. These descendants developed Archaic cultures during the early post-Pleistocene because they had to replace bygone big game hunting to survive. At this critical juncture, necessity forced them to shift from their former focus and begin to adapt to a wider and less specialized subsistence base. They wisely diversified their livelihood and took increasing advantage of the different ecosystems they encountered by developing the necessary tool kits to exploit the various foodstuffs each offered. Lacking large game, Archaic Indians divided their livelihood into three distinct major food sources, or subsistence modes: small game hunting, gathering, and fishing. Small game hunting involved primarily deer, along with bear, elk, rabbit, raccoon, squirrel, beaver, and other small mammals; wild birds, such as turkey and migratory waterfowl; and turtles. They used an array of weapons: spears with notched and stemmed projectile points made from stone and bone, deadfalls, snares, and decoys. Most likely they practiced occasional collective hunting. When gathering they chose wild fruits, seeds, nuts, tubers, roots, bark, leaves, berries, honey, maple syrup, and eggs on land; and shellfish, snails, and mussels in the rivers and lakes. To fish they used spears, hooks and lines, gaffs, nets and seines, weirs, and perhaps poison, which we know was used in later times.

Archaic Indians successfully developed and utilized a much wider and more intimate working knowledge of their environment than had the Paleo-Indians. Given a number of available ecosystems to exploit, Archaic Indians adapted to ecological and geographical variation accordingly. As a result, each group evolved some cultural differentiation, according to the particular nature of its local resource base. In Ohio, archaeological evidence revealed that Archaic Indians usually optimized their small game hunting, gathering, and fishing livelihood by selecting the riverine environment for habitation. The Raisch-Smith site in Preble County is representative. Choosing to live at least part of the year alongside rivers and lakes enabled these people to easily exploit both terrestrial and aquatic ecosystems. Riverine sites also provided abundant water for refrigeration, cooking, drinking, washing, and transportation. The oldest watercraft found in North America is a dugout canoe about 3,600 years old (1600 B.C.) recovered near the head of the Vermilion River in Ashland County. This wise placement of village sites near bodies of water enabled the Archaic peoples to systematically harvest the bounty of one ecosystem and then another as the changing seasons made different foodstuffs available. One strategy was to settle during the spring and summer along waterways where they could easily fish, hunt turtles and birds, and gather shellfish. In the fall they camped in forests to gather nuts and other forest plant foods, then dispersed during winter in small mobile units—such as extended families—to hunt deer and birds. This Archaic seminomadic and more-territorial lifestyle thus replaced the nomadic Paleo-Indian existence.

The rise of seasonally inhabited villages—evidenced by thicker, excavated cultural strata, along with other stationary features such as cemeteries and heavy stone utensils and vessels—reflected greater locational stability that constituted a major step away from nomadism and towards sedentarism. Creating a larger inventory and supply of food generated more leisure time, resulting in a more diversified division of labor and occupational specialization. Food supply not only became more varied and abundant in the Archaic, it also became more regular and dependable. As a consequence of this more assured subsistence base, life expectancy, population, and social-group size all increased. Without the constant movement and specialized hunting skills associated with the Paleo-Indians, both the very young and the elderly no longer constituted major societal burdens.

Archaic Indians continued to learn from and exploit their environment, as evident from the many innovations revealed in their artifacts of flaked stone, polished stone, bone, antler, horn, wood, shell, freshwater pearls, and occasionally copper and meteoritic iron. The quantity and variety of tools, weapons, utensils, ornaments, and ceremonial objects excavated by archaeologists increased throughout the Archaic, showing higher levels of economic success and social well-being. Long-distance trade arose, evidenced by some artifacts made of materials unavailable in the areas where they were used. The existence of long-distance trade indicates a complex social structure had evolved based on food surpluses, a stable political climate, occupational specialization, denser populations, and leisure time. More people with greater amounts of material and nonmaterial culture reflected the existence of wealth on a significant scale for the first time. Such cultural elaboration—in addition to much larger, more numerous, more sedentary sites—constituted major differences for the Archaic stage, as compared to the preceding Paleo-Indian stage.

Woodland

The climax of aboriginal culture east of the Rocky Mountains was reached during the Woodland stage. Beginning about 3,000–3,500 years ago (1500–1000 B.C.), it is presently subdivided into three horizons: Early, Middle, and Late Woodland. In Ohio these are referred to as Adena, Hopewell, and Fort Ancient–Monongahela, respectively. The Woodland stage was marked by the use of ceramics. Although earlier containers constructed of bark, wood, woven and dried plant materials, skin, bone, shell, and stone continued to be produced, they were now supplemented by pottery. Clay was tempered for strength using sand or shell grit, then rolled into coils, and fired for durability. Decorations to pottery vessels in Ohio were applied by incising, stamping, punctating, and impressing or wrapping with cords or occasionally some other textile.

While the Woodland stage continued the Archaic trimodal livelihood of small game hunting, gathering, and fishing, a new food source was slowly being adopted: horticulture. Though its impact was for a long time limited, the cultivation of domesticated plants continued to increase in significance. Woodland Indians became more dependent on garden cultivation by making food staples from plant domesticates. Horticulture expanded food variety and supply and helped create a more dependable subsistence base by providing its practitioners with greater and more direct control over their natural environment. This reflects both the wisdom of the aborigines and their willingness to experiment with their surroundings and adopt new subsistence forms. The increase in horticultural practices was accompanied by an increase in life expectancy. This led to expansion of population and social group size, more leisure time, greater division of labor, and occupational specialization. The material and nonmaterial aspects of Woodland culture multiplied in variety, quantity, and sophistication. Although horticulture was adopted throughout Ohio and was widely practiced elsewhere, it never became an Indian cultural universal. Horticulture occurred only where it was both environmentally feasible and needed. In regions where climate was hostile to horticulture (such as cold northern coniferous forests and tundra, and arid intermontane) or where bountiful and easily exploited natural resources were available (such as Pacific Northwest coast fishing and marine mammal hunting, and California acorn gathering), the Archaic stage persisted until the coming of the Europeans.

Gardening arose in Ohio primarily in major river valleys, which were geographically and ecologically optimal locations to maximize horticultural success. They contained large, wide, level floodplains that made gardening relatively simple and rewarding. Although huge forests covered 95 percent of present-day Ohio, the floodplains did not require much clearing before being planted. Compared to the laborious turning-over-of-sod necessary in the prairie grass regions tackled unsuccessfully by early white settlers, the floodplains were easily worked. In addition, these floodplains were covered with fertile alluvial soil renewed naturally with the annual spring floods. Woodland Indians utilized sophisticated conservation techniques. They practiced *intertillage* by raising several different crops together. Intertillage maximized available space and maintained soil fertility by reducing (1) soil loss due to the erosion common in fallow fields, (2) nutrient loss common in monoculture, and (3) crop loss due to plant diseases, also common in monoculture. They also used fish as a fertilizer when necessary. Domesticated plants included maize (corn), beans, squash, pumpkins, gourds, sunflowers, tobacco, goosefoot, pigweed, smartweed, and probably other edible plants that archaeologists and palynologists have not yet found or recognized as domesticates.

Horticulture as a significant subsistence component necessitated the adoption of a relatively sedentary lifestyle; crops had to be planted, tended, harvested, preserved, and stored. Permanent villages arose as Indian groups remained in an area year-round while maintaining access to periodically exploitable ecological zones nearby. "Permanent" as used here is not equivalent, however, to the settlements of today, nor does it indicate settlements of multigenerational duration. Due to the gradual diminution of local supplies of firewood, game, and perhaps productive soil, Woodland villages existed for only several decades at most. Eventually, faced with the effects of environmental depletion, including declining crop yields, the group migrated en masse to a different site and erected a new village. When economic prosperity resulted in a settlement growing beyond a manageable size, the inhabitants commonly split up, creating two new villages.

A fascinating development of the Woodland stage that occurred in particular areas was the rise of numerous and impressive religious structures, artifacts, and activities in eastern North America. These extensive artificial landscapes display the greatest impact that aboriginal culture made in transforming the natural environment in this region. Ohio was the center of this ceremonial activity in Early and Middle Woodland times and here were constructed the most spectacular and advanced cultural landscapes that had yet occurred. There were three major related components: (1) thousands of artificial earthworks, many of which were of monumental size and effort (including burial, temple, and effigy mounds; walled enclosures; and paved walkways), (2) multitudes of specially manufactured ceremonial grave goods made from exotic raw materials and of high-quality and workmanship, and (3) a nearly continental network of long-distance trade or tribute that imported the desired resources used for the funerary objects.

Such cultural florescence required a complex and stratified society. Continuing food surpluses were essential to sustain large numbers of laborers and other skilled workers, all of whom were nonfood producers, over long building periods. Construction of grandiose ceremonial centers, manufacture of lavish burial offerings, maintenance of a long-distance trade network, and related ritual activities required a level of coordination typically associated with a class structure. A small religious elite likely formed a nobility, or ruling class, supplemented by a larger professional class of skilled workers—including architects, surveyors, engineers, artisans, foremen, and traders—with a commoner class of ordinary drafted laborers constituting the majority. The many examples of monumental architecture completed over a lengthy period attest to continuous local and regional political stability, prosperity, and social acceptance or acquiescence.

The mechanism that successfully integrated these various culture elements was religious rather than social. Religion more easily lent divine justification to large-scale, long-term, and far-reaching human endeavors which became even more ambitious and imposing through time. Although advanced levels of art, architecture, astronomy, economics, engineering, and technology were attained, these disciplines did not exist for their own sakes, or even as discrete entities as we know them. Instead, they were purposefully developed for religious-political reasons. The monumental earthworks, sophisticated ceremonial artifacts, long-distance trade networks, and accompanying mortuary rituals all specifically related to the special and lavish treatment afforded the important dead and their needs in the afterlife. This state religion was narrowly and obsessively directed, unlike the much simpler utilitarian culture and folk religion. The religious mechanism was a burial cult successfully superimposed on the preexisting, broader society.

It is not known if Ohio's ceramics, horticulture, and burial cult activities arose in situ or were the result of outside innovation diffusion. If the later occurred, as is commonly suggested, these Woodland traits probably originated in the Mesoamerican civilizations of coastal Mexico and diffused up the Mississippi River and its major tributaries, reaching as far north as southern Ontario, Canada. Another possibility is that Ohio's relative location presented an interface between two distinctive and established culture regions located respectively to the north and south, and Ohio peoples ably exploited their fortuitous geographical situation. Regardless of origins, the Early and Middle Woodland cultural climax in Ohio was not indicative of a new culture or cultures. Rather, the elaborate Adena and Hopewell complexes were burial cult manifestations accepted by or imposed on various local cultures that otherwise remained intact. Ohio was part of an interactive circum-Caribbean region of advanced culture approaching civilization. Several hallmarks of civilization realized in Ohio included monumental architecture, state religion and government, long-distance trade, social stratification, division of labor, food surpluses, and wealth. However, other significant defining criteria, especially writing and the keeping of records (i.e., literacy), either were not achieved or not common, such as the wheel, currency, smelting, and urbanism. Only in Late Woodland times would the Mississippian complex that eventually extended well into the continental interior, in the Southeast and the lower Mississippi Valley, surpass the earlier Adena and Hopewell to become the most advanced aboriginal culture ever to have existed in eastern North America.

Adena and Hopewell. Adena and Hopewell sites—whose inhabitants are also known as the Mound Builders—are famous for their spectacular earthworks and grave offerings. They collectively cover a period of over 1,500 years that began with the Adena

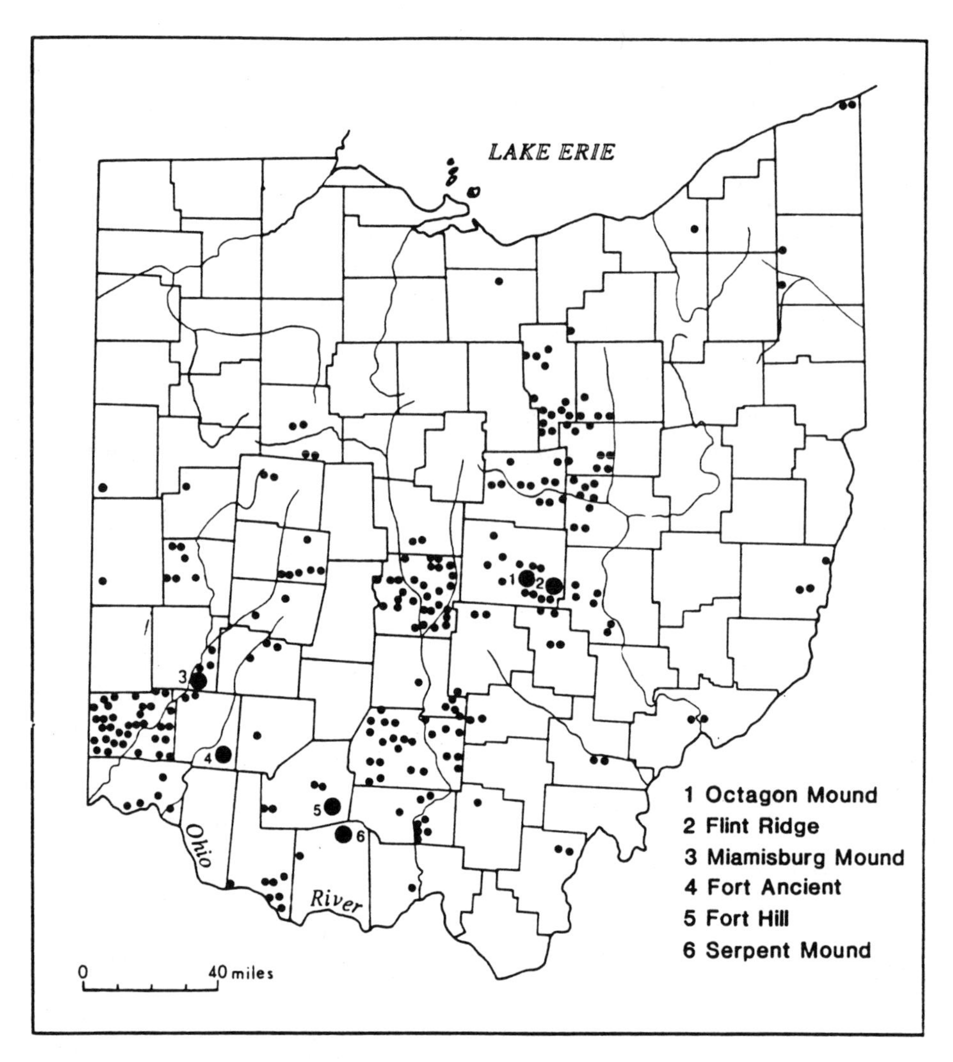

Fig. 5.1. Adena and Hopewell Mound Sites (From George Knepper, *Ohio and Its People,* 1989. Map by Margaret Geib)

about 2,800 years ago, incorporated the Hopewell about 1,900 years ago, and ended with the Hopewell about 1,200 years ago. Adena and Hopewell peoples coexisted for several centuries. Differences in cranial morphology revealed from burials suggest that the Hopewell elite were recent immigrants to Ohio who practiced a burial cult similar to the local Adena. The Hopewell cult was heavily influenced by, and continued many traits of, the Adena, but in general its artifacts were more elaborate, abundant, and sophisticated—especially in ceramics. Adena and Hopewell sites were concentrated in approximately the same geographical region in central and southern Ohio, specifically in the Miami, Scioto, and Muskingum River Valleys (Figure 5.1). Given their temporal and spatial overlap, as well as similar artifact inventories, both Adena and Hopewell are treated here as a single "tradition."

The extremely successful and extralocally influential Adena and Hopewell were aided in part by an augmented resource base, both natural and domesticated. Their regional core in central and southern Ohio contained various microenvironmental zones due to its greater local relief. The Adena and Hopewell wisely exploited this ecologically diverse natural setting and generated an intensified subsistence base. Although all of Ohio, except the Great Black Swamp in the northwest, was inhabited by peoples of the same Woodland-stage culture, the Adena and Hopewell tradition did not significantly affect peoples in northern Ohio. Probably the poorer natural environment was inadequate to provide the high subsistence threshold necessary to sustain the burial cult. Evidence of this cult is widespread throughout the state, especially in the form of mounds and grave goods.

Mounds

Ohio has a greater variety and number of Indian mounds than any other state; they appear either as effigies or as mounds with a specific burial or religious connotation. The few effigy mounds were constructed in the shape of animals as well as birds and reptiles. One example is a raptorial bird—probably an eagle, hawk, or vulture—at Newark in Licking County; it extends 200 feet between wingtips. Another very impressive example is the Great Serpent Mound at Locust Grove in Adams County. It is approximately 1,330 feet long (1/4 mile), 3–6 feet high, with a body width of 20 feet and represents a partially coiled snake (Figure 5.2).

The fact that few effigy mounds are located in Ohio suggests that they did not directly relate to the Adena and Hopewell burial cults. Burial mounds, on the other hand, constituted the most numerous mound type. Serving as tombs and places of worship (temple mounds), they were constructed in many different forms and usually housed elaborate log tombs containing cremations and inhumations. These remains only of important persons were provided with many special ceremonial and utilitarian mortuary offerings. Their graves were adorned with red ocher and the entire tomb was covered with earth. This procedure—repeated through time on a given site—often included burials that, without tombs, were placed on the surface of the earlier mound. The eventual result was multiple tombs and burials within conical or domed mounds of great size: 160–470 feet long and 20–32 feet high. Burial mounds have been found both in unprotected clusters and within massive earthen-walled enclosures. The largest mound in Ohio is the Miamisburg Mound at Miamisburg in Montgomery County. It is a cone-shaped mound, 850 feet in circumference and 68 feet high. A second example is the Seip Mound southwest of Chillicothe in Ross County. It is 240 feet long, 160 feet wide, and 30 feet high. A third example is the Harness Mound, south of Chillicothe in Ross County. It has a rectangular base and measures 160 feet long, 80 feet wide, and 20 feet high.

Grave Goods

Grave goods recovered from burial mounds usually differed in a number of significant respects from those found in commoner-class burials. First, burial mound artifacts of the elite were often produced in large quantities and often of rare materials. The copious amount sometimes lavishly and literally blanketed the remains. Examples are cut mica and freshwater pearls. Second, burial mound artifacts show high levels of dexterity, care, and technical expertise. These goods took considerable time and effort to produce and were doubtless made for the deceased by skilled, probably full-time, craftspeople. Third, burial mound artifacts were aesthetically appealing; many examples are exquisite. Often combining several techniques, they reflect an overall level of sophistication and ingenuity that represents a well-developed sense of professional artistry. Fourth, burial mound grave goods were commonly constructed of extra-local raw materials. These exotic materials were imported from distant locations and would otherwise probably never have been seen by the local populace. Examples include alligator skulls and conch shells. Fifth, burial mound artifacts were often ceremonial in form or function. They were not meant for utilitarian or daily use as they appear to be of symbolic importance and often display recurrent, probably religious, motifs. In many cases, such as monolithic stone axes, they could not even have been used in the traditional sense. Sixth, many grave goods were intentionally broken at the time of interment. Archaeologists refer to such objects as "killed," meaning that they were deliberately mutiliated so they also could journey to the afterlife and there aid the deceased.

These factors taken together indicate that significant levels of commerce, wealth, and manufacturing existed for religious reasons, most specifically for rituals surrounding the dead. Tombs, mounds, and funerary offerings were obviously created for the benefit of important persons. Although many artifacts are labeled "ceremonial," archaeologists often apply this misleading designation when the function of an artifact is not apparent. Burial offerings included tools and weapons of flaked flint, polished stone, and copper; utensils and vessels of bone, pottery, and shell; figurines of pottery; ornaments of bone, copper, mica, and freshwater pearls; gorgets of bone, shell,

Fig. 5.2. Serpent Mound, Adams County (Courtesy of the Ohio Historical Society)

and stone; apparel of cloth and leather; birdstones, smoking pipes, and engraved tablets, all made of stone; and amulets, talismans, and medicine bundles, including drilled animal teeth and animal jaws. These objects were made in a variety of ways, often showing multiple techniques that included flaking, polishing, hammering, molding, cutting, carving, drilling, incising, weaving, and inlay. Many objects were further decorated, again in different ways; these included geometric, zoomorphic, and human patterns—for example the commonly found raptorial bird and hand-eye motifs. However, these specialized representations served as religious icons rather than as mere art.

Long-Distance Trade

Although long-distance trade arose in the Archaic, during the Woodland stage it became far more spatially expansive and systematic, covering most of North America. A wide-ranging commercial network of natural waterways and trails existed primarily to import certain exotic raw materials used to manufacture many of the grave goods needed for the lavish, and no doubt spectacular, funerals and ceremonies of the burial cult. Although this extensive trade was vital to acquire materials for religious needs, it also became a major vector, intentional or not, that helped diffuse the Adena and Hopewell burial cult.

Table 5.1
Exotic Raw Materials Found in Ohio Adena and Hopewell Burials and Their Source Regions

DIRECTION FROM OHIO	SOURCE REGION	RAW MATERIAL
North	Lake Superior	copper
	Ontario	silver
East	Appalachian Mountains	mica
	E. Pennsylvania, Delaware	steatite, chlorite
Southeast	Florida	alligator teeth & skulls, pottery
South	Gulf of Mexico	conch shells, shark teeth
West	Missouri	galena
	Rocky Mountains	obsidian, grizzly bear, canines

Exotic raw materials excavated from Adena and Hopewell burials were imported into Ohio from source regions across the United States. For example, these materials included minerals such as copper, silver, mica, steatite, chlorite, and obsidian as well as teeth and skulls from sharks, alligators, and grizzly bears (Table 5.1 and Figure 5.3). A comparable set of Ohio exports is more difficult to ascertain. Perhaps many were perishable (food) or made from perishable materials (textiles). Understandably, Ohio materials have not been found in concentration in a given burial or group of related burials beyond the Adena and Hopewell region. This was probably due to a lesser emphasis on cult burials, as practiced in Ohio, by peoples living elsewhere. However archaeologists have determined at least two major Ohio exports that were highly valued by distant peoples: Flint Ridge flint and Ohio pipestone. Though Flint Ridge is located in southern Licking County, this type of flint was also quarried at sites in Licking and Muskingum Counties. Both unfinished "blanks" of Flint Ridge flint and the flaked objects made from it, such as ceremonial knives and spear points, were exported (Figure 5.4). Ohio pipestone was quarried in the hills of the Scioto River Valley north of Portsmouth. This pipestone and the polished objects made from it, such as tubular pipes, were also exported (Figure 5.5). These two Ohio exports have been uncovered as far east as West Virginia, Maryland, New York, and Vermont, as far west as Iowa, and as far south as Florida.

Excavated scraps of exotic raw materials and artifacts made from them that were probably judged to be of substandard quality and discarded represent evidence suggesting that some structures within the earthwork enclosures specifically housed artisans who manufactured funerary goods. This manufacturing also included the export of Ohio raw materials and finished goods that comprised very desirable complementary commodities to Indians elsewhere. Further, artisans at certain earthworks specialized in the export of a single Ohio resource. For example, the Newark site specialized in Flint Ridge flint, while the Portsmouth site specialized in Ohio pipestone. Thus regional production, specialization, and perhaps even monopolistic practices appear to have existed in the Adena and Hopewell export trade.

Enclosures

Enclosures were earthworks constructed of long earthen and stone walls either surrounding an area or fencing in those portions located on easily accessible topography. These enclosed areas seem to have been erected as gathering places or community ceremonial centers for the local populace. They were not permanently inhabited, except perhaps by some clergy, maintenance workers, or artisans. Enclosures were used periodically for various religious activities (especially funerals) and for long-distance commerce,

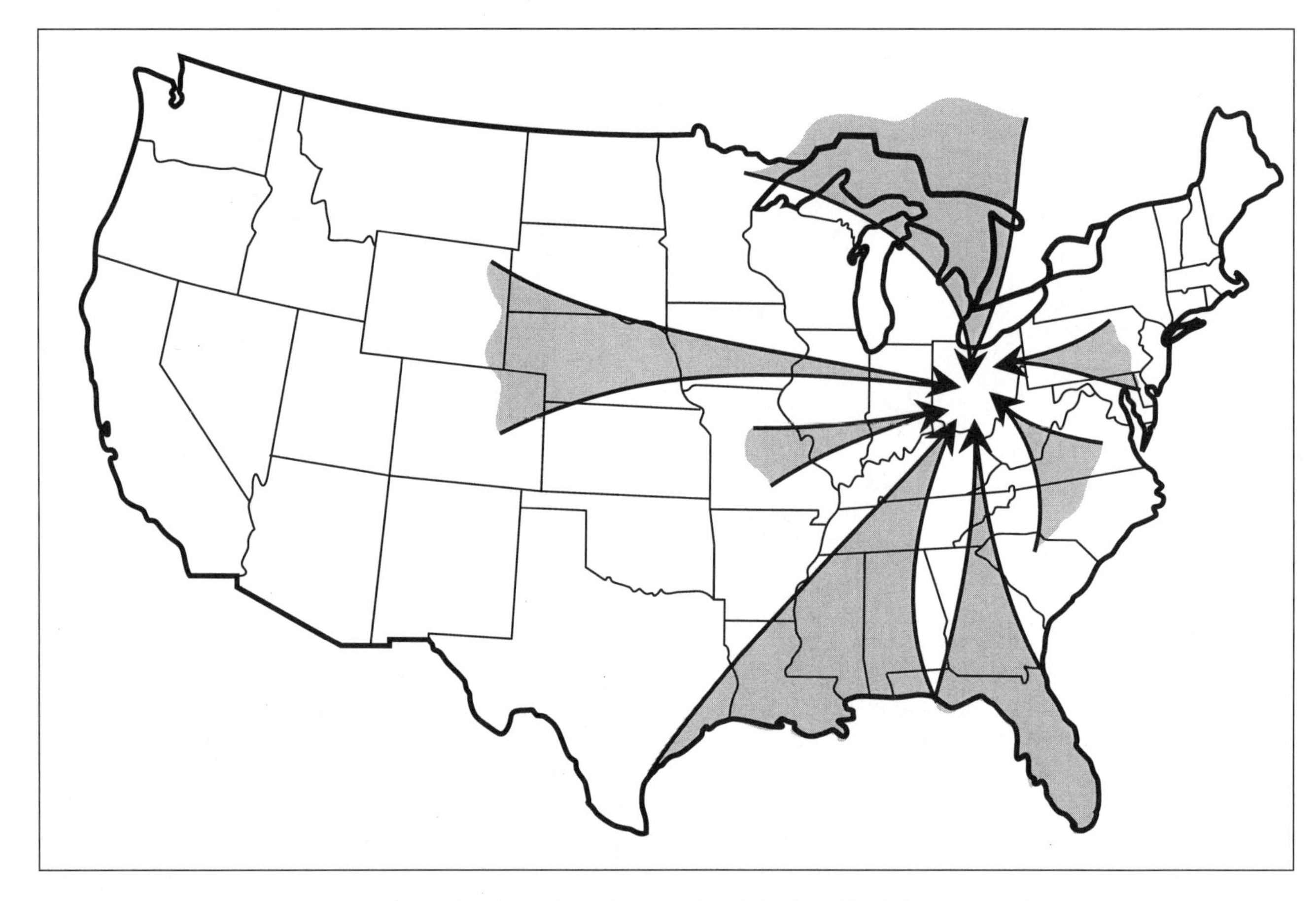

Fig. 5.3. Source Regions of Exotic Raw Materials Found in Ohio Adena and Hopewell Burials

and their solar and lunar alignments suggest astronomical observations enabled a calendrical scheduling of rituals, and as refuges in troubled times. Their walls were massive: up to 50 feet wide at the base and 25 feet high. They had circumferences of several miles and enclosed areas ranging from 20 acres to over 100 acres (the equivalent of 50 city blocks). Contained within the enclosures were burial mounds, effigy mounds, paved walkways, and other structures. Even the smaller and simpler enclosures must have required a minimum of several years and possibly as long as several generations to complete.

There were two major types of enclosures, based on location. Enclosures built on the flat bottomlands of river valleys were geometrical, usually circular, square, pentagonal, octagonal, rectangular, or oval in shape. A few notable examples are: (1) the Newark Earthworks at Newark in Licking County, an immense complex originally covering almost 4 square miles (Figure 5.6); (2) Seip Mound near Bainbridge in Ross County; and (3) Mound City at Chillicothe in Ross County.

Another enclosure type, identified initially as forts, was built on high elevations. One example is Fort Ancient near Lebanon in Warren County. Situated on a plateau overlooking the Little Miami River, it covers an area of more than 100 acres enclosed by 3 1/2 miles of earthen and stone walls 10–25 feet high, and contains burial mounds, crescent-shaped mounds, and paved stone walkways and plazas. A second example is Fort Miami in Hamilton County. Situated on a cliff overlooking the Ohio River near the confluence of the Ohio and Little Miami Rivers, it covers 12 acres surrounded by massive earthen and stone walls 50 feet wide at the base. A third example is Spruce Hill in Ross County whose walls enclose an area exceeding two miles in length and 100 acres in area.

Fig. 5.4. A Scraper Made from Ohio Flint (From Martha Potter Otto, The First Ohioians, in *Ohio's Natural Heritage,* 1979)

Fig. 5.5. Adena Pipe (From Martha Potter Otto, The First Ohioans, in *Ohio's Natural Heritage,* 1979)

Late Woodland. It is not clear why the thriving burial cult tradition was replaced in Ohio by the Late Woodland horizon that began around 1,300 years ago (700 A.D.). Perhaps climatic change occurred as had happened at other times in the past and significantly decreased the biomass available to the Hopewell for subsistence. Perhaps the local peoples rebelled against ever-increasing Hopewell demands and the undemocratic stratified society it imposed. Perhaps the long-distance trade network established to obtain vital raw materials for burial goods was usurped by envious peoples elsewhere—the early Mississippians, for example—or was disrupted by regional unrest. Without exotic materials and the impressive objects made from them—certainly vital components of mortuary rituals—the viability of the burial cult may have severely weakened.

Some archaeologists believe that hilltop ceremonial centers represented a major situational change or strategic retreat from the open floodplains. Perhaps under pressure, the Hopewell and their remaining adherents withdrew to more inaccessible and defensible higher terrain. The introduction of stockades or moats (or a combination of both) around these hilltop enclosures suggests a defensive posture and thus potential con-

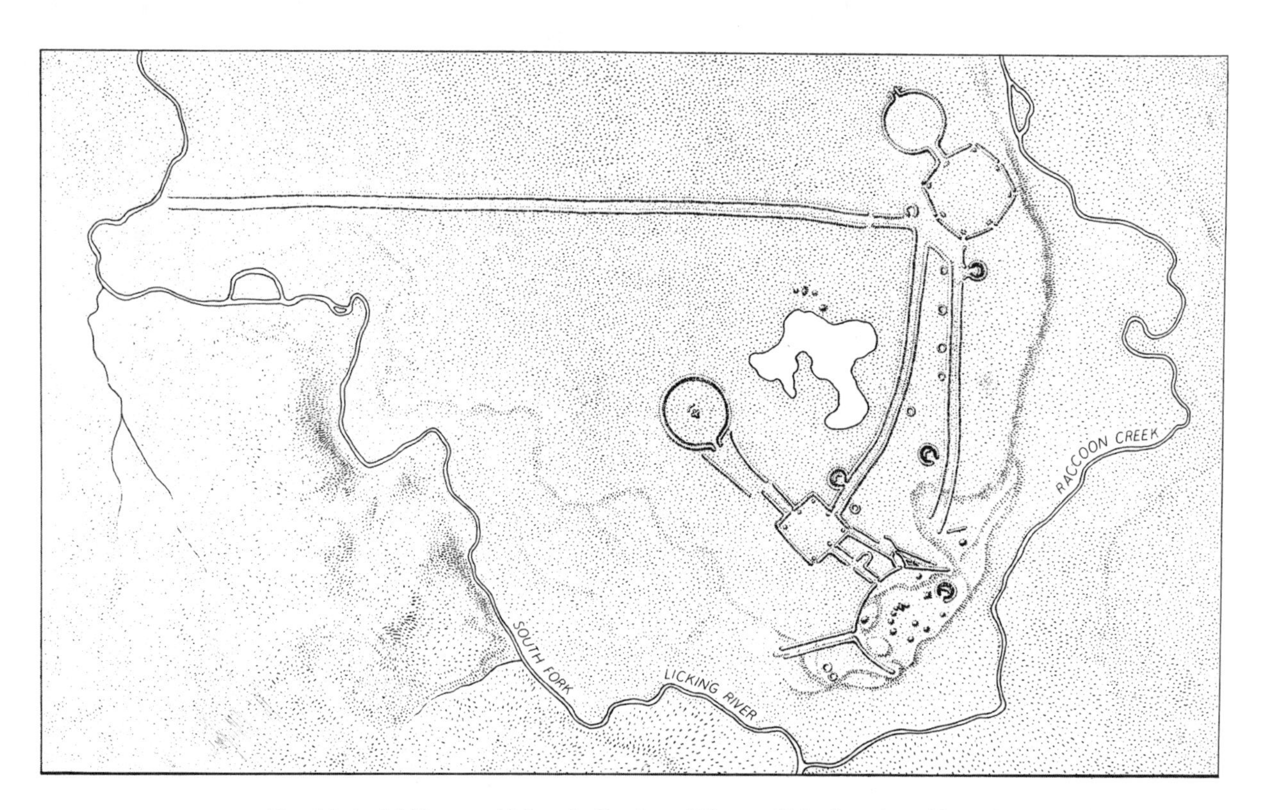

Fig. 5.6. Aerial Photo and Map of a Portion of Hopewell Earthworks at Newark, Ohio (From Olaf H. Prufer, The Hopewell Cult, in *Scientific American,* December 1964. Top illustration courtesy of the Smithsonian Institution; map courtesy of *Scientific American.*)

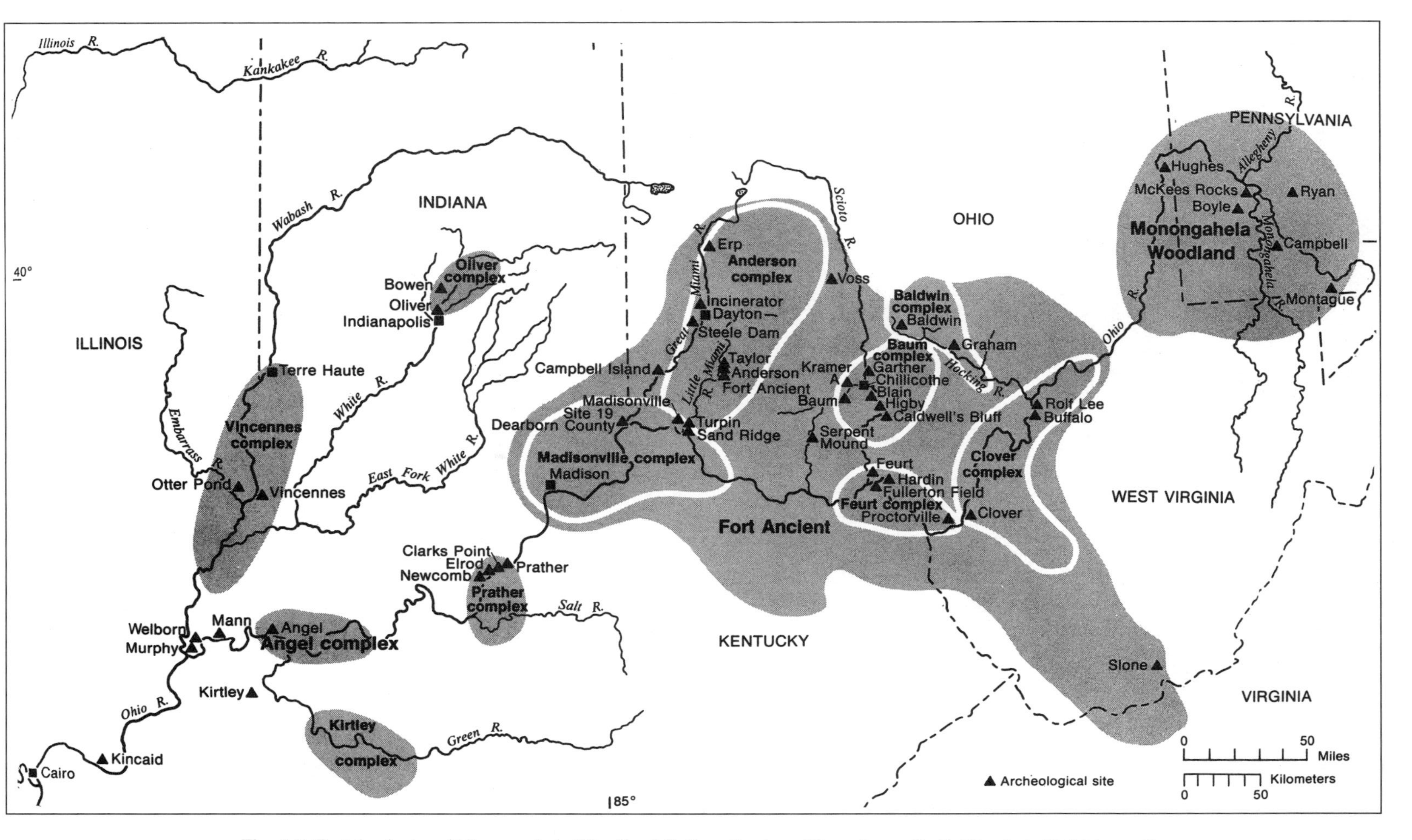

Fig. 5.7. Fort Ancient and Monongahela Woodland Culture Regions (From James B. Griffin, Late Prehistory of the Ohio Valley, in *Handbook of North American Indians: Northeast,* 1978. Courtesy of the Smithsonian Institution.)

flict. The introduction about this time of the bow and arrow, a weapon using multiple projectiles tipped with small triangular stone points, created a quantum leap for potential mayhem and destruction, and suggests another cause for the toppling of the Hopewell complex.

The Late Woodland horizon in Ohio is viewed by many archaeologists as a period of cultural decline. They base the lower postclimax culture level on the premises that monumental earthworks, such as mounds and enclosures, were no longer being built and that the long-distance trade system, with its manufacture of elaborate funerary goods, also ceased. The Turpin Farm site near Newton in Hamilton County and the Merion site in Franklin County are characteristic Late Woodland examples. Subsequent artifacts, made mostly of local stone, were of cruder construction and intended for utilitarian more than ceremonial purposes. This major reversion suggests either a cultural regression or the arrival of new immigrants with a simpler culture. However, it may be correct to view these changes not as cultural decline but within the context of the fall of an elaborate, narrowly focused burial cult that did not affect the basic Woodland culture of the local peoples in lasting ways. Unfortunately, due to the lack of "exciting" artifacts and landscape structures that attract substantial scholarly attention, knowledge of post-Middle Woodland peoples in Ohio is scanty. Interestingly, Adena and Hopewell earthworks, mortuary artifacts, culture, and peoples were as much a mystery to later Ohio Indians as they were to the early European explorers who inquired about them.

Fort Ancient–Monongahela Woodland. This last prehistoric Indian culture phase in Ohio began around 1000 A.D. and lasted until European contact in 1654 A.D. It is divided into two complexes, or geographical regions, with different cultures: Fort Ancient and the Monongahela Woodland. A major characteristic of this period was the fortified site. Villages were usually protected with either encircling earthen ramparts, or stockades of upright pointed stakes, or a combination of both. This new condition indicated conflict, if not actual aggression. Villages with their wealth of material goods, stores of food, and fields of standing crops may have represented tempting targets for other peoples, especially those armed with a new "automatic" weapon of great destruction power—the bow and arrow. Another major diagnostic of this time was the central courtyard. Villages were constructed so that dwellings and other structures encircled an open area, or plaza. This innovation was used as a public square and diffused from the Mississippian peoples.

The Fort Ancient complex occurred in the same central and southern Ohio region as did the preceding Adena and Hopewell. It reflected advanced or reinvigorated culture, probably influenced by the Mississippian peoples. Notable examples include the Baum Village site on Paint Creek in Ross County, the Gartner site in Ross County, and the Feurt site in Scioto County. Fort Ancient peoples may be the ancestors of the historic Shawnee tribe, or perhaps the Shawnee happened to settle on the earlier Fort Ancient sites.

The Monongahela Woodland complex occurred in the southeasternmost portion of Ohio. It reflected the culture of peoples who had lived in the east and had migrated westward. Like the Fort Ancient peoples, the Monongahela peoples lived during the same time span, built stockades and central courtyards, and probably experienced instability in the form of aggression from marauders. The Monongahela peoples may be the ancestors of the protohistoric Erie tribe.

European materials excavated from sixteenth-century Indian sites in Ohio—consisting of such European trade goods as glass beads and iron axes, or of Indian artifacts fashioned from brass and iron European objects—signify that a protohistoric era had begun. Ohio Indians, although not yet in direct contact with the Europeans living to the east on the Atlantic seaboard, were nevertheless in receipt of European influence in the form of material culture that diffused successively inland as one Indian tribe in turn traded with another. In the seventeenth century, when direct contact was finally established between Ohio Indians and Europeans who recorded these contacts, the transitional protohistoric era became the historic era in Ohio. In 1654, the Indians whose ancestors had ventured into Ohio at least 13,000 years prior to the coming of the Europeans entered yet another culture stage.

REFERENCES

Anderson, Douglas D. 1968. A Stone Age Campsite at the Gateway to America. *Scientific American* (June). Also published in Ezra Zubrow et al., 1974.

Associated Press. 1995. Spear Point Found Near Findlay. *Bowling Green Sentinal-Tribune.* (2 August).

Begley, Sharon. 1991. The First Americans. *Newsweek* (fall), special issue.

Carter, George F. 1975. *Man and the Land: A Cultural Geograpy.* New York: Holt, Rinehart and Winston.

Claiborne, Robert et al. 1973. *The Emergence of Man: The First Americans.* New York: Time-Life Books.

Collins, Williams R. 1974. *Ohio: The Buckeye State.* Englewood Cliffs, N.J.: Prentice Hall.

De Blij, Harm J., and Peter Muller. 1988. *Geography: Regions and Concepts.* New York: John Wiley & Sons.

Drexler, Michael. 1992. Farm Owners Get Thrill Out of Prehistoric Find. *Cleveland Plain Dealer* (12 March).

Durham, Michael S. 1995. A Walk in the Steps of the Real Mound Builders. *Toledo Blade* (7 May).

Fellman, Jerome, Arthur Getis, and Judith Getis. 1990. *Human Geography: Landscapes of Human Activities.* Dubuque, Iowa: William Brown.

Gabriel, Mary. 1988. Artifacts Place Humans in Americas 30,000 Years Ago. *Bowling Green Sentinel-Tribune* (10 March).

Glover, James L. 1984. Mastodons Along Glacial Border. In Hansen, Michael C. *Ohio's Glaciers.* Columbus: Ohio Dept. of Natural Resources. Educational Leaflet No. 7.

Griffin, James B. 1978. Late Prehistory of the Ohio Valley. In *Handbook of North American Indians: Volume 15, Northeast,* Bruce G. Trigger, ed. Washington, D.C.: Smithsonian Institution.

Hansen, Michael C. 1984. *Ohio's Glaciers.* Columbus: Ohio Dept. of Natural Resources. Educational Leaflet No. 7.

Haynes, C. Vance, Jr. 1966. Elephant Hunting in North America. *Scientific American* (June). Also in Zubrow et al., 1974.

Jackson, W. A. Douglas. 1985. *The Shaping of Our World: A Human and Cultural Geography.* New York: John Wiley & Sons.

Jennings, Jesse D. 1974. *Pre-History of North America.* New York: McGraw-Hill.

Kendall, Henry M. and Robert Glendinning. 1976. *Introduction to Cultural Geography.* New York: Harcourt Brace Jovanovich.

Knepper, George. 1989. *Ohio and Its People.* Kent, Ohio: The Kent State University Press.

Lindsey, David, Esther Davis, and Morton Biel. 1960. *An Outline History of Ohio.* Cleveland: Howard Allen.

Melvin, Ruth W. 1970. *A Guide to Ohio Outdoor Education Areas.* Columbus: Ohio Dept. of Natural Resources and Ohio Academy of Science.

Noble, Allen G., and Albert J. Korsok. 1975. *Ohio: An American Heartland.* Bulletin 65. Columbus: Division of Geological Survey, Dept. of Natural Resources.

Otto, Martha Potter. 1979. The First Ohioans. In *Ohio's Natural Heritage,* Michael B. Lafferty, ed. Columbus: Ohio Academy of Science

Prufer, Olaf H. 1964. The Hopewell Cult. *Scientific American* (December). Also in Zubrow et al., 1974.

Roberts, Carl H., and Paul R. Cummins. 1956. *Ohio: Geography, History, Government.* N.p.: Laidlaw Bros.

Scripps Howard News Service. 1989. Prehistoric Man May Have Used "Refrigerators." *Toledo Blade* (19 November).

Sloat, Bill. 1991. Park May Save Works of Ohio's Mound Builders. *Cleveland Plain Dealer* (6 October).

Snyder, Sarah. 1983. Map in England Called New Key to Area Past. *Toledo Blade* (14 November).

Spencer, Robert F., and Jesse D. Jennings. 1977. *The Native Americans.* New York: Harper & Row.

Van Fossan, William. 1937. *The Story of Ohio.* New York: Harvey Macmillan Co.

Walker, Byron H. 1973. *Indian Culture of Ohio: A Resource Guide for Teachers.* Columbus: Ohio Historical Society.

Wheat, Joe Ben. 1967. A Paleo-Indian Bison Kill. *Scientific American* (January). Also in Zubrow et al., 1974.

Zubrow, Ezra B. W., Margaret C. Fritz, and John M. Fritz. 1974. *New World Archaeology: Theoretical and Cultural Transformations.* San Francisco: W. H. Freeman & Co.

CHAPTER SIX

Ohio's Settlement Landscape

Hubert G. H. Wilhelm and Allen G. Noble

The Ohio Country, a term coined during the earliest years of settlement and later applied to the new state, was part of a vast area lying west of the Pennsylvania line and north and west of the Ohio River. Ceded to the United States in the Peace Treaty of Paris in 1783, which ended the American War of Independence, it became known as the Northwest Territory. Occupying and using this territory was considerably different than the legal process of mapping a huge new piece of real estate. In 1785, in response to action taken by the Congress of the United States, Thomas Hutchins, the official geographer of this country, and his crew of surveyors began the arduous task of accurately measuring the first small segment of the area, which later would be subdivided into the states of Ohio, Michigan, Indiana, Illinois, Wisconsin, and a part of Minnesota. Beginning at the intersection of the Pennsylvania line with the Ohio River, they laid down a survey line due west for a distance of 42 miles, or seven ranges. When their work was completed, Hutchins had charted the first federally surveyed part of the Northwest Territory; it became known as the Seven Ranges (Figure 6.1). Today, it is difficult to imagine the hardships that Hutchins and his surveyors faced in accomplishing their mission. One can be sure that Indians were included among their concerns.

Ohio's Indians offered stiff resistance to white settlement. Their numbers and technology, however, were no match for the increasing pressure exerted by settlers from the Northeast, East, and South, as well as by large numbers of European immigrants. The Indians made their last stand during the Battle of Fallen Timbers on August 20, 1794, but were routed by Major General Anthony Wayne's Kentucky Volunteers. Their defeat led to the Treaty of Greenville in June 1795, when the Indian tribes ceded most of their Ohio lands to the federal government. The Greenville Treaty line (Figure 6.2) formed a temporary division between native Americans to the north and new Americans to the south. Its location influenced settlement dispersal by whites and, in part, explains the path of the frontier.

Though the year 1795 was crucial to the settlement and development of Ohio, settlements were in place on the Ohio River, in disputed territory, several years before the Greenville Treaty. Both Marietta and Cincinnati were founded in 1788. The main settlement thrust did not occur until after 1795, when an avalanche of humanity descended on the Ohio Country. The first census conducted in the area in 1800 counted 45,365 residents. By 1850 Ohio's population was nearing 2 million. The dispersal of the new settlers was favorably influenced by the Ohio River and its northern tributaries and their valleys, most notably the Muskingum, Hocking, Scioto, and Little and Great Miami Rivers. By 1797 a crude overland road called Zane's Trace had been blazed between Wheeling, West Virginia, and Limestone (now Maysville), Kentucky (Figure 6.1). In 1803 statehood was granted, and in 1843 the last of Ohio's organized, surviving Indian tribes, the Wyandots, was forced to emigrate from their ancestral hunting grounds for less well-known lands to the west.

Ohio's Cultural Base

It would be an error to negate the impact of native Americans on Ohio's landscape. One need only consider the hundreds of burial mounds that dot this state to be reminded that people predated written accounts.

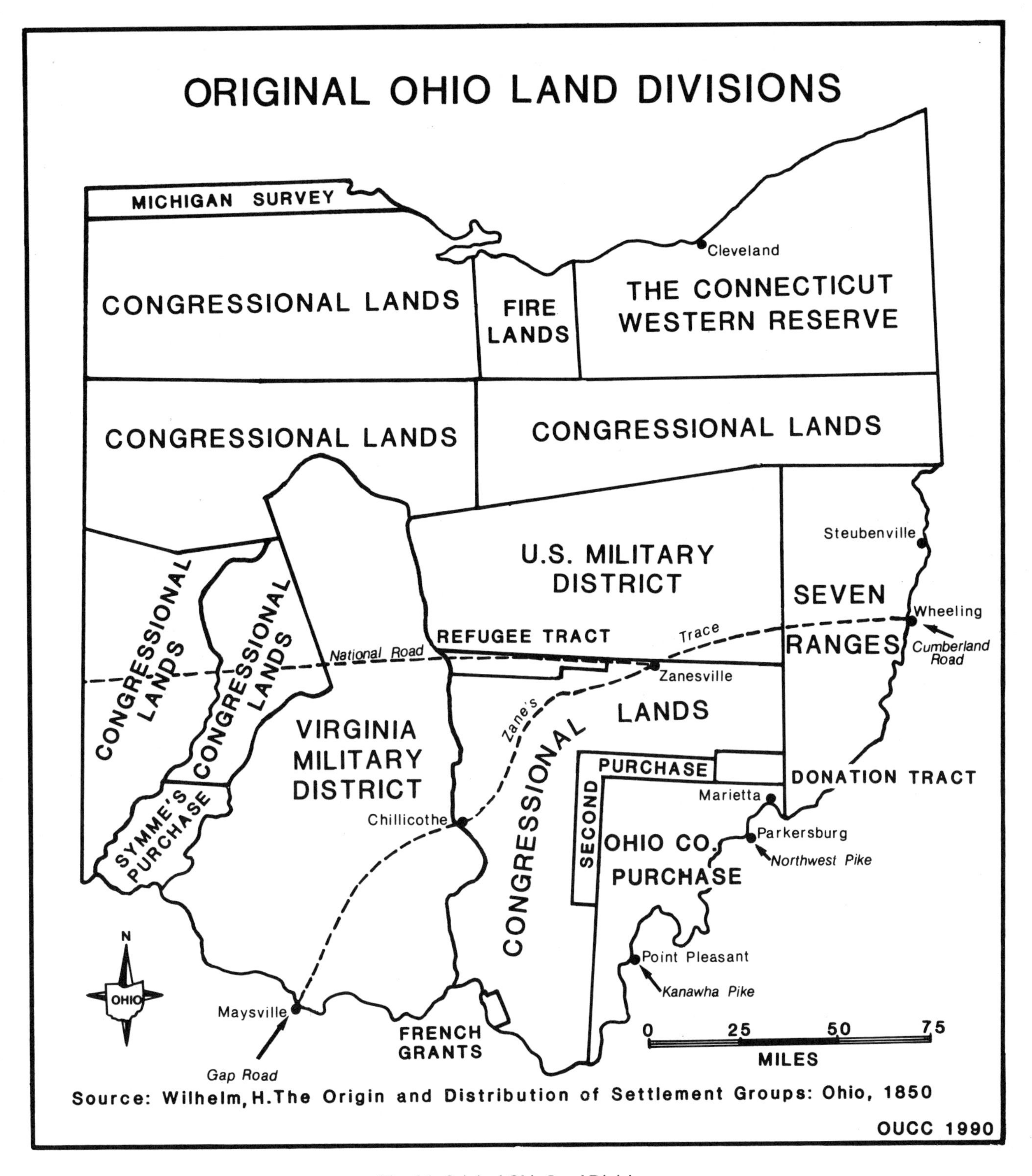

Fig. 6.1. Original Ohio Land Divisions

These earliest occupants also left an intricate network of trails that crossed Ohio from north to south and from east to west and was later incorporated into the modern route system. Indians also contributed to the development of Ohio's prairies through their use of fire to drive animals, clear areas for crop raising, or improve grazing conditions. Unintentionally they functioned as ecological agents, disturbing the

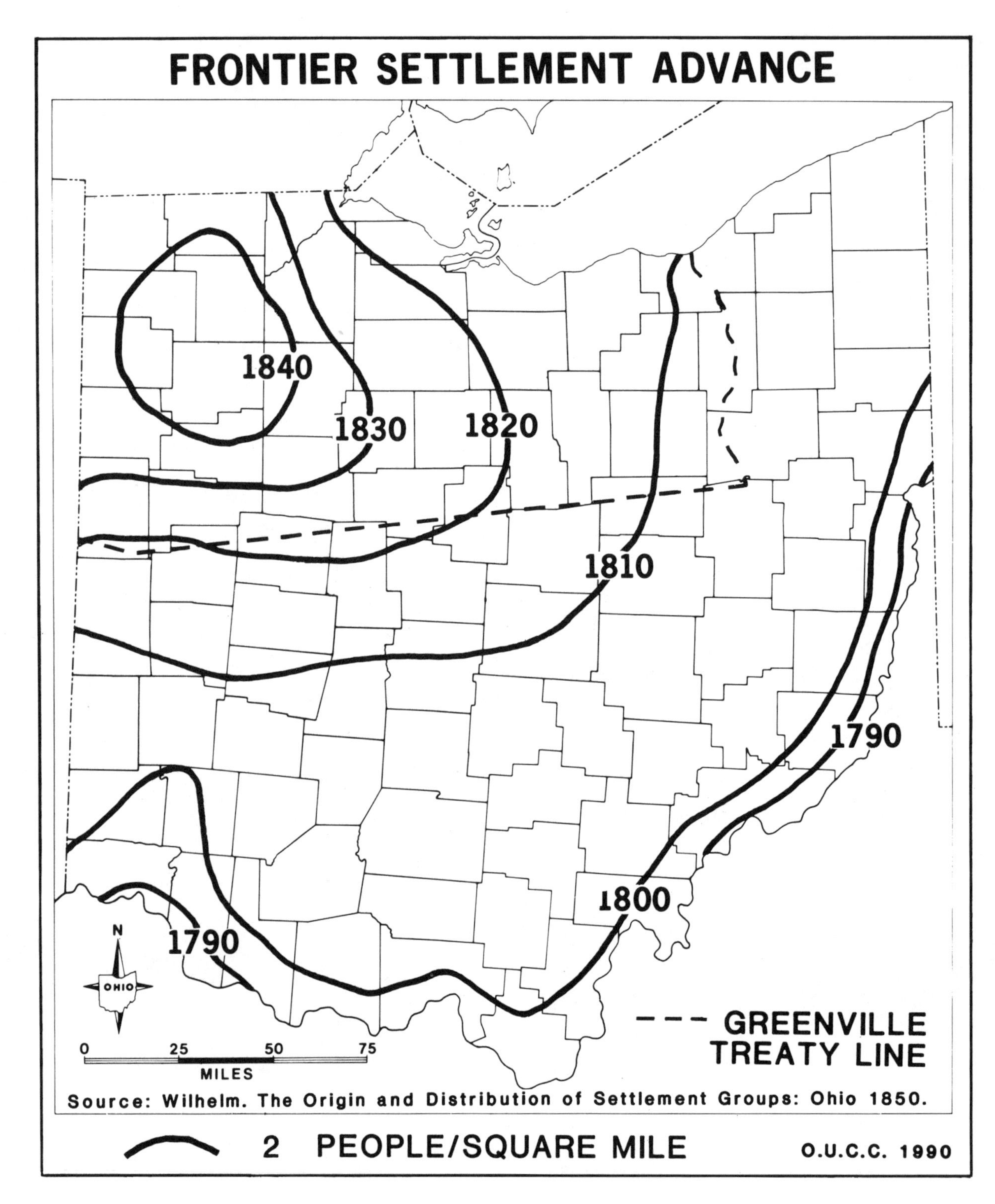

Fig. 6.2. Frontier Settlemernt Advance

forested habitat and assisting in the evolution of one of Ohio's more unusual natural environments—the prairie. Another obvious Indian legacy is the great number of Indian place-names. The frequency of such names may relate to the fact that surveyors often were assisted by Indian scouts who identified places by

Table 6.1

Principal States of Origin of Non-Ohio Born Residents in 1850

STATE	NO.	%
Pennsylvania	190,396	38
Virginia	83,300	17
New York	75,442	15
Maryland	34,775	7
New Jersey	21,768	4
Connecticut	20,478	4
Massachusetts	16,437	3
Vermont	13,672	3
Kentucky	11,549	2
Indiana	5,059	1
	472,876	94

Source: Wilhelm. 1981. *The Origin and Distribution of Settlement Groups: Ohio, 1850.*

their native names. The surveyors then recorded these names, thereby assuring their legal entry. One of the most satisfying exercises in cultural geography is the study of place-names. More than simply intriguing—with such names as Chillicothe, Wapakoneta, or Kinnikinnik (the Shawnee word for tobacco)—these names can be a key to understanding cultural origins and diffusion.

Although some of the native Americans' landscape imprints remain, their recognition requires close, often archaeological scrutiny. The present-day landscape of Ohio is essentially the result of both traditional and popular influences combined with the state's development as a principal agricultural, industrial, and commercial area. In order to understand today's Ohio one must step back into history to the days of Hutchins, Ebenezer Zane, and "Mad" Anthony Wayne, to times of the Greenville Treaty and the opening of the Ohio Country to settlement by Anglo-Americans and European immigrants.

Because the Ohio territory was located between Lake Erie and the Ohio River it became the spatial focus that attracted large numbers of settlers of varying cultural backgrounds (Tables 6.1 and 6.2). For example, from the northeast came migrants of largely English extraction who held to the New England traditions of settling in small towns, attending Congregational churches, and raising dairy cattle and sheep; while from the east came the wheat and livestock farmers largely rooted in the Pennsylvania-German culture of southeastern Pennsylvania. Characteristically, settlers from Pennsylvania belonged to the Lutheran, Evangelical, Reformed, Brethren, Mennonite, and Amish denominations. The distribution of these religious groups remains important today as a key to Middle Atlantic or Eastern settlement in Ohio.

Southerners entered Ohio primarily from either Appalachian Virginia or a greater cultural region known as the Upland South. "From there, Ohio received its hill-culture traditions, including a goodly proportion of Scots-Irish ethnicity, erecting buildings from logs, following Presbyterian religious ideas, and raising corn and tobacco" (Wilhelm 1982). European immigrants added their cultural uniqueness to Ohio's diverse population. Especially important were Germans; the settlements of Berlin, Hamburg, Bremen, Fryburg, Gnadenhutten, and others attest to their cultural influence. Later, as Ohio began its urban and industrial development, immigrants from eastern and southern Europe enhanced the state's cosmopolitan character. However, these immigrants, in contrast to the earlier ones, became localized in specific cities—Cleveland and Cincinnati, for example—or in mining districts such as the Hocking Valley coalfield in Athens, Hocking, and Perry Counties.

Ohio's cultural patterns were formed during pre-1850 settlement. During that time, settlement was primarily in response to the pull of virtually free and superior land and the push of ever-increasing population pressures in other parts of the Union. Ohio functioned as the smaller end of a huge funnel through which America's migrating masses were channeled on their way to the broader interior. This early corridor function was greatly aided by Ohio's natural routeways—especially the Ohio River, Lake Erie, and the numerous interior river valleys—and the glacial ridges or moraines that offered dry ground across often waterlogged terrain (Wilhelm 1982).

Ohio's Settlement Groups and Their Distribution

The distribution of settlers in the Ohio Country was controlled by several factors, including existing pioneer roads that terminated on the Ohio River (Figure 6.2), specific land claims or purchases by individuals, land set aside by companies or states for certain migrant groups, and cultural affinity (the desire of settlers to move into areas already settled by like-minded people). The great variety of methods by which the land was divided reflected the different groups who laid claim to the state's lands. There were the Western Reserve lands and the Virginia Military District, where Connecticut and Virginia respectively retained titles; the Ohio Company Purchase, nearly 1.2 million acres in the southeast which the Ohio Company of Associates, a Massachusetts-based land company, acquired for settlement by New Englanders; in the southwest the Symmes Purchase; and in the east-central region, the United States Military District, 2.5 million acres which Congress set aside to satisfy land warrants issued after the Revolutionary War. Two of the smaller special grants were the Refugee Tract, offered to Canadians who had sided with the American rebels during the Revolution, and the French Grant in Scioto County. This latter parcel consisted of 25,200 acres granted to the unlucky French who arrived in southeastern Ohio (at Gallipolis) only to find that the land they had purchased, in France, was already in the hands of the Ohio Company of Associates (Sherman 1972).

The largest parts of Ohio were controlled by Congress. These parcels varied according to time and method of survey (Figure 6.3). The United States Land Ordinance of 1785, which stipulated that the public domain should be surveyed into a rectangular system of townships and sections, was first implemented in the Seven Ranges (Figure 6.2), the eastern part of Ohio that was, theoretically, the first area opened for settlement. Unfortunately, there were few takers. The government had difficulty selling Ohio Country land due to conditions set forth in the Land Ordinance of 1785. According to this ordinance, the minimum purchase was one section of 640 acres at a minimum price of $2 per acre. Thus a settler wishing to purchase land in the Seven Ranges needed to meet the "threshold" price of $1,280. Not surprisingly there were few settlers who could afford to do so, and the way was left open for speculators. Eventually, to prevent excessive land speculation, the government stepped in, gradually lowering its minimum purchase area and price. By 1832 Congress set the minimum purchase area at 40 acres and the minimum price at $1.25 per acre, resulting in a more affordable threshold price of $50 (Hart 1975). The federal government's different and constantly changing approach to land sales had considerable impact on Ohio's rural landscape because of its effect on farm sizes and land use practices.

One valuable source of information about land purchases is the Ohio land office records. These offices conducted the day-to-day business of land sales. Legal sales of the territory's public land began as early as 1796, through an office in Pittsburgh, Pennsylvania (Peters 1930). By 1800 the Marietta and Steubenville offices had opened in Ohio. The small structure where the early land sales in Marietta were conducted still survives as a part of the Campus Martius historical exhibit. Other land offices followed: Cincinnati and Chillicothe in 1801, Zanesville in 1804, Canton in 1808, Delaware and Piqua in 1820 (Figure 6.3). In 1876 the Chillicothe land office was the last to close its doors, bringing the era of public land sales in Ohio to an end (Burke 1987). The closing of the Chillicothe office corresponds with the end of frontier settlement in the state (Figure 6.1).

A second and more accessible source is the Manuscript Population Schedules of the United States Population Census. The most applicable of the schedules is the one for 1850, which was the first to list residents on a township-by-township basis as well as by place of birth, thus providing insight into the geographical and cultural origins of the settlers. In 1850 Ohio consisted of eighty-seven counties; Noble County was the last, formed in 1851. One should keep in mind that 1850 is rather late to gain a sufficiently comprehensive picture of non-native Ohioans because by that time the massive thrust of migration into the state had subsided. For example, Ohio's population in 1850 was 1,980,329. Of these, 1,215,876 (62 percent) were native Ohioans, leaving slightly over one-third of the population non-Ohio born. A large proportion of the in-state born would have been younger children of families where one or both parents and any older children were born out of state (Wilhelm 1982). The maps (Figures 6.4, 6.5, and 6.6) represent the principal migrant source regions. These regional cultural contrasts, including differences in language, religion, land tenure, architecture, and agriculture, were part of the cultural baggage brought in by settlers. Here, the persistence of tradition assured the survival of these cultural variations through time.

Analysis of these three maps reveals areas of concentration of regional migrants and the overlap of settlers on the margins of their respective settlement territories. One can surmise that in these contact areas,

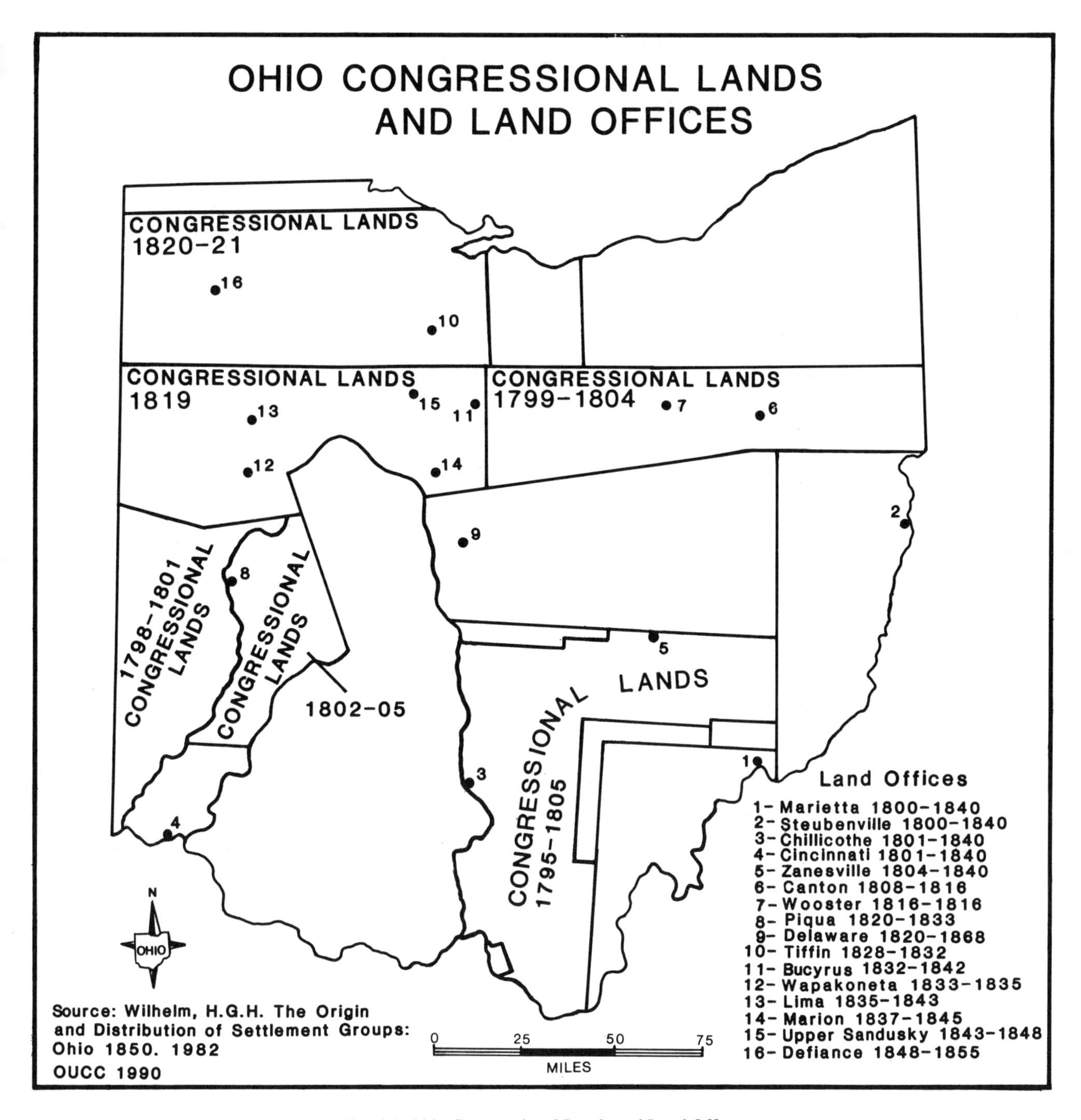

Fig. 6.3. Ohio Congressional Lands and Land Offices

cultural mixing, or *acculturation,* took place. Nevertheless, the maps do indicate regional centers of Northeastern, Middle Atlantic (Eastern), and Southern settlement in Ohio. These centers reflect the relative location of migrant source areas, the presence of early trans-Appalachian roads, and land set aside for specific regional settlement groups. Thus settlers from New England became concentrated in the Western Reserve and in the southeast between Marietta and Athens. The presence of Ohio University in Athens is representative of one New England cultural trait, the importance of education.

Southern migrants, who came principally from Virginia, entered the state at two points: the first near Point Pleasant, West Virginia, and Gallipolis, Ohio; and the second at Limestone (Maysville), Kentucky,

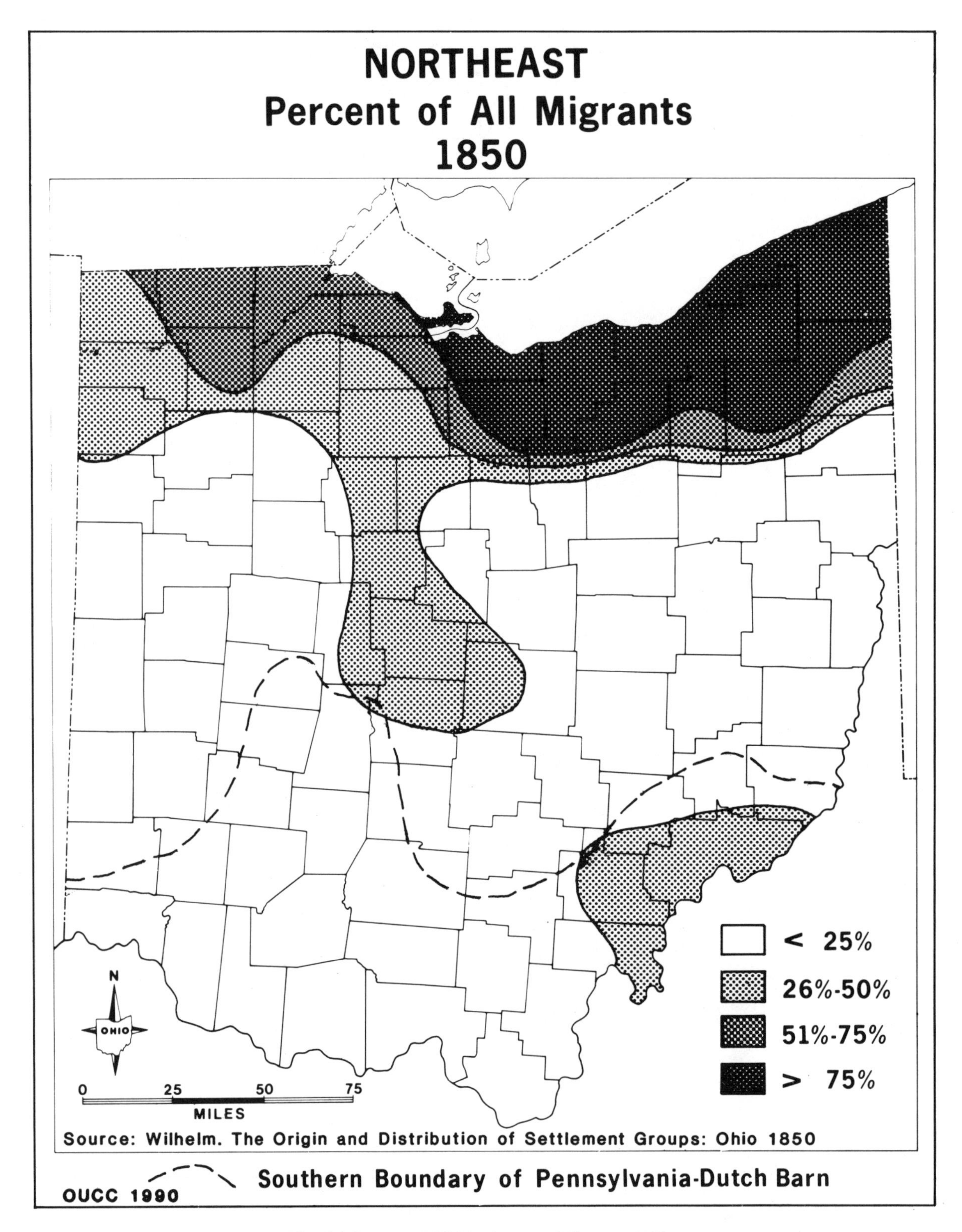

Fig. 6.4. Percent of All Northeastern Migrants, 1850

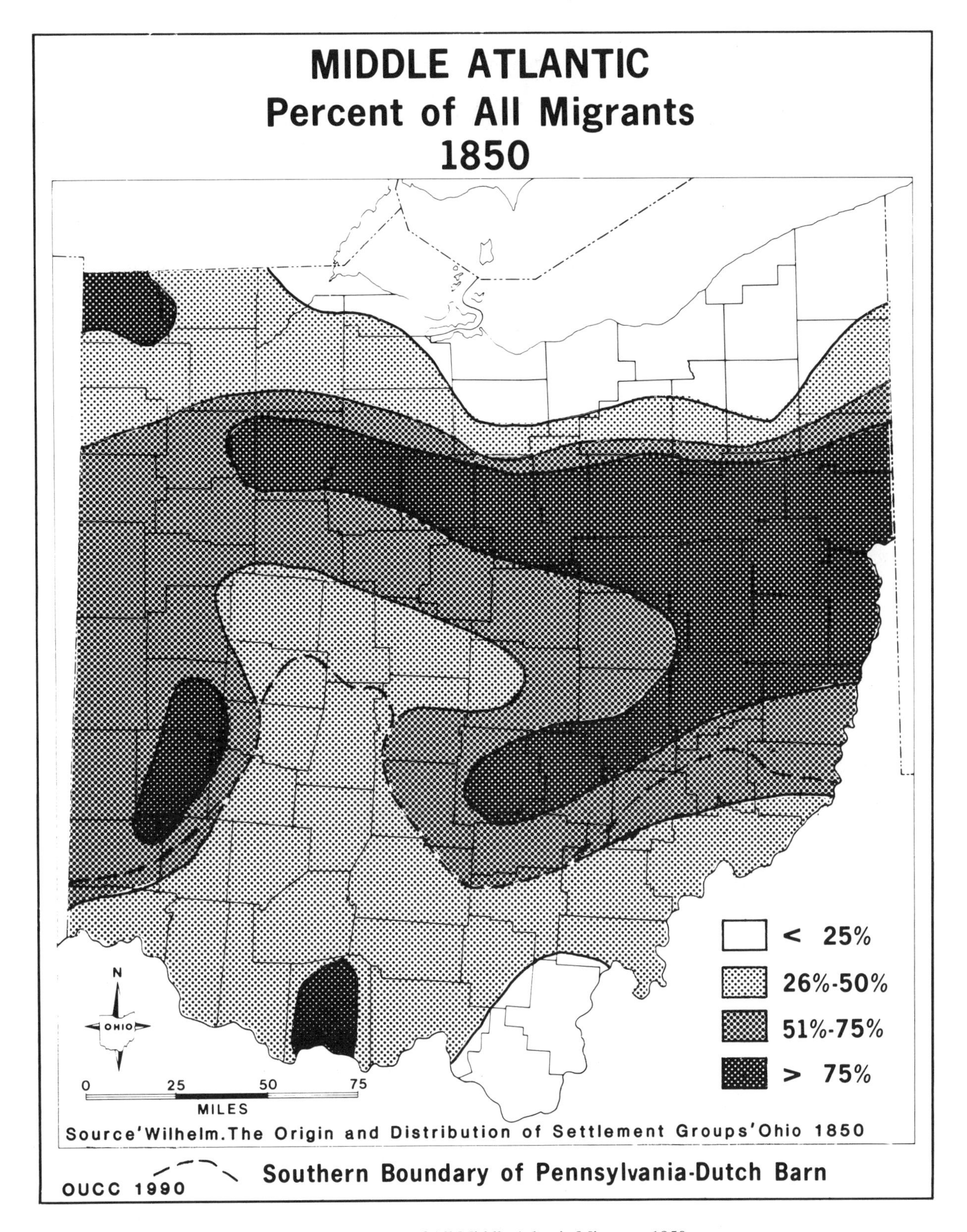

Fig. 6.5. Percent of All Middle Atlantic Migrants, 1850

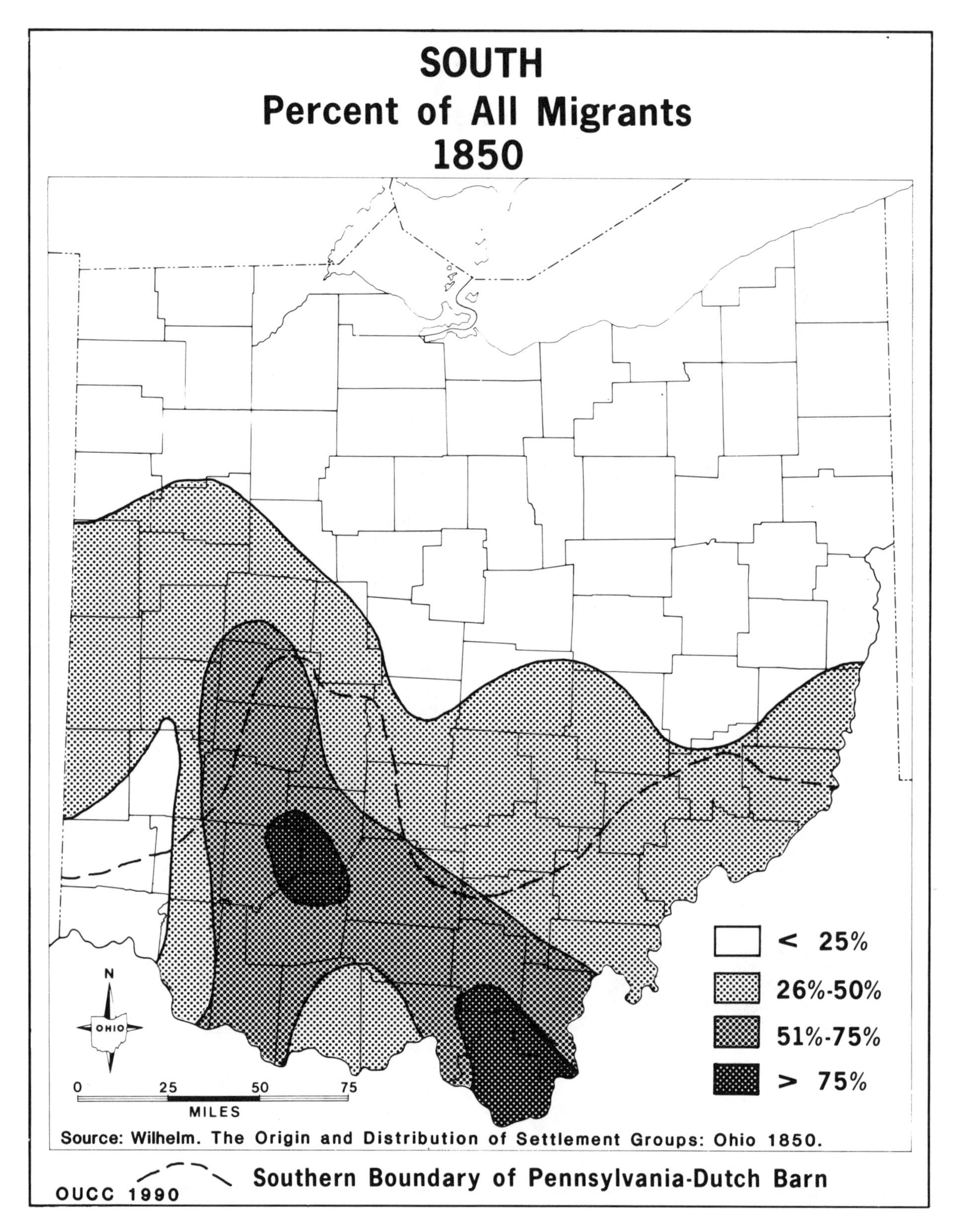

Fig. 6.6. Percent of All Southern Migrants, 1850

opposite from Zane's Trace in Ohio. The majority of these settlers concentrated in the center of the Virginia Military District, particularly Clinton, Fayette, and Highland Counties. To this day, Ohio's largest farms are located in the Virginia Military District, reflecting the large number of Revolutionary War officers and their families who took advantage of Virginia's liberal land policies for former military men choosing to settle in Ohio. A major general could, for example, claim as much as 15,000 acres in the Virginia Military District, and a common soldier or sailor at least 100 acres (Sherman 1972). This area's indiscriminate pattern of settlement, which will be discussed later in greater detail, is unique to the Virginia Military District and survives as a Southern cultural settlement imprint on our landscape.

As the 1850 census reveals, the greatest number of out-of-state born were from the Middle Atlantic states. In fact, Ohio's 190,396 Pennsylvanians outnumbered settlers from the Northeast and the South combined, thus exerting a high impact on Ohio's landscape. Middle Atlantic settlers pushed across the state in two major prongs (Figure 6.5). One prong coincides with the principal drainage divide of the state, sometimes referred to as Ohio's Backbone. The land here is rolling, consisting of low hills and wide, open valleys, much of it affected by glaciation. It was the type of land, well-drained and productive, these farm folk from the East desired. Of importance when considering the distribution of the Eastern settlers is the location of the Connecticut Western Reserve to the north and the United States Military District to the south. Both of these areas represented barriers to settlement from the East, the former because it was set aside for New Englanders, and the latter because it was reserved for veterans of the Revolutionary War and the War of 1812. Many of the migrants from the East were members of pacifist religious groups, therefore ineligible to acquire land in the Military District.

The second prong of Eastern settlers was aligned with Zane's Trace, a logical routeway for people arriving from Pennsylvania. Extending southeastward through the Congressional Lands, this prong ends abruptly at the eastern margin of the Virginia Military District, whose border lies in Franklin, Pickaway, and Ross Counties. On the western side of this district, in Montgomery and Miami Counties, is a large concentration of Middle Atlantic settlers, while to the east of the Virginia Military District, settlers from the Middle Atlantic region were concentrated in Fairfield, Hocking, Perry, and Pickaway Counties. Fairfield County, especially, became the center of Pennsylvania-German settlement. The county seat, Lancaster, was originally named New Lancaster after Lancaster, Pennsylvania, located in the heart of Pennsylvania's Dutch settlement.

The distribution of the three principal American migrant groups in Ohio represents an expected internal division (Figure 6.7). It must be remembered that these divisions are not rigid; considerable settlement overlap and acculturation always occurs on settlement margins. In addition both short and longer distance movements continued, characterizing the periods after the introduction of canals and railroads, and the growth of cities. Nevertheless, the three-part division of Ohio based on the cultural origins of its early settlers is evident when specific cultural traits—such as language, folk traditions, and in particular, building practices—are considered. One of the several cultural regions shown in Figure 6.8 is suggested by the southern limit in Ohio of the Pennsylvania barn, a uniquely distinctive structure because of its cantilevered second story, which creates a pronounced overhang. It is significantly different from barns built by Southerners in Ohio and clearly delineates the Pennsylvania settlement imprint.

Immigrants

Typical of pre-1850 immigration in the United States, Ohio's immigrant populations came predominantly from northwestern European countries, especially Germany and Ireland (Table 6.2). Germans were numerically predominant and continued as Ohio's single major immigrant group through the remainder of the nineteenth century and on into the twentieth. Germans flocked to Ohio because of serious economic problems at home, especially unemployment in urban areas and landlessness in the countryside. Many also had strong ties to Pennsylvania-German settlers already in Ohio.

European settlement in Ohio became a patchlike pattern of local concentrations. This was true of the large German contingent and becomes particularly striking when some of the smaller immigrant groups such as the French, Swiss, and Welsh are considered.

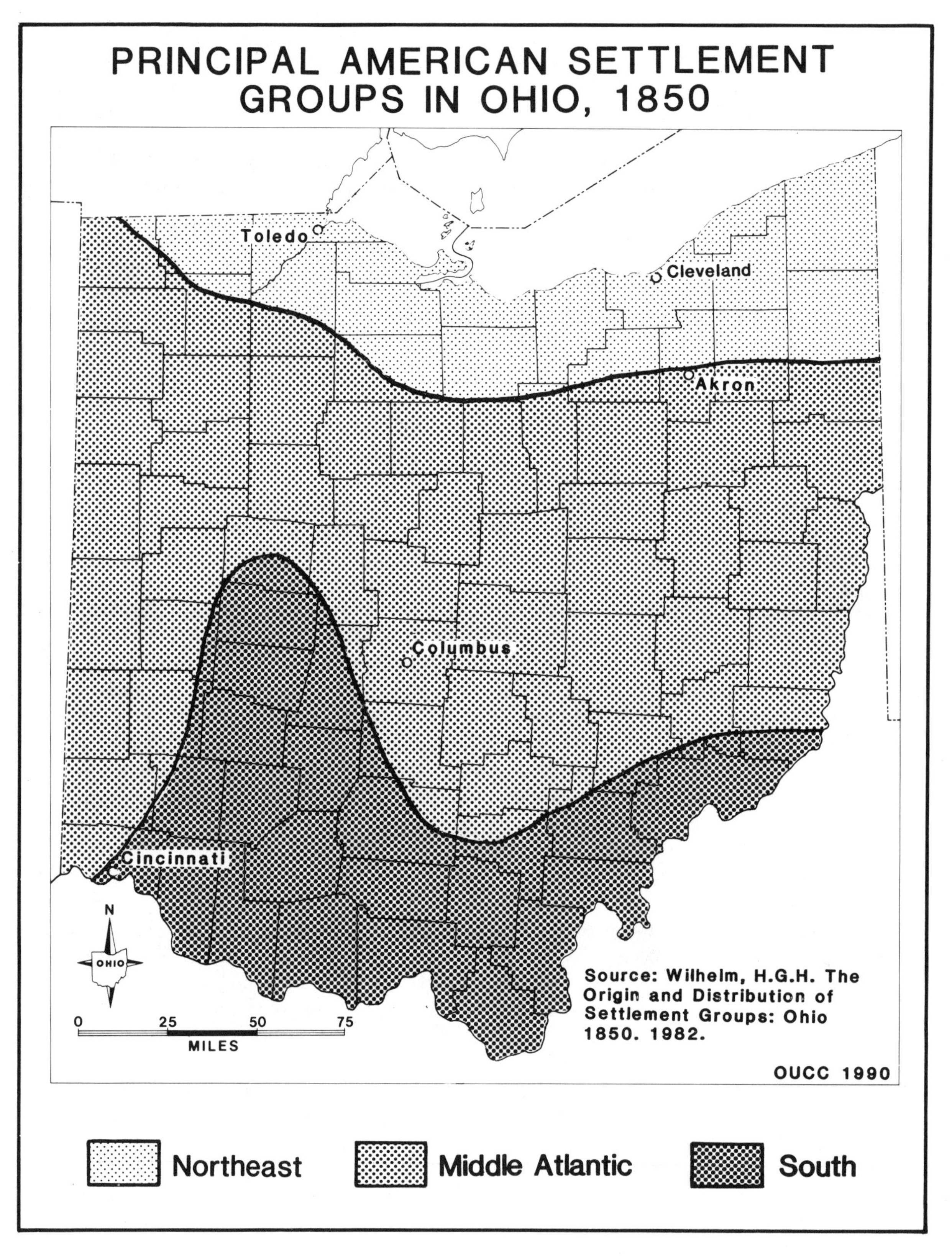

Fig. 6.7. Principal American Settlement Groups in Ohio, 1850

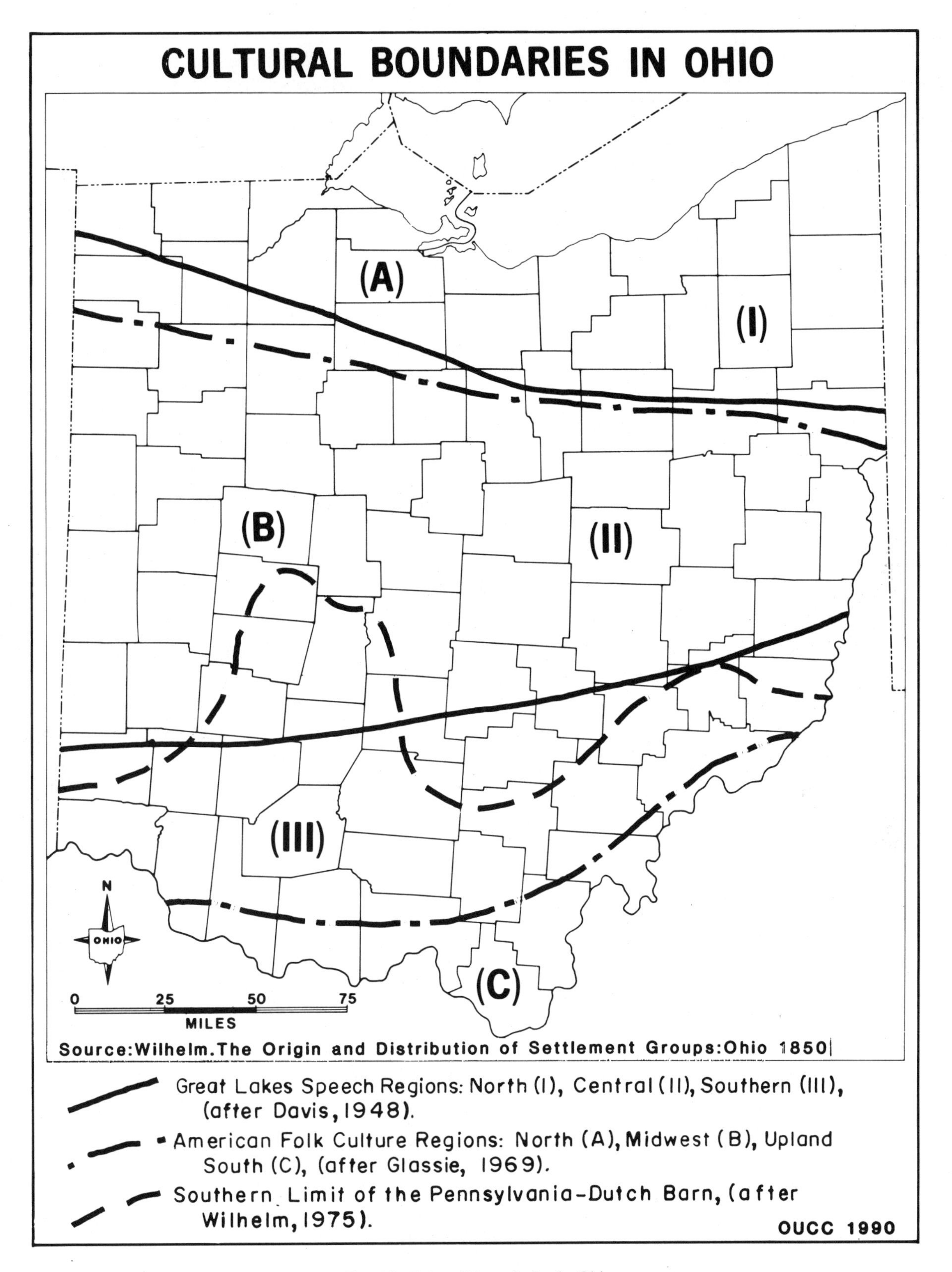

Fig. 6.8. Cultural Boundaries in Ohio

Table 6.2

Origin of Immigrants in Ohio, 1850

COUNTRY	NO.	%
Germany	70,236	48
Ireland	32,779	22
England	19,509	13
France	6,326	4
Wales	5,045	3
Canada	4,606	3
Scotland	4,003	3
Switzerland	3,000	2
Others	488	—
	145,992	100

Source: Wilhelm. 1981. *The Origin and Distribution of Settlement Groups: Ohio, 1850.*

For example, in 1850 most of the 5,045 Welsh were concentrated in six counties: Delaware, Gallia, Jackson, Licking, Meigs, and Portage. Jackson County had the largest contingent of Welsh settlers, and it is here that the Welsh cultural legacy, expressed most notably in language and religion, remains apparent. This localization of immigrants is best explained on the basis of cultural affinity. When settling in a new and strange land, foreign settlers sought contact with their own kind, in the process establishing centers of ethnicity rather than contiguous areas (Wilhelm 1982). Figure 6.9 shows the seemingly haphazard dispersal of local concentrations of European ethnic groups, which exist like islands in a sea of Anglo-American settlement.

The Ohio Settlement Imprint

Ohio's historical settlement by Anglo-American migrants and European immigrants produced geographical convergence from a number of culturally diverse source areas. Although cultural mixing occurred among migrant and immigrant populations, centers of American regional and foreign national dominance were established. Settlers continued their traditional practices, imprinting different regions with contrasting traditional traits. For example, it is widely known that Cincinnati's reputation as an early major beer producer is directly related to the city's strong ethnic German background. Similarly, the presence in southeastern Ohio of such place-names as Barlow, Chester, Plymouth, and Waterford is representative of the early New England settlement influence. The survival of numerous contrasting cultural traits in Ohio's landscape allows the geographer to map these traits and to interpret the state's cultural geography. Prominent in this cultural geographic pattern is the effect of the three Anglo-American settlement groups: those from the Northeast, East, and South.

Land Subdivision and Survey Patterns

One readily apparent settlement imprint is the way in which people distribute the land they occupy. The state of Ohio stands out as a region where both public and private attempts to create order on the land abound. It became the first state to be subjected to the rectangular survey system, better known as the *township and range system,* adopted by the United States Congress in 1785. This system's first creation was the Seven Ranges of eastern Ohio, and its use spread from there throughout the various congressional areas of the state. Between the Seven Ranges and the northwestern regions that were surveyed last, the rectangular survey underwent several modifications, including a running of section lines on the ground; an adjustment to base lines that overcame the problem of meridian convergence; and a change in the section numbering system (Sherman 1972). The impact of rectangular survey on the landscape can be observed by anyone who has used road maps or has traveled Ohio's secondary roads, particularly those on the glaciated terrain of western Ohio. Based on that system, section and township lines run north-south, usually parallel to surface roads. Individual properties, including field rows and their associated fences, houses, and barns and other outbuildings, are similarly oriented. Among the Pennsylvania folk, it had been standard practice to orient a barn so that its front, with the overhang and related cattle yard, would face southeast. This traditional practice ceased when farm structures were aligned with the prevailing north-south and east-west survey and boundary lines.

The Virginia Military District. The Virginia Military District is one large area of Ohio where the rectangular survey is totally absent, supplanted by Virginia's traditional way of subdividing land, the indiscriminate *metes and bounds system.* This system varies both in

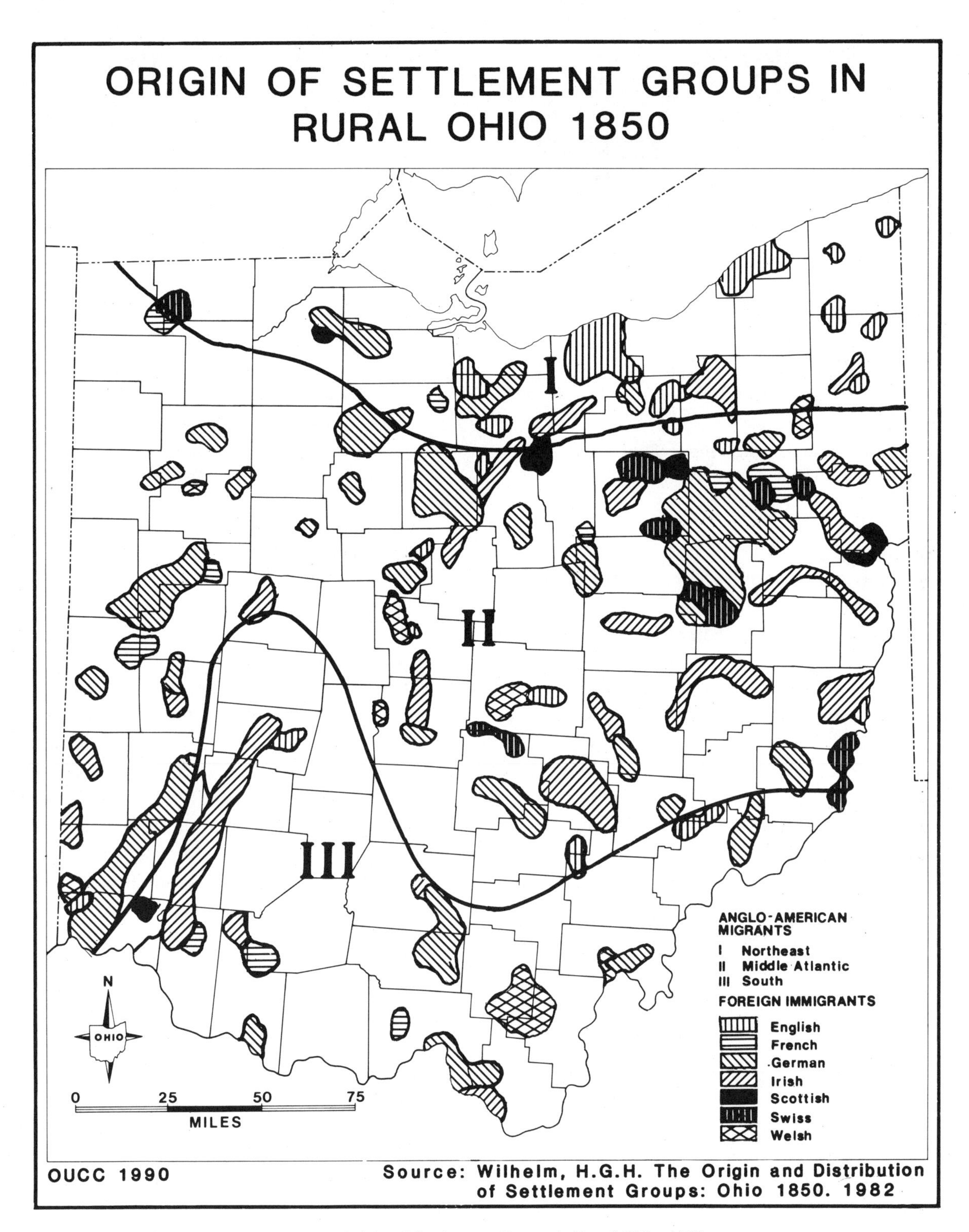

Fig. 6.9. Origin of Settlement Groups in Rural Ohio, 1850

principle and results from the township and range system. It is based on the concept of "subsequent" survey, which means that individuals could make a claim against land before it was surveyed. Often such claims were based on tomahawk marks made on trees, hence the phrase "tomahawking a claim." Between four such

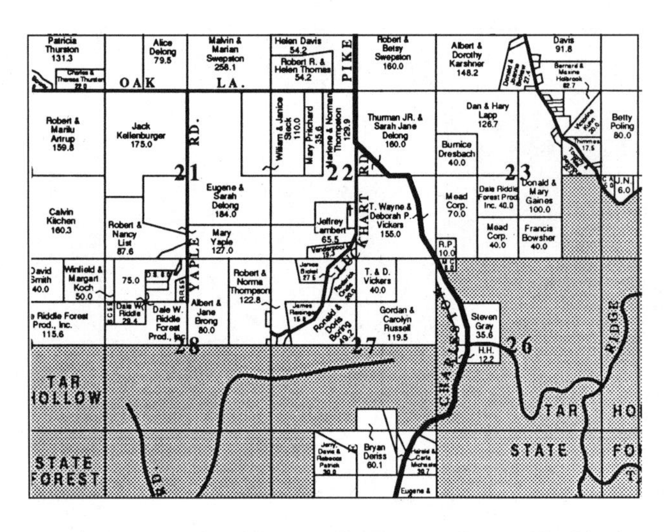

Fig. 6.10a. Example of Rectangular Property Pattern Used in the Congressional Lands, Colerain Township, Ross County

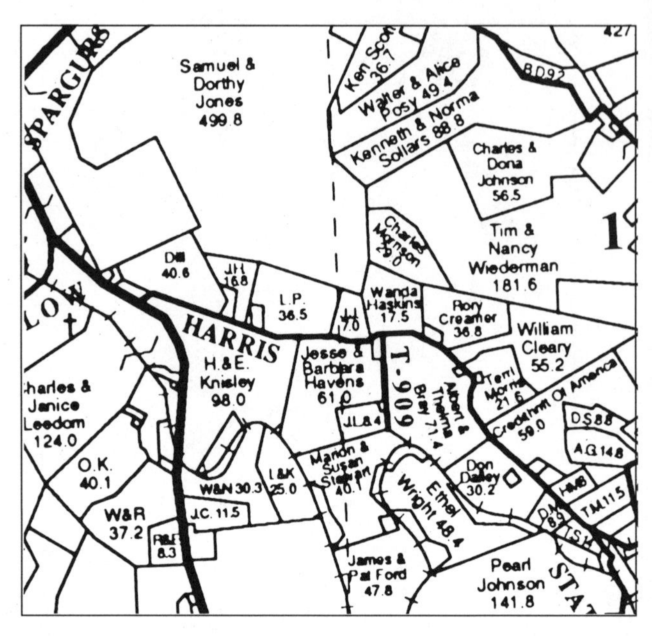

Fig. 6.10b. Example of Metes and Bounds Property Pattern Used in the Virginia Military District, Twin Township, Ross County

prominent reference points (metes) lay the bounds. Understandably, all land in the Virginia Military District is irregularly subdivided, giving that region a character all its own. Because reference markers such as trees are not permanent, the original shape and size of land tracts are often disputed, leading to frequent feuds and real estate suits between neighbors. An excellent example of the contrasting patterns of properties is found in Ross County, which straddles two survey areas—the Congressional Lands on the east side and the Virginia Military District on the west side (Figure 6.10a and b). There, Colerain Township, in the Congressional Lands, is divided into neat rectangular parcels, while Twin Township, in the Virginia Military District, has tracts of all shapes and sizes—the effect of metes and bounds. (Maps of the two townships were taken from the 1989 county plat book [Wilhelm and Mould, 1991].)

New England Settlement Patterns. New England settlement patterns and local surveys reflect New England traditions. In southeastern Ohio where the Ohio Company of Associates controlled early settlement, the partitioning of land followed the congressional plan. These surveys, however, contained two exceptions: fractional sections and river lots. *Fractional sections* were rectangular lots approximately 260 acres in size situated in the center of a township. Their presence and location reflect these New Englander's preference for compact, central settlements. By locating smaller segments of land in the center of a township, the plan for centralized settlement was realized because the smaller fractional sections were easier to sell than regular 640-acre sections. One also finds *river lots*, long lots which border on the Hocking and Ohio Rivers, in southeastern Ohio. They were introduced into the area by the principal surveyor of the Ohio Company of Associates, Rufus Putnam, who argued that the congressional system of sections was not an equitable system in an area dominated by slope. Putnam correctly surmised that without these long lots, too few people would have access to the flat terrain along the principal rivers of the purchase area.

The Connecticut Western Reserve, in the northeastern part of the state, also exhibits a special survey landscape. Here, the square townships measure five by five miles (25 square miles) rather than the common six by six miles, and they were never subdivided into sections. A likely reason for this survey pattern is the traditional New England desire for centrally located and compact towns or villages. A similar plan was suggested for southeastern Ohio. By limiting the size of the townships in the Western Reserve and excluding sections, greater choice for centralized, clustered settlement was attained.

Unlike Southerners from Appalachian areas, who preferred individualistic, isolated locations, New Englanders were above all interested in town settlements. The individual town plans throughout Ohio vary according to several factors, including the locally prevailing surveys and certain regional traditions. For example, New Englanders approached town layout in a similar manner to that practiced in the Northeast. Both regular and irregular plans exist, but the majority of towns founded by New Englanders in Ohio share several common internal geographical characteristics. These town were laid out around a central green or parade ground, with houses facing the open space, and roads radiating outward. There was usually an area set aside as a commons, to be used by community members for such purposes as cutting wood, grazing animals, or establishing gardens. Beyond the immediate town were the cropping fields, or out-lots. Several excellent examples of these town plans are to be found in the northeast, with Tallmadge, Ohio, located between Akron and Kent, among them. Tallmadge's green is surrounded by a circular drive and the characteristic starlike pattern of radiating roads. On the green itself are the Greek Revival structures common to New England settlements, the Congregational Church and the meeting house.

Eastern and Southern Settlement. Eastern and Southern town settlement patterns bear considerable similarity, both displaying variations of the Philadelphia plan involving a regular grid of streets and residential blocks with a central square. Southerners, especially, preferred to locate the county courthouse within the square. Easterners, reflecting practices established in Pennsylvania, laid their street grid around a diamond or square formed by taking land from surrounding blocks. Lancaster, Somerset, and Cambridge, Ohio, are all examples of a "Pennsylvania" town.

Although there were numerous other conditions that influenced the patterns of villages and towns—terrain being a dominant factor—many towns were aligned with the rectangular land survey grid that imposed a rigid north-south and east-west street pattern, influencing the direction of Main Street, which runs east-west in most Ohio towns. When the railroads began to crisscross Ohio, town patterns accommodated the direction of the rails. This is especially true in the flatter parts of the state where rail lines were not confined by terrain. The influence of transportation is most apparent today in urban centers that have grown and expanded in response to the large, four-lane highway systems known as the Interstates.

Buildings on the Landscape

Preindustrial, or folk, buildings are a "key to cultural diffusion" (Kniffen 1965). Certainly buildings of all kinds are as much a part of the cultural fabric of a people as religion and diet. Ohio's early rural settlers consistently adhered to certain traditional types when building houses or barns. These architectural forms developed over a period of approximately 200 years in the Northeast, Middle Atlantic, and South, and were continued by Ohio's new settlers. Because these houses, barns, outbuildings, and churches are relatively permanent, many have outlived their original function and still remain, indicating regional and cultural origins. To the cultural geographer the built landscape becomes quite literally a living map of regional settlement patterns, and houses are probably the most important in revealing regional variations. A famous early German geographer, August Meitzen, sagely referred to the house as "the embodiment of a people's soul" (1882). More than simply wood, brick, or stone, a house is a statement of who we are, what we are, and, in the case of folk houses, where we originated.

In its broadest interpretation, the settlement geography of Ohio involved three macroregions, as depicted in Figure 6.7. Each of these regions produced a distinctive ensemble of folk buildings that provide a particular character to the landscape. Within each macroregion, finer differentiation caused by the ethnic associations of local settlers may be possible.

Throughout northern Ohio, folk buildings reflect a New England or Yankee (New York) provenance. New Englanders were not only drawn into this area because of the land reserved for some of them in the Connecticut Western Reserve and the Firelands, they also were following the route of topographical least resistance, located just south of the eastern Great Lakes along a plain where glacial beach ridges offered firm footing and easy pathways. The early houses they favored were the New England One-and-a-Half Cottage (especially prevalent in northeastern Ohio), the Gable Front house, and the Upright and Wing (concentrated in northwestern Ohio) (Figure 6.11). These

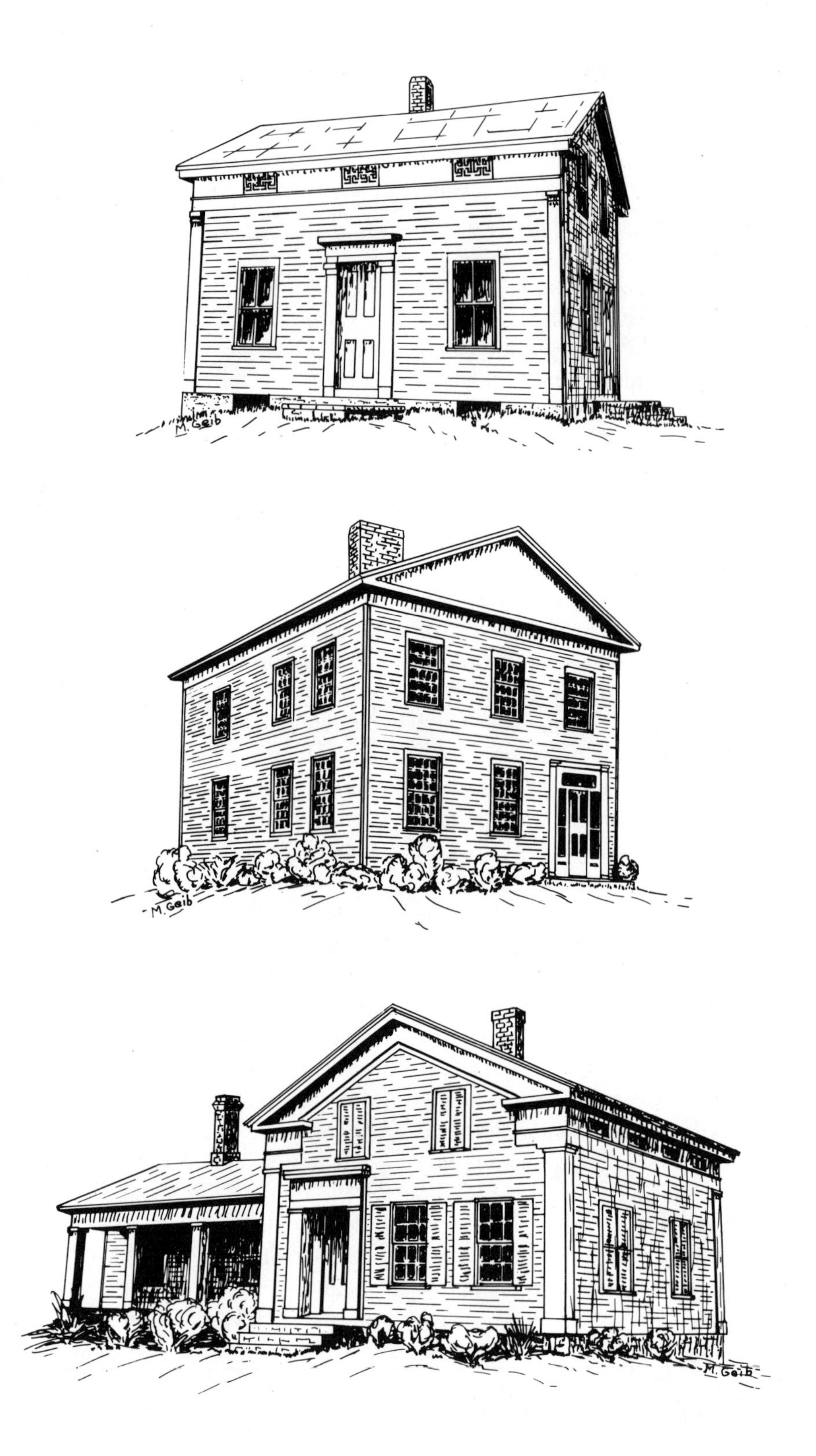

Fig. 6.11. Vernacular Houses of Northern Ohio. From top to bottom: the New England One-and-a-Half Cottage, the Gable Front House, and the Upright and Wing House. (Drawings by M. Margaret Geib)

three houses still characterize the countryside and small towns of northern Ohio.

Common to them all was Greek Revival styling, popular in New York and New England. The earliest of the houses, the New England One-and-a-Half Cottage, maintained the traditional orientation in which the long side of the house faced the street. Here, Greek Revival styling was evident in a lower roof pitch, a balanced facade, and especially in an entablature usually pierced by small half-windows. The Gable Front house was turned ninety degrees, so that its gable end looked out on the street. This allowed for full classical treatment, including pilasters, columns, pediment, and cornice returns. Upright and Wing houses initially continued the tradition of Gable Front structures, with the main entrance on the gable end. Later on, when the larger kitchen wing began to dominate as the center of household activities, the house's main entrance often shifted to the side wing. The large farm kitchen of these houses provided the model for the popular mid-twentieth-century family room.

The barn which Yankee–New Englanders brought with them was a simple English Three-Bay structure (Figure 6.12). This structure was divided into three roughly equal parts: a central threshing floor, and two bays, one on either side. The bays were used to store both threshed and unthreshed grain, farm equipment and machinery, and sometimes to stable farm animals. Up above, a loft provided for the dry storage of hay. As farm operations grew and these smallish English barns proved inadequate, new barns were built raised up on a stone foundation, providing a basement to house the steadily increasing numbers of farm livestock. These Raised or Basement Barns became typical not only of northern Ohio but of much of the eastern Midwest (Noble 1984).

The middle third of Ohio bears the influence of Middle Atlantic settlement, with its strongly Germanic character. The most common early habitat of these settlers was the simple, one- or two-room log house. Log building techniques came into Ohio from both the East and the Upland South. Introduced by Swedish and German immigrants who settled in the Delaware Valley and in southeastern Pennsylvania, they spread both westward and southward, becoming the common method of construction in frontier America, including Ohio. Because they fell out of fashion sometime in the late nineteenth or early twentieth century when their inhabitants were disdainfully viewed as rough, uncultured folk, log structures were either replaced with more "refined" wooden frame houses or sided over with clapboards. A surprisingly large number of these disguised log houses survive, although only the practiced eye finds them today.

The most important and widely distributed vernacular dwelling is the I-house, so named because it is characteristic of Indiana, Illinois, and Iowa (Figure 6.13a). It is easy to identify because of its narrow, one-room depth and two-story facade. The placement of its windows, doors, and chimneys ranged widely, depending on whether the house incorporated classical styling. More often than not, an L-shaped addition was built in the rear to house the kitchen.

The most easily recognized barn of central Ohio is the Pennsylvania-Dutch, Switzer, or Forebay Barn, brought into the area mostly by Pennsylvania-German settlers. The barn is also widely called a Bank Barn because its lower level was built into the slope or bank of a hillside, with the entrance to this lower level from a feedlot on the downslope side. The upper level, which resembles that of the English Barn in both form and function, was entered directly from the upper part of the hill. The barn's chief diagnostic feature is its overhang, or *forebay* (Figure 6.13b). Cantilevered over the feed lot, the forebay offered protection for stock and permitted direct gravity feeding.

The southern part of the state shows the Upland South cultural characteristics most strongly. Throughout this area, Virginia influences are pervasive. As was true in central Ohio, the I-house is the most widespread vernacular dwelling in southern Ohio. The Southern or Virginia I-house usually has a double porch and a lower roof pitch (Figure 6.14). The chimneys are located on the gables, and the orientation is with the long side toward the road. These houses were usually built of timber frame, but some were constructed of brick (Wilhelm 1989).

The I-house was a symbol of rural prosperity and wealth. Smaller, simpler dwellings, such as the Saddlebag house and a double-room house with gable chimneys, housed less affluent families (Figure 6.14b and c). This latter dwelling may have been an expansion of the basic log cabin to two rooms, or a derivation of the Virginia Hall and Parlor house. In any event, it was widely built throughout southern Ohio, both in the countryside and in towns. The Saddlebag house, characterized by a central chimney stack with

Fig. 6.12. Barns of Northern Ohio. The English Three Bay Barn (top) and the Raised or Basement Barn (bottom). (Drawings by M. Margaret Geib)

back-to-back fireplaces, arose out of the log building traditions of the Upland South. When a family wished to expand a single log pen house, one of the easiest ways was to build an identical pen abutting the gable chimney. Although the earliest attempts were rough, often without an interior connection between the two rooms, gradually the idea and design evolved so that two-room houses were built as complete units.

In the portions of southeast Ohio initially settled by New Englanders, an ancient New England house known as the Saltbox is common. This is usually a two-story house on which the rear roof extends to the level of the first floor. The Saltbox house was frequently built in New England in the 1600s and early 1700s and was reestablished among some of the early settlers in southeastern Ohio, where it gained some popularity.

Southerners brought their own barn design, one quite different from that of Yankee or Pennsylvania barns, to southern Ohio. Called the Transverse Frame Barn, it was an outgrowth of earlier, simpler log crib barns, a very few of which still survive as obscure outbuildings. Unique to the Transverse Frame Barn was the placement of its door on the gable end of the structure rather than on the side (Figure 6.15). Some barns have shed-roofed side wings (Montell and Morse 1976). The barn was used for animal shel-

Fig. 6.13. Common Vernacular Buildings of Middle Ohio. The I-House (top) and the Pennsylvania German Bank Barn (bottom). (Drawings by M. Margaret Geib)

ter and as storage for corn and equipment. Another Southern barn found in parts of southern Ohio is the elongated Tobacco Barn (Figure 6.15). For the most part these are relics of earlier, more widespread tobacco agriculture. Clearly related to Transverse Barns, their length and frequent presence of roof ventilation, which helped to dry the tobacco leaves, makes these barns distinctive and easily recognizable.

Language Patterns

Ohio is crossed by two speech boundaries, indicating the convergence of different speech forms (Figure 6.9). The east-west orientation of these boundaries reflects the general westward drift of migration, with northern Ohio retaining predominantly New York–New England traditional dialect forms and central and southern Ohio developing greater word

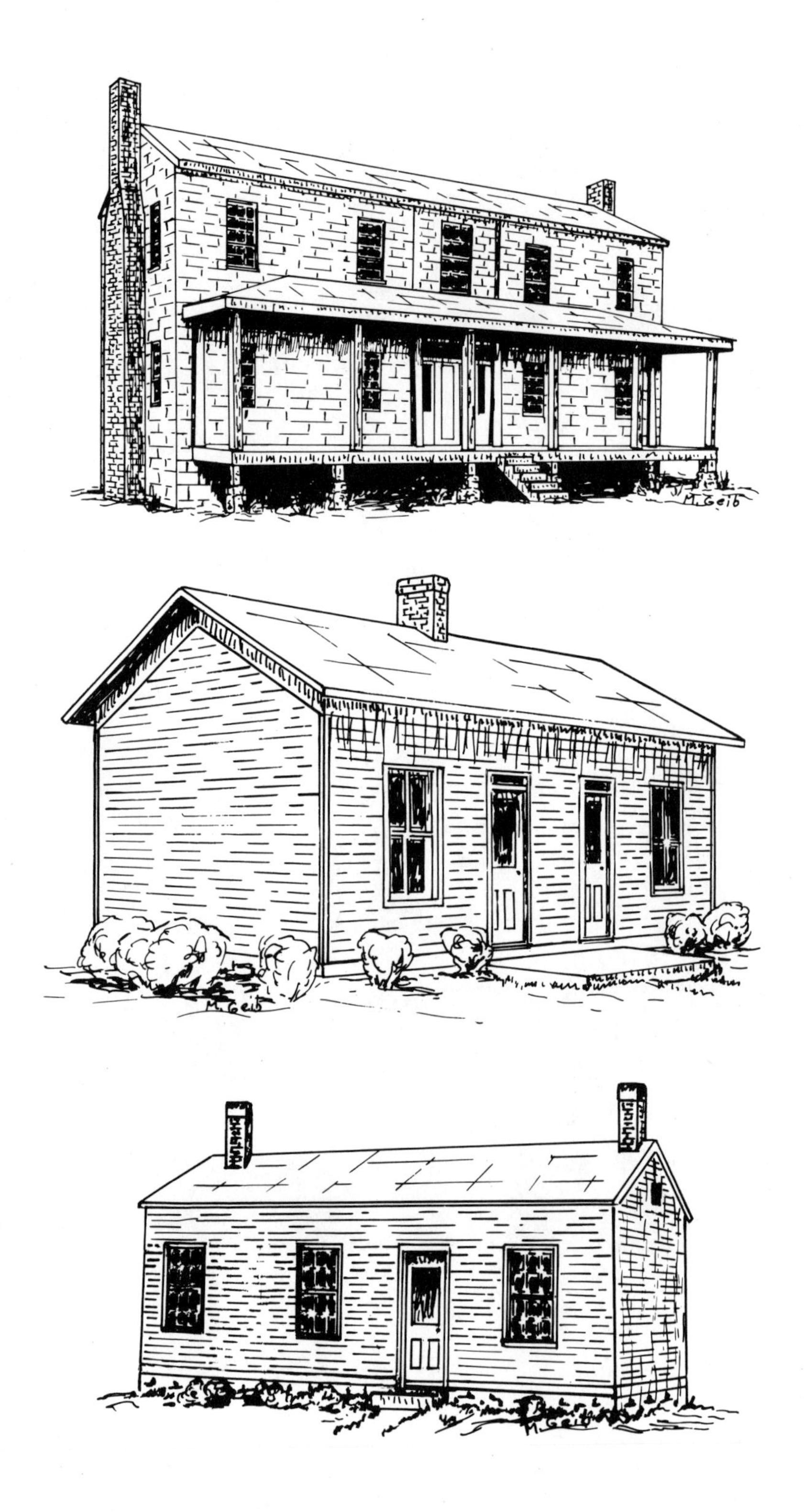

Fig. 6.14. Vernacular Houses of Southern Ohio. From top to bottom: the Southern or Virginia I-House, the Saddlebag House, and the Double Pen House. (Drawings by M. Margaret Geib)

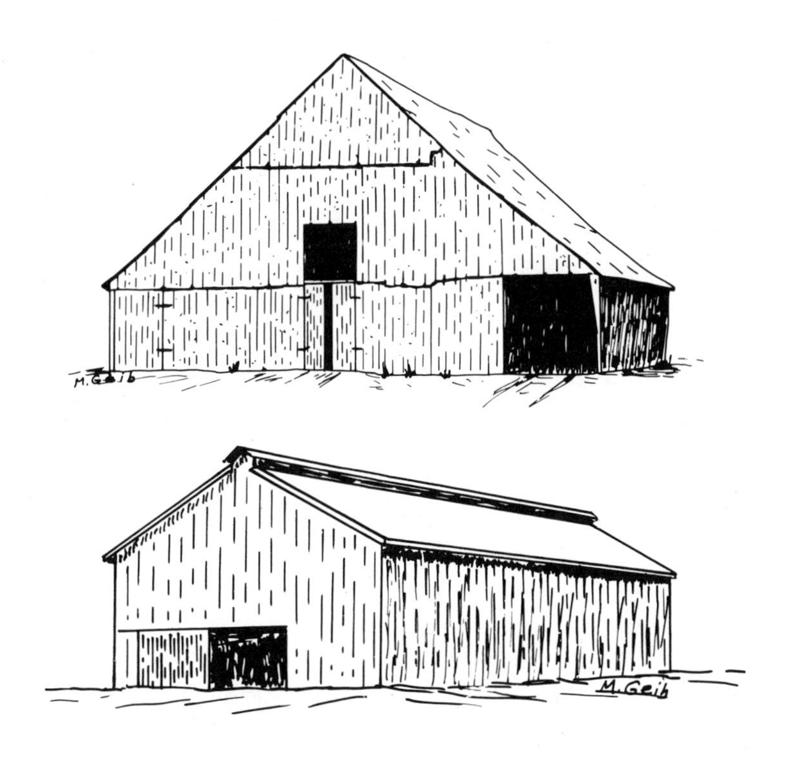

Fig. 6.15. Barns of Southern Ohio. The Transverse Frame Barn (top) and the Ohio Tobacco Barn (bottom). (Drawings by M. Margaret Geib)

complexity, a probable result of these lower regions' increased settlement diversity. For the most part, Ohio south of the great Northern speech divide was dominated by Midland dialect forms. The geographical core area of Midland speech coincides roughly with the Middle Atlantic states, extending from northern Pennsylvania to just beyond the James River in Virginia (Kurath 1966). Because the bulk of the settlers in central and southern Ohio came from the Midland speech area, that area's word forms became widely distributed. Samples of once-common Midland terms or phrases encountered in the southern half of Ohio include quarter till, blinds (shades), skillet, cling peach, belling (celebration after a wedding), sookie (call to cows), fishing worm, snake feeder (dragon fly), poison vine, and belly buster (riding a sled on one's stomach). True Southern speech forms are rare in Ohio; however, those classified as South Midland words from parts of Virginia are quite common, among them fire board (fireplace screen), clabbered milk (sour milk), trestle, favors, sunup, light bread, snack, snap beans, and hay shocks (Davis 1951). In areas primarily settled by Pennsylvania Germans, words or phrases such as clook (*Glucke,* "setting hen"), smear case (*Schmierkaese,* "cottage cheese"), school leaves out, and make the door to (*mach die Tuer zu,* "close the door") closely resemble the original German (Kurath 1966).

Examples of New England speech survive in southeastern Ohio between Marietta and Athens, where place-names illustrate early Yankee settlement influence. A diligent researcher working in that area can still come across such Northern words as pail (bucket), swill (pig slop), Dutch cheese (cottage cheese), clingstone (wet stone), nigh horse (saddle horse), bossi (call to cows), angle worm, and Johnny Cake (corn bread) (Davis 1951).

Today, most words used to illustrate regional language contrasts would be unknown or, at best, perceived as archaic. With time and formal education, the old words and phrases that expressed a specialized, rural way of life have been replaced by forms enjoying uniform usage, therefore revealing nothing about geographical variations.

Religious Patterns

Religion, which, as a conservative indicator of culture changes slowly over time, can serve as an ideal tool to reveal regional variations in settlement. Before proceeding to an analysis of several specific denominations and their distribution, one should turn to the general pattern of religion in the state. Quantitative information pertaining to participation in formal religions is no longer as readily available as when it was included in the decennial censuses. Today this information is published at lengthy and irregular intervals. The results of the most recent survey were published in 1971 (Johnson, Picard, and Quinn 1971).

Figure 6.16's map of church membership reveals some of Ohio's cultural geography. The north, especially the northwest, has above average church membership. Mercer County, on the Ohio-Indiana line, has the single highest participation rate, that of 90.1 percent. On the other hand, the majority of southern Ohio counties show below average church membership, with Pike County the lowest at 19.8 percent. An explanation of this contrasting pattern rests with the social and economic characteristics of each area as conditioned by the cultural background of its people. Northwestern Ohio is predominantly rural-agricultural, with strong attachment to traditional religions, especially Lutheran and Catholic. In contrast, the cultural background of the state's southern half is primarily from the Upland South, a region where people favored independence and tended to dislike formal social organizations. Another contributing factor was the area's hilly nature, which offered isolation and therefore less uniform spatial development.

The distribution of the two principal religious groups, Protestants and Catholics (Figure 6.17), shows the state to be overwhelmingly Protestant, especially in its rural counties. The Protestant majority of 67.5 percent indicates that the state's settlement background was rooted in the non-Catholic areas of northwestern Europe, especially Great Britain and Germany. German Catholics, nonetheless, are well represented in several western Ohio counties, from Hamilton County northward, and including Montgomery, Shelby, Mercer, Auglaize, and Putnam Counties. The most Catholic counties in Ohio are Putnam and Mercer, with 74 percent and 72 percent respectively. Both of these counties are rural-agricultural, and the huge spires of the Catholic churches distributed throughout the countryside stand out as striking features on the landscape. Any traveler in that part of Ohio can appreciate the visual impact of these imposing brick structures.

In urban northeastern Ohio the Catholic Church dominates, a result of heavy immigration after 1850, especially from predominantly Catholic eastern Europe, into the rapidly growing industrial centers. Mahoning County, which includes Youngstown, has a 65 percent Catholic majority. The greater ethnic diversity in the major urban-industrial centers of Cleveland, Akron, Canton, and Youngstown lends a particular cosmopolitan flavor expressed by religion and other cultural indicators.

The distribution of three typical regional churches epitomizes settlement patterns in the state. They are the Congregational Christian Church, whose roots are in New England; the Evangelical United Brethren Church, with origins in Pennsylvania and Maryland; and the Disciples of Christ Church, whose adherents came originally from Kentucky and West Virginia. Each one of these churches is representative of the regional American settlement groups who populated Ohio. Figure 6.18 shows the largest concentrations of members of these three denominations. Because of church mergers during the ecumenical movement of the last twenty to thirty years, all three of these churches have merged with other denominations and their names have largely disappeared from the landscape; hence membership data was derived from the 1952 census figures found in *Churches and Church Membership in the United States* (1957). The Congregational churches and members were mostly in northeastern Ohio, the region settled predominantly by New Englanders. Southern Ohio, on the other hand, became the focus for the Disciples of Christ Church, whose origins were often immediately across the Ohio River; some of the Disciples pushed farther north into central Ohio, wherever work or cheap land was readily available. The Evangelical United Brethren Church was strongly aligned with Pennsylvania-Dutch culture, and its distribution falls primarily within the Eastern or Middle Atlantic settlement area. The E.U.B. Church, as it was commonly known, was related philosophically with the Methodist Church but differed in ethnic makeup: unlike the Methodists, initially E.U.B. members were German immigrants.

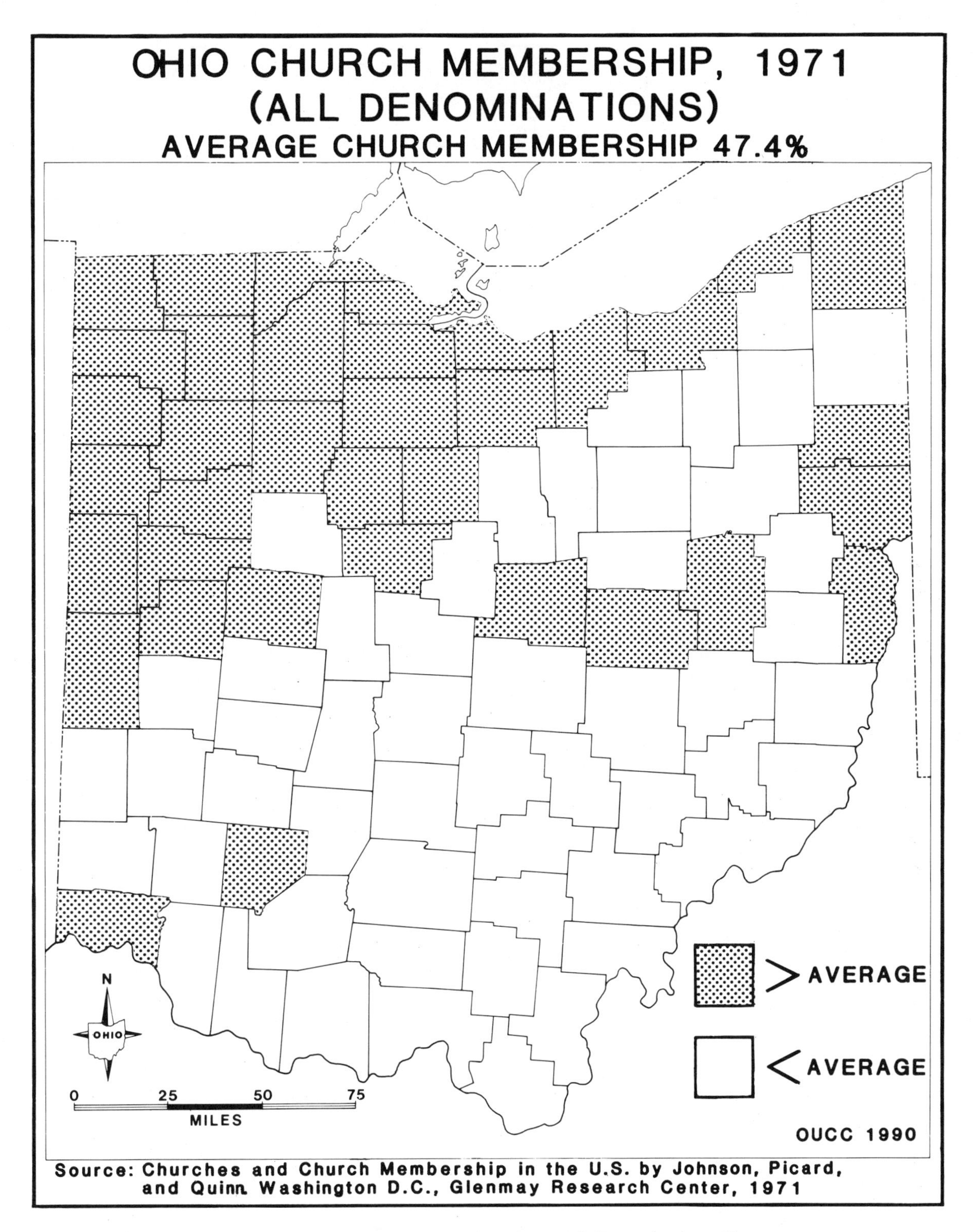

Fig. 6.16. Ohio Church Membership, 1971—All Denominations (Average Church Membership, 47.4 Percent)

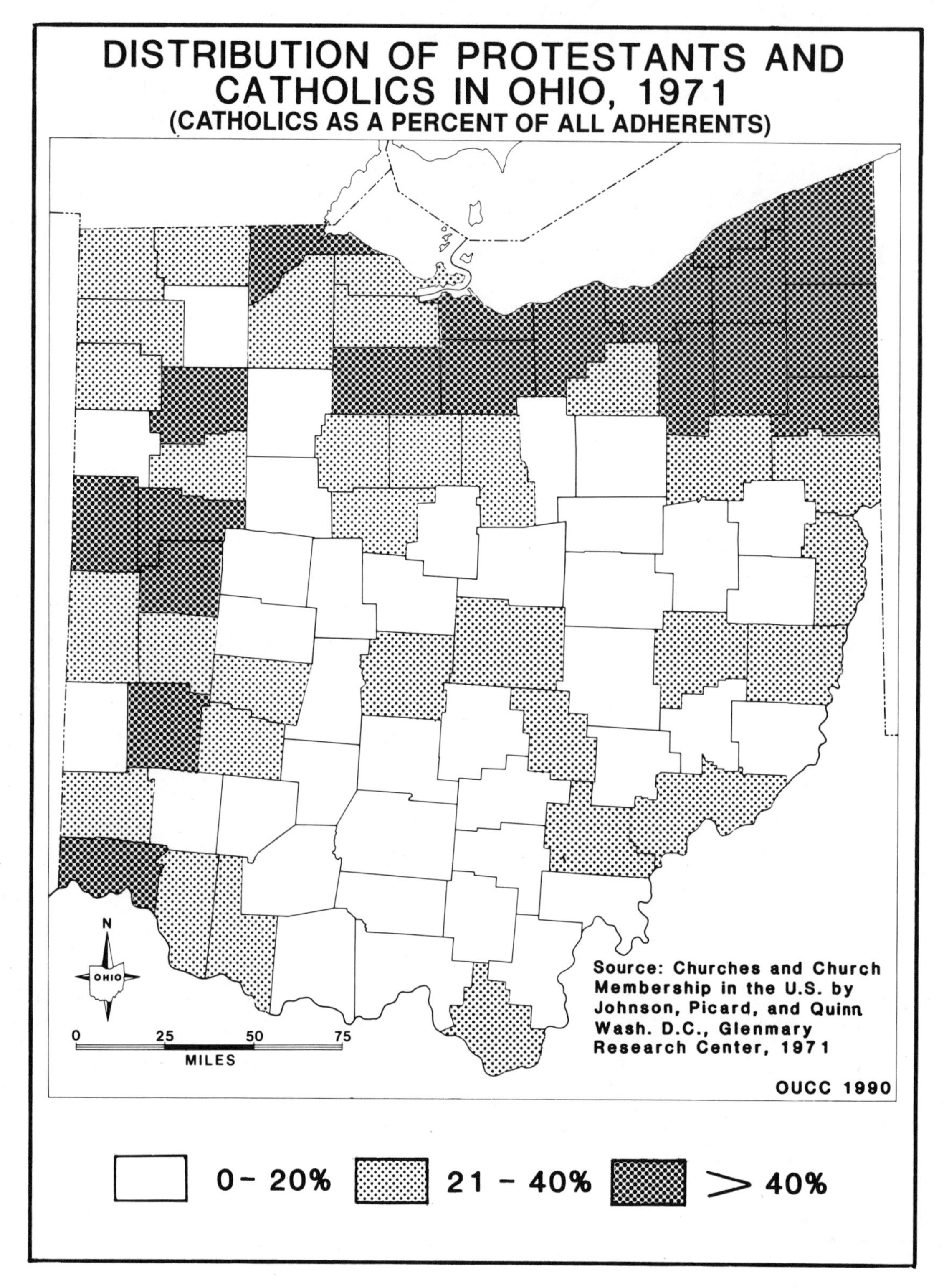

Fig. 6.17. Distribution of Protestants and Catholics in Ohio, 1971 (Catholics as a Percentage of All Adherents)

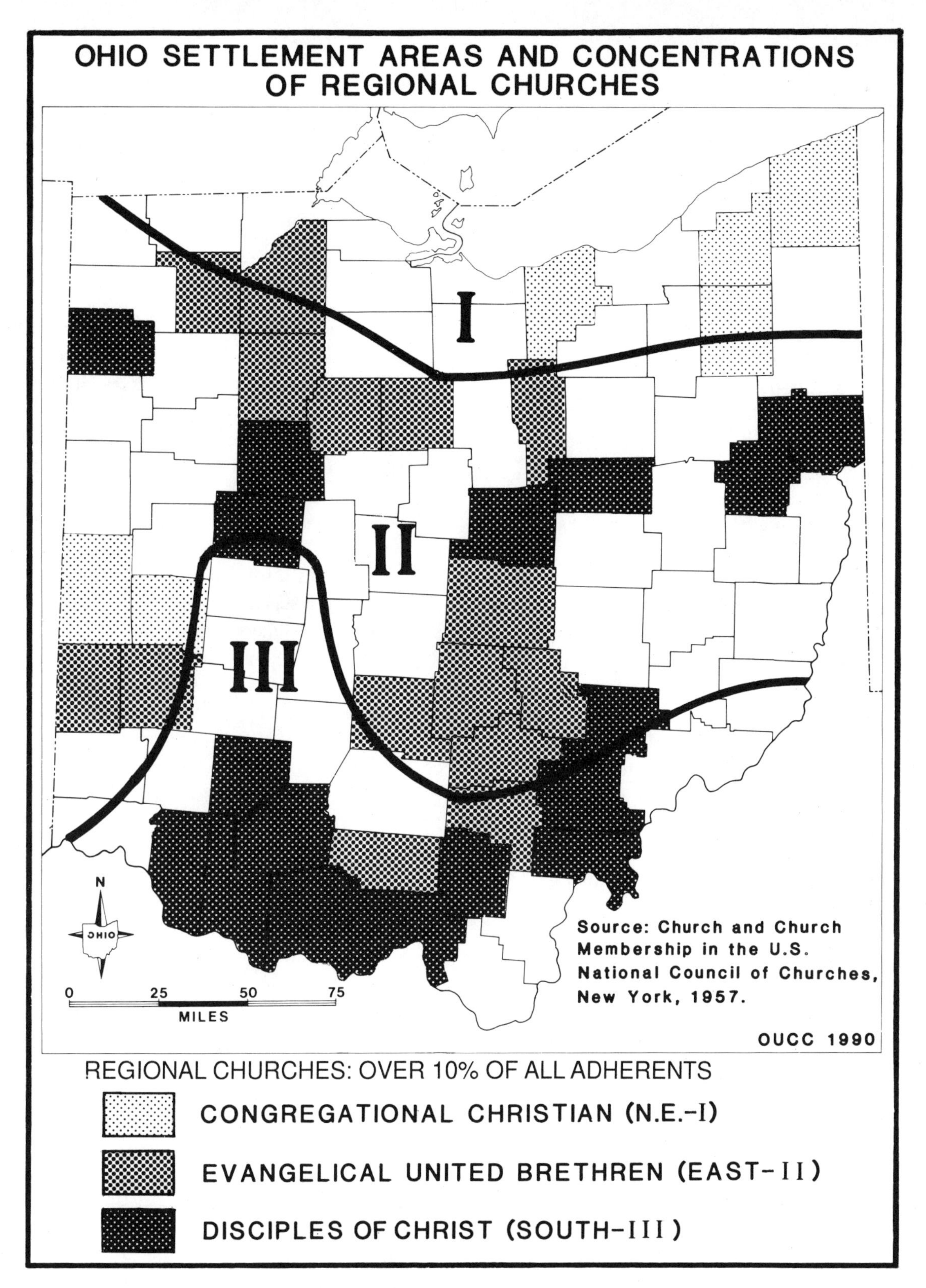

Fig. 6.18. Ohio Settlement Areas and Concentrations of Regional Churches

Ohio's Amish

Among Ohio's great variety of religious denominations, the Amish religion (Old Order Amish) is one of the most fascinating. More than simply members of a particular sectarian church, the Amish are people with a distinctive history and unique culture. The origin of the Amish church dates from the late seventeenth century Anabaptist or Swiss Brethren movements of Europe. Today's Amish are the religious descendants of Jacob Amann, who, sometime during 1693–1697, broke with the Mennonites to form the Amish church. Subsequent persecution dispersed the Amish from Switzerland into the Rhineland in Germany (Palatinate), Alsace-Lorain in France, and the Netherlands. Eventually forced from these areas, they migrated during the eighteenth and nineteenth centuries to America, initially settling in southeastern Pennsylvania among other German-speaking peoples. The Amish are one of the numerous subgroups that constitute the Pennsylvania-Dutch (Mook and Hostettler 1957).

As farmers, the Amish spread westward with the frontier, always in search of good land. By the end of the first decade of the nineteenth century, they had arrived in Ohio, settling first in Wayne County (Schreiber 1962). Today, this county along with several of its neighbors, including Holmes, Tuscarawas, Stark, and Coshocton Counties, are known as Amish Country. Among the many settlements in other counties (Figure 6.19), those in Morgan, Noble, Pickaway, and Licking Counties are quite recent, having been founded during the past 10–20 years. This recent immigration is a consequence of the conflict between the rapid urbanization of the northern counties and the rural existence so integral to Amish culture. Ohio and Pennsylvania are the two states with the largest number of Amish church districts. Amish are also found in Canada and some Central American countries.

Land use among Ohio's Amish centers on a five-year rotation of corn, wheat, oats, and grass. Farms average 80–100 acres; larger areas could not easily be handled by their horse teams. Amish farmsteads always include two houses: the granddaddy or "grossvader" house and the main house. The former may be attached to the main house and usually shelters the retired farmer and his wife. There are also many outbuildings, including a large barn. Amish residences are usually distinguished by the lack of wires connecting the house to poles along the road. For religious reasons they decline to use machinery unless sanctioned by the church as necessary to carry out their trade.

As in other parts of the country, Ohio's Amish are known for their dairy operations. The farms usually keep 10–15 active milkers. Because milking is done by hand, members of the extended family, from the older children to the grandparents, help with that task. Amish farmers get their income primarily from the sale of milk to local cheese makers. In areas close to centers of urbanization, many Amish farms have begun to use refrigeration so that the milk will conform to USDA regulations and thus can be sold to large dairies for bottling.

The Amish are socially organized in church districts, which they also call settlements. Between twenty-five and thirty farms may belong to a single district. Church services are held every two weeks and alternate between the various farms of the district. Although very community oriented, the Amish are not a communal people. In fact, they are very good capitalists who save primarily to purchase land for their children. Amish families are quite large, and the youngest child usually inherits the homestead, a traditional practice known as *ultimogeniture.* In Ohio, as in other settlements, the Amish suffer from two main problems: insufficient land and outside cultural pressures. These problems often force families to move, leading to the recurring establishment of new settlements.

Ohio's Amish Country, especially Holmes and Tuscarawas Counties, has become an important tourist region. Along Route 39 between Sugarcreek and Millersburg, numerous tourist services have sprung up, among them restaurants, craft and antique shops, and the usual "authentic" Amish farms, complete with buggy rides. The local Amish appear to have adjusted to all this as inevitable. For some it actually presents a financial boon because tourists are all-important purchasers of their homemade quilts, furniture, and baked and canned goods. Each year, Ohio's Swiss Cheese Festival, which attracts huge crowds, is held in Sugarcreek in Tuscarawas County.

This chapter has examined Ohio's settlement geography as part of the cultural landscape. The cultural geography of any area was not shaped yesterday, but rather over a long period of time. Because of the state's location, situated between the Ohio River and Lake Erie, Ohio became the focus of three important

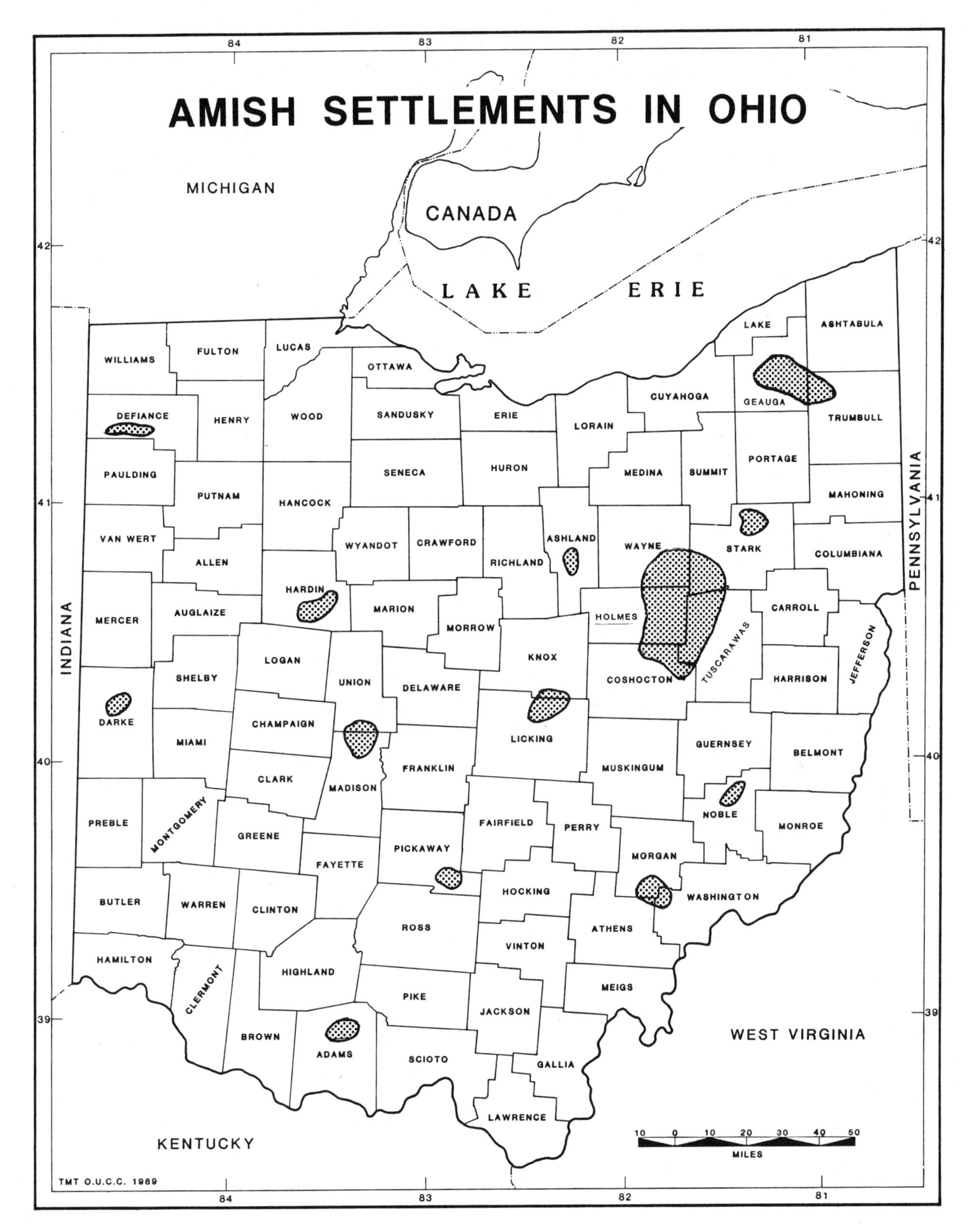

Fig. 6.19. Amish Settlements in Ohio

American cultural streams, the Northeast, Middle Atlantic, and Upland South. The peoples of each one of these areas had developed, over time, distinct material and nonmaterial cultural traits, including survey systems, agricultural land use, language forms, religious practices, and others. Because those various traits were traditional among these regional groups, they were transferred into Ohio during the migration and settlement period of the nineteenth century. Many of these traits remain present today in the landscape of the state and form the basis for regional differentiation. Because settlement in Ohio advanced from the east westward, the three cultural influences were spread across the state as three distinctive east-west regional belts. Any observant traveler, driving across the state from south to north or in the opposite direction, would become aware of the contrasting landscape images which reveal Ohio's diverse cultural background.

REFERENCES

Burke, Thomas A. 1987. *Ohio Lands: A Short History.* Columbus, Ohio: Auditor of State Office.

Churches and Church Membership in the United States: An Enumeration and Analysis by Counties, States and Regions. 1957. New York: Bureau of Research and Survey, National Council of Churches.

Davis, Alva L. 1951. Dialect Distribution and Settlement Patterns in the Great Lakes Region. *Ohio State Archaeological and Historical Quarterly* 60:48–56.

Glassic, Henry. 1969. *Pattern in the Material Folk Culture of the Eastern United States.* Philadelphia: University of Pennsylvania Press.

Hart, John F. 1975. *The Look of the Land.* Englewood Cliffs, N.J.: Prentice Hall.

Johnson, Douglas W., Paul R. Picard, and Bernard Quinn. 1971. *Churches and Church Membership in the United States.* Washington, D.C.: Glenmary Research Center.

Kniffen, Fred B. 1965. Folk Housing: Key to Diffusion. *Annals of the Association of American Geographers* 55: 549–77.

Kurath, Hans. 1966. *A Word Geography of the Eastern United States.* Reprint, 1949. Ann Arbor: University of Michigan Press.

Meitzen, August. 1882. *Das deutsche Volk in seinen volkstuemlichen Formen.* Berlin: Verlag von Dietrich Reimer.

Mook, Maurice A., and John A. Hostettler 1957. The Amish and Their Land. *Landscape* 6:3, 21–28.

Montell, William Lynwood, and Michael Lynn Morse. 1976. *Kentucky Folk Architecture.* Lexington: University Press of Kentucky.

Noble, Allen G. 1984. *Wood, Brick, and Stone: The North American Settlement Landscape.* Vol. 2: Barns and Farm Structures. Amherst: University of Massachusetts Press.

Peters, William E. 1930. *Ohio Lands and Their History.* Athens, Ohio.

Schreiber, William I. 1962. *Our Amish Neighbors.* Chicago: University of Chicago Press.

Sherman, Christopher E. 1972. *Original Ohio Land Subdivisions.* Ohio Cooperative Topographic Survey, final report, vol. 3. 1925. Reprint, Columbus, Ohio: Ohio Geological Survey.

Wilhelm, Hubert G. H. 1974. The Pennsylvania-Dutch Barn in Southeastern Ohio. M. J. Walker and W. G. Haag, editors, *Geoscience and Man* 5:155–62.

———. 1982. *The Origin and Distribution of Settlement Groups: Ohio, 1850.* Athens: Cutler Service Center, Ohio University.

———. 1989. *The Barn Builders: A Study Guide.* Athens: Cutler Service Center, Ohio University.

Wilhelm, Hubert G. H., and David Mould. 1991. *Log Cabins and Castles: Virginia Settlers in Ohio.* A Study Guide. Athens: Ohio Landscape Productions.

CHAPTER SEVEN

The Development of the Economic Landscape

Richard T. Lewis

This chapter examines the past geographies of Ohio and the spatial processes which, through time, converted the state from one geography to another. Certain highlights of the state's development are briefly described. The focus is on post-Revolutionary Ohio, and the organization is chronological. Most of the chapter is a review of a series of 40-year time periods, with just one or two main topics discussed in each period. These topics are almost exclusively economic, since ethnic and cultural aspects of the population are discussed in another chapter of this book. Population size and location form a continuing theme carried through each of the periods. Five subdivisions of the state are used in this review of population change.

Ohio Territory During the Colonial Period: 1600–1788

The land of the future state of Ohio was rather remote from the sites of initial European intrusion in North America (Figure 7.1). Although closer in miles to the English at Jamestown and the Dutch at New Amsterdam, the region was most accessible to the French at Quebec. For the British and Dutch, the Appalachians intervened physically, and the British would use these highlands as a political boundary for the next century and a half. But French settlers in the St. Lawrence Valley had access via the river and lake system, and using it entered the Ohio Country quite early. Furthermore, the French laid claim to the lands of both the Great Lakes–St. Lawrence basin and the Mississippi basin, thus asserting sovereignty over the whole of the future state of Ohio.

French-British Conflict

The French, motivated by their desire to exploit the fur resources of the continental interior, developed a river-based system to accomplish their goals. Their principal thrust was directly westward out of the St. Lawrence core, up the Ottawa River, and into the upper Great Lakes. A major interior base was established at Michilimackinac, which became the focus of operations in the region of Lake Michigan and the upper Mississippi Valley. Ohio Country was marginal to this scheme, for it was less productive of furs and too near the rival British and their allies, the Iroquois.

As the British-French rivalry intensified, the French sought to assert themselves more effectively south of Lake Erie. They encouraged their Algonkian allies to be more active here, with the result that Miamis, Ottawas, and Wyandots occupied large tracts of the region. Shawnees entered from the South, and Delawares and Senecas came in from the East, pushed by the disruption of territorial occupancy patterns along the Atlantic coast. These were the Amerindian groups whom the Euro-American settlers would encounter when westward migration began in the 1790s.

On four occasions the French-English rivalry broke into open warfare. While the Ohio Country was important in the competition, the outcomes were decided largely through fighting in the settled areas on the Eastern seaboard, particularly in the Hudson-Champlain lowland corridor. The ultimate effect of the English victory in 1763 was the removal of the French from political control in North America. The British were without European rival throughout nearly all of the eastern half of the continent. The

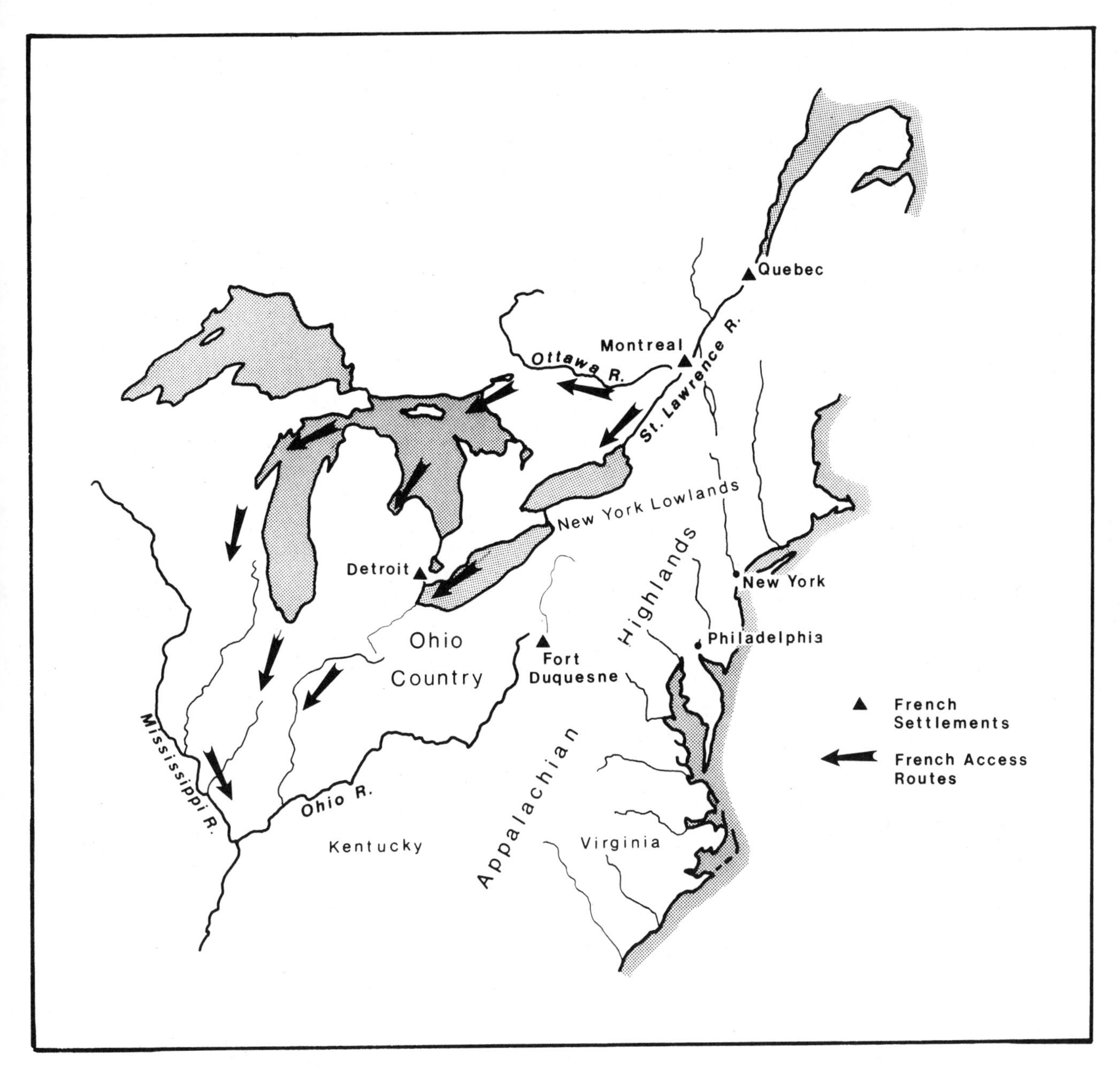

Fig. 7.1. Ohio's Colonial Setting

question was then raised of jurisdiction over the trans-Appalachian region within the British colonial scheme. Eastern coastal colonists looked on the interior with great expectations for expansion. Several of the colonial charters actually authorized extension of territorial control into the region that, though previously impossible due to the French occupation, might now be accomplished. However, British authorities chose to maintain the interior as an extension of the St. Lawrence Valley settlement, declaring it to be Indian territory and, in 1774, actually designating it a part of their Quebec colony. Resentment, instilled among the American settlers, added to the colonists' accumulating problems with the Crown that finally culminated in the War for Independence.

The Revolutionary War, too, was largely fought and won in the East. The major exception was the attacks undertaken by Britain's Indian allies on American pioneer settlements along the frontier, some of which had developed as far west as the Bluegrass region of Kentucky. In response to these attacks, American leaders authorized a punitive expedition that was undertaken in 1778–1779 by George Rogers Clark and troops of the Virginia militia. This mission had limited success in weakening the Indians of the region, but in the treaty negotiations that ended the conflict it provided further support for American claims to the interior. By 1783 the newly independent United States extended from the Atlantic to the Mississippi and from the Great Lakes nearly to the Gulf of Mexico. Within two decades the national identity of the Ohio territory shifted from the French to the British to the Americans.

Ohio Country and the New Land Policies

This expansion of the territory under American control raised three fundamental questions whose answers formed the basis of American national land policy. The first concerned jurisdiction over the interior. Britain ceded this territory to a "united states," which at the time constituted a poorly defined entity—thirteen victorious colonies trying to decide how to relate to one another in their new condition. Should the west belong to all of them or only to those with charters specifying certain western territories as being within their jurisdiction? Should states' rights or central authority prevail? The decision in favor of the latter was a notable step in the direction of national unity in the Confederation period of the 1780s. The result was the creation of the National Domain, the body of public lands accumulated by the national government through acquisitions from this time on. Nearly all of the future Ohio was part of this domain, but there were two major exceptions. These western lands had become "national" through the yielding of claims by individual states. Two of the states chose to withhold tracts for their own benefit. Virginia maintained jurisdiction over land between the Scioto and Little Miami Rivers to meet the claims of state militia veterans of the Revolution who were paid in land warrants. Connecticut reserved a territory south of Lake Erie, expecting to gain profit from its sale with which to recompense citizens whose properties had been destroyed during the war. Thus, the Virginia Military District and the Connecticut Western Reserve, the latter with its Firelands subdivision, were created (Figure 6.2).

The second and third land questions followed from this creation of the National Domain. How would the western lands be sold to farmer-settlers who would make them productive? And, once settled, how would statehood be achieved?

Responding to the land sale question, Congress opted for a procedure of direct sale to individuals. The well-known subdivision system called the Township and Range Survey, a system of 36-square-mile townships and 1-square-mile sections, was developed to create farm-size units suitable for sale. It was applied for the first time along the Ohio River just west of the Pennsylvania line, in the Seven Ranges. From there it was extended across the state, then across the continent to the Pacific. But there were exceptions. The national government made some large-scale land sales to companies that then assumed the duties of subdividing and selling the land to settlers. There were two such cases in Ohio's territory: the Ohio Company Purchase west of the Seven Ranges and the Symmes Purchase between the Little and Great Miami Rivers. A third such sale, involving lands between the Ohio Company Purchase and the Scioto River, was approved by Congress but never carried through. Another exception occurred just west of the Seven Ranges in the United States Military District, a part of the National Domain where land was set aside for veterans of the Continental Army. Here, for practical reasons related to the land-granting process, the Township and Range Survey System was rejected in favor of a different form of rectangular survey.

The result of these varied activities was a patchwork of land subdivisions within the state: lands outside the National Domain where Eastern states established their own settlement procedures; lands within the National Domain that were sold to wholesaling companies which assumed settlement responsibilities; lands both in and out of the Domain that were granted to Revolutionary War veterans; and lands within the Domain that were subjected to the Township and Range Survey and sold directly to settlers. Eventually, this last procedure would account for the largest share of the state lands, including all of the northwestern quadrant. In a significant way, Ohio was a practice field for the national government's land disposal policies, especially the Township and Range

System, which had attained a high degree of refinement by the time the western parts of the state were surveyed.

The question of state formation was answered by Congress through passage in 1787 of the Northwest Ordinance. It established two stages of territorial status prior to statehood, with each stage attained under specified conditions of development. In 1803, when Ohio entered the Union as the seventeenth state, it was the first to have done so under this Territorial System. As with the Township and Range Survey, Ohio pioneered a process that was to be followed through nearly all of the westward expansion that followed.

Pioneer Ohio: 1788–1830

While policies concerning the Northwest were being formulated, settlement was already progressing in trans-Appalachian lands to the south—the future states of Kentucky and Tennessee. They had been opened to settlers in the 1770s and had a combined population exceeding 100,000 by 1790. Both would join the Union prior to Ohio, unaffected by the land policies that Congress was creating.

Kentucky's attractiveness enhanced the importance of the Ohio River as a transportation corridor. Whereas the early pioneers went directly westward from Virginia through the mountains, the route along the Ohio became a viable alternative for Virginians as well as a more direct approach for Pennsylvanians and others going to Kentucky from the Middle Atlantic region. Water traffic on the river and land movement along its southern shore grew steadily in the 1780s. Pittsburgh developed as the great "jumping off" place, primarily for those coming directly overland across Pennsylvania. Wheeling also held a strategic position in this process as the point of access to the river for many who came up the Potomac Valley.

The Northwest Territory

Passage of the Northwest Ordinance ended the restriction on settlement north of the Ohio. Once the region was opened to settlement, much of the Ohio Valley traffic began to turn northward into this new and virgin land. Settlement of Ohio thus began from the Ohio River, along the southern and eastern boundaries of the state. The subdividing described above resulted in various forms of governmental control over different parts of the territory. This impacted the settlement process in many ways. Among these were differences in the timing of settlement and variations in the sources of immigration into the new lands. The Ohio Company and the Symmes Purchase territories experienced very rapid opening to development, due to quick preparation and efforts to attract settlers. Marietta and Cincinnati, both founded in 1788, and adjacent pioneer farms were the immediate results. In the Virginia Military District, however, administrative complications related to land titles had a retarding effect on development. Its first town, Manchester, dates from 1791. Movement into the Western Reserve was also delayed when New Englanders found intervening opportunities for settlement in the newly opened lands of western New York State. Fear of problems with Ohio's Indians was another inhibition to settlement. The Greenville Treaty of 1795 extinguished Indian claims only to lands east of the Cuyahoga River. Ten years would pass before the western part of the Reserve was opened. Generally, early Reserve settlement came about a decade after the beginnings in the Ohio Valley.

Once in-migration finally came about in these areas, their settlers often came from a limited range of Eastern origins. This was especially so for the Virginia Military District and the Western Reserve. Though the Ohio Company, being a New England enterprise, did create a secondary concentration of settlers from the Northeast, its location made a strong Virginia presence inevitable. As in later western states, Ohio showed evidence of the latitudinal phenomenon characteristic of the migration process: northern counties drew heavily from New England, southern counties from the South, and the middle section from the Middle Atlantic region. The principal effect of these processes was the fact that, at this time, Ohio's population represented a greater diversity of "Americans" than could be found in any other state.

By the start of the new century, Ohio's population growth had been sufficient to attain statehood, and the political process toward that end was undertaken. By then, Ohio had been part of Quebec (both the French and British versions) and the Northwest Territory. It had missed the chance, under a rejected plan developed by Thomas Jefferson, of becoming, in whole or in part, states called Washington, Metropotamia, and Saratoga. Or it might have been wholly or partially incorporated within Virginia, New York, or Connecticut. Slight changes in the course of events could have made permanent any of these situations. As it happened, the course history took in 1803 brought a little over 40,000 square miles of land into the national fold as Ohio, the seventeenth star in the flag.

Ohio's founders chose to establish a political geographic structure similar to that of the northeastern states. This included the creation of both townships and counties to act as extensions of the state government into local areas and as units of local self-government. All parts of the state were provided with both levels of government. In addition, the incorporation of municipalities was authorized by the founders, providing the opportunity to develop even more specialized governing capabilities. Thus were created the bases for forming numerous territories of governmental jurisdiction. Through a process that not only subdivided new settlement areas but often resubdivided older regions, the eighty-eighth and final county was established in 1851.

Also a part of the state's political creation was the process of its own boundary formation. Obviously these boundaries, generally taken for granted today, have had much significance for the state's development. The Pennsylvania line had been established in 1785 as a northward continuation of the boundary between, and jointly agreed to by, Virginia and Pennsylvania. Ohio's southern boundary, the Ohio River, is unlike most river boundaries in that it runs along the northern shore of the river rather than through the middle. This came about when Virginia ceded the lands north and west of the river while retaining all of the river itself. On the west as on the east, the boundary is a meridian. The mouth of the Great Miami River was chosen as the starting point for this line, which was surveyed in 1799 as the first principal meridian in the Township and Range Survey. On the north, the state line coincides with the national boundary in the middle of Lake Erie. Only the land boundary on the north was problematic at the time of statehood, and it was not defined satisfactorily until more than 30 years later. The original line, based on an errant map, was drawn in such a way as to exclude Maumee Bay, including the site of Toledo, from the state's territory. This was considered unsatisfactory by Ohioans who successfully delayed the statehood of Michigan until the issue was resolved in Ohio's favor by Congress in 1835.

Early Development of the New State

In these early years of the nineteenth century, Ohio displayed three characteristics that would be substantially transformed during the ensuing 100 years: it was primarily rural but would become primarily urban; it was principally agricultural but would become principally industrial and commercial; and its developmental focus was in the south but would shift to the north (Figure 7.2). The vast majority of early Ohioans were farmers, engaged in the difficult task of creating productive acreage out of nature's wilderness. They came with great differences—in attitudes and goals, in levels of skill, and in their supply of the goods or wealth needed to start farming. But many became quite successful in creating a pioneer farm that at least provided a good supply of the things their families needed for survival. Many quickly moved beyond that level and began to grow surpluses of various products. These surpluses were initially sold to new neighbors who were not yet able to supply all they needed for themselves, to migrants passing through on their way to lands farther west (or returning east), or to the small but growing populations of nearby towns (Figure 7.3a). However, in many areas production rapidly outstripped all of these forms of local demand, creating a desire for transportation connections to external markets, especially those formed by the cities of the East.

Internal improvements in transportation came quite early with the opening of Zane's Trace in 1796 (Figure 7.3b). Another important highway was the National Road, intended to be a transnational route from Baltimore and Washington to the Mississippi. Though started in 1807, the crossing of Ohio was not completed until the 1820s. Both roads had important impacts on the state's development, especially in the movement of people and in communication. But their usefulness in agricultural marketing was seriously limited.

By the late 1810s a strong interest in canals had developed. The primary stimulus was the construction of New York's Erie Canal, completed in 1825, and the opportunity it offered to Ohioans living near Lake Erie to ship their products to the growing

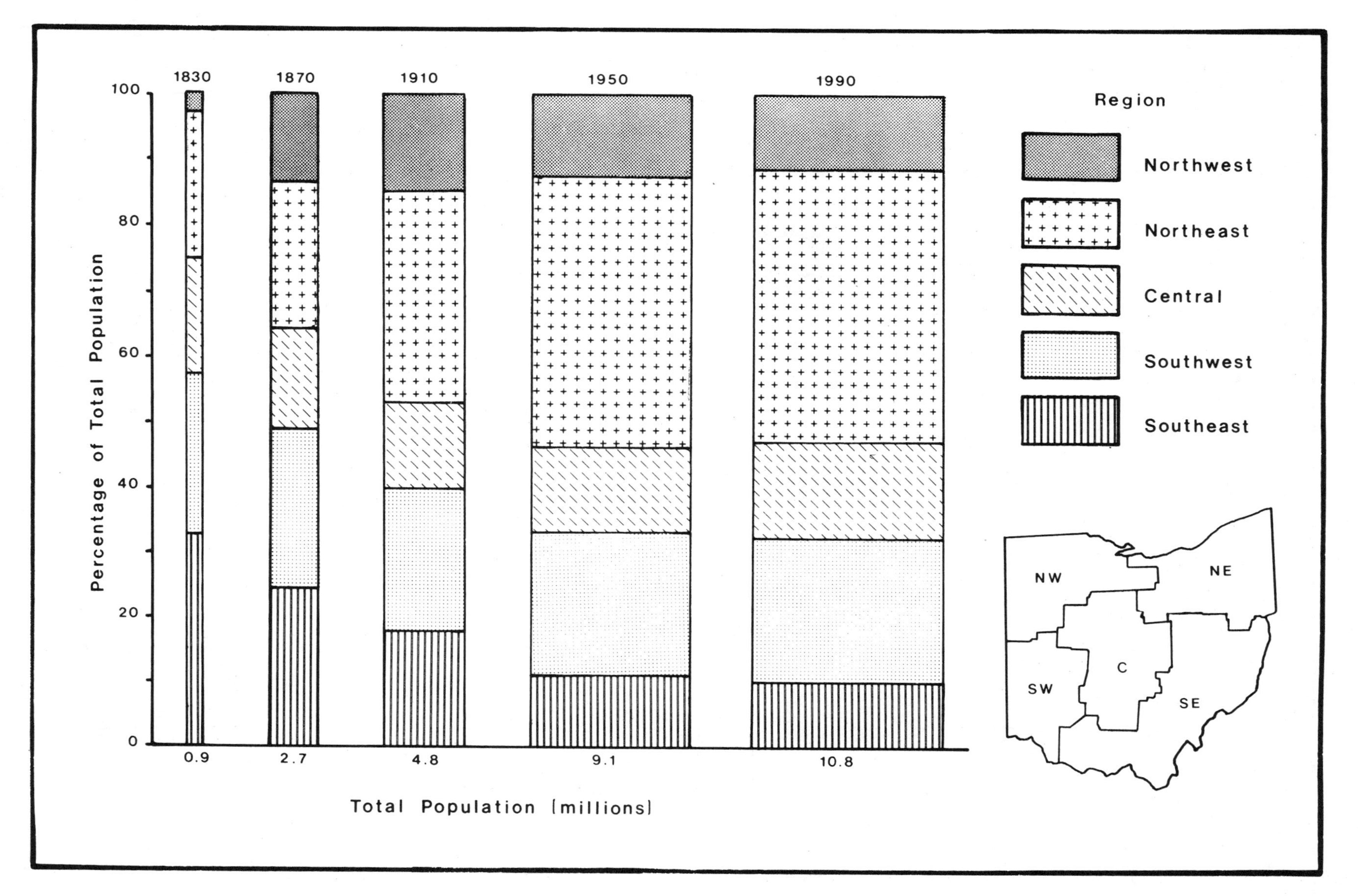

Fig. 7.2. Regional Population Shares, 1830–1990

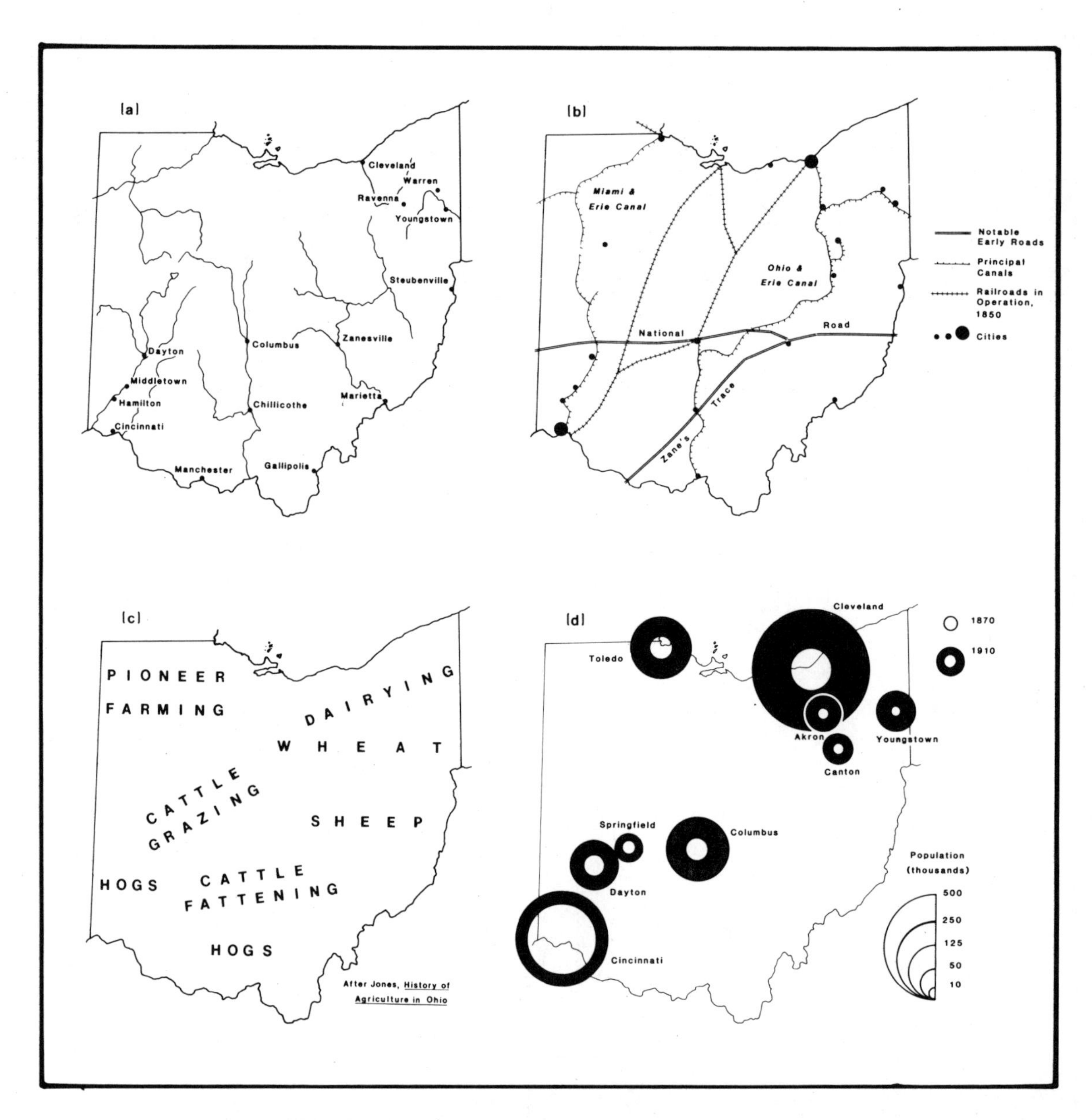

Fig. 7.3. Selected Features of Ohio's Historical Geography. (a) Pre-1800 Settlements (b) Early Transportation Routes (c) Agricultural Regions, 1850 (d) Growth of Major Cities, 1870–1910

markets of the Atlantic states, and often on to Europe. Leaders of Ohio responded favorably to the idea of investing in a similar facility, or facilities, to extend these marketing conditions as broadly as possible across the state. Studies of potential canal routes were undertaken and advice was sought from those involved in the New York enterprise. It was clear that there were several possible routes provided by the several pairs of rivers that rise close to each other in the upland divide that extends across the state and separates Lake Erie drainage from the Ohio River basin. Three of these pairs were given the strongest endorsement: the Cuyahoga-Muskingum (Tuscarawas), the Sandusky-Scioto, and the Maumee (Auglaize)-Miami. After further study indicated a limited water supply in the Sandusky headwaters, that segment was abandoned. But the Scioto was chosen, and it was decided to connect its canal with that built in the Cuyahoga and Tuscarawas Valleys, creating a long artery from Cleveland to Portsmouth (the Ohio and Erie Canal). Construction was begun immediately. This route selection meant abandoning the lower Muskingum Valley route, but residents there were placated by a "river improvement," a deepening of the river channel, in lieu of a canal. The western Maumee-Miami route (the Miami and Erie Canal) was also accepted, but its construction was delayed for several years.

Population, 1830, and Labor Force, 1840

By 1830 the population of the state had reached 937,903, which was 7.3 percent of a national population of over 12.8 million. (Data related to this, and to later discussions of population patterns, are given in Table 7.1.) Having overtaken North Carolina in the 1820s, the state now ranked fourth and was closing rapidly on number three: Virginia. The two largest states, New York and Pennsylvania, have continued to exceed Ohio's numbers up to the present day. The state's growth had been rapid, at a rate of 61.3 percent for the 1820s, about twice that of the nation as a whole.

Within the state, the southern regions dominated, with 32.3 percent of the population in the southeast and 25.2 percent in the southwest, a total of 57.5 percent. Central Ohio, the upper Scioto Valley, had another 17.6 percent. These three regions would each experience continuous decline in its share of the state's total from 1830 until 1950. The northeast, with 22.2 percent in 1830, would increase its share during the same period.

Town development in the state generally coincided with, or even preceded, farm settlement. The establishment of Cincinnati and Marietta in 1788 and Manchester in 1791 on the Ohio River accompanied the opening of their respective regions. When pioneer farming expanded up the tributary valleys, more towns were created in the process. The founding of Hamilton, Middletown, and Dayton in the Miami Valley, Chillicothe and Franklinton (Columbus) in the Scioto, and Zanesville in the Muskingum occurred in the mid 1790s.

In the Western Reserve, Cleveland was founded as the base for the surveying process in 1796, about the same time as the interior Ohio River basin towns. Cleveland's location was determined by the need for a harbor at the mouth of the Cuyahoga River and the access to the interior which the river valley offered. It was also the westernmost site available along the lakeshore due to the Greenville Treaty. Other early Reserve settlements were Youngstown and Warren on the Mahoning and also Ravenna, one of the few early towns not located on a river.

While Cincinnati and Marietta had begun life at the same time, Cincinnati occupied a more strategic riverbend location where the travelers' various needs were greater than at Marietta. Cincinnati, too, became the servant to a more productive hinterland that included the Miami Valley lands to the north and much of Kentucky's Bluegrass region southward, to which the Licking River gave access. These factors supported a more substantial growth at Cincinnati than occurred at any other point in all of the Ohio Valley. By 1830, Cincinnati, with nearly 25,000 people, was far and away the largest city west of the Atlantic seaboard. Pittsburgh, with about 15,000, and Louisville, with 10,000, were the nearest regional competitors. Within the state, Zanesville, Dayton, Steubenville, Chillicothe, and Columbus, all mainly trade centers in the southern interior, ranged in population from 2,400 to 3,100. Cleveland (1,076), left far behind, was smaller than Canton (1,257), and only slightly larger than Wooster (977). Many other towns in the Ohio River tributary valleys—including Urbana, Springfield, Portsmouth, Marietta, Hamilton, Circleville, and Lancaster—all exceeded 1,000 residents.

During this early period of development, several cities had assumed the unique public function of territorial or state capital. After Cincinnati had been the seat of territorial government, state offices had been located in Chillicothe and, briefly, in Zanesville. Then Columbus was designated in 1816, having been built for that purpose. While Chillicothe had been central to the southern half of the state, Columbus was more central to the whole of Ohio, occupying an intermediate point between the two territorial extremes of the southwest and the northeast.

While the 1830 census reported no direct information on labor force characteristics, the 1840 census revealed that three-fourths of the workers in the state were in the primary sector (extractive activities including agriculture, mining, forestry, and fishing). This figure declined constantly through the ensuing years, as mechanization and increasing farm size reduced the labor needed for production. The secondary (or manufacturing) sector, which then included artisans and small-scale factories, accounted for 18.5 percent, while the tertiary (wholesale and retail trade and

Table 7.1

Ohio Population Characteristics, 1830–1990

	1830	1870	1910	1950	1990
POPULATION					
Total (000s)	937.9	2,665.3	4,767.1	7,946.6	10,847.1
% of US	7.3	6.9	5.2	5.3	4.4
Rank	4	3	4	5	7
Larger states	NY PA	NY PA	NY PA	NY CA	CA NY
	VA		IL	PA IL	TX FL
					PA IL
% by Region					
Northwest	2.8	13.7	14.8	12.4	11.8
Northeast	22.2	22.3	31.6	41.4	38.8
Central	17.6	15.6	14.0	12.9	16.2
Southwest	25.2	24.0	21.6	21.8	23.8
Southeast	32.3	24.4	17.9	11.6	9.4
% Urban	3.9	25.6	55.9	70.2	74.1
Main Cities	Ci 25	Ci 216	Cl 561	Cl 1,384	Cl 1,677
(000s)	Za 3	Cl 93	Ci 364	Ci 813	Ci 1,213
	Da 3	To 32	Co 182	Co 438	Co 945
	St 3	Co 31	To 168	Ak 367	Da 613
	Ch 3	Da 30	Da 117	To 364	Ak 528
	Co 2	Sa 13	Yo 79	Da 347	To 489
	La 2	Sp 13	Ak 69	YW 298	YW 362
		Ha 11	Ca 50	Ca 174	Ca 245
		Po 11	Sp 47	Sp 82	LE 224
		Za 10		Ha 63	
		Ak 10			
LABOR FORCE					
Total (000s)	357.9	840.9	1,919.1	3,059.6	4,931.4
% by sector					
Primary	76.3	48.6	24.7	8.1	2.3
Secondary	18.5	22.1	36.6	42.4	28.3
Tertiary	5.1	29.4	38.7	49.5	69.4

Notes: Labor force data are for 1840, not 1830.

Incorporated city population 1830–1910; Urbanized Area population for 1950 and 1990.

Cities: Ak=Akron, Ca=Canton, Ch=Chillicothe, Ci=Cincinnati, Cl=Cleveland, Co=Columbus, Da=Dayton, Ha=Hamilton, La=Lancaster, LE=Lorain-Elyria, Po=Portsmouth, Sa=Sandusky, Sp=Springfield, St=Steubenville, To=Toledo, Yo=Youngstown, YW=Youngstown-Warren, Za=Zanesville. Cincinnati Urbanized Area is partly in Kentucky; Toledo Urbanized Area is partly in Michigan.

all forms of services) sector had only one in 20 Ohio workers.

Transportation Change and Agricultural Revolution: 1830–70

The canals had positive effects on the state. In the east the Ohio and Erie Canal was begun in 1825—the same year New York's Erie Canal was completed—and finished in 1832. While the southern segment of the western Miami and Erie Canal, between Cincinnati and Dayton, was also opened in the early 1830s, financing problems delayed its completion through the sparsely settled west to Lake Erie until 1845. In addition, Indianans built a canal through the Wabash Valley that connected with the Miami and Erie. Fed by both these western channels, the young port of Toledo became a striking example of the many cities brought to life by the system. By giving direct access to eastern markets, canals improved the fortunes of Ohio farmers. They also contributed to improved internal communications within the state, fostered the growth of both transportation and marketing enterprises, and, in the beginning, required a long-lasting construction program that provided business opportunities for those who supplied the needs of the building process. The employment opportunities were important both for already settled Ohioans and for immigrants to the state from the East and from Europe.

The canals turned out to be a temporary expedient in the evolution of an integrated transportation system connecting Ohio with the rest of the country. Though the Ohio canals were destined to die out within a few decades of their completion, the era itself produced two canals, both external to the state, that benefit Ohio even today. One is the Welland Canal across the Niagara Peninsula in Ontario. Originally built in 1829, it was replaced by a new facility in 1932. More significant in terms of the volume of shipments destined for Ohio is the pair of "Soo Canals" between Lake Superior and Lake Huron: the American canal, which opened in 1855 and was later reconstructed, and the Canadian channel, which opened in 1895. Their role in the shipping of, first, copper and then iron ore achieved huge proportions. Within Ohio, however, canal traffic had declined substantially by the Civil War.

Early Railroads

It is ironic that in 1825, the year in which the Erie Canal was completed and in which the Ohio and Erie Canal was begun, the first American steam locomotive was patented, for it was the development of the railroads that led to the decline of the canals. Railroads ran faster, were cheaper to build, and could be constructed in far more environments, thereby serving more people in more places. While the costs of shipment per mile were somewhat higher by rail, their many advantages outweighed the higher costs, and the days of the canals were numbered.

When railroads began to appear on the Ohio landscape in the 1840s, they captured the imagination of many who saw the opportunity to establish direct contact with the Atlantic coast. It was this connection that many felt would assure the economic success of any community at the Ohio end of the line. The initial rail lines, however, were links between the water arteries. Sandusky sought to make up for losing out in the canal building process by supporting the construction of a railroad, called *Mad River and Lake Erie,* to Springfield. A second line was soon built to Mansfield, then an important agricultural center. For a while Sandusky's port activity increased, but the eventual destiny of the railroads was to obviate the use of the water system and provide direct cross-country, high-speed service that eliminated the costs of intercarrier transfer of goods at port cities.

By 1860, nearly 3,000 miles of railroad tracks had been built within the state, and these rails were linked directly to the Atlantic coast by several routes. After the Civil War, railroad building continued to such an extent that by 1900 Ohio became the state with the highest density of lines, as it continues to be.

Agricultural Regionalization

Integration of the national transportation system led to great improvement in the marketing opportunities for Ohio's farmers. An important effect of this process was to allow for regional specialization in agriculture within the state (Figure 7.3c). Farmers concentrated on the particular product that brought them the greatest return, and usually this meant that nearly all in a particular area grew the same product. By midcentury, distinctive regions of agricultural specialization had developed throughout most of the state. The Western Reserve was strongly oriented to dairy farming and the manufacture of cheese, while farmers to the south in the "backbone counties"—from Columbiana west to Richland—turned to wheat production. Still farther south, in the Muskingum basin, sheep raising and wool production became possible in a location somewhat less accessible to the markets than the northern areas. In the Scioto Valley and over much of the southwest, the production of grain-fed cattle and hogs emerged as an early stage in the development of what would become the Corn Belt long associated with the American Midwest. The northwestern section of the state, where swampiness had interfered with effective occupation, was finally subjected to large-scale drainage and opened to farming during the mid 1800s.

Population and Labor Force: 1870

While farming still prevailed as Ohio's major economic activity, by 1870 its share of the state's total employment had already begun to decline. Though in 1840 the primary sector accounted for more than three-fourths of the labor force, by 1870 its share was less than half. The total number employed in all sectors continued to increase, but with the primary sector growing much slower than the others. Ohio's 1870 population numbered 2,665,260, nearly three times its 1830 size but down to 6.9 percent of the national total. Ohio now ranked third among the states, the highest rank it would achieve. But as Ohio's rate of growth slowed to levels below that of the national rate, other parts of the country, both older and newer, experienced faster growth.

Within the state, the spatial distribution was more balanced than it ever had been or would be again. Only 11 percentage points separated the largest region, the southeast (24.9 percent), from the smallest, the northwest (13.7 percent). Striking changes since 1830 had affected those two areas, as the former declined by 9.1 percent in its share of the state's total, while the latter gained 12.3 percent of that total. The 23.1 percent share in the southwest meant that nearly half of all Ohioans were still in the two southern regions. The southwest, the center (at 16.5 percent), and the northeast (at 21.8 percent) each changed by only 2–4 percentage points in the four decades from 1830 to 1870; for the northeast it was a positive change, while the southwest and center declined slightly in their shares of the total. Labor force characteristics reflected corresponding changes in the economy. With many former farm residents moving into cities and city-based activities, the tertiary sector now accounted for nearly 30 percent of the labor force, while the remaining 22 percent were in secondary industries.

Cincinnati continued to be the state's dominant city. Ohio's major center for southwest meat production, this "Porkopolis" had reached a population in 1870 of 216,000, more than twice the size of Cleveland. But the "Queen City," as Cincinnati was more appealingly known, no longer reigned supreme in the Midwest, since first St. Louis in 1860 and then Chicago in 1870 rose to be the leading city of the region.

As had occurred in 1830, a cluster of three Ohio cities followed the two leaders. Included this time were Toledo, Columbus, and Dayton, each with just over 30,000 people. In the 10,000–15,000 range were six towns: in the Lake Erie basin were Sandusky and Akron; in the Ohio River basin were Springfield, Hamilton, Portsmouth, and Zanesville.

Population redistribution clearly favored the north side of Ohio in the mid 1800s. The canals and the early railroad system both helped to shift development northward away from the Ohio River. The draining of the Black Swamp opened a vast farmland to settlement for the first time and was a significant component of the overall population change. As will be seen, however, this process of change was just getting under way in 1870. The next decades would see even more dramatic growth in the north.

Industrializing Ohio: 1870–1910

As the United States embarked on a period of strong industrial growth in the decades after the Civil War, Ohio was a leading participant. The northeastern region of the state, especially, presented significant advantages for manufacturing, and several cities emerged as national industrial leaders.

But these changes did not occur overnight. There was small-scale manufacturing in most of the state's cities before the Civil War. Especially notable was an iron-manufacturing region deep in southern Ohio, stretching from Lawrence County northward into Perry, called the "Hanging Rock District." It had been an important producer in the 1830s and 1840s but had fallen on difficult times, partly due to the destruction of nearby forests from which its charcoal fuel had been derived and also because the local iron ores were nearly depleted. Manufacturing elsewhere within the state was primarily devoted to the needs of farmers, either supplying their tools and equipment or processing their crops. Cincinnati's meat processing industry was the most notable example. The Civil War had been a strong stimulant to fledgling manufacturing industries that probably would have spent a longer time in infancy had the war not occurred.

New Manufacturing Industries

The manufacturing developments of the late 1800s were overwhelming in comparison to what had occurred before. Many varied industries took root after 1870, the single most significant being the steel industry, which entered the northeastern region. Improvements in the processes of making steel had been developed in the mid 1800s and put into operation in the Pittsburgh area in the 1860s. Coal, convertible to a purer form of carbon called "coke," was mined in the Allegheny Plateau around the city. Iron ore, while present locally in limited quantities, was acquired mainly from newly opened mines in Upper Michigan and Wisconsin and carried by boat to Lake Erie ports. It was not long before the railroad cars that had carried iron ore to the Pittsburgh area were used to transport coal to Cleveland, Canton, and Youngstown, which had possessed iron furnaces since early in the century. Ohio's steel industry blossomed, and by the end of the century this area, along with the centers in western Pennsylvania, formed the world's leading steel-producing district.

Petroleum processing was another industry that encouraged the development of the northeast, especially Cleveland. Again, the plateau country of western Pennsylvania was involved, for it was there, at Titusville, that the first successful oil well was operated in 1858. A boom in the drilling of wells followed and substantial production ensued. The Clevelander John D. Rockefeller moved quickly to gain control of the processing and distribution of petroleum products, creating the Standard Oil Corporation with its headquarters and much of its processing activity located in Cleveland. A second oil boom occurred in northwestern Ohio, near Findlay, in the 1880s. Ohio's significance in the industry dwindled rapidly after the opening of the vast Texas-Oklahoma fields a decade later.

The array of diversified manufacturing that evolved in the state in these last decades of the nineteenth century is enormous. Agricultural processing continued—including the Civil War–stimulated oatmeal production in Akron—and other resource-oriented manufacturing flourished, including the potteries of East Liverpool. Agricultural machinery production was important in many cities. Soap making in Cincinnati (an offspring of the meat industry), glass making in Toledo, and rubber production in Akron were prominent industries of the age that would eventually bring substantial wealth and worldwide fame to their cities. But in addition to these large, notable enterprises, there were numerous smaller operations that often evolved because of them. For example, centers of steel production became attractive to entrepreneurs who devised newer and better things to do with the steel, and the resulting products and machinery became important segments of the total.

As an industry develops, it is often quickly joined by another that either supplies its needs or uses its output in some productive way. Thus the presence of manufacturing itself became a significant attraction to new manufacturing growth, and the region was filled with many intermediate-stage manufacturers, processors not of raw materials but of materials previously processed by someone else, and not makers of goods for the household consumer but of producer goods, destined to be formed or worked into another product by another manufacturer. These symbiotic relationships led to intensive concentrations of industrial activity within industrial districts within industrial cities. An entirely new landscape was created in the cities of the state. A steel mill was the sort of structure that, operating day and night, with its size, sound, smell, and sparking illumination, had never been witnessed before.

Nowhere are interfactory linkages so apparent as in the automobile industry, which began to develop in the state soon after 1900. Elaboration of the production of motor vehicles would eventually reach the point where hundreds of parts, each with its own developmental process, would be passed from one factory to another until reaching an assembly plant, which would bring forth the finished product. While Detroit and southeastern Michigan became the heart of this industry, northeastern Ohio became a major concentration.

Steel, through its impact on transportation, also affected manufacturing. Its use in the making of rails, engines, and rolling stock brought about a substantial increase in the capacity of the railroad system. In turn, this increased capacity in the shipping of coal, iron ore, and finished products made large steel mills possible. During the last decades of the 1800s, Ohio's rail mileage continued to expand, reaching nearly 10,000 by 1910, when the state led the nation with a density of 23 miles of railroad track for every 100 square miles of territory, a fairly good indication of the effectiveness of rail service for the people and businesses of the state.

The development of the Cleveland-Pittsburgh steel district was one of a series of such events occurring almost simultaneously in Europe and the United States. Other major world steel centers were the English Midlands, the German Ruhr Valley, the Donets basin in Russia, and the Chicago-Milwaukee area in the United States. In each instance the critical local resource supporting the development of steel manufacturing was coal, and the availability of this fuel caused rapid growth of the mining industry in the plateau country of east-central Ohio, in and around Belmont County. While not as pure as that found in other parts of the plateau, Ohio coal was easily exploited, with production reaching a peak of about 50 million tons in 1908. Then a period of decline followed, as other coalfields and other fuels came into use, but in the 1930s, production began increasing once more.

A Diversity of Spatial Changes

While the railroads continued to expand and improve during this period, two other related innovations had a very important impact on the state's geography. Within cities, the electric streetcar was introduced in the 1890s and quickly exerted an influence on locations, allowing homes to spread outward along the lines while the downtown area, at the heart of the radial system, blossomed as a retailing center. Similarly, interurban electric railroads appeared at about the same time, providing improved linkages not only between individual cities but also between a city and its rural environs. The first of these in the United States was built between Granville and Newark in 1889.

While employing fewer and fewer Ohioans, agriculture continued to be an important element. Among the new patterns emerging at this time was the shift on dairy farms, especially in the northeastern counties, to fluid milk production: rapid rail service now made possible the marketing of this very perishable commodity. Furthermore, the northwest saw the development of Corn Belt meat production, combined with Lake Plain vegetable farming, that became characteristic of this region.

Geographic change took many other forms during this period. Factors of long-term significance, for example, were the increasing importance of the state government and the creation of a state university, enhancing both the importance and the growth capabilities of Columbus. These developments would place that city in a fortuitous situation several decades into the next century. With the creation of state-funded colleges in Athens, Oxford, Bowling Green, and Kent, higher education spread to the four corners of the state.

Population and Labor Force: 1910

By 1910, the 4.8 million people in the state comprised 5.2 percent of the national total, down slightly from 1870. Illinois had grown more rapidly and moved up to the third rank as Ohio dropped to fourth. A greater change occurred in internal regional population patterns, which were strikingly different from those of 1870. While the northeast rose to include nearly one-third of the state's residents, and the northwest increased slightly, the other three regions—southeast, southwest, and central—though still growing in total population, all declined in their share of the state's total, with the southeast continuing to experience the greatest reduction.

Ohio had parlayed tremendous advantages in transportation, market access, materials, fuel supplies, and labor into strong manufacturing growth. This in turn produced urban growth, making the state by 1910 more than 50 percent urbanized, a doubling of the level of 1870 urbanization. Among the cities, Cleveland had, in 1900, passed Cincinnati for the first time and by 1910 led with 560,663 to Cincinnati's 363,591. Following these two leaders were Columbus, 181,511; Toledo, 168,497; and Dayton, 116,577; with Akron, Canton, and Youngstown ranging 50,000–100,000.

These eight cities had thus established their preeminence in the state and would maintain that status throughout the twentieth century.

During the early years of this industrialization, labor was drawn from rural areas, where demand for workers was declining, as well as from the growing immigrant population that reached Ohio via a relatively short journey from Atlantic ports of entry. This latter group, discussed at length elsewhere in this book, became an increasingly large part of the work force, bringing great diversity and a distinct ethnic geography to the cities. Urban growth was reflected in the characteristics of the state's labor force in 1910. The largely non-urban primary sector was only half as significant as it had been in 1870, declining to 24.7 percent of the total. Secondary and tertiary employment divided the remainder fairly evenly, with 35–40 percent each.

As proposed earlier in this historical discussion, the Ohio that had been forged by 1910 was remarkably different from the state that embarked on the nineteenth century. The north, especially the northeast, gained predominance over the south, especially the southwest. While agriculture continued to be a significant element in the state's economy, its relative position was declining, and both manufacturing and services had become more significant. From being mainly a rural state in its infancy, most Ohioans in 1910 were urban. But many of these trends would also shift as the new century unfolded. The northeastern counties would gain no greater dominance than they possessed in 1910, whereas the southwest would eventually see its status grow. Services would increase their share of the state's employment while manufacturing would stabilize and, by the end of the 1900s, join agriculture in decline. Even the level of urbanization (the percentage of the population living in cities) would start to go down before the twentieth century came to a close.

Geographic Change in Times of Stress: 1910–50

The broad spatial patterns evident in Ohio at the end of the nineteenth century changed only gradually during the twentieth. The regional arrangements of agriculture, mining, manufacturing, urbanization, and population in 1950 strongly resembled those of 1910, and this resemblance continues through to the present. Once such patterns become established, they do not change very quickly. Local modifications within the broad patterns occurred largely because of the auto age, which drastically reshaped the locational patterns of all sorts of human endeavors. In addition, and more in Ohio than in most places, the production of these motor vehicles became an important function and, as such, an important component in the state's economic geography.

The Auto Impact

As with all innovations in transportation, new locational patterns emerged because of the new efficiencies and new cost structures of movement. Motor vehicles were initially most effective for short trips and thus made their greatest impact in and around those places where economic activity was already concentrated—that is, the cities. Of great immediate effect was the replacement of the horse and wagon by the truck. As cities had grown in the preceding decades, so had the horse population, and their care and feeding had become quite costly. Trucks appeared on Ohio streets much more quickly than automobiles, since people could still use the trolleys or the interurbans for local trips. Cars were used primarily as recreational vehicles, for Sunday drives around town or in the country, before they became a part of the regular economic life of cities. But gradually, as the use of automobiles broadened, their impact was felt in the further separation of home and workplace, the outward expansion of development into the countryside, and the sharper segregation of residential areas from other land uses.

The use of cars and trucks required improved highways. Unlike the privately owned trackage used by the railroad companies, roads were built and maintained by the government, primarily at the local level. In the 1910s, the state government moved into this highway development effort in earnest, and it has been a major state function ever since. In the process, a new form of place-name was created to identify these long strips of paved surface, and names like *S.R. 7* and *S.R. 235,* and later, *U.S. 30* and *U.S. 23,* grew common, providing a sense of connectedness between people living in towns and cities spread out across the state and the nation.

Auto and truck manufacturing blossomed in the first decade of this period. Ohio generally benefited from the development of motor vehicle production in at least two ways. The steel industry, especially in the Cleveland and Youngstown areas, grew because of the great quantities of the metal needed for autos. In addition, large parts of the auto industry itself, especially components manufacturing, put down roots in Ohio, becoming a substantial part of the industrial diversity of the state. Akron became the most specialized manufacturer of auto-related products. Tire manufacturing had developed locally in response to the bicycling craze of the 1890s, and so Akronites were ready when the automobile began to appear in large numbers. The city was Ohio's leading growth center in the 1910s and continued to expand throughout the first half of the century. Toledo, close to Detroit, also became an auto center, and Dayton evolved in tune with twentieth century developments, making contributions to the auto industry, to the growing tertiary sector through cash register manufacturing, and, as a result of the presence of the Wright brothers, to the budding aviation industry. It enjoyed the greatest sustained growth among Ohio's large cities during this period.

In these years, Ohio continued to be exceptionally diversified in its manufacturing. Though steel and its products, including transportation equipment, led all others, chemicals, food products, glass, paper, ceramics, and rubber remained significant in a state that was truly the heart of the American manufacturing belt.

Some New Developments

With the beginning of World War I, the process of immigration into the United States declined dramatically, changing population geography nationwide and putting the maintenance of American industrial growth, thus far dependent on immigrant labor, at risk. But a new source of factory labor emerged from the American South, as whites, especially from the Appalachian highlands, and African Americans began to enter northern cities in large numbers, often through the recruiting efforts of northern corporations. For the first time in several decades, the proportion of African Americans in Ohio began to grow, from 2.3 percent in several prior decades, to 3.2 percent in 1920, reaching 6.5 percent by 1950. Nearly all of this growth was in the cities, especially Cleveland.

A greater interest in the relationships with the natural environment emerged during this period, with the technological age taking a part in new expressions of these relationships. One such new expression was the use of technology, such as flood control, to control the environment. In 1913, enormous flooding ravaged Ohio, taking over 400 lives. The Dayton area, hit especially hard, responded with a strong movement toward control, and a regional organization of private and governmental interests, the Miami Conservancy, was granted strong powers by the state to undertake control projects. It built earthen dams and reservoirs in the upper parts of the Miami basin to effectively prevent future deluges. A similar and larger-scale conservancy was developed in the 1930s in the Muskingum watershed.

While flood control was the direct aim of these projects, recreational uses of the reservoirs and surrounding lands were important results, forming a second expression of the new human-environment relationship. As the growth of manufacturing put many people to work on a rigid time schedule, enlightened management and strong unions increasingly saw to it that this schedule included two-day weekends and annual vacations. With the ease of travel provided by the automobile, a demand evolved for places in which to spend leisure time at reasonable cost, thus fueling the development of parks, particularly by the state. It was during the Depression that much important, long-term environmental development occurred in the state, a benefit of the many and varied tasks to which the unemployed were assigned by New Deal programs.

Population and Labor Force: 1950

During the four decades following 1910, worldwide and national upheavals were reflected within the basic patterns of population change within Ohio. During the two war decades of the 1910s and 1940s, Ohio's growth rate exceeded that of the nation as a whole—an indication of the importance of the state's industries to national military mobilization. But during the other two decades, the 1920s and 1930s, the growth rate of the state fell below that of the rest of the nation. This was most notably the case during the Depression years of the 1930s, when, after growing by at least 13 percent in each decade throughout its history, Ohio's growth dropped to just 3.9 percent at a time when the national rate was 7.2 percent. Recovery, in an economic sense, characterized the 1940s, largely because of World War II, which required many things that Ohio could provide. By 1950 the state had experienced a decade of population growth at 15 percent, comparable to pre-Depression times. With nearly 8 million people, Ohio was now fifth among the states, having been passed by the West Coast upstart, California. Since 1910, the proportion of the national total had risen slightly and now claimed 5.3 percent. Northeast Ohio continued to hold the greatest portion: 41.4 percent. The urbanizing southwest also increased its share over the 1910 figure and now reached 21.8 percent. But the other three regions declined: the southeast hitting 11.6 percent, and the northwest and center down only slightly to 12.4 and 12.9 percent, respectively.

The 1950 census reported populations for city-suburban totalities called *urbanized areas,* which were, in comparison with incorporated cities, more realistic expressions of the magnitude of the economic and cultural concentrations which the term *city* implies. The Cleveland urbanized area contained over 1,380,000 people at midcentury. The other major concentrations were Cincinnati with 813,000 and Columbus with 438,000. With around 350,000 each were Akron, Toledo, and Dayton, while Youngstown had just under 300,000 and Canton about 175,000. Tertiary employment accounted for half of the labor force in 1950. Its share would continue to increase, but the proportion in the secondary sector would soon join the primary share in decline.

From 1950 to the Present

The postwar period saw industrial growth continue in the state for some time and population growth sustained through the 1950s. But seeds of change had been sown. Interregional shifts had been stimulated during the war years; industries had expanded to the balmier climates, especially to the West to sustain the war in the Pacific, and masses of personnel from the North, trained in Southern military camps, found the South attractive. New technologies, especially air conditioning for the South and water supply in the West, as well as the benefits of ever-expanding transportation, provided the capabilities for major relocations of the American people and their economic activities. The state's total population reached a plateau of just under 11 million in 1970 and sustained it for the next 20 years, as other parts of the nation gained the momentum of growth. By 1990, Ohio's total of 10.8 million was only 4.4 percent of the national population, the lowest it had been since 1810, and six states were larger: Texas had passed Ohio in the 1960s; Florida did the same in the 1980s.

During the previous 40 years, the regional distribution of the state's population was marked by gains in the shares of the center and southwest and reduced shares in the other three regions. Decline in the northeast's portion ended the pattern of increase that had prevailed in that region since the beginning of statehood, while the southeast's lower percentage marked continuation of a constant decline there.

Primary employment dropped to only 2.3 percent of the labor force, but, more notably, secondary employment—the manufacturing sector with which Ohio had been so closely identified for a century—also declined. Only the tertiary sector increased its share of the total, reaching nearly 70 percent. The age of services, or what many call the *postindustrial age,* had arrived. In addition, the proportion living in urban settings was at 74.1 percent, higher than in 1950.

Among the urbanized areas, Cleveland remained the population leader at 1.7 million, with Cincinnati next at 1.2 million. But the focus of urban growth had shifted to Columbus, the government-education-trade center still third with over 900,000 but growing faster than the others. And in fact, during the 1980s the central city of Columbus became the state's largest incorporated place. Dayton meanwhile moved from sixth to fourth place, ahead of both Akron and Toledo. And following Youngstown-Warren and Canton, Lorain-Elyria placed ninth due to strong growth in the postwar decades.

These geographic patterns are now evolving into the future through diverse processes. Developing transportation technologies, changing understanding and treatment of the natural environment, evolving patterns of energy production and use, and modified social relationships among population groups, particularly in cities, are among the forces most likely to transform today's geography of Ohio for tomorrow. Changes are likely to come slowly, but they will surely come. Knowing the geographies of the past, and of the present as it becomes the past, should help Ohioans create a more livable geography of the future.

REFERENCES

Jakle, John. 1977. *Images of the Ohio Valley: An Historical Geography of Travel.* New York: Oxford University Press.

Jones, Robert L. 1983. *The History of Agriculture in Ohio to 1880.* Kent, Ohio: Kent State University Press.

Knepper, George. 1989. *Ohio and Its People.* Kent, Ohio: Kent State University Press.

Noble, Allen G., and Albert J. Korsok. 1975. *Ohio: An American Heartland.* Bulletin 65. Division of Geological Survey. Columbus: Ohio Department of Natural Resources.

Schreiber, Harry N. 1969. *Ohio Canal Era: A Case Study of Government and the Economy: 1820–1861.* Athens: Ohio University Press.

Smith, Thomas. 1977. *The Mapping of Ohio.* Kent, Ohio: Kent State University Press.

Wilhelm, Hubert G. H. 1982. *The Origin and Distribution of Settlement Groups: Ohio, 1850.* Athens: Cutler Service Center, Ohio University.

Wittke, Carl, ed. 1943. *The History of the State of Ohio.* 6 vols. Columbus: Ohio State Archaeological and Historical Society.

CHAPTER EIGHT

Population Patterns

David T. Stephens

Since people give character to places, geographic understanding can be enhanced through awareness of the distribution of and variety among an area's population. The demographic characteristics of a place are greatly influenced by the nature of the locale's economy. Given the dismal performance of Ohio's economy in the past thirty years, one would expect to see changes in the nature of the state's population. Examination of Ohio's population reveals a story of growth, offers a picture of current conditions, and provides a basis for making projections about the future.

Historic Trends

A historic perspective on population offers insights into Ohio's quick rise to prominence and helps to explain its recently declining position within the national power structure. Table 8.1 is a summary of the state's population growth. Buoyed by a heavy wave of young, land-seeking immigrants and high birth rates, Ohio's population rose rapidly in the first three decades following statehood. From 1803 until 1840, Ohio's growth rate exceeded that of the nation, making Ohio the third most populated state. Since then, though natural increase has overshadowed immigration as a source of growth, the state has experienced significant immigrant influxes since 1840, including north and west Europeans up to 1880; east and south Europeans from 1880 to 1920; African Americans from 1900 to 1920; people from Appalachia from 1945 to 1960; and most recently, Hispanics and Asians.

The state's growth rate exceeded that of the nation during three periods: from statehood to 1840; from 1910 to 1920, when the influx of European immigrants and African Americans combined with natural increase to boost Ohio's growth above the national average; and during the early years of the baby boom. Since 1960, Ohio, like other Rust Belt states, has suffered from deindustrialization. The steel, automobile, and rubber industries, long Ohio's strength, have proven to be an Achilles' heel. Decline in these and related industries has ushered in an era when Ohio's population growth has been well below that of the nation. For the decade of the 1980s, Ohio grew only 0.5 percent; in the 1970s the growth was only 1.5 percent. In light of these numbers, the Ohio State Data Users Center's projection of a 3.3 percent growth in population in the 1990s seems too optimistic. Ohio will be fortunate if it does not lose population by the turn of the century.

Ohio retained its rank as the third most populated state until displaced by Illinois in 1890. In the twentieth century, first California, then Texas, and most recently Florida eclipsed the Buckeye State in population. Even if it should lose population in the 1990s, it appears Ohio will keep its seventh-place rank through the year 2000. The recent pattern of either slow growth or no growth has some important implications that are explored later in this chapter. One consequence is an impact on Ohio's status in the national political arena.

Apportionment of seats in the United States House of Representatives is based on a state's population. The number of representatives allocated to Ohio rose during the state's youth and maturation but in recent decades has declined. Reflecting the rapid popula-

Table 8.1

Ohio's Population Growth, Projections, and Changing Congressional Strength

DATE	POPULATION	% CHANGE	NATIONAL RANK	NUMBER OF REPS.	PERCENT OF THE HOUSE
1800[a]	45,365				
1810	230,760	408.7[b]	13	6	3.2
1820	581,434	152.0[b]	6	14	6.6
1830	937,903	61.3[b]	4	19	7.9
1840	1,519,467	62.0[b]	3	21	9.1
1850	1,980,329	30.3	3	21	8.9
1860	2.339,511	18.1	3	19	7.8
1870	2,665,260	13.9	3	20	6.8
1880	3,198,062	20.0	3	21	6.5
1890	3,672,329	14.8	4	21	5.9
1900	4,157,545	13.2	4	21	5.4
1910	4,767,121	14.7	4	22	5.1
1920	5,759,394	20.8[b]	4	22	5.1
1930	6,646,697	15.4	4	24	5.5
1940	6,907,612	3.9	4	23	5.3
1950	7,946,627	15.0[b]	5	23	5.3
1960	9,706,397	22.1[b]	5	24	5.5
1970	10,657,423	9.8	6	23	5.3
1980	10,797,630	1.3	6	21	4.8
1990	10,847,115	0.5	7	19	4.4
2000	11,188,300[c]	3.3			
2010	11,552,000[c]	3.0			

Sources: Censuses of Population and Ohio Data Users Center

[a] Population of the Northwest Territories

[b] Exceeded the national growth rate

[c] Projected estimate from the Ohio Data Users Center

tion growth in the first part of the nineteenth century, Ohio's congressional delegation grew from six in 1810 to twenty-one by 1880. In 1840 Ohio reached a peak in the proportion of House seats it controlled, with just over 9 percent. Although proportional representation has not grown since, the number of Ohio seats increased to a high of twenty-four in 1930 and again in 1960. Since 1960 the number of seats has declined significantly, reaching nineteen through the 1990s. This has meant fewer federal dollars have made their way back to Ohio, and in an era of budget cutting, with less protection Ohio's federal facilities are more likely to end up on closing lists, and Ohio is less likely to be considered when new federal facilities are constructed. Given past population trends, Ohio will be fortunate to avoid further erosion of its influence when the House is reapportioned in 2000.

Spatial Patterns

Where do Ohioans live? Figure 8.1 offers some insights into this question. Most can be found in a band of thirty-four counties that crosses the state diagonally from northeast to southwest. This band—the *urban axis,* which contains Cleveland, Akron, Canton, Youngstown, Columbus, Dayton, and Cincinnati—holds 73.4 percent of the state's population. Outside this concentration lies only one noteworthy outlier, in the northwest around Toledo. The southeast and the balance of the northwest are sparsely settled. Figure 8.2 suggests Ohio counties vary dramatically in their population destiny. Cuyahoga County, with more than 3,000 persons per square mile, is one of the most densely settled counties in the United States, while Noble County, in the southeast, has less than thirty persons per square mile.

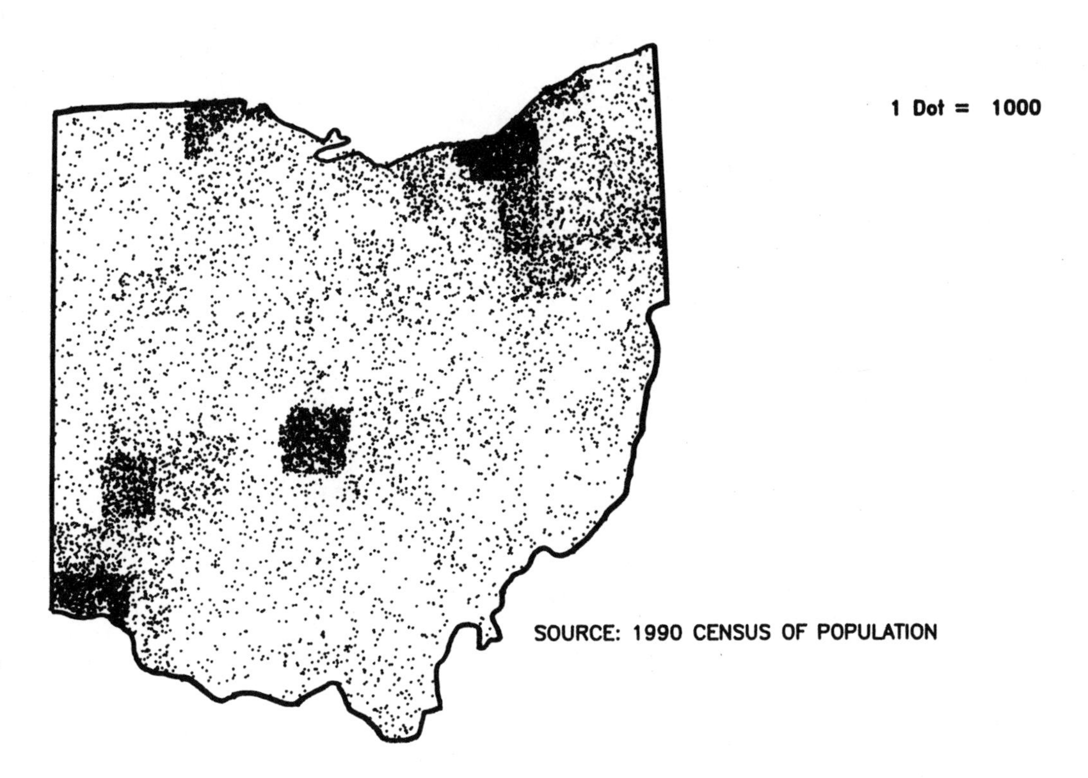

Fig. 8.1. Population Distribution, 1990

A popularly held image of Ohio is that of a densely settled, highly urbanized state, but the reality is a densely settled urban axis paralleled by two zones of very modest densities. In 1990, 74.1 percent of the state's population was classed as urban. The rural population concentrated outside the urban axis. Figure 8.3 shows that even within the axis there are several counties where over 75 percent of the residents are rural. What may be unexpected is that most Ohio counties have less than half their population living in urban places. The image of a very urbanized Ohio is not as correct as might be thought. As Figure 8.3 indicates, the most rural areas of Ohio are the southeast followed by the northwest.

Ohio's urban population can be better understood by examining metropolitan populations. The Office of Management and Budget defines U.S. metropolitan areas. Ohio has fifteen metropolitan areas of various types. These areas are shown in Figure 8.4 and their populations and component counties are shown in Table 8.2. In terms of growth, these metropolitan areas mirror certain trends. Those areas in the southwest and central, Cincinnati-Hamilton, Dayton-Springfield, and Columbus, grew in the last decade, while the balance did not. The Steubenville and Wheeling MSAs (Metropolitan Statistical Areas) fared the worst, experiencing declines of more than 10 percent. Closer inspection of the table reveals that growth occurred in suburban counties, while those counties having central cities lost population.

From 1960 to 1990, Ohio's population increased, but not all counties shared in these increases. For that period the average increase in a county's population was 21.3 percent. Figure 8.5 shows the spatial variation in population changes within the state during those thirty years. Six counties—Belmont, Cuyahoga, Harrison, Jefferson, Mahoning, and Scioto—declined from their 1960 totals. The erosion of their economic bases—coal mining and steel industry—and significant out-migration are the primary contributors to this decline.

Those counties experiencing slow growth, less than the average for all counties, fall into several categories. One group includes most of the counties

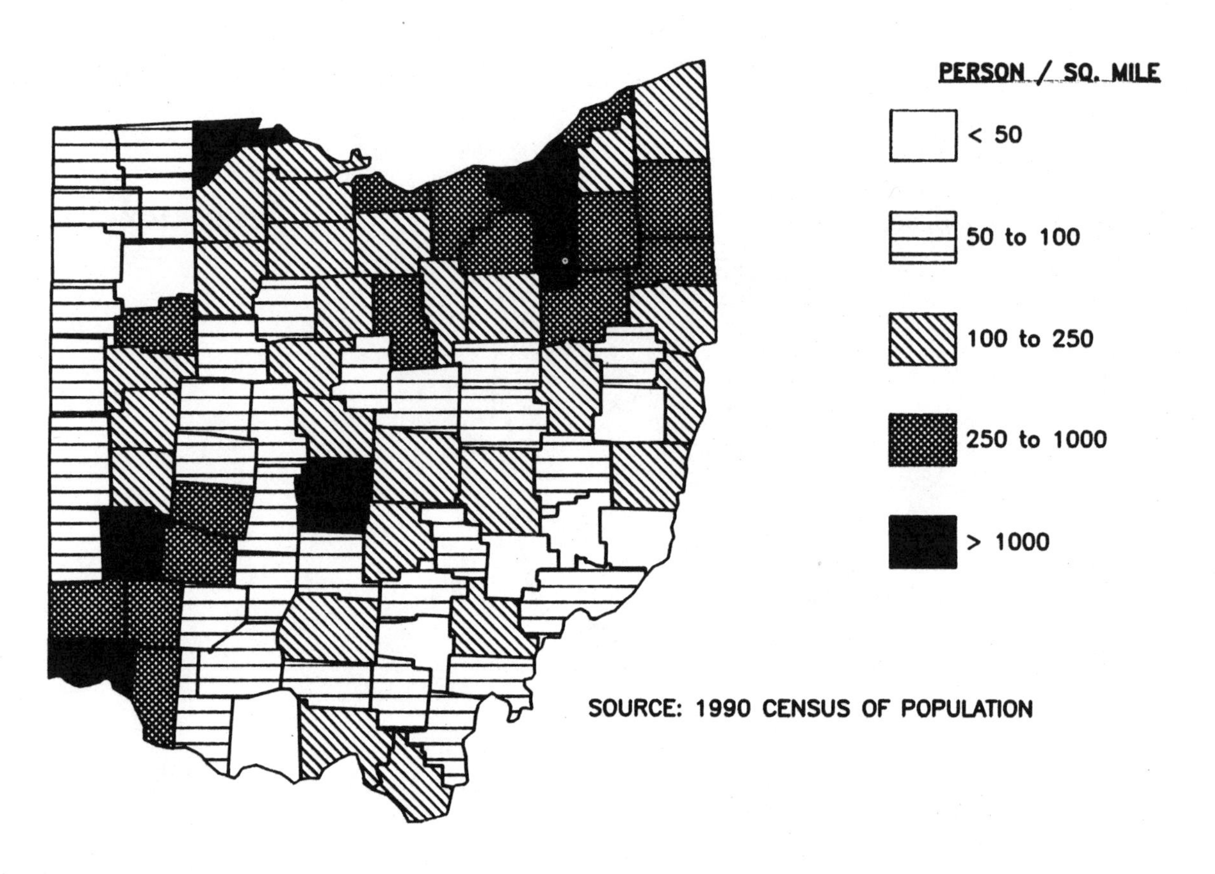

Fig. 8.2. Population Density, 1990

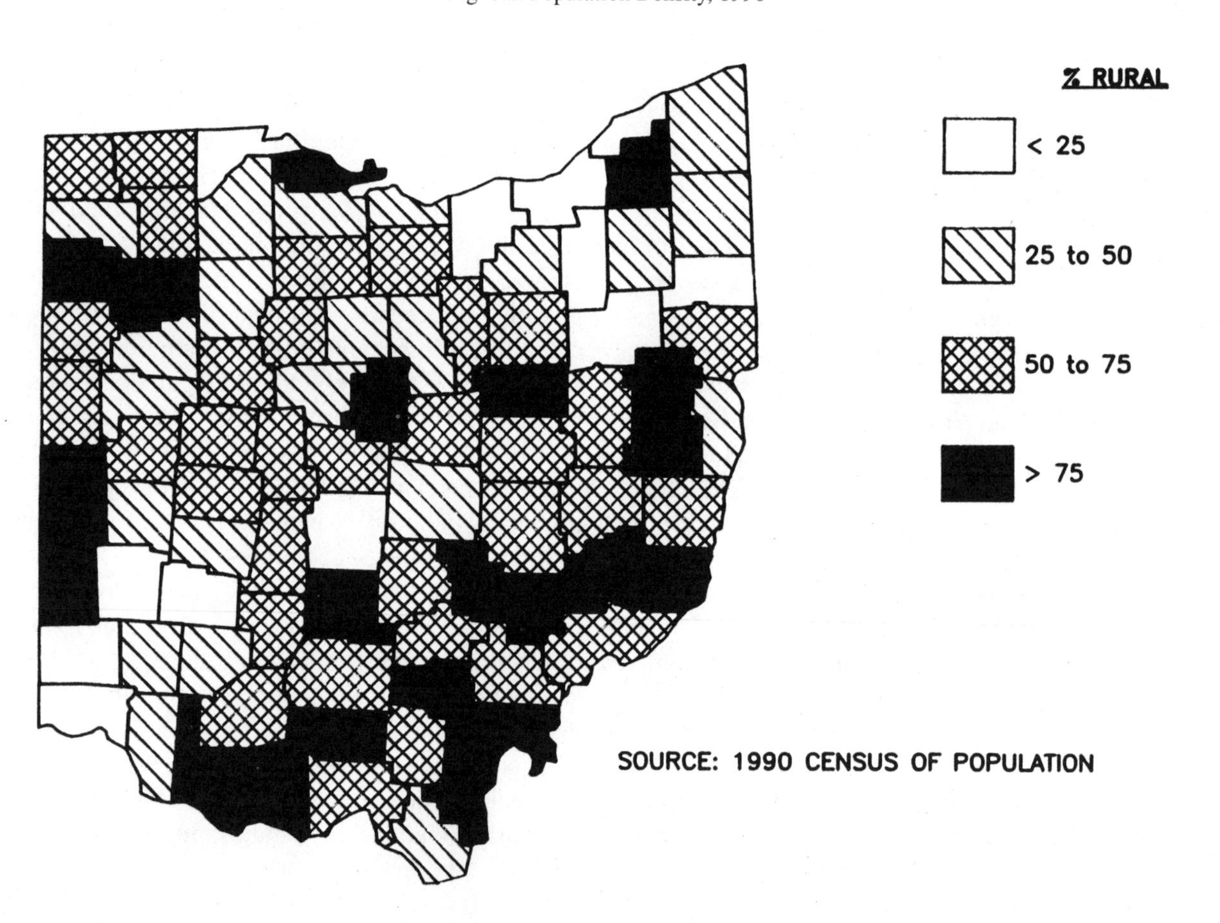

Fig. 8.3. Rural Population, 1990

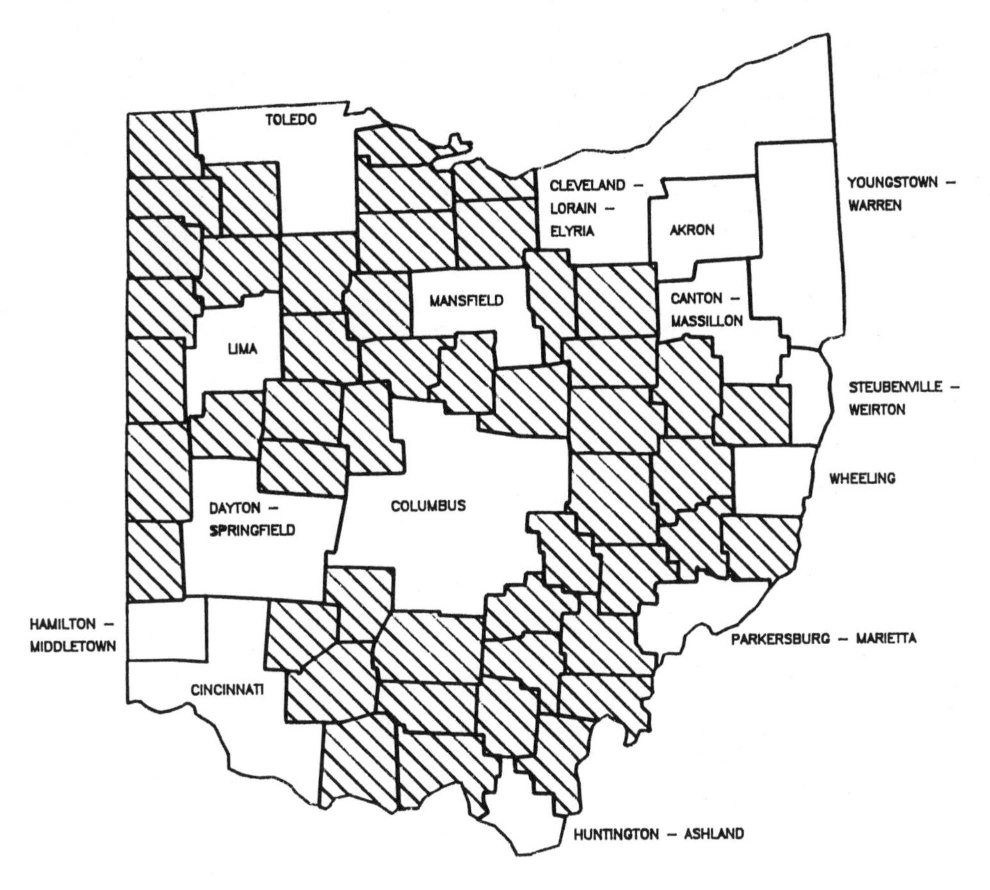

Fig. 8.4. Metropolitan Areas, 1992

that have the state's largest cities: Allen, Clark, Hamilton, Lucas, Montgomery, Stark, Summit, and Trumbull Counties all contain large central cities. These cities have experienced plant closings, layoffs, and out-migration—all contributors to population declines. A second group is composed of predominantly rural western counties which, with their declining farm populations and small service centers, not only have not attracted much new economic activity but have lost some of what little they had, thereby slowing population growth. A third group of slow growers is found in eastern Ohio. These include much of the state's coal country, Appalachian counties characterized by chronic poverty, and counties like Ashtabula and Columbiana that have suffered erosion of their manufacturing bases.

Counties growing above the average have been divided into two classes: rapid growth (greater than the all-county average) and very rapid growth (more than twice the all-county average). The rapid growers fall into three groups. The first is composed of the suburban counties that surround Cleveland, Columbus, and Cincinnati. These counties have been the destinations of many fleeing central-city residents and functions. Often this flight has been hastened and abetted by the development of outerbelts that divert traffic and development from central cities. A second group is made up of western counties that have benefited from the dispersal of the automotive industry in the last ten years. The third group consists of counties in the interstate corridors along Interstate 71 and Interstate 75 that have profited from the increased connectivity afforded by access to the interstate network.

Very rapid growth counties house the sites of most of the suburban growth of Ohio's major cities: Cleveland, Columbus, Cincinnati, Dayton, and Toledo. Two counties that do not fit that description are Morrow and Holmes. In the former's case, a small 1960 population, a 1970s energy boom, and Interstate 71 seem to have been major factors in its growth. Holmes County is frequently a demographic anomaly because of its large Amish population; high fertility levels among the Amish explain Holmes's rapid growth.

Table 8.2

Ohio's Metropolitan Areas

COUNTY	1990 CENSUS	PERCENT CHANGE 1980–90
Canton-Massillon MSA	396,958	–2.6
Carroll	26,521	3.6
Stark	367,585	–3.0
Cincinnati-Hamilton CMSA	1,817,571	5.3
Cincinnati PMSA (OH-KY-IN)	1,526,092	4.0
Brown	34,966	9.5
Clermont	150,187	16.9
Hamilton	866,228	–0.8
Warren	113,909	14.7
Hamilton-Middletown PMSA	291,479	12.6
Butler	291,479	12.6
Cleveland-Akron CMSA	2,859,844	–2.7
Akron PMSA	657,575	–0.4
Portage	142,585	5.0
Summit	514,990	–1.8
Cleveland-Lorain-Elyria PMSA	2,202,069	–3.3
Ashtabula	99,821	–4.2
Cuyahoga	1,412,140	–5.8
Geauga	81,129	8.9
Lake	215,449	1.3
Lorain	271,126	–1.4
Medina	122,354	8.1
Columbus MSA	1,345,450	10.8
Delaware	66,929	24.3
Fairfield	103,461	10.4
Franklin	961,437	10.6
Licking	128,300	6.0
Madison	34,068	12.3
Pickaway	48,255	10.5
Dayton-Springfield MSA	951,270	1.0
Clark	147,548	–1.8
Greene	136,731	5.4
Miami	93,182	3.1
Montgomery	573,809	0.4
Huntington-Ashland MSA (WV-KY-OH)	288,189	–7.4
Lawrence	61,843	–3.2
Lima MSA	154,340	–0.3
Allen	109,755	–2.2
Auglaize	44,585	4.8

Table 8.2 (*cont.*)

Ohio's Metropolitan Areas

COUNTY	1990 CENSUS	PERCENT CHANGE 1980–90
(Lorain-Elyria MSA: See Cleveland-Lorain-Elyria PMSA)		
Mansfield MSA	174,007	–4.0
Crawford	47,870	–4.4
Richland	126,137	–3.9
Parkersburg-Marietta MSA (WV-OH)	149,169	–5.5
Washington	62,254	–3.1
Steubenville-Weirton MSA (OH-WV)	142,523	–12.6
Jefferson	80,298	–12.3
Toledo MSA	614,128	–0.4
Fulton	38,498	2.0
Lucas	462,361	–2.0
Wood	113,269	5.5
Wheeling MSA (WV-OH)	159,301	–14.2
Belmont	71,074	–13.9
Youngstown-Warren MSA	600,895	–6.8
Columbiana	108,276	–4.7
Mahoning	264,806	–8.5
Trumbull	227,813	–5.8

Source: Ohio Data Users Center
Note: December 1992 definitions

What will be the nature of changes in Ohio's population in the future? The Ohio Data Users Center has prepared population projections for the year 2015 (1993b). Figure 8.6 shows the expected changes in population based on these projections. The Center expects population losses in the major urban areas of northern Ohio, Toledo, and Cleveland. Declines are also projected for eastern Ohio's coal counties and the predominantly agricultural counties in the northwest. The expansion of suburban populations around Cleveland, Cincinnati, and Columbus is expected to continue, as is a continuation of increases among the Amish in Holmes County.

What is producing these changes? Erosion of economic bases has had a significant impact, but there are other factors at work. To gain a better understanding, we need to examine the two components of population change: migration and natural change. Figure 8.7 offers a perspective on spatial variations in recent migration rates. During the 1980s, Ohio had a net migration rate of –5.5 as over 600,000 Ohioans left the state. Most, but not all, counties suffered this out-migration. Eleven counties, all located in the central and southwest regions, countered the trend and had more arrivals than departures. They can be grouped into three clusters: one in the southwest, another north and east of Cincinnati, and a third around Columbus and in west-central Ohio. The attraction in both the Cincinnati and Columbus areas is suburbanization of many former central-city functions, while the magnetism of Logan and Champaign Counties reflects the dispersion of the automotive industry into parts of western Ohio.

The heaviest out-migration is associated with the decline of the steel and coal-mining industries in eastern Ohio. Mahoning and Jefferson Counties are representative of the former, whereas in Belmont, Harrison, and Monroe Counties, coal mining has been a leading employer. Counties not suffering the greatest losses, but still exceeding in the state average, include most of the state's major urban areas (excepting Columbus), the predominant agricultural counties of western Ohio, and some of the poorer counties of southeast Ohio. Unfortunately for Ohio's future, those leaving are often the "best and brightest"—the young and well educated. This does not bode well in terms of both future social services needs and Ohio's attempts to attract high-tech industries, which require a well-trained work force.

The natural element of population change depends on births and deaths. With the national birth rate at 16.2 in 1991, Ohio's birth rate of 15.7 is low. The map of birth rates, Figure 8.8, shows the spatial pattern of fertility within the state. Low rates are common in southeastern Ohio, with the lowest rates found in the extreme eastern and western parts of the state. The most likely explanations for these low rates are an older population, significant out-migration, and depressed local economies. Not surprisingly, the highest rate is found in Holmes County. Other high rates appear in counties having major urban centers; there is also a high-rate cluster in west-central Ohio. High rates in inner-city areas and the attractiveness of large hospitals may help to explain the higher rates in urban counties. An explanatory variable for a group of counties in the west-central region is probably its significant Catholic population (Wilhelm 1990).

Ohio's 1991 death rate of 9.1 exceeded the national norm of 8.5. Mortality in the state is influenced by age structure, poverty level, quality of life, and the availability of medical care. As Figure 8.9 reveals, the highest rates—those over 11—are found mainly in the southeast. This part of the state consistently ranks low in measures of the quality of life. Low rates are most likely to occur in suburban counties and in the west, as a result of younger populations and a better quality of life.

By combining information on these two demographic variables, the rate of natural change can be computed. The spatial pattern of this variable is depicted by Figure 8.10. In 1991, Ohio had two counties where deaths exceeded births. These two, Belmont and Harrison, have experienced severe economic stress created by declines in the coal industry. Historically, workers in this industry have a high incidence of accidental death. Both counties have aging populations, and high unemployment and poverty levels. Might these two counties be indicators of Ohio's future? If so, in that future Ohio, the need for funds to educate a shrinking young population might be reduced, but those gains could be more than offset by increased demands for Medicaid and other programs to assist a growing elderly population.

The highest rates of natural increase occur in suburban counties, the Catholic enclave in the northwest, and the Amish strongholds in Holmes and Wayne Counties. None of these rates represent very rapid growth. Holmes County, with its rate of increase at 1.5 percent, will require more than forty-five years to double its population. And if Ohio's 1991 rate of about 0.6 percent natural change remains constant, the state's population will not double for well over 100 years.

One consequence of out-migration, dropping birth rates, and increasing death rates is an aging population. Figures 8.11 and 8.12 are population pyramids for 1960 and 1990. The 1960 pyramid with its widening base shows a population that had been expanding for fifteen years. This growth represents the baby

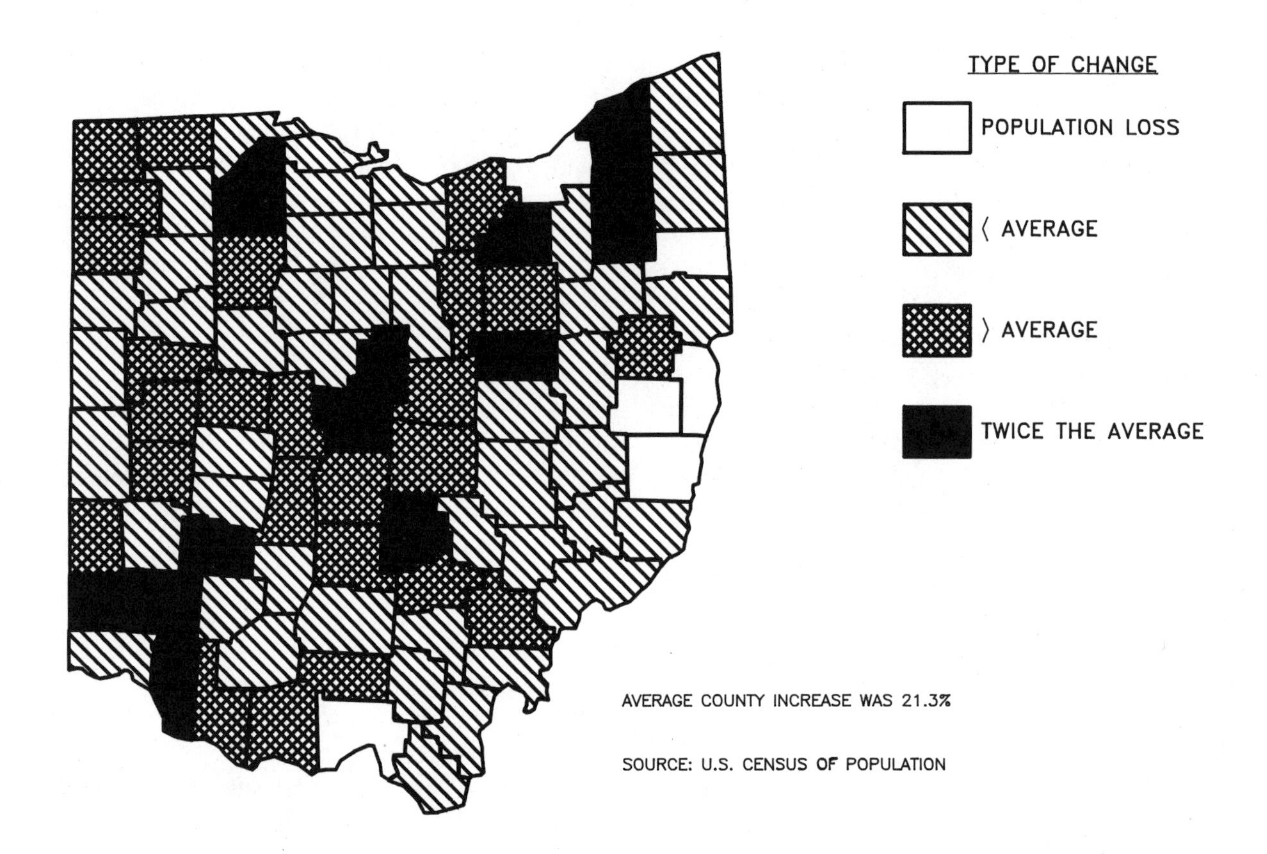

Fig. 8.5. Population Change, 1960–1990

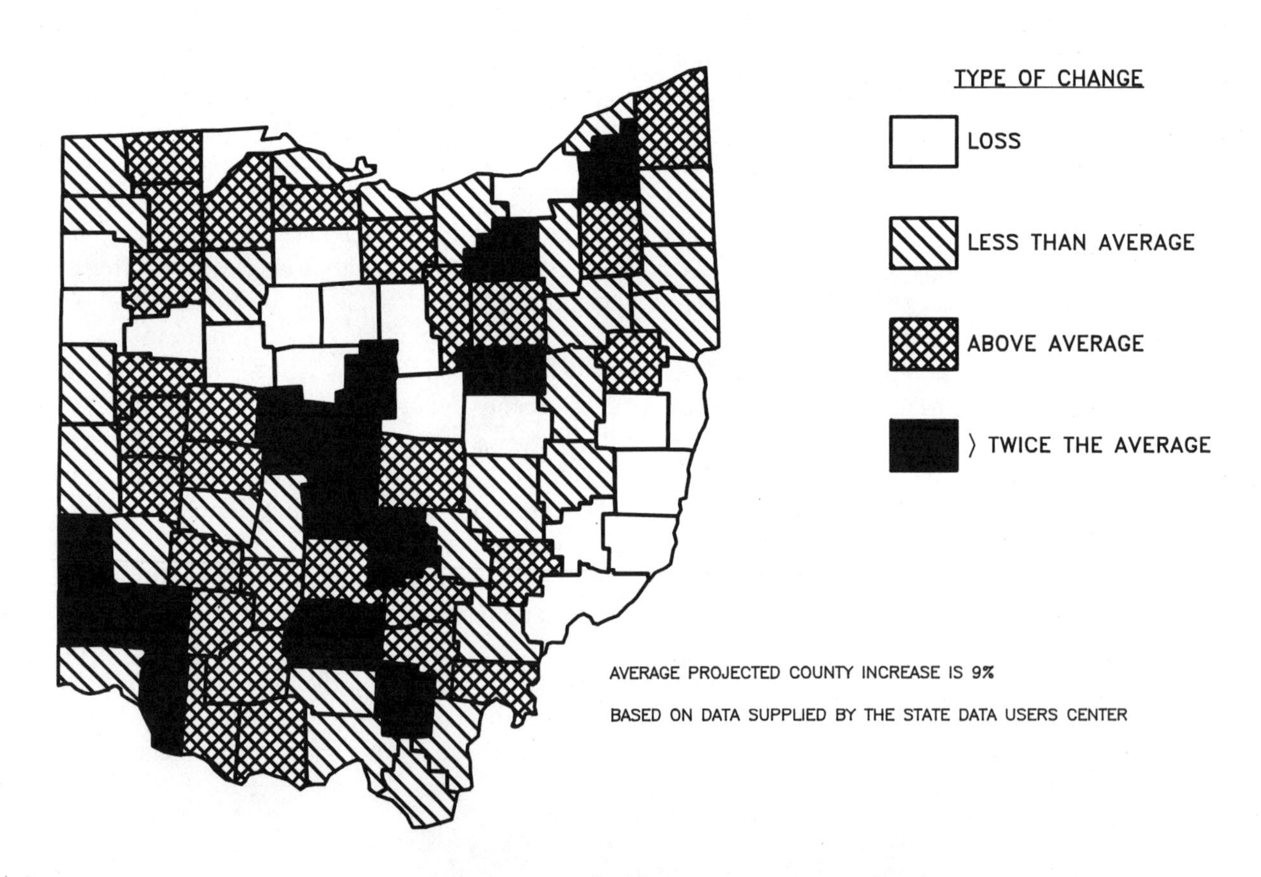

Fig. 8.6. Projected Population Change, 2015

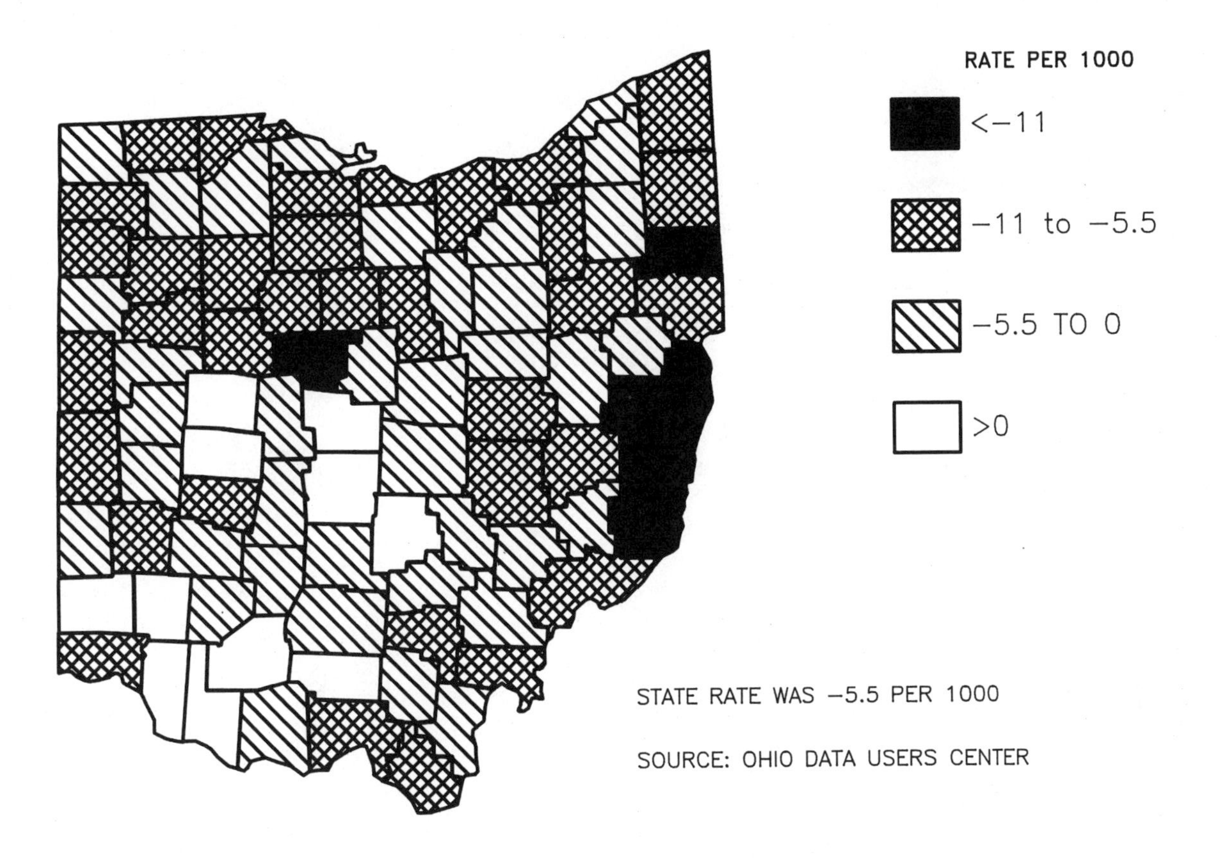

Fig. 8.7. Migration Rate, 1980–1990

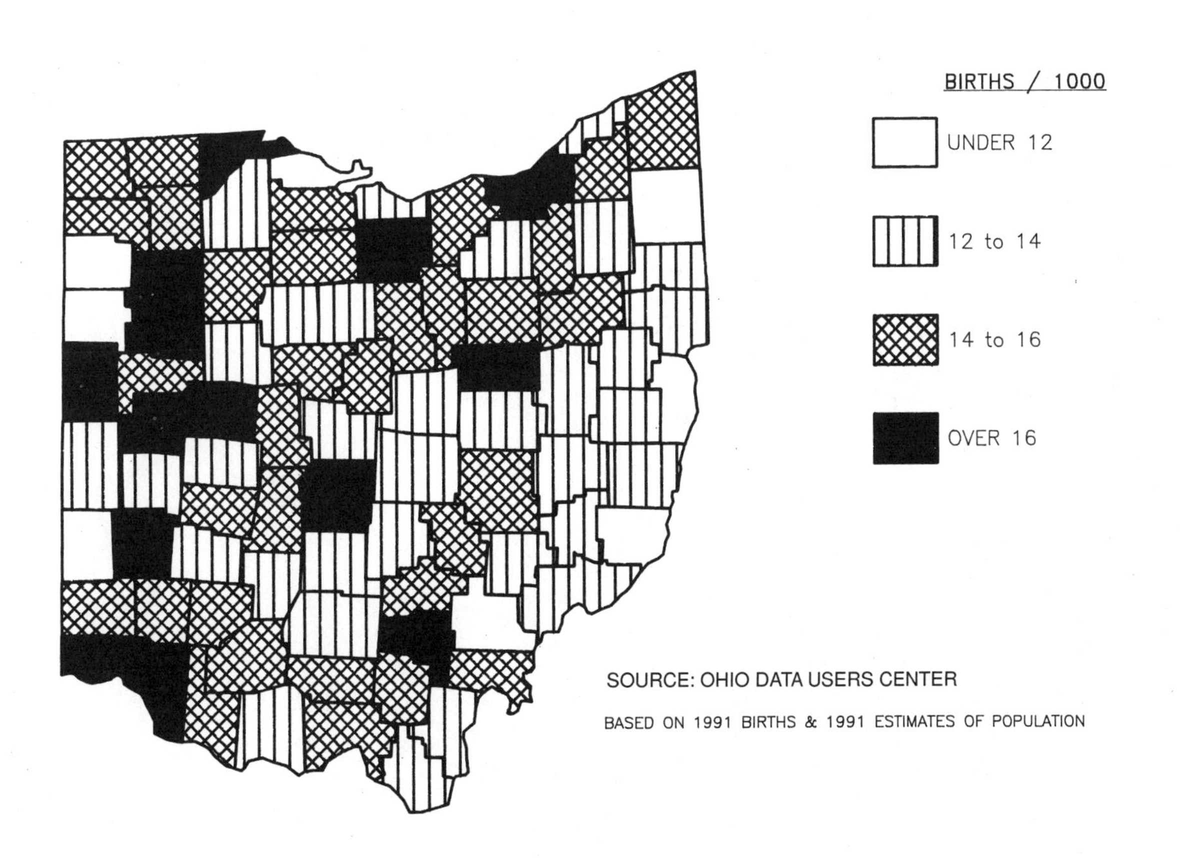

Fig. 8.8. Birth Rates, 1991

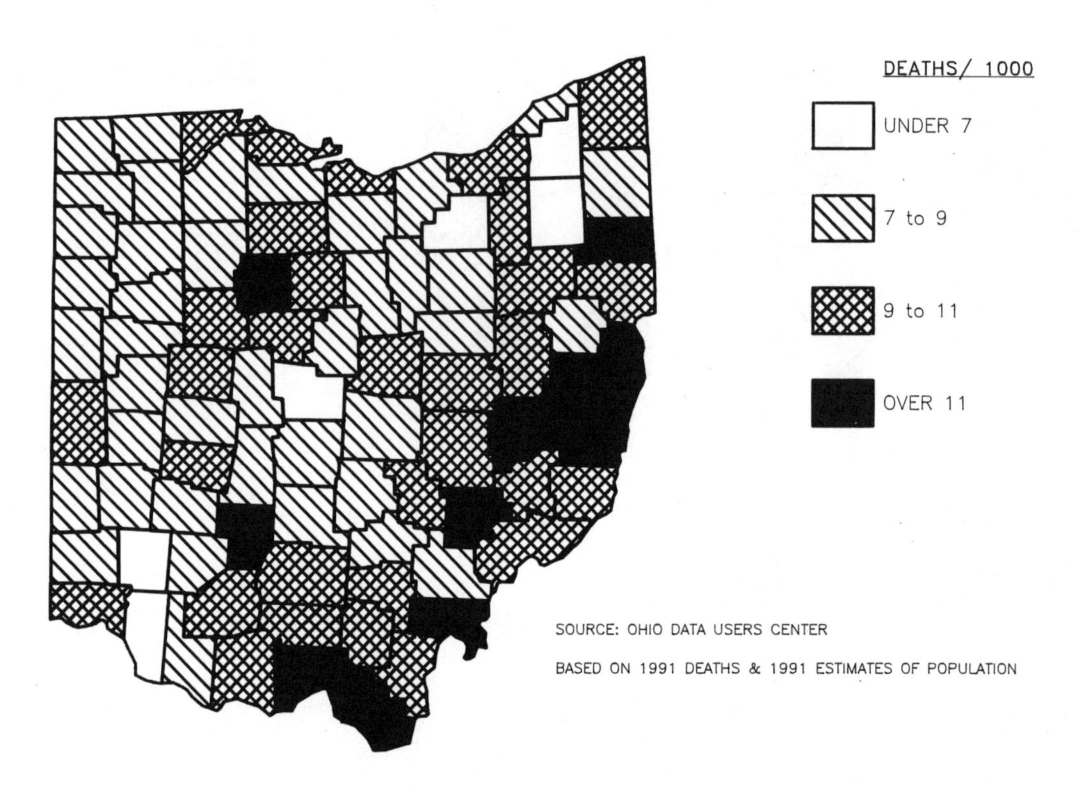

Fig. 8.9. Death Rates, 1991

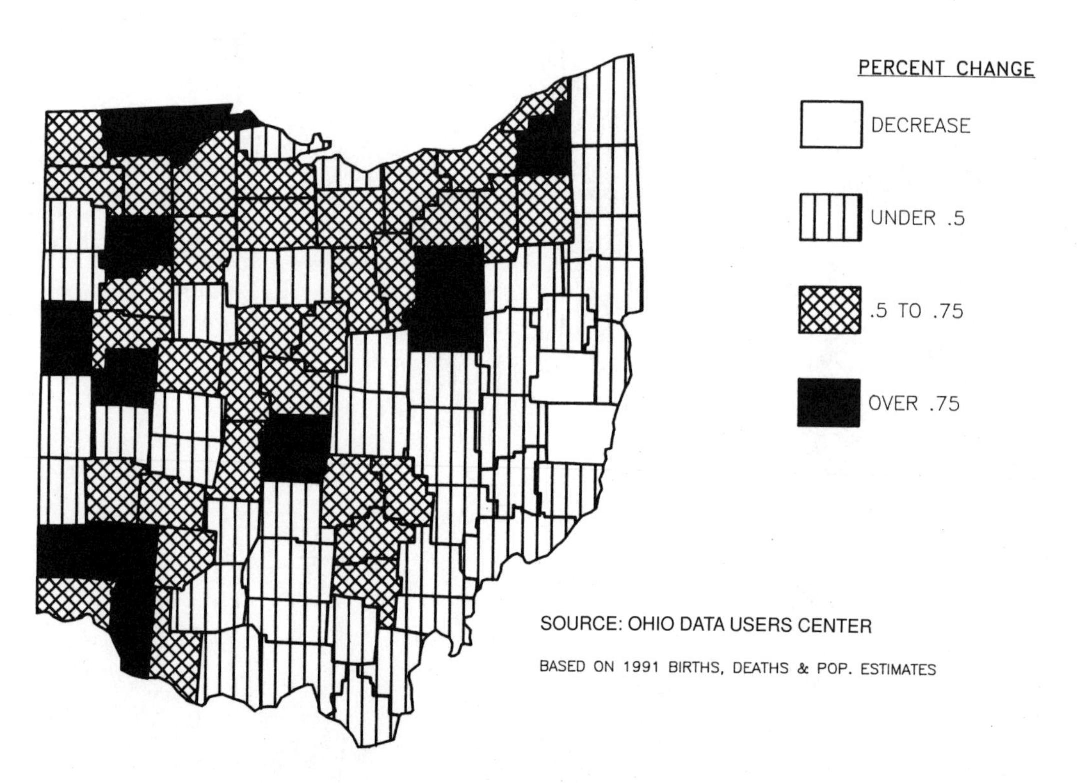

Fig. 8.10. Natural Change, 1991

boom. By 1990 the boom was over; the base of the pyramid had contracted, and the baby boomers had moved into the age 30–45 cohort. The pyramid for 1990 reflects little growth.

The pyramids also indicate a shift in the segments of the population below 15 and over 65. In 1960, 31.7 percent of the population was under 15. By 1990 that percentage had dropped to 21.5. Based on those figures, one could assume that Ohio's public education needs should have declined, while, given the rise in population of those over sixty-five, up from 9.3 to 13.2 percent in a thirty-year period, the costs of programs for older persons, such as Medicaid, should have increased. As the population has aged, school districts find it harder to pass bond issues to supplement state funding of education, and legislators are reluctant to cut programs for older citizens because of their power at the ballot box. This, coupled with declining tax revenues due to plant closings and the state's general economic malaise, plus the need for programs to meet social problems caused by the economic downturn, has meant declining resources for education. Ohio has never been known for largess in its support of public education, and there is little to suggest that this pattern will change in the future. It should be remembered that states which do not invest in their future may not have a future.

Ethnic Diversity

Ohio is a place of considerable ethnic diversity. Early immigrants were from western Europe, followed by east and south Europeans, African Americans, and most recently Hispanics and Asians. The presence of economic opportunity at the time of a particular group's arrival greatly influenced where they took up residence. Patterns established long ago persist today, creating a mosaic of ethnic diversity across the state. Figures 8.13–19 show the distribution of selected ancestries.

When asked by the U.S. Bureau of the Census in 1990 about their ethnicity, German was the response of most Ohioans. Many Germans arrived early in the state's settlement process and took up farming. Even today there is a very strong spatial association between good farmland and German ethnicity in Ohio. While there are a significant number of persons claiming German ancestry in every Ohio county, several areas house important concentrations, most notably Holmes County with its German-speaking Amish and the agricultural counties of the northwest. There are, however, two areas—the urbanized northeast and the extreme southeast—where few persons claiming German ancestry are found.

Other European groups have different patterns. The Irish, who came a bit later than the Germans, were not farmers. Instead they took an important role in the construction of Ohio's canal and railroad systems. The Irish are concentrated in the southern half of the state. Poles and Italians are representatives of groups that arrived in Ohio at the height of industrialization—the beginning of the twentieth century—drawn by jobs in the growing industrial cities of northern Ohio. A spatial dichotomy exists in the settlement pattern of east and south Europeans. This dichotomy is exemplified by Toledo, where Poles are a significant population group and Italians are present in much smaller numbers.

Ohio's population also springs from non-European roots. African Americans were present in Ohio at an early date and became a more significant element with the recruitment of Black workers from the South during World War I. As Figure 8.17 shows, there is a high degree of spatial association between African Americans and the urban axis, with this group making up more than 10 percent of the population in each of Ohio's seven largest cities. Here they tend to be found near city centers; outside urban areas African Americans are not common. In the northwest, excluding Lucas and Allen Counties, their number is few, and in Appalachian Ohio they comprise less than 5 percent of the population, with many counties in this region having populations that are less than 1 percent African American.

Like the rest of the nation, Ohio's Hispanic population has been rising, though with only 1.3 percent of the state's population, Hispanics remain a small minority group. They have, however, a very distinctive distribution, as Figure 8.18 shows. Historically, northwest Ohio's vegetable-producing regions were serviced by Spanish-speaking migrant workers, and some of those workers opted for permanent residence in the region. There is also an interesting spatial dichotomy in the state's Hispanic population. From Lorain County eastward the Hispanic population is dominantly Puerto Rican, whereas to the

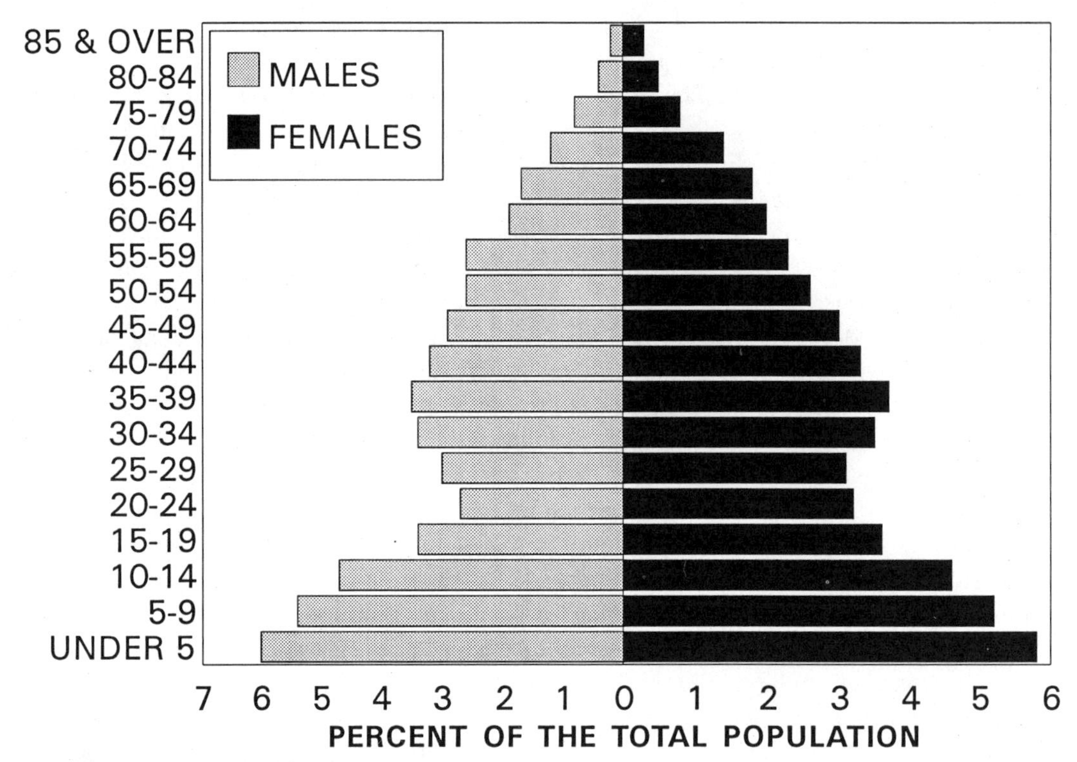

Fig. 8.11. Ohio Population, 1960

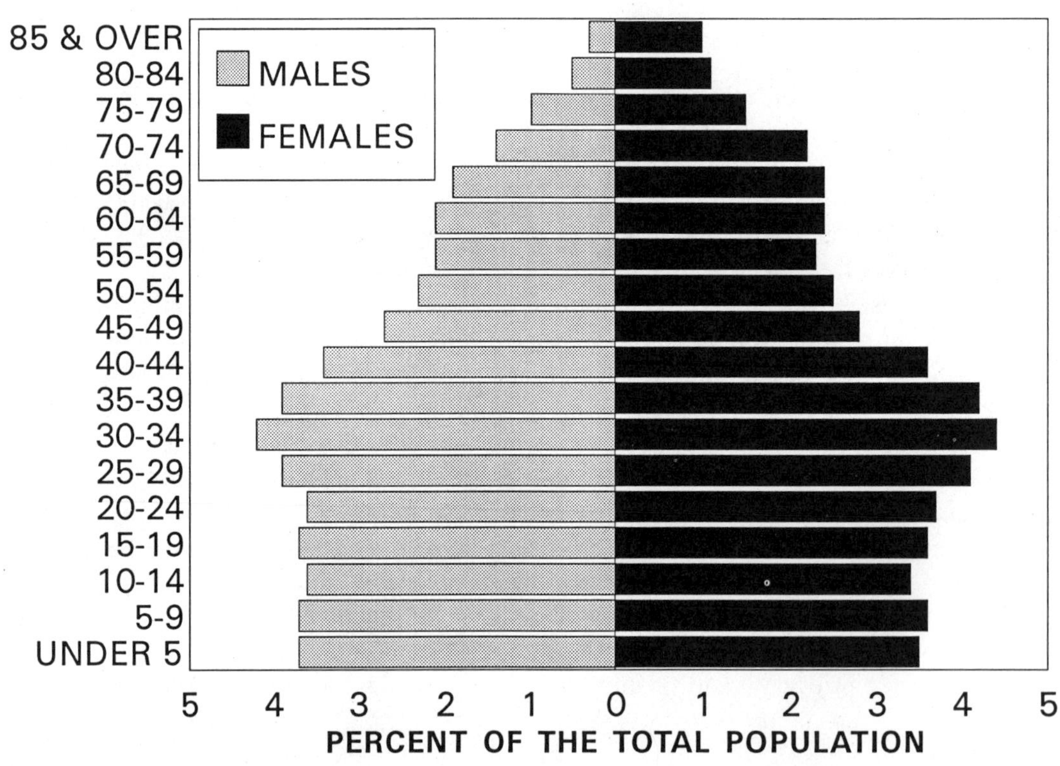

Fig. 8.12. Ohio Population, 1990

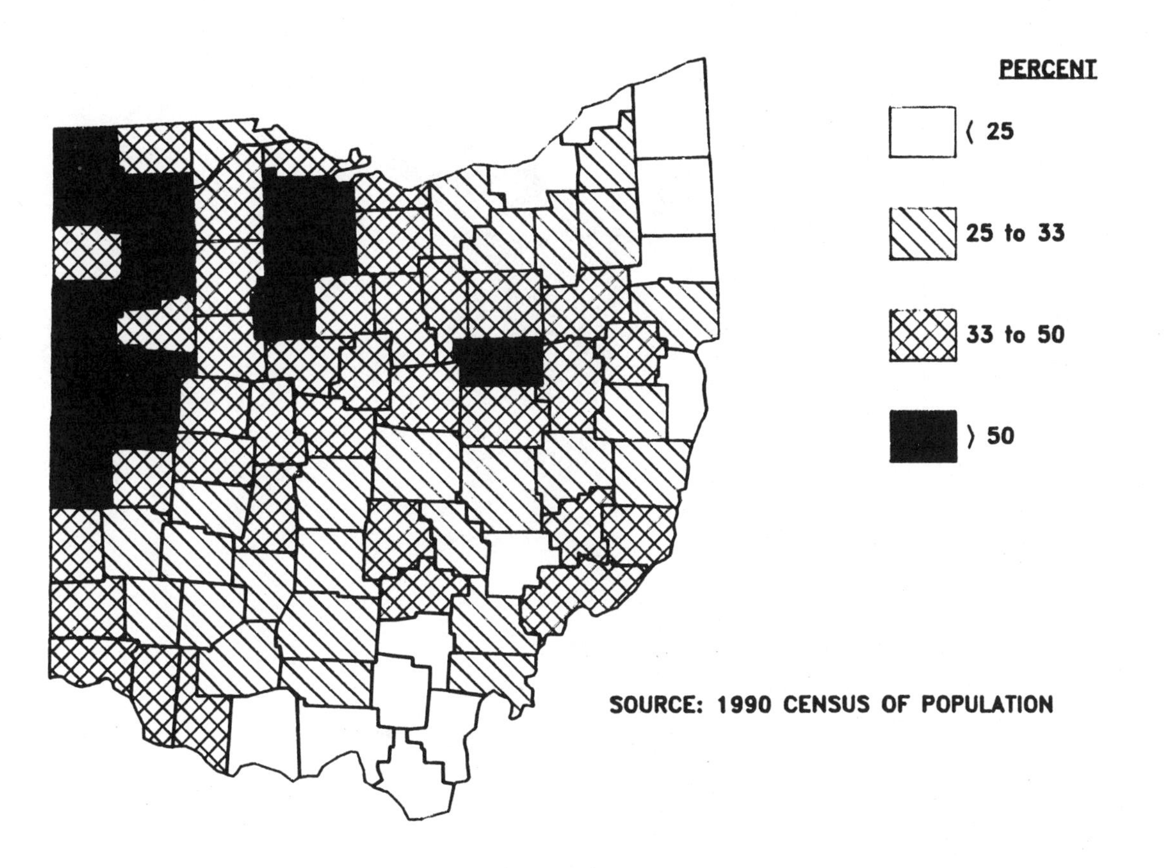

Fig. 8.13. Percent German, 1990

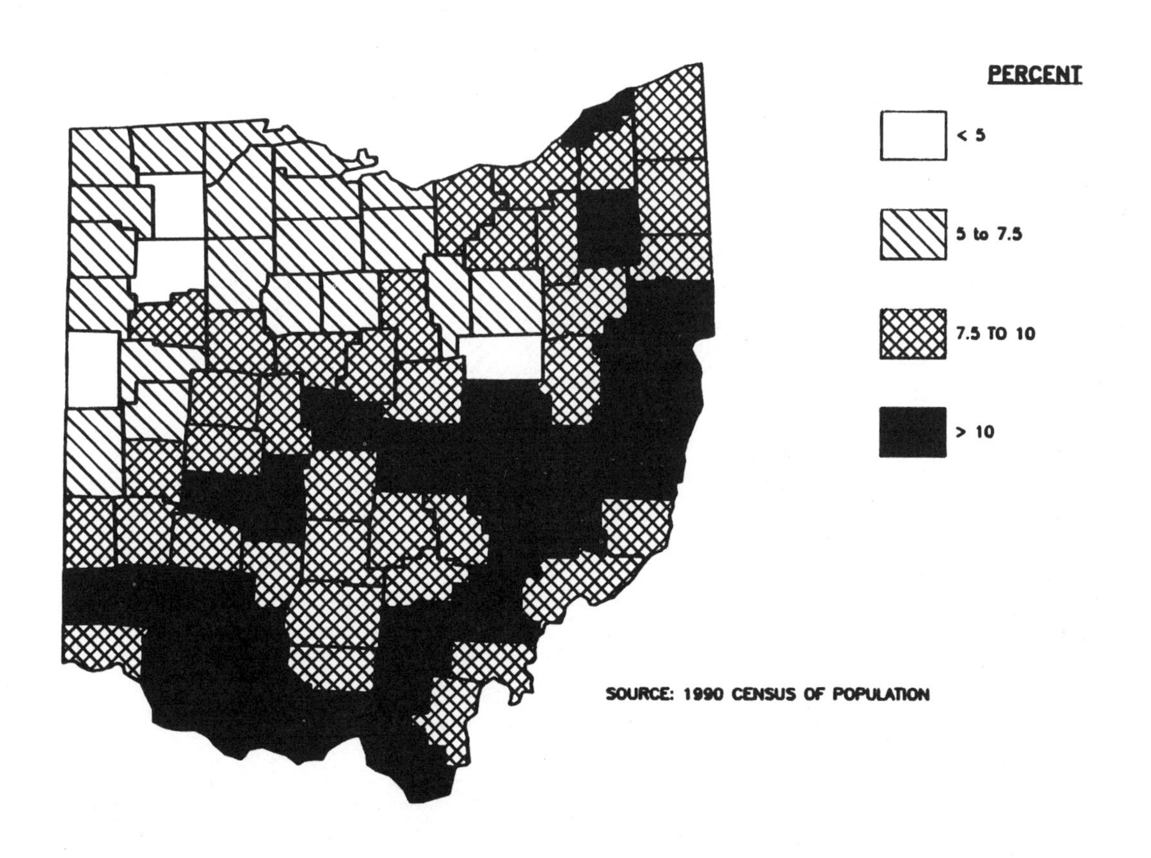

Fig. 8.14. Percent Irish, 1990

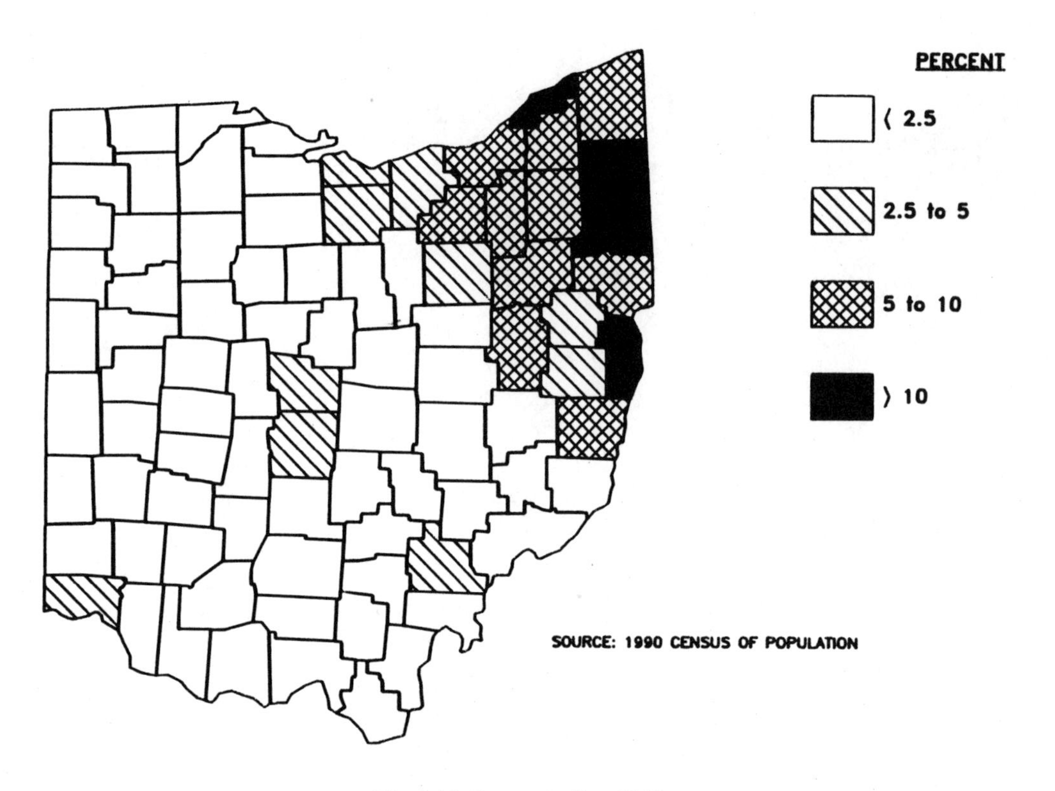

Fig. 8.15. Percent Italian, 1990

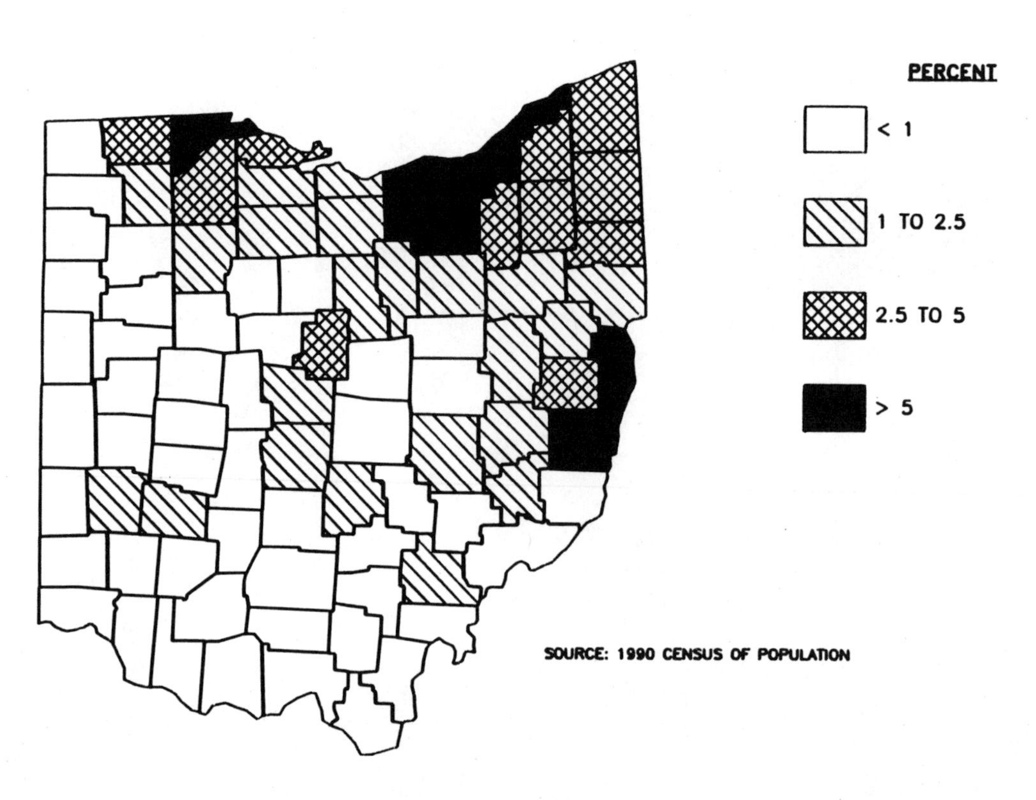

Fig. 8.16. Percent Polish, 1990

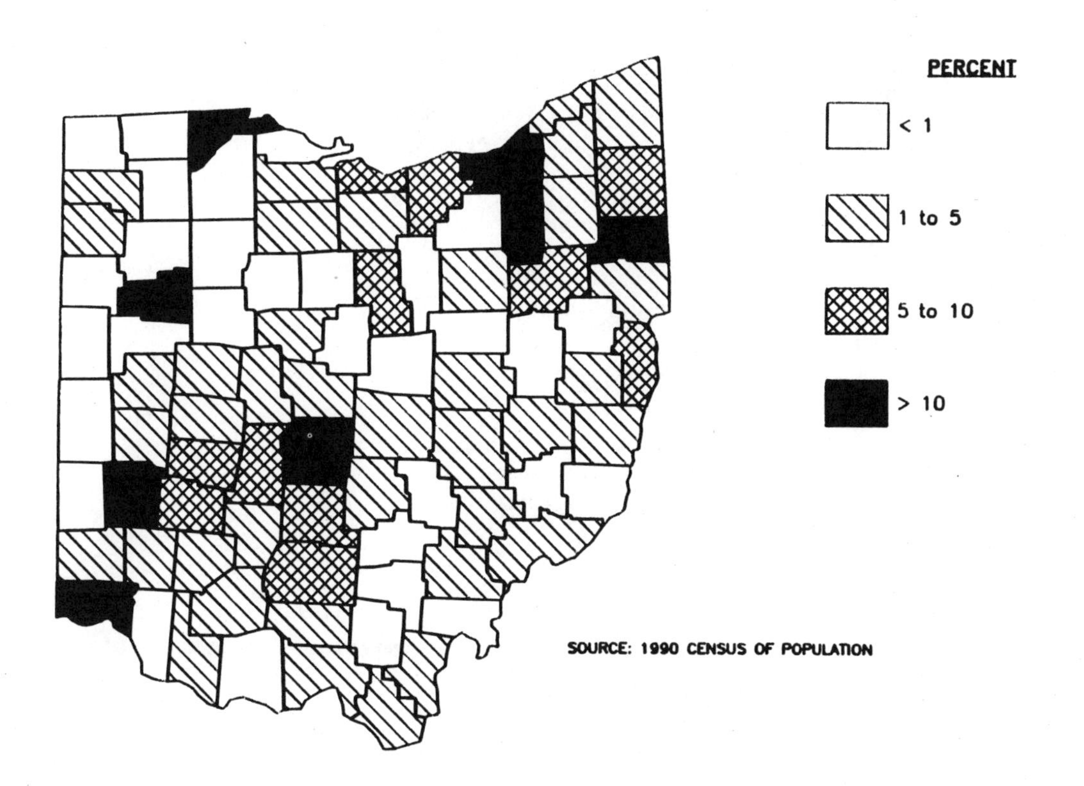

Fig. 8.17. Percent Black, 1990

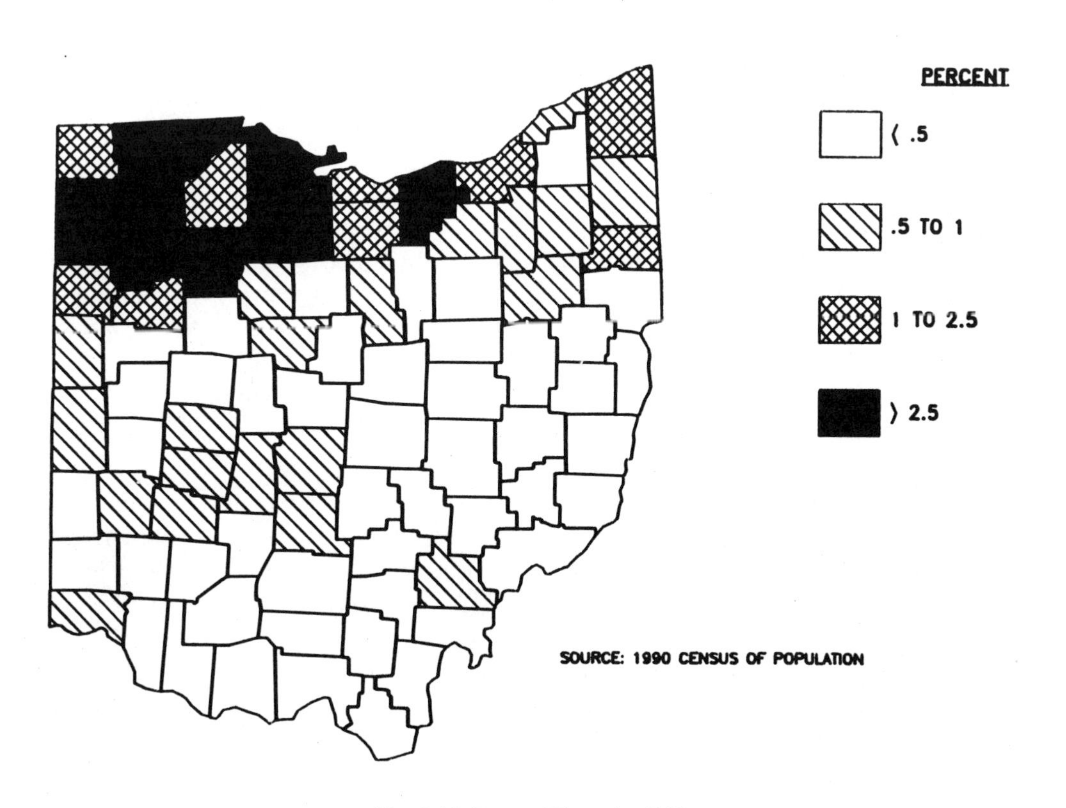

Fig. 8.18. Percent Hispanic, 1990

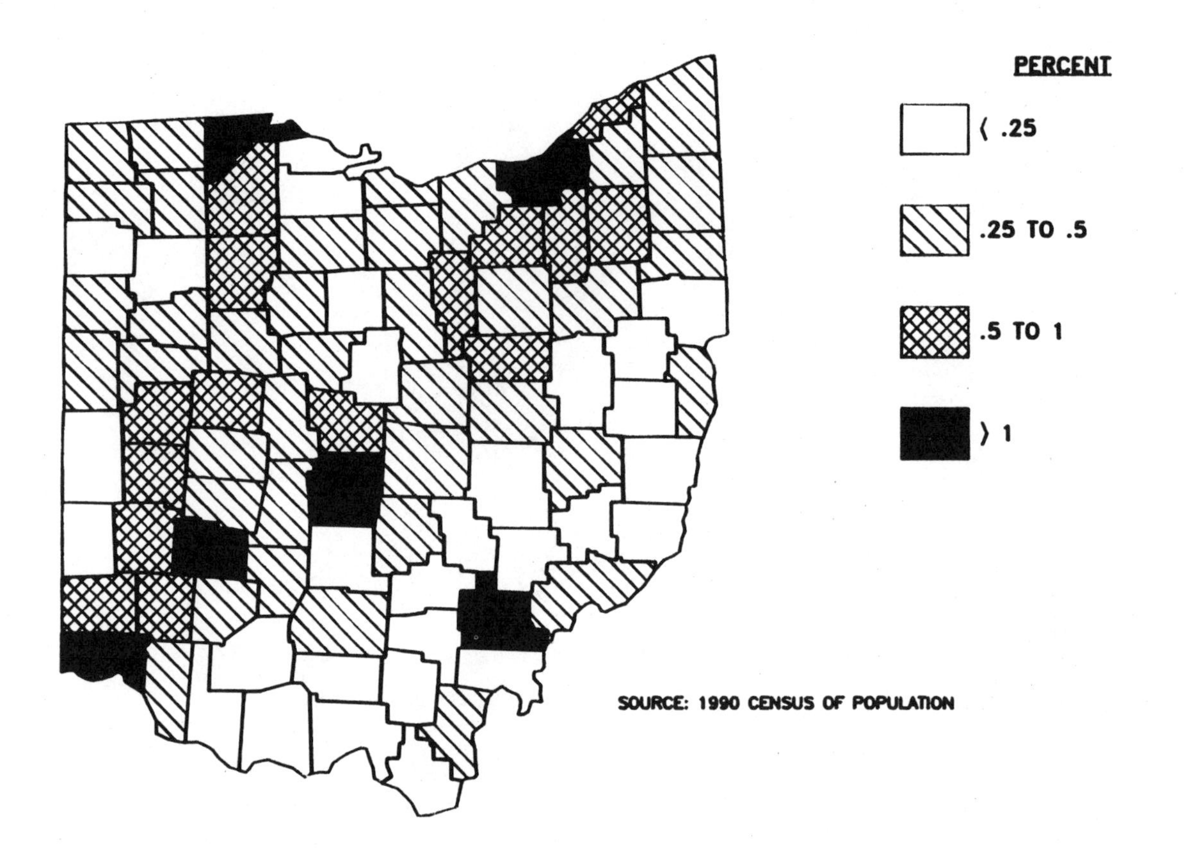

Fig. 8.19. Percent Asian, 1990

west the origin of most Hispanics is Mexico. Lorain is an especially interesting case. More than 5 percent of the county's population is Hispanic, and most of these—more than 70 percent—are Puerto Rican. This significant concentration is explained by the recruitment in 1947 of groups of Puerto Ricans to work at National Tube Company (Allen and Turner 1988).

Asians represent a recent immigrant group in Ohio. They are a very small proportion of Ohio's population, accounting for less than 1 percent, but their numbers are increasing. Like Hispanics they have an interesting distribution (shown in Figure 8.19), but unlike Hispanics, Asians are much more likely to settle in Ohio's major cities: Cleveland, Columbus, Cincinnati, Toledo, and Dayton. The attraction is economic opportunity, especially in the medical and scientific fields. Outside urban areas Asians are drawn to places where universities are present, such as Athens County in southeast Ohio. This county is also home to one of the state's major universities: Ohio University. Those areas with few Asians are rural, poor, and lack many opportunities in the fields of medicine and science.

Poverty

The economic maladies of Ohio have limited opportunities for many Ohioans. In 1990, 12.5 percent of the population was at the poverty level. As Figure 8.20 shows, the economic misery is not evenly distributed. The poorest part of Ohio is the southeast, where more than one-fifth of the population lives below the poverty threshold. This area has long been plagued by chronic poverty. It is a region of limited economic opportunity, marginal farms, and long dependence on coal. With the country more environmentally conscious, Ohio coal—with its high sulfur content—has not been able to compete effectively with low-sulfur coals from the West and other alternative fuels. The result has been the devastation of the economic base of many Appalachian counties. It is not surprising to

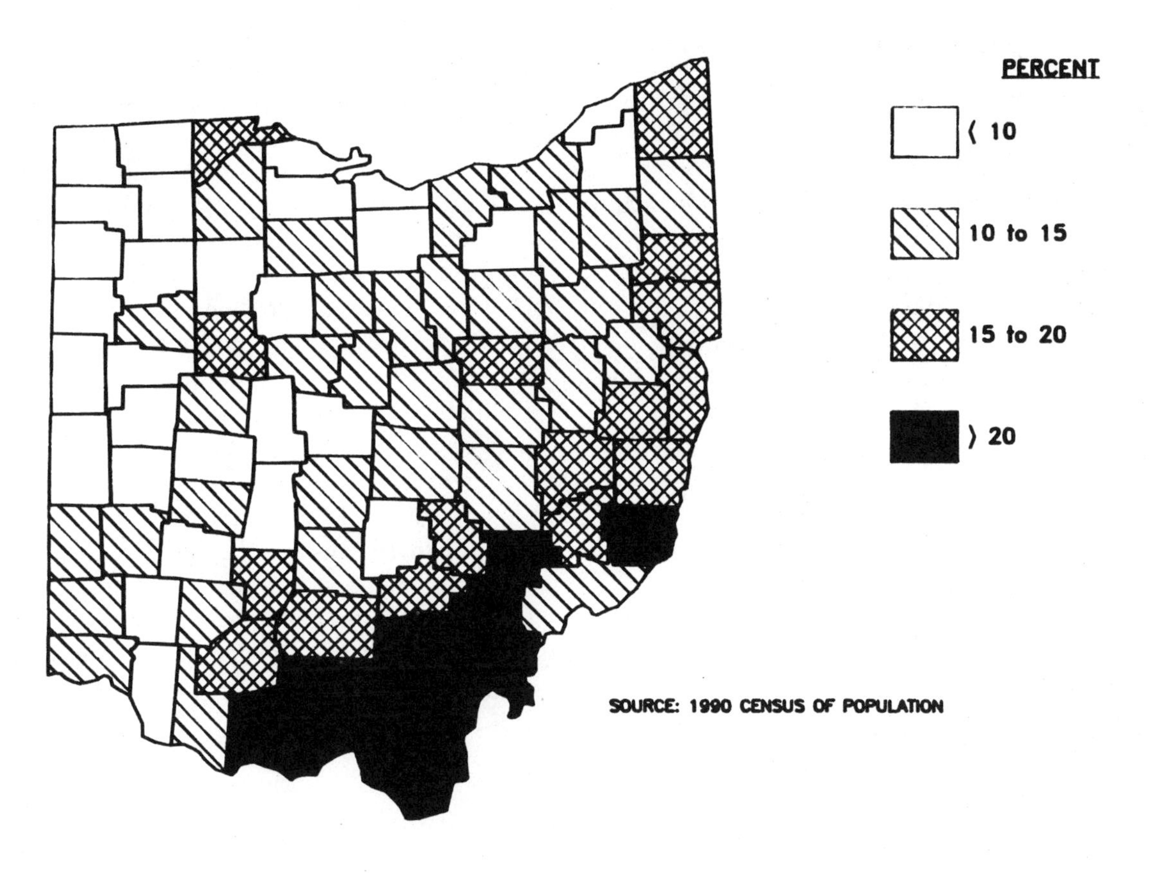

Fig. 8.20. Persons Below Poverty Level, 1990

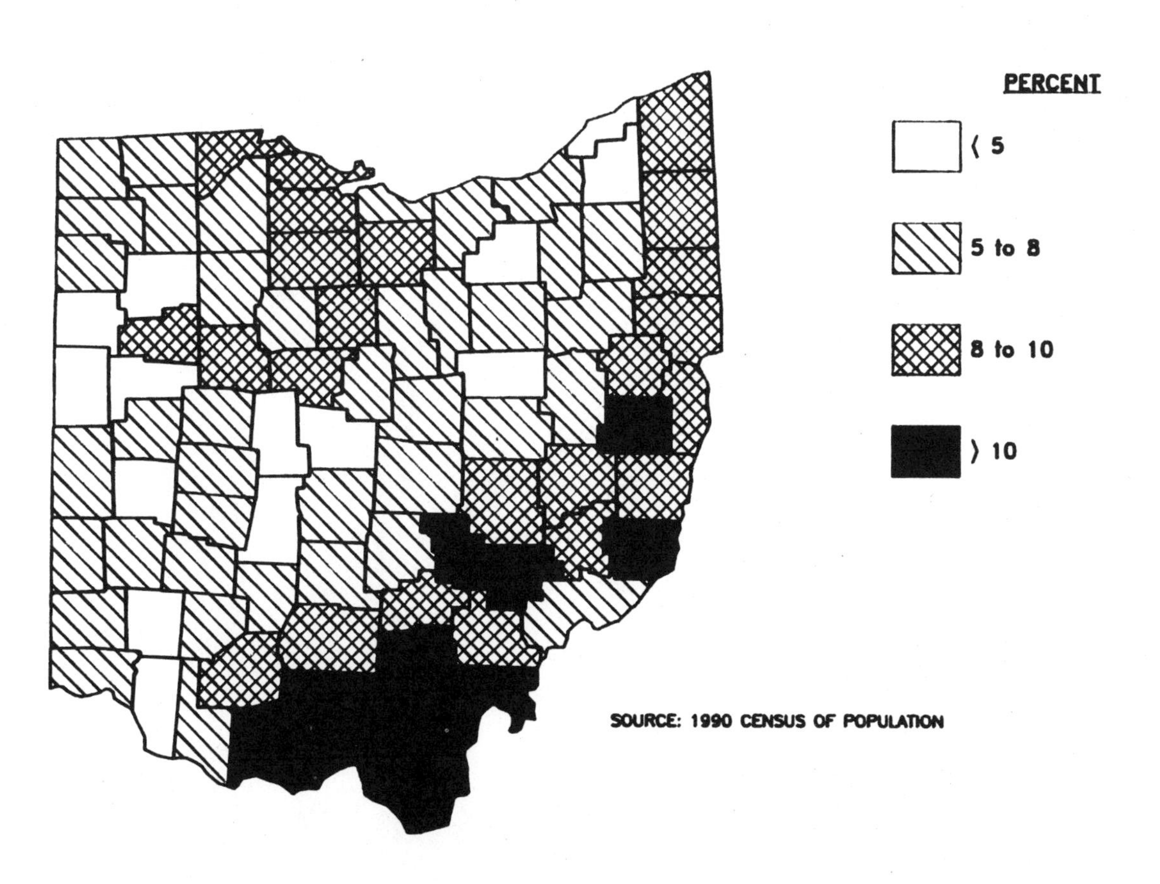

Fig. 8.21. Percent Unemployed, 1990

find that Figure 8.21, the map of unemployment, mimics the pattern of poverty. The two are unfortunately inextricably linked.

Numerically speaking, the greatest poverty in the state occurs in the urban counties. Cleveland and Cincinnati, because of their size, have more persons below the poverty level than any other area. The inner-city neighborhoods of Ohio's metropolitan areas are beset by a host of social and economic ills. These same regions paradoxically contain the wealthiest neighborhoods, and these wealthy neighborhoods with their high levels of affluence tend to offset the poverty of the central cities, creating a distorted picture. There are in reality two parts of Ohio where poverty and unemployment are critical problems: the southeast and the inner cities.

Conclusions

In the last thirty years Ohio has undergone tremendous economic upheaval, producing important demographic changes that are likely to continue. Barring some miraculous economic upswing, the state is likely to see its population decline as its best and brightest are attracted elsewhere, leaving behind a population that is growing older, poorer, and less well educated. These are not conditions that attract economic opportunities, nor are they likely to produce innovations. Given the erosion of the state's power at the national level, it is not practical to expect a bailout from Washington.

Not all of Ohio is doomed, but it is hard to be optimistic about the metropolitan inner cities and the rural southeast. Northern Ohio seems destined to suffer from continued deindustrialization, as heavy industries keep disappearing. The southwest, however, with its more diversified economic base, may be able to hold its own. And Columbus, with its hold on state government, will probably continue to outperform the rest of the state.

REFERENCES

Allen, James Paul, and Eugene James Turner. 1988. *We the People: An Atlas of America's Ethnic Diversity.* New York: Macmillian Publishing.

Northern Ohio Data and Information Service. 1991. *1990 Population of Ohio Counties and General Assembly Districts.* April. Cleveland: The Urban Center, Maxine Goodman Levin College of Urban Affairs, Cleveland State University.

Ohio Data Users Center. 1993a. *Ohio Metropolitan Areas and County Populations: Census Counts and Intercensal Estimates.* January. Columbus: Ohio Department of Development.

Ohio Data Users Center. 1993b. *Population Projections: Ohio and Counties By Age and Sex: 1990 to 2015.* January. Columbus: Ohio Department of Development.

Ohio Department of Industrial and Economic Development. 1960. *Statistical Abstract of Ohio, 1960.* Columbus.

U.S. Department of Commerce, Bureau of the Census. 1990. *1990 Census of Population, Summary File Tapes, 1A and 3A.* Washington, D.C.

Wilhelm, Hubert G. H. 1990. Cultural Settlement Patterns. In *The Changing Heartland: A Geography of Ohio.* Ed. Leonard Peacefull. Needham Heights, Mass.: Ginn Press.

CHAPTER NINE

Cities in Ohio

Allen G. Noble and Leonard Peacefull

Ohio, along with most of modern America, has a greater proportion of its residents in urban areas. With an urban population of 8,039,409, or 74 percent of the state's total population, Ohio is on a par with most other heavily populated states in the nation, and like other states, Ohio has seen its larger, older industrial centers depopulated while its medium-sized towns have grown. This chapter will focus on the larger cities; chapter 10 will consider smaller towns.

In the previous chapter, Ohio's urban axis—which forms a line from Cleveland in the northeast to Cincinnati in the southwest—was identified as the focus of population distribution. This concentration of people shows up clearly as on a county level in Figure 8.1. Although this map does not show the distribution of the larger cities, it is easy to postulate where they will be found. The populations of the fifty largest cities, all over 25,000 inhabitants, are shown in Figure 9.1 and listed in Table 9.1, in descending order. The axis, essentially a broad band from the Lake Erie shore to the Ohio River, includes the satellites around Cleveland, Akron-Canton, Columbus and its environs. Springfield-Dayton-Kettering, and Hamilton and Cincinnati. There are several notable outliers to this axis, namely Toledo and Youngstown-Warren. This axis reflects the industrial heartland of Ohio, where both traditional and more recent high-tech industries are located.

The Development of the Ohio City

It is difficult to say with absolute certainty just why Ohio cities have originated where they have. There is often a measure of chance or human whim in urban location that is overlooked, simply because this factor is so difficult to assess (Johnson 1972). In discussing the rationale for the location of urban centers, two terms are often used: site and situation. *Site* refers to the immediate physical conditions which influence the positioning and settlement of a town, addressing itself to the reasons for the community's establishment. *Situation* is a broader concept, referring to the relationship of the urban site to nonlocal geographic, economic, and other considerations and addressing itself to the reasons for growth or nongrowth in urban areas.

In many instances the platting of a town site with no special locational advantages was the result of the purchase of newly surveyed frontier lands by energetic speculators eager to turn a profit. Among sites that became cities one finds Canton (Stark County), Mansfield (Richland County), Marion (Marion County), Xenia (Greene County), Ashland (Ashland County), Salem (Columbiana County), and Wadsworth (Medina County). The site of Bellefontaine (Logan County) is somewhat more fortuitous, located as it is near abundant clear spring water so significant that it gave the settlement its name.

Certain Ohio cities recapitulate the pattern of earlier Indian settlements. Among the best-known examples are Akron (Summit County), Wooster (Wayne County), Defiance (Defiance County), Conneaut (Ashtabula County), and Van Wert (Van Wert County). Each of these cities is situated along a well-used former Indian trail, with Akron lying along the famous portage between the Cuyahoga and the Tuscarawas Rivers and the others at points where the trails cross streams.

Accessibility always had a great deal to do with city origin. Perhaps the clearest expression of this factor is centrality. Columbus, which was founded as

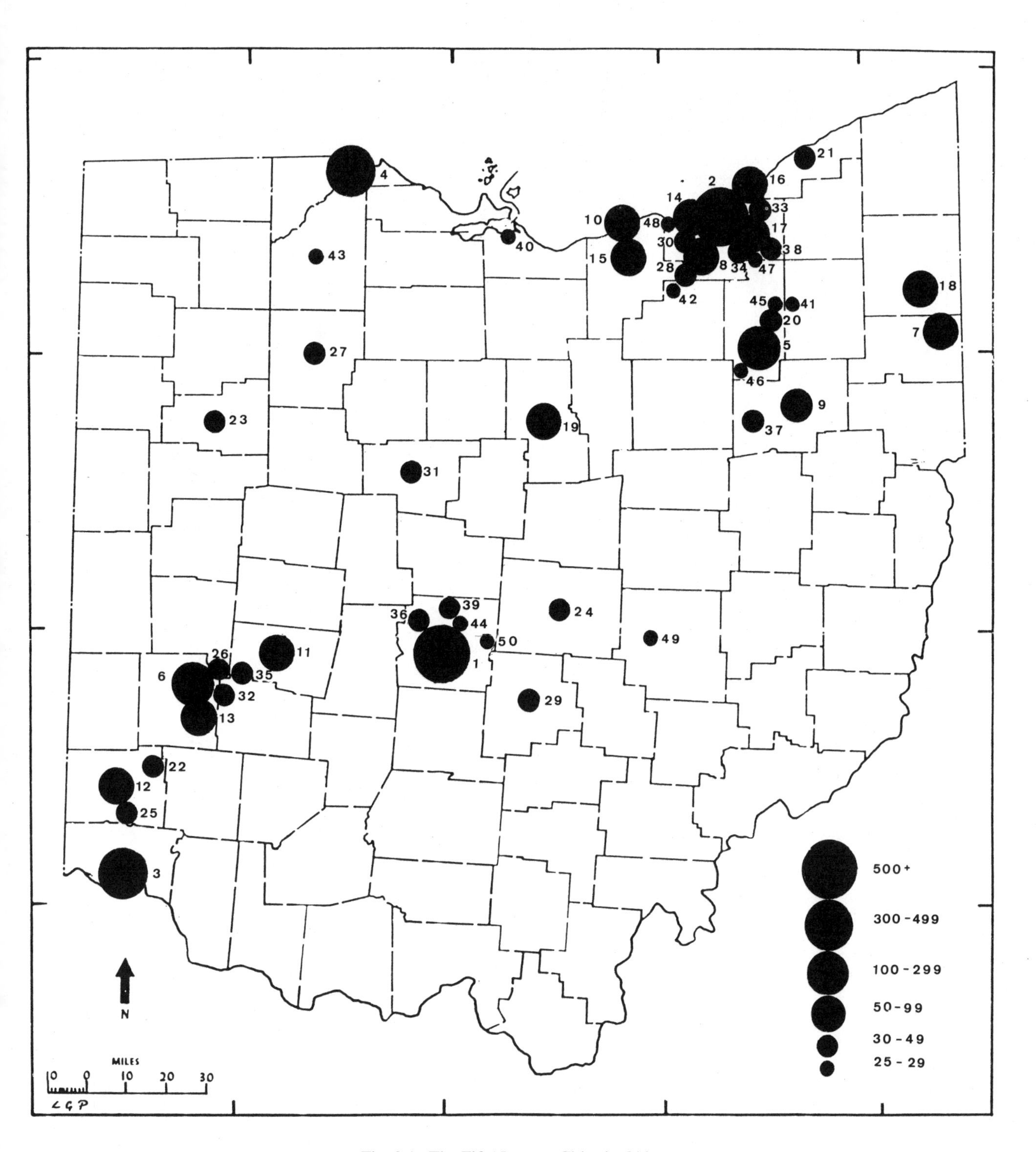

Fig. 9.1. The Fifty Largest Cities in Ohio

Franklinton in 1797, owes its location to a crossing of the Scioto River. Its continued importance and growth is the result of its selection as state capital, a selection that was made in large measure because of its central position in the state. Similar criteria were used in several cities organized and platted as county seats, including Lima (Allen County), Troy (Miami County), Sidney (Shelby County), Washington Court House (Fayette County), Ravenna (Portage County), Urbana (Champaign County), Medina

Table 9.1

The Fifty Largest Cities in Ohio in 1990

	AREA				POPULATION				
CITY	1980	1990	CHANGE	% CHANGE	1980	1990	80–90	% CHANGE	DENSITY PER SQ. MILE
1 Columbus	180.9	191.0	10.1	5.6	565,032	632,910	67,878	12.01	3313.66
2 Cleveland	79.0	77.0	–2.0	–2.5	573,822	505,616	–68,206	–11.89	6566.44
3 Cincinnati	78.1	77.2	–0.9	–1.2	385,409	364,040	–21,369	–5.54	4715.54
4 Toledo	84.2	80.6	–3.6	–4.3	354,635	332,943	–21,692	–6.12	4130.81
5 Akron	57.5	62.2	4.7	8.2	237,177	223,019	–14,158	–5.97	3585.51
6 Dayton	48.4	55.0	6.6	13.6	193,536	182,044	–11,492	–5.94	3309.89
7 Youngstown	34.5	33.8	–0.7	–2.0	115,511	95,735	–19,776	–17.12	2832.40
8 Parma	19.7	20.0	0.3	1.5	92,548	87,876	–4,672	–5.05	4393.80
9 Canton	19.5	20.2	0.7	3.6	93,077	84,161	–8,916	–9.58	4166.39
10 Lorain	23.8	24.1	0.3	1.3	75,416	71,245	–4,171	–5.53	2956.22
11 Springfield	18.1	19.5	1.4	7.7	72,563	70,487	–2,076	–2.86	3614.72
12 Hamilton	19.4	20.0	0.6	3.1	63,189	61,368	–1,821	–2.88	3068.40
13 Kettering	18.4	18.7	0.3	1.6	61,168	60,569	–599	–0.98	3238.98
14 Lakewood	5.6	5.5	–0.1	–1.8	61,963	59,798	–2,165	–3.49	10872.36
15 Elyria	19.4	19.4	0.0	0.0	57,538	56,746	–792	–1.38	2925.05
16 Euclid	10.3	10.7	0.4	3.9	59,999	54,875	–5,124	–8.54	5128.50
17 Cleveland Heights	8.1	8.1	0.0	0.0	56,438	54,052	–2,386	–4.23	6673.09
18 Warren	14.8	16.0	1.2	8.1	56,629	50,793	–5,836	–10.31	3174.56
19 Mansfield	25.5	27.9	2.4	9.4	53,927	50,627	–3,300	–6.12	1814.59
20 Cuyahoga Falls	8.9	25.5	16.6	186.5	43,890	48,950	5,060	11.53	1919.61
21 Mentor	28.3	26.8	–1.5	–5.3	42,065	47,358	5,293	12.58	1767.09
22 Middletown	20.0	20.2	0.2	1.0	43,719	46,022	2,303	5.27	2278.32
23 Lima	12.3	12.7	0.4	3.3	47,827	45,549	–2,278	–4.76	3586.54
24 Newark	16.8	18.0	1.2	7.1	41,200	44,839	3,639	8.83	2491.06
25 Fairfield	20.2	20.6	0.4	2.0	30,777	39,729	8,952	29.09	1928.59
26 Huber Heights	20.9	20.8	–0.1	–0.5	35,480	38,696	3,216	9.06	1860.38
27 Finlay	13.8	13.6	–0.2	–1.4	35,594	35,703	109	0.31	2625.22
28 Strongsville	25.1	24.6	–0.5	–2.0	28,577	35,308	6,731	23.55	1435.28
29 Lancaster	15.1	15.7	0.6	4.0	34,953	34,507	–446	–1.28	2197.90
30 North Olmstead	11.5	11.5	0.0	0.0	36,486	34,204	–2,282	–6.25	2974.26
31 Marion	8.0	8.0	0.0	0.0	37,040	34,075	–2,965	–8.00	4259.38
32 Beavercreek	25.7	25.7	0.0	0.0	31,589	33,626	2,037	6.45	1308.40
33 East Cleveland	3.2	3.1	–0.1	–3.1	36,957	33,096	–3,861	–10.45	10,676.13
34 Garfield Heights	7.6	7.2	–0.4	–5.3	34,938	31,739	–3,199	–9.16	4408.19

35	Fairborn	10.2	11.2	1.0	9.8	29,702	31,300	1,598	5.38	2794.64
36	Upper Arlington	9.6	9.6	0.0	0.0	35,648	31,128	–4,520	–12.68	3242.50
37	Massillon	9.6	13.2	3.6	37.5	30,557	31,007	450	1.47	2349.02
38	Shaker Heights	6.2	6.3	0.1	1.6	32,487	30,831	–1,656	–5.10	4893.81
39	Westerville	7.2	10.5	3.3	45.8	22,893	30,269	7,376	32.22	2882.76
40	Sandusky	9.1	10.0	0.9	9.9	31,360	29,764	–1,596	–5.09	2976.40
41	Kent	7.4	8.7	1.3	17.6	26,164	28,835	2,671	10.21	3314.37
42	Brunswick	12.2	11.5	–0.7	–5.7	28,104	28,238	134	0.48	2455.48
43	Bowling Green	7.5	7.9	0.4	5.3	25,728	28,176	2,448	9.51	3566.58
44	Gahanna	10.1	13.2	3.1	30.7	17,391	27,791	10,400	59.80	2105.38
45	Stow	18.0	17.1	–0.9	–5.0	25,303	27,702	2,399	9.48	1620.00
46	Barberton	7.5	7.6	0.1	1.3	29,751	27,623	–2,128	–7.15	3634.61
47	Maple Heights	5.1	5.2	0.1	2.0	29,735	27,089	–2,646	–8.90	5209.42
48	Westlake	16.5	15.9	–0.6	–3.6	18,783	27,018	8,235	43.84	1699.25
49	Zanesville	9.7	10.4	0.7	7.2	28,655	26,778	–1,877	–6.55	2574.81
50	Reynoldsburg	7.6	9.3	1.7	22.4	20,428	25,748	5,320	26.04	2768.60

Source: Bureau of the Census 1988 & 1992

(Medina County), and Wilmington (Clinton County). The site of each of these cities has little to commend it except centrality of location.

In the pioneer period, access was normally via water, and thus a very large number of Ohio cities have grown up along the state's waterways. Most settlements occurred only at certain types of geographic situations. Along the Ohio River, and to a much lesser extent smaller rivers such as the Miami, many settlements developed on river terraces out of reach of all but the highest floods, and placement on the gradually sloping insides of river bends gave easy accessibility and thus facilitated growth. Cincinnati (Hamilton County), Steubenville (Jefferson County), East Liverpool (Columbiana County), and Portsmouth (Scioto County) all fall within this category. Although Cincinnati is somewhat limited in terms of land suitable for easy development, its location at the nexus of three major Ohio River tributaries gave it an early strategic significance, which made it Ohio's largest city from 1810 to 1900. The fertile lands of the Miami Valley contributed to keep Cincinnati secure in its position as "Queen City of the West."

Found along the courses of many streams are rapids and falls that blocked early navigation but also offered the possibility of waterpower. The interruption to navigation forced handling of passengers and/or cargo and promoted establishment of facilities to handle this activity. This dual impetus resulted in the siting of countless settlements, some of which have grown into thriving cities. Elyria (Lorain County), which now is included within the Lorain urbanized area, originated at the falls of the Black River. Kent (Portage County), Piqua (Miami County), Fremont (Sandusky County), and Zanesville (Muskingum County) are other cities platted adjacent to falls or rapids.

Zanesville also illustrates another typical urban site: the junction of two or more streams. River junctions are significant because of the increased accessibility at such locations, for each valley offers a route. As a confluence site par excellence, Dayton boasts a junction of no fewer than four streams: the Great Miami, Mad, and Stillwater Rivers, and Wolf Creek. Other Ohio cities with confluence locations are Columbus, Defiance, Coshocton (Coshocton County), and Newark (Licking County).

The mouth of a river or estuary is still another typical settlement site; Lorain, Toledo (Lucas County),

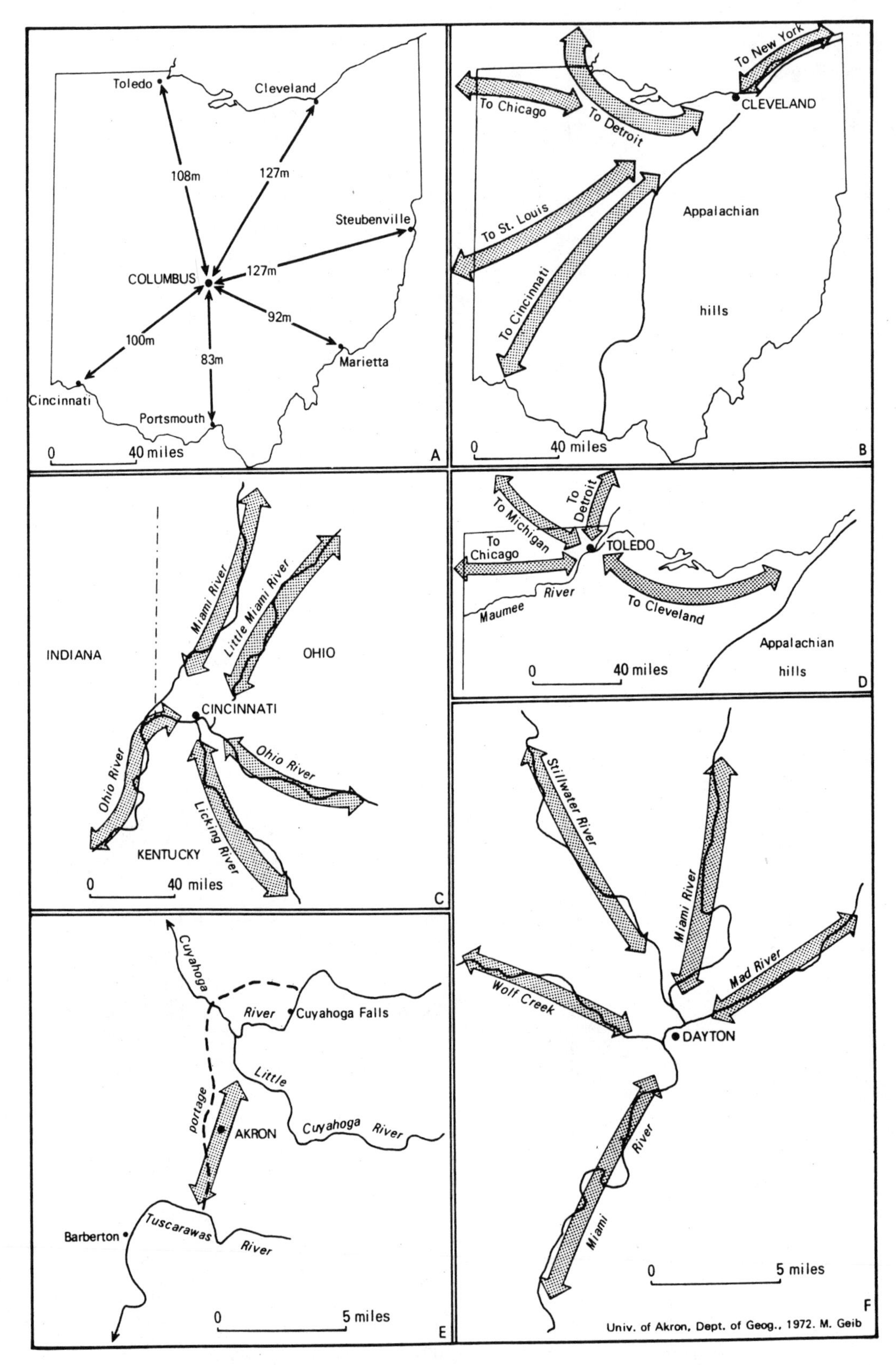

Fig. 9.2. Geographic Situation of Six Major Ohio Cities. A, Columbus (showing distances from selected boundary cities); B, Cleveland; C, Cincinnati; D, Toledo; E, Akron; and F, Dayton. (After Noble and Korsok 1975)

and Cleveland are examples of sites established along Lake Erie. The great size of Toledo and Cleveland today is related to geographic situation rather than site characteristics (Fig. 9.2). Not only does Toledo function as a port at the mouth of the Maumee River, but in its position at the western extremity of Lake Erie, it straddles the major transportation routes to the entire Upper Midwest. The latitudinal position and the elongated shape of the lake, together with the nearby United States–Canada political barrier, which restricts traffic north of the lake, serve to focus many of these transportation routes on Toledo. Essentially similar is the situation of Cleveland, which functioned in the last half of the nineteenth and the first half of the twentieth centuries as a focal point for rail and highway routes from the entire Midwest. The great barriers of Lake Erie on the north and the Appalachian hills on the southeast come together at Cleveland. Looking eastward, the major transportation arteries of the northeastern United States are trapped in the narrow corridor south of Lakes Erie and Ontario and in the Mohawk and Hudson River Valleys. These corridors have long served as the major access routes between the magnificent harbor of New York and the Midwest, with Cleveland as the great collecting point.

Ashtabula and Conneaut, two cities which now function as lake ports, did not originate on Lake Erie, although both have expanded until today they partially front on the lake. In each instance the original settlement was established some two or three miles inland at a point where a glacial-lake beach ridge, used as an Indian and pioneer trail, crosses the river. U.S. 20 follows this trail today. The elongated beach ridges of Ohio were sandy and dry and occurred several feet above the adjacent swampy forest. Serving first as Indian trails, these ridges, because of their natural advantages, commonly became the major stage and highway routes. Especially important as town sites were places such as Bowling Green (Wood County) and Norwalk (Huron County), where ridges join, and Defiance, where ridges approach streams. At both Norwalk and Defiance, important waterway routes crossed major paths of land travel. In addition to Ashtabula and Conneaut, cities which have grown up where beach ridges meet stream crossings include Painesville (Lake County), Tiffin (Seneca County), Fostoria (Hancock and Seneca Counties), and Van Wert. The site of Findlay (Hancock County) was especially favorable not only because beach ridge and river intersect but also because two separate ridges meet there.

Streams can be crossed at many other places besides ancient beach ridges, and such crossing sites have provided another representative settlement location. The significance of the site is normally a function of both the importance of the land route and the size of the stream. Two routes—Zane's Trace and the National Road—produced several early cities at the sites of fords, ferries, and bridges: Cambridge (Guernsey County) at the Wills Creek crossing, Zanesville at the Muskingum River, Lancaster (Fairfield County) at the Hocking River, Columbus at the Scioto River, and Springfield (Clark County) at the Mad River. Although Chillicothe (Ross County) lies adjacent to Zane's Trace at its crossing of the Scioto, the settlement of the city actually preceded the establishment of the trail. The significance of the Chillicothe site lies in the relationship between topography and drainage. The city lies on the river at the boundary between the Appalachian Plateau and the Central Lowlands Province, between the rough hill country to the southeast and the smooth plains to the northwest.

The city of Akron deviates somewhat from the major pattern of stream crossing sites, for Akron is located between streams (see the following vignette). In this instance the elements are reversed. The movement has been along the streams, with the portage the barrier to be crossed. Cuyahoga Falls (Summit County) grew at one end of the portage, assisted by its potential for waterpower, while Akron developed along the portage.

A handful of Ohio cities have site locations that do not fit conveniently into the principal categories. Both Athens (Athens County) and Oxford (Butler County) owe their locations to considerations other than site, for both cities were founded by commissioners more concerned with survey boundaries and land revenues than with geographic features. It is true that Oxford sits on a commanding rise and that Athens had early connections by the easy water route of the Hocking River, but these conditions were secondary factors.

Another pair of cities—Massillon (Stark County) and Dover (Tuscarawas County)—are the only canal-period relics that have continued to prosper. Both communities originated along the Ohio and Erie Canal. Although Dover was actually laid out in 1807, it was an insignificant unincorporated village until the canal reached it in 1830. Massillon was laid out shortly after the canal was built. All other Ohio cities along the routes of both the Ohio and Erie and the Miami and Erie Canals developed much earlier than either waterway.

The city of Sandusky (Erie County) is unique in its placement as the only Ohio lake port not at a stream mouth. Secure behind the barrier of Cedar Point and adjacent to the protected anchorage of Sandusky Bay, the city of Sandusky has been an important fishing and shipping port since 1817. Bypassed by the Miami and Erie and the Ohio and Erie Canals, it never reached the significance of either Toledo or Cleveland, which became the terminal cities of the two canals.

The Centrality of Ohio Cities

Another important consideration bearing on the growth and distribution of cities is the relationship these cities have with the surrounding countryside and with each other. Just as transportation access is the key to understanding growth in five of Ohio's six largest urban areas, central location is the major factor explaining the growth of the remaining large city, Columbus. Centrality was not only the raison d'être for the location of many county seats but also accounted for much of their growth. In a large number of instances, the county seat functions as both the administrative capital and as the market, service, and shopping center for the county.

Because the size of Ohio's counties is relatively uniform, it is hardly surprising to find that many county seats are roughly the same size. While a number of these centrally located county seats function as small city trade centers, others, by virtue of manufacturing or special service activities, have reached a larger size.

For all cities, whether county seats or not, their relationship with the surrounding countryside is of utmost importance. A variety of terms have been employed to describe this hinterland: umland, catchment area, tributary region, urban field, or zone of influence. As such terms imply, the city exerts some control over the surrounding area.

The identification of the spheres of influence of Ohio cities has received scant attention. Studies by Dutt and Harrity (1970, 1971), using newspaper circulation as the index, noted four distinct levels of cities grouped on the basis of size. Northwest and southeast Ohio fall outside the influence of the three so-called first-order cities: Cleveland, Cincinnati, and Columbus. To include the influence of Columbus as far south and east as the Ohio River Valley, as was done in these studies, is questionable and is doubtless due to the lack of consideration of any large city to the southeast. If, on the other hand, cities outside Ohio—such as Charleston, West Virginia, or the Ashland, Kentucky–Huntington, West Virginia, urban area—were considered, then it is doubtful if the area of Columbus's influence would extend to many of the Appalachian counties of Ohio.

Of the second-order cities, the influence area of Akron is smallest and that of Toledo largest, a function of the presence or absence of competing centers. The only serious nearby competitor for Toledo is Detroit, whereas Akron must content not only with Cleveland but also with the third-order cities of Youngstown and Canton.

Current Evolution in Ohio's Cities

There is a change taking place in Ohio's cities, as people move out of the larger cities and into the rural surroundings and small towns found within commuting distance. This transformation is paramount in cities with the largest population (Table 9.1). Of the nineteen largest cities, only Columbus grew between 1980 and 1990. Youngstown, Cleveland, and Warren, all centers of early steel making, suffered declines over 10 percent each. Of the fifty largest cities in Ohio, only twenty increased in population between 1980 and 1990. Many of the most rapidly growing cities were residential suburbs. Gahanna, Westlake, Westerville, Reynoldsburg, and Strongsville typified these cities, all located on the expanding periphery of major metropolitan centers. At the same time, many of the older inner suburbs (Maple Heights, Barberton, Shaker Heights, Upper Arlington, East Cleveland, Cleveland Heights, Lakewood, and Parma) began to experience the population out-migration which has long been a phenomenon of central cities. Educational centers, such as Westerville, Kent, and Bowling Green, also increased in size. During the decade, Columbus grew by more than 12 percent to become the largest city in the state (see the previous section for a discussion of the importance of centrality in explaining growth). Figure 9.3 shows the thirteen cities that have experienced a population increase greater than 10 percent and the

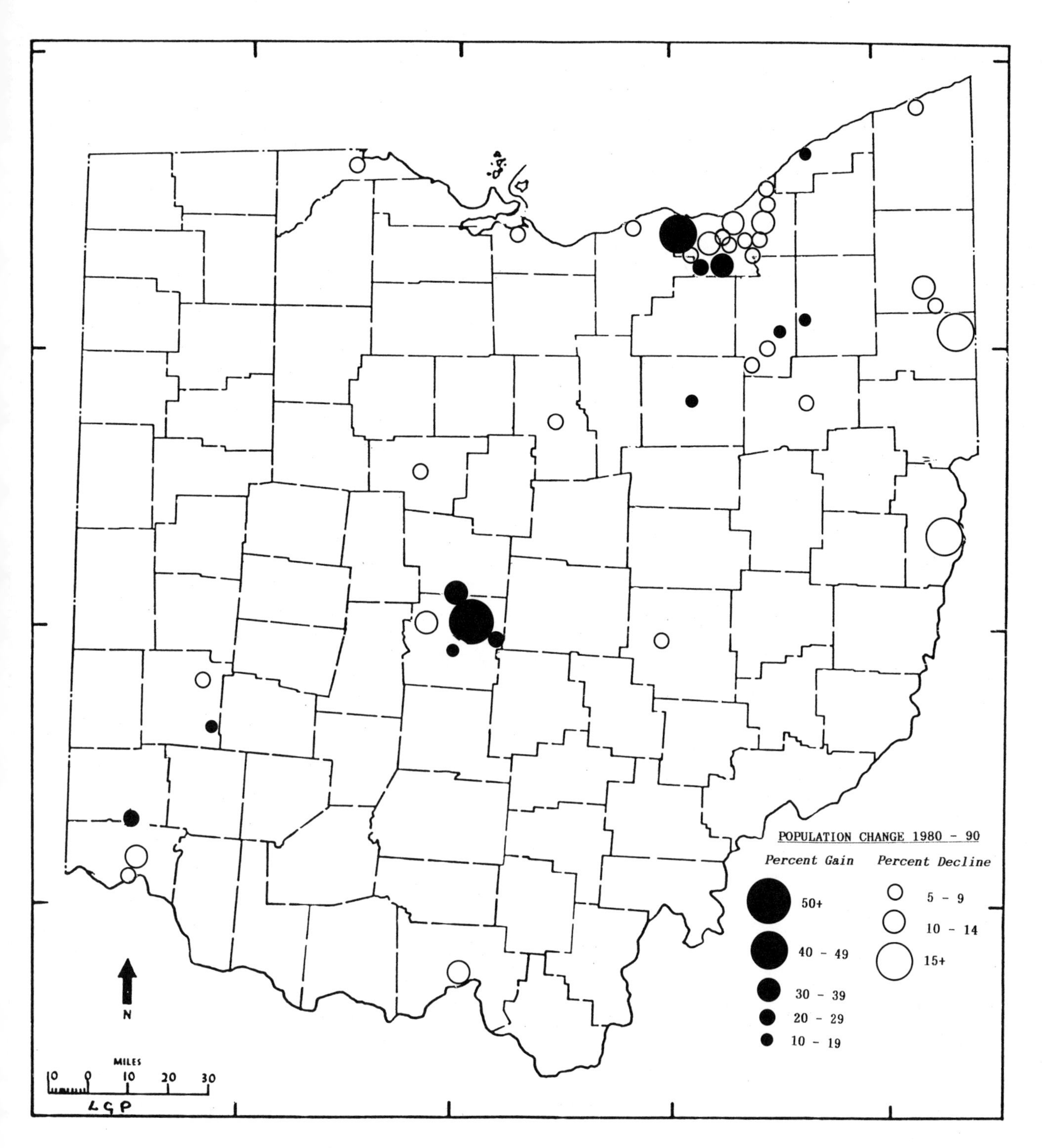

Fig. 9.3. Changes in Urban Population

thirty cities that have suffered a decline of greater than 5 percent. The mapped cities are also listed in Tables 9.2 and 9.3.

How important is the physical growth of a city on its population's stability? During the 1980s, boundary changes brought on by the incorporation of surrounding tracts of township real estate have gone on apace, as sharks swallow up their neighboring minnows. The reasons behind these incorporations are varied, ranging from a desire to increase the tax base to opportuni-

Table 9.2

Cities with a Growth Greater than 10%

CITY	1980	1990	CHANGE	% GROWTH
Gahanna	17,997	27,791	9,794	54.42
Westlake	19,478	27,018	7,540	38.71
North Royalton	17,671	23,197	5,526	31.27
Westerville	23,406	30,269	6,863	29.32
Fairfield	30,777	39,729	8,952	29.09
Reynoldsburg	20,660	25,748	5,088	24.63
Strongsville	28,577	35,308	6,731	23.55
Wooster	19,297	22,191	2,894	15.00
Mentor	42,065	47,358	5,293	12.58
Columbus	565,032	632,910	67,878	12.01
Centerville	18,879	21,082	2,203	11.67
Cuyahoga Falls	43,890	48,950	5,060	11.53
Kent	26,164	28,835	2,671	10.21

Source: Bureau of the Census 1980 & 1990

ties to improve the quality of the facilities of the incorporated area. More important in many cases is the protection size offers against incorporation by an even larger neighboring city. For example, Cuyahoga Falls and Westerville both grew to protect themselves from the designs of Akron and Columbus, respectively.

Does incorporation lead to either population increase or population stability? There certainly is no correlation between the two variables. Table 9.4 shows the ten cities with the largest percent population growth in relation to their change in area. Apart from Gahanna and Westerville, where a change in area does reflect a population increase, the argument is inconclusive. According to the census, Westlake, with a 43.84 percent population growth, shrank by 3.6 percent (Bureau of the Census 1992). Massillon, with the third largest area gain, saw a population growth of a mere 1.47 percent, while Cuyahoga Falls's massive 185 percent increase in area, a result of its annexation of the whole of Northampton Township, netted only an 11.5 percent growth in population. Dayton, which grew by 13.6 percent, lost almost 6 percent of its 1980 population.

There are, of course, many variables responsible for these figures, and these variables should be considered alongside the data. The acquired land may have a variety of attributes attractive to the acquiring city; it is not necessary that it have a substantial population. Also significant is the fact that most of the area changes recorded by the census are less than 1 square mile and thus could be accounted for by statistical rounding. Table 9.4 lists the cities that grew by more than 1.5 square miles.

The older industrial cities and inner suburbs are declining, and Ohioans are moving to the outer suburbs from the congested and socially disrupted city centers. A larger, longer distance migration is going on, as noted by Stephens in chapter 8, who suggests that there is a major migration from the state, and this suggestion is confirmed by the overall lack of growth in Ohio's population. Those who stay behind are moving to more rural areas, forcing small towns to face increases in population which, in some cases, they may not be able to handle.

Another consequence of these population shifts is change in the layout, or *morphology,* of major Ohio cities. It has been stressed earlier in the chapter that the nature of Ohio's cities, their location and form, was determined by the people who first settled the area. Constraints of physical geography—a river crossing, confluence, or lakeshore—played important parts in the original location. Added to the various physical constraints were the economic and social pressures created by such attributes as a nearby route or mineral site, or even the proximity of a market for trade. The internal form of the city was usually decided by the founders, who laid out the plan for their "city." The square—later to become the heart of the

Table 9.3

Cities with a Population Decline more than 5%

CITY	1980	1990	CHANGE	% DECLINE
Youngstown	115,511	95,735	–19,776	–17.12
Steubenville	26,400	22,125	–4,275	–16.19
Brook Park	26,195	22,865	–3,330	–12.71
Upper Arlington	35,648	31,128	–4,520	–12.68
Portsmouth	25,943	22,676	–3,267	–12.59
Cleveland	573,822	505,616	–68,206	–11.89
East Cleveland	36,957	33,096	–3,861	–10.45
Warren	26,342	23,674	–2,668	–10.13
Norwood	93,077	84,161	–8,916	–9.58
Canton	93,077	84,161	–8,916	–9.58
Garfield Heights	34,938	31,739	–3,199	–9.16
Maple Heights	29,735	27,089	–2,646	–8.90
Euclid	59,999	54,875	–5,124	–8.54
Niles	23,021	21,128	–1,893	–8.22
Marion	37,040	34,075	–2,965	–8.00
Ashtabula	23,356	21,633	–1,723	–7.38
South Euclid	25,713	23,866	–1,847	–7.18
Barberton	29,751	27,623	–2,128	–7.15
Parma Heights	23,025	21,448	–1,577	–6.85
Zanesville	28,655	26,778	–1,877	–6.55
North Olmstead	36,486	34,204	–2,282	–6.25
Mansfield	53,927	50,627	–3,300	–6.12
Toledo	354,635	332,943	–21,692	–6.12
Akron	237,177	223,019	–14,158	–5.97
Dayton	193,536	182,044	–11,492	–5.94
Cincinnati	385,409	364,040	–21,369	–5.54
Lorain	75,416	71,245	–4,171	–5.53
Shaker Heights	32,487	30,831	–1,656	–5.10
Sandusky	31,360	29,764	–1,596	–5.09
Parma	92,548	87,876	–4,672	–5.05

Source: Bureau of the Census 1980 & 1990

city's central business district (CBD)—dominated city life as it still does in many of today's small towns.

The CBD became the heart of the industrial city, where commerce, finance, and retail trade flourished. Public transport in the form of buses, streetcars, or interurban lines focused routes on the downtown, and street patterns reflect that area's preeminence in many of Ohio's cities. This design is epitomized in Cleveland's Public Square, where all the major routes into or out of the city once met. Along these arteries the basic parts of the city developed, with industries, retail operations, warehouses, residential developments, and cultural sites all focusing on downtown. As the CBD thrived, potential occupants vied for a place, space became a premium, and the city center grew. Land values rose too, and with them, buildings; skyscrapers towered over city centers, which now became concrete histograms reflecting the market values of the properties. High-density development reflected the importance of the industrial city in Ohio. Consider Cleveland, where the higher buildings all cluster around Public Square, dominated by Terminal Tower—which for a long time was the tallest building in the region between New York and Chicago. Today both the BP Tower and the Society Corporation Building, which has replaced the Terminal Tower

Table 9.4

Cities with an Increase in Area Greater than 1.5 Square Miles

CITY	1980	1990	AREA Change	% Change	POPULATION 1980–90 Change	% Change
Cuyahoga Falls	8.9	25.5	16.6	186.52	5,060	11.53
Columbus	180.9	191.0	10.1	5.58	67,878	12.01
Dayton	48.4	55.0	6.6	13.64	–11,492	–5.94
Akron	57.5	62.2	4.7	8.17	–14,158	–5.97
Massillon	9.6	13.2	3.6	37.50	450	1.47
Westerville	7.2	10.5	3.3	45.83	7,376	32.22
Gahanna	10.1	13.2	3.1	30.69	10,400	59.80
Mansfield	25.5	27.9	2.4	9.41	–3,300	–6.12
Reynoldsburg	7.6	9.3	1.7	22.37	5,320	26.04
Wooster	10.2	11.8	1.6	15.69	2,896	15.01

Source: Bureau of the Census 1980 & 1990

as the tallest building between New York and Chicago, continue to reflect the predominance of the city center.

Now, in the last decade of the twentieth century, this concentration of business in the city center is fading; in some cities this process began a decade or two earlier. In many Ohio cities the center does not reflect the city's former prosperity. Stores are closed and business has moved elsewhere. Where there once was a thriving main street, functionality has changed and premises lie empty. Mayors and managers are today trying to refocus and thereby revive the CBD.

No single influence emerges as greater than any other in the demise of the city center. In fact, the cause may be different in each case. A decline in manufacturing jobs around the city center may account in one case, while another is brought on by the greater mobility of the city's inhabitants. Reliance on the automobile allows movement further and further afield. People are able to travel greater distances to work and to perform the necessary rituals of life, like shopping and leisure activities. Businesses in the service sector were able to snap up cheaper land outside the city center and so develop attractive locations—such as shopping malls, large stores, movie theaters, motels, and restaurants—for the mobile population to visit. These developments continued along the arteries of communication into the city, but at a distance from the CBD. As a result the area around the center became a fragmented fringe of blighted residences and wholesaling districts encircled by a zone of older properties occupied by low-income groups and ethnic minorities (Clark 1985). The prosperity of a city is no longer reflected by the scale of the downtown properties but by the grouping of shopping malls, plazas, and low-rise office complexes, all with extensive parking facilities on the city's outskirts. As trade and jobs have moved out, so too have people. *Greenfield* housing developments—suburban developments with open, green space—have become the norm on the fringe of the cities.

Belden Village Mall, north of Canton, is an example of this form of fringe development. The area was designed with the expectation that it would attract patrons from Akron located to the north. Yet the accelerated growth around the mall during the 1980s was not wholly the result of visitors from the neighboring cities but a consequence of people both visiting and moving to the area from further afield. Even commuters from the suburbs of Cleveland now flock to the new housing developments that are flourishing in greenfield sites in northern Stark and southern Summit Counties. It may not be long before the area surrounding Belden, Jackson Township, has city status; Parma and Kettering, once small suburbs of Cleveland and Dayton, respectively, now rank in the top ten cities. Other examples of "edge cities," such as Gahanna and Westerville, are likely to be even more common in the future.

Will the traditional morphology of the Ohio city continue into the next century, or will the central business district be overshadowed by growing service centers and satellite cities around the margins of the old city? In the major cities of Ohio—Columbus, Cincinnati, and Cleveland—there will always be an attraction for businesses to locate in the downtown, producing a concomitant high cost to the real estate. But in other cities—Toledo, for instance—where the downtown area is but a skeleton of its former self, the city managers have a major task of rejuvenation. It is a matter of conjecture as to whether the older, established cities can maintain themselves into the next century or whether the new-growth cities, which have little or no manufacturing base but provide an overwhelming mass of services, can grow to be dominant urban centers.

REFERENCES

Clark, D. 1985. *Post Industrial America.* London: Methuen.

Dutt, A. K. and P. Harrity. 1970. Newspaper Circulation and a City's Sphere of Influence. Quoted in Allen G. Noble and Albert J. Korsok. *Ohio: An American Heartland.*

———. 1971. Newspaper Circulation and City Hierarchy. Quoted in Noble and Korsok, *Ohio: An American Heartland.*

Johnson, J. H. 1972. *Urban Geography: An Introductory Analysis.* Oxford: Pergamon Press.

Noble, Allen G., and Albert J. Korsok. 1975. *Ohio: An American Heartland.* Columbus: Ohio Department of Natural Resources, Division of Geological Survey.

U.S. Bureau of the Census. 1992. *General Population Characteristics: Ohio.* Washington, D.C.: Supt. of Documents.

City Vignettes

The following section contains short vignettes of seven of Ohio's largest cities: Columbus, Cleveland, Cincinnati, Toledo, Akron, Youngstown, and Dayton. Each of these cities has a different character and history; consequently the authors have all taken different perspectives for their narratives. Providing only enough detail to give the reader a flavor of the community, the authors portray their respective cities as they see them in the final decade of the twentieth century and discuss how these cities might develop in the years ahead.

COLUMBUS
The Capital City

Henry L. Hunker

Columbus is the largest political city in Ohio and, in many ways, the state's most dynamic city. During a period when Ohio's large industrial centers were adjusting to the structural shift in the U.S. economy, Columbus emerged as a major center of growth in the northeastern quadrant of the nation. The city and its metropolitan region are among the more rapidly expanding in Ohio in terms of population and economic development. Much of this expansion is related to Columbus's function as a sophisticated service center whose growth reflects the nature of an evolving economy in the region. The basis for this expansion lies in the city's historic role as Ohio's capital and, in a sense, the historical development of the city holds the key to understanding its present stature.

Historic Overview

Columbus is a planned state capital. It was created by an act of the General Assembly in 1812 to be located on a site on the high banks of the Scioto River. Physically the newly designated capital stood alone on the relatively flat Central Ohio Till Plain, a trait that was not to change until the mid-twentieth century. In due course, the site was platted by Joel Wright into a rectangular grid with plots set aside for a capitol, arsenal, and penitentiary. Shortly, settlement and commercial development were under way, with the state government function and the city's geographic location major factors affecting its development.

These initial actions set the stage for future development in at least two ways. At the local level, Wright's plat established the basic plan of the city. The rectangular grid, on an approximate north-south bias, and the related street layout and orientation not only established the community's internal pattern, but provided the basic elements that linked it to the interregional transportation system that was to evolve. At the broader level, the basis for Columbus's interaction with other parts of the state and beyond was established both through its role as Ohio's center of government and through the city's physical access to other regions. These conditions also sparked the growth of a commercial or service economy linked to the government function through a primitive transportation network and such services as inns, restaurants, and government suppliers. A commercially based economy remains an important factor shaping the present society.

Columbus grew steadily and by 1834, when it was incorporated as a city, had 3,500 residents. By 1850, there were approximately 18,000 persons in the city deriving their living from government functions and related activities. Early manufacturing was limited by the local resources and restricted market. With improved transportation—early turnpikes and the National Road, the opening of a feeder to the Ohio and Erie Canal, and, later, the coming of the railroads—a manufacturing economy based on the agricultural productivity and the forest and mineral resources of southeastern Ohio emerged. This evolved into an active industrial economy by 1870. During the last two decades of the nineteenth century, population more than doubled, reaching over 125,000 by 1900. The expansion of Columbus in this period was reflective of the rapid urban industrialization occurring elsewhere in Ohio and the nation during the "industrial revolution" of the late nineteenth century.

From this time on into the early years of the twentieth century, Columbus evolved a distinctive industrial economy based, in part, on local inventions, innovations, and initiative, as well as local capital. What resulted was a number of highly successful manufacturing enterprises that were unique to Columbus and, in time, grew to be among the largest such operations in their fields. This characteristic of Columbus manufacturing remained a dominant feature until World War II and the postwar boom in manufacturing. It gave rise to firms such as Jeffrey Manufacturing (mining machinery), Buckeye Steel Castings, Columbus Coated Fabrics (oil cloth), Jaeger Machinery (cement mixers), and Seagraves (fire engines), each reputed to be the largest company of its kind in the nation, if not in the world.

The Evolving Economy

The expansion of the industrial economy was encouraged locally as a force designed to shape the economic growth of the region. But a major change occurred with the decision by the federal government to locate the Curtis-Wright aircraft plant in the city in 1941. That decision, and the resulting expansion of the labor force through in-migration, primarily from Appalachia and the South, set the stage for postwar industrial growth.

By the mid 1950s, Columbus had taken on a number of new manufacturing activities reflecting national investment in the region and further recognition of its locational advantages. The Ohio geographer, Alfred J. Wright, writing in 1930 about Columbus's locational advantages, stated that its "geographical position . . . augurs most for the ultimate development of Columbus. . . . [It is] the one inexhaustible resource which gives stability and permanence to Columbus" (Wright 1930). Certainly, good rail and highway transportation, market accessibility, and the pool of reasonably well-trained industrial labor helped to shape locational decisions that brought new industry to the city. Aircraft production at the large Curtis-Wright plant during World War II was replaced by that of North American Aviation in the 1950s. In addition, large branch plants of other national firms came to Columbus in the postwar years: for example, General Motors (Ternstedt Division), Westinghouse, and Western Electric. Later, and again responsive to the city's continuing locational advantage and excellent interstate highway network, Columbus developed as a major distribution center, with a large Anheuser-Busch brewery and a number of sizable distribution facilities including huge Sears and JCPenney warehouse and distribution centers, and numerous other distributors located in the city's Outerbelt.

These characteristics continue to help explain business location and expansion in Columbus's growing Service economy. National organizations, such as American Electric Power and Bordens, have relocated their headquarter offices to Columbus. The spectacular expansion of The Limited retail organization, with its roots firmly established in Columbus and with corporate offices and extensive warehousing and distribution operations in central Ohio, is another example. And this activity extends to the financial and insurance sectors, with dynamic growth in several major area banking institutions and the city's prominent national insurance industry. Critical to these latter businesses is the development of the modern computer-based communications system. In this context, Columbus has proven to be a leader, not in the production of computers and software, but in their imaginative use in the expansion of extensive information systems. Key among these are the Online Computer Library Center (OCLC) and CompuServ, major centers for worldwide information distribution, and other major information-generating and -using institutions—such as The Ohio State University, Battelle Institute, Chemical Abstracts, and the banks and insurance industries of the area. More recently, the United Nations identified Columbus as the North American site for a new international electronic data network that will connect with more than fifteen sites throughout the world. These activities give credence to the assertion that Columbus is one of the major information centers in the world.

Employment

The Columbus Metropolitan Statistical Area (MSA) is a seven-county area composed of Franklin, Delaware, Fairfield, Licking, Madison, Pickaway, and Union Counties. Currently, it is the fastest growing metropolitan area in Ohio in employment and will likely continue to be during the 1990s. Estimates of employment indicate approximately 700,000 workers in the Columbus MSA, with Central City Columbus—its principal economic center—accounting for approximately 74 percent of the total. Consistent with the recent past, most of the growth is expected to occur in the Service sector. The goods-producing manufacturing sector experienced an approximate 12 percent loss in employment during the 1980s, in the Columbus MSA and, by

Table 9.1.1

Employment by Sector, Columbus MSA 1980–1992

SECTOR	1980	SHARE OF TOTAL	1992	SHARE OF TOTAL	ACTUAL CHANGE	% CHANGE (1980–92)
Manufacturing	96.7	18.80%	101.7	14.41%	5.0	5.17%
Construction	21.1	4.10%	22.8	3.23%	1.7	8.06%
Transport and Utilities	24.8	4.82%	30.1	4.26%	5.3	21.37%
Wholesale and Retail	122.8	23.88%	174.0	24.65%	51.2	41.69%
F.I.R.E.	38.0	7.39%	61.1	8.66%	23.1	60.79%
Service	105.9	20.59%	180.7	25.60%	74.8	70.63%
Government	105.0	20.42%	135.4	19.18%	30.4	28.95%
Total Service	396.5	77.10%	581.0	82.32%	184.5	46.53%
TOTAL EMPLOYMENT	514.3	100.00%	705.8	100.00%	191.5	37.24%

Source: Ohio Bureau of Employment Services

1990, accounted for less than 15 percent of all workers in the area's labor force (Table 9.1.1).

The dominant role of the Service economy is well-illustrated by a review of the major employers in the central Ohio region. The two largest are the state government, with 26,400 employees, and The Ohio State University, with 21,000. Government and education dominate the group of top ten employers, although Honda of America, with 7,800 workers, Nationwide Insurance, with 6,900, and AT&T with 6,500, are among the ten leading business firms. Of the approximately sixty firms employing more than 1,000 workers in the MSA, only seventeen are in manufacturing; all others are representatives of the Service economy.

The growth of employment in the Service (or Tertiary) economy has been a key factor in the region's development over the last two decades. Wholesale and retail trade, general services, and the finance–insurance–real estate (FIRE) sectors report approximately 60 percent of all current employment. This growth, combined with the decline in manufacturing, helps to explain the differences between the experiences of the Columbus economy and that of Cleveland, Youngstown, Akron, Toledo, Dayton, and other cities of the industrial Midwest, which are more heavily committed to an industrial base.

The Columbus experience is reasonably consistent with national trends, however. The region has the base on which to evolve an economy consistent with the projected growth economies of the future—those involved in the generation, distribution, and consumption of information and knowledge. Associated with this growth has been relatively low unemployment in the Columbus MSA counties, with rates that are among the lowest in Ohio.

Population

The Columbus MSA ranks twenty-ninth in the nation in population, with a 1990 total in excess of 1.3 million. As such it ranks third among Ohio's metropolitan areas. The Cleveland Consolidated MSA leads with over 2.7 million and the Cincinnati Consolidated MSA has over 1.7 million. From 1980 to 1990, the Columbus MSA increased in population by 10.7 percent (Table 9.1.2), whereas the Cleveland area had a 2.6 percent loss, and the Cincinnati area grew by only 5.1 percent. Growth throughout the MSA was by no means evenly distributed. Delaware County had an increase of 24.3 percent, the highest increase in population of any Ohio county in the decade; at the same time, Licking County posted the lowest increase in the MSA: 6.0 percent. It is expected that Delaware will remain the growth leader in the 1990s. Clearly the Columbus MSA is one of the dynamic regions of the state in terms of population growth.

At the heart of this expansion is Columbus, the Central City of the MSA. Political Columbus is now the largest city in Ohio; its population numbers 633,000. Its increase of 12.0 percent from 1980 to 1990 was

Table 9.1.2
Population Characteristics, 1980–1990

AREA	1980	1990	CHANGE	% CHANGE
Franklin County	869,109	961,437	92,328	10.62%
Columbus	564,826	632,910	68,084	12.05%
Delaware County	53,840	66,929	13,089	24.31%
Fairfield County	93,678	103,461	9,783	10.44%
Licking County	120,981	128,300	7,319	6.05%
Madison County	33,004	37,068	4,064	12.31%
Pickaway County	43,862	48,255	4,393	10.02%
Union County	29,536	31,969	2,433	8.24%
Seven County MSA	1,244,010	1,377,419	133,409	10.72%

Source: Bureau of the Census

the highest percentage increase of a city outside the Sun Belt among the nation's twenty-five largest cities. Columbus is now larger than Cleveland and Cincinnati; and the political city continues to thrive, whereas Cleveland and Cincinnati have sustained losses or had only modest growth.

Closely linked to population growth in Columbus has been the aggressive annexation program launched by the city in 1952. This expansion into suburban areas was brought about by using access to sewer and water services as spurs to induce annexation. The principal consequence was an increase in the city's area from 52 square miles in 1952 to approximately 193 square miles by 1990. (In contrast, both Cleveland and Cincinnati had little opportunity to annex during this period and have not added significantly to their corporate areas.) Related to land acquisition was the acquisition of a growing population in the suburban and rural fringe of political Columbus. The historic observation that Columbus "stands alone" in central Ohio helps to explain why the city was able to annex growth areas while Cleveland and Cincinnati, hemmed in by political suburbs, could not. Perhaps more than any other factors, the annexation policy of the city of Columbus in the 1950s and the expansion that resulted explain its position today. Continued growth in both the Central City and the Metropolitan Statistical Area is expected.

Along with other major urban centers in the U.S., there is concern regarding the polarization of the population between the Central City and the suburban areas of the Columbus MSA. In effect, the suburban areas remain overwhelmingly white while the Central City is home to the prevailing minority members of the community. Only about 15 percent of Columbus's minority population has located in the suburbs. This problem is compounded by the fact that among the African Americans, Asian, and Hispanic minority groups, the African American minority, the largest, has been the slowest to move into suburban communities.

A Great Place to Live

It has often been suggested that like many Midwestern cities, "Columbus is a great place to live, but you wouldn't want to visit it!" The city's residents would ardently support the initial phrase—it *is* a great place to live and to raise a family. The city and its suburban neighbors are characterized by relatively low density, single-family housing of considerable variety relative to amenities, neighborhoods, and housing costs. It is well served by interregional transportation, such as Interstates 70 and 71, which pass through Columbus, and, until recently, with an internal transportation network that had been relatively free of major traffic problems: the journeys-to-work and -to-shop were not major chores. Low unemployment has typified the region, resulting in a fairly wide range of employment opportunities, increasingly in the Service economy. Wage rates are competitive: Columbus has both lower overall earnings and less unionization than its more industrialized urban neighbors. Cultural amenities have improved significantly in recent years, with a good symphony and ballet in place, expanded theatrical events, and enriched museums. Closely linked is the important role that The Ohio State University and other area colleges and universities—Capital, Franklin, Ohio Dominican, and Ohio Wesleyan, among others—play in the metropolitan area and the expanded research and development environment to which Battelle, Chemical Abstracts, OCLC, and others contribute.

In a sense, much of what is happening in Columbus comes from the continued expansion of the city's economy and its amenities. The early success of the City Center retailing complex is yet one more example; whether it will serve to promote core area (Central Business District) revitalization or not remains to be seen. A number of high-rise office buildings constructed by the state and by private enterprise in the 1980s lend a distinctively different feel to the city's skyline and to the core area. Principal examples include the Huntington Center, the expanded Nationwide Insurance complex, the Riffe State Office Building, and the workers' compensation facility. Projected is a new high-rise planned for Capitol Square, representing the cooperative efforts of two of Columbus's major family interests, the Galbreaths and Wolfes. With all of this activity, there is no reason to question at this time the belief that the city remains a good place to live for most of its residents. This point is well illustrated by initially skeptical newcomers who have relocated to Columbus from larger urban centers and who eventually settled in comfortably in the somewhat slower paced and more livable central Ohio environment.

Columbus appears a bit more dubious as a place to visit. Lacking the historical experiences that so often provide the basis for tourism, like many urban downtown core areas there is little that is either unique or exciting. On the other hand, Ohio State continues to fill its stadium on football Saturdays in the fall; and the city's Museum of Art (with its Sirak Collection), the Center for Science and Industry, the Ohio Historical Society Museum and Village, the Columbus Zoo, and the State Fair in August are key attractions. In 1992, the city held AmeriFlora '92, an international floral and cultural festival to help celebrate the 500-year anniversary of Columbus's "discovery" of the Americas. Renovation and restoration of Ohio's capitol building and square, recognized by scholars as one of the finest Greek Revival public buildings in America, is nearing completion. It is clearly the focal point of state government. As always, the city's centrality makes it readily accessible to convention visitors. Columbus's failure to capture these, in part due to inadequate facilities, has been addressed with the construction of the Columbus Convention Center. Conservatism in public spending and the lack of consistently strong leadership is reflected in the major public institutions and infrastructure. Nevertheless, with a dynamic decade of growth behind it, Columbus stands ready to meet the challenges of the future.

REFERENCES

U.S. Bureau of the Census. 1981. *General Social and Economic Characteristics 1980.* Washington, D.C.: U.S. Department of Commerce.

———. 1991. *General Social and Economic Characteristics 1990.* Washington, D.C.: U.S. Department of Commerce.

Wright, A. J. 1930. *Columbus, Ohio: An Analysis of a City's Development.* Columbus: Industrial Bureau of Columbus.

CLEVELAND
Building the Great Lakes City of the Future?

Harry L. Margulis

Cleveland, the blue-collar corporate industrial metropolis, is a city in economic and political transition. As economic restructuring eliminates manufacturing jobs, service sector employment replaces traditional smokestack industries, and global restructuring fosters labor migration and resettlement, this often-chosen All American City is rebuilding its image and redefining its role in the rapidly changing global, regional, and national economy. Confronted by decades of economic disinvestment, continuing population losses, fiscal distress, and real estate abandonment, the city is beginning to strategically alter its direction of development. A city of pluralistic politics and social redistribution is giving way to a corporatist elite whose focus on downtown redevelopment and public-private partnerships markedly contrasts with ravaged neighborhoods and contentious biracial political coalitions. Surrounded by noncooperating, often hostile, go-it-alone suburbs, the city is swiftly becoming one of many multifoci centers in a larger metroplex—a decentralizing metropolitan area with growing suburban core cities containing emerging downtown areas.

Racial divisiveness has created a political vacuum into which has stepped a corporatist business elite whose constancy remains downtown redevelopment. Nonetheless, exogenous centrifugal global forces are interjecting influences into the political economy over which local leadership has little control. Together with race polarization and the difficulties of consensus building, the social fabric of the city is becoming increasingly frayed. Chronic unemployment, welfare dependency, crime, drugs, homelessness, and teenage pregnancy are but a few of the incessant problems haunting the city's future. Whether the city's economy can be truly turned around depends not only on locational circumstances but also on the ability of enlightened leadership to position the city to move beyond economic stalemate and to capture the next Kondratiev cycle of technological growth—the 'flex-tech' 'throughware' economy.

The Industrial City: The City of the Past

Almost thirty years elapsed between 1795—the time the federal government granted the state of Connecticut rights to the Western Reserve in the Northwest Territory, which made possible the speculative efforts of the Connecticut land companies to quickly sell off land in the tract—and the decision in 1825 to locate the northern terminus of the Ohio and Erie Canal at the mouth of the Cuyahoga River. The completion of the canal in 1832—linking Portsmouth on the Ohio River to Cleveland, with connections to New York State's Erie Canal, to the densely settled Atlantic seaboard, and to overseas markets—transformed a commercial frontier village into a commercial city (Miller and Wheeler 1990). By improving transportation access and reducing freight rates, the economic ascendancy of Cleveland was assured.

The Flats-Oxbow area at the mouth of the Cuyahoga River became the transportation hub of an emerging metropolitan area, collecting produce as it flowed north and sending manufactured goods south into Ohio's hinterland. On-loaded at the docks were flour, butter, and cheese; off-loaded were salt, lumber, and manufactured merchandise. Both migrants and immigrants were drawn to the city by local commerce and the rise of small manufacturing industries, but the great influx of immigrants that would forever ethnically, racially, and socially alter the city would await the coming of the railroads.

Industrial supremacy came in the late 1850s, when railroads constructed along the lake plain provided all-season access to Pittsburgh, New York City, and Chicago. By the outbreak of the Civil War, Cleveland had become a preeminent rail center—the terminus of five railroads and numerous canal and steamship lines. Industrial development quickly spread up the Cuyahoga River Valley and along the rail lines. The war created an insatiable appetite for war materials, a demand that fueled industries producing kerosene, lubricants, machinery, castings, bar iron, structural iron, railroad equipment, and stoves.

As industrial growth accelerated, the city's population rapidly grew from 43,417 in 1860 to 92,829 in 1870 (Miller and Wheeler 1990). German, Irish, Czech, Hungarian, and Italian immigrants, and African American laborers sometimes comingled with each other but remained residentially segregated from their prosperous Protestant New England neighbors. The lavishly built homes of the wealthy on Euclid and Superior Avenues starkly contrasted with the modest cottages of artisans and mill owners and the immigrant working-class domiciles found in the burgeoning ethnic enclaves.

Fueled by a growing iron and steel industry, as well as associated metal fabricating, petroleum refining, shipping, shipbuilding, materials handling, and consumer goods industries, Cleveland's products poured into the expanding national and regional markets. In 1890 the population of the city was 261,353, a population largely composed of foreign-born or the children of foreign-born parents. Drawn by rapidly multiplying factory jobs and other economic opportunities, large numbers of unskilled and semiskilled southern and eastern Europeans came to work in the city's industries. For over 100 years the industrial city would be an economic magnet, drawing in migrant and immigrant workers. Only structural economic change—in the form of suburbanization, the decentralization of business and industry, and the resulting thinning and decanting of the city's population—would substantially alter Cleveland's locational advantages.

The Crisis of Economic Restructuring

Cleveland is a city confined to fixed boundaries (Figure 9.2.1), a winter city that is slowly responding to the forces of decentralization. Table 9.2.1 shows the population of Cleveland, suburban Cuyahoga County, and the Primary Metropolitan Statistical Area (PMSA)—consisting of Cuyahoga, Geauga, Lake, and Medina Counties—from 1870 to 1990. From 1870 to 1930 Cleveland had an average population growth rate of 47.3 percent. In the decade from 1930 to 1940 the growth rate declined by 2.5 percent, but population recovery occurred in the 1950s, when a population apogee of 914,808 was reached. Over the next three decades Cleveland experienced a steady population decline averaging 14 percent, with population falling from 914,808 to 573,822 (a loss of 340,986 people). Although the rate of population decline has slowed, the city lost another 72,474 people from 1980 to 1990. Population projections for the next two decades estimate a further population decrease of 197,700 people.

In contrast, the population of surrounding suburban Cuyahoga County increased steadily for 100 years, reaching a peak of 970,397 in 1970. Of the four counties in the PMSA, Cuyahoga County remains the most populous and in 1990 housed 77.1 percent of the PMSA's total population. However, Cuyahoga County's share of the PMSA's population is currently declining, while the proportion for Geauga, Lake, and Medina Counties is increasing. Population projections suggest that Cuyahoga County will suffer further decline and by the year 2000 will have only 73.6 percent of the PMSA's total population.

Cleveland's long-term relative and absolute population decline is also reflected in the redistribution of households within the PMSA (Table 9.2.2). In 1940, 72.0 percent of the households in Cuyahoga County resided in the city, but only 35.2 percent of the county's households resided there in 1990. Projections for the year 2000 indicate that household decentralization should continue. By the end of the century only 28.8 percent of the PMSA's households are expected to reside in Cleveland.

The deindustrialization of production and distribution functions makes older industrial cities vulnerable to structural shifts in the nation's economy. This phenomenon is evident in the dramatic changes taking place in the city's employment and occupational characteristics. Historic industrial employment trends reveal that the traditional manufacturing base of Cleveland is eroding. The future geography of the advanced service sector economy of the city and region will be largely deter-

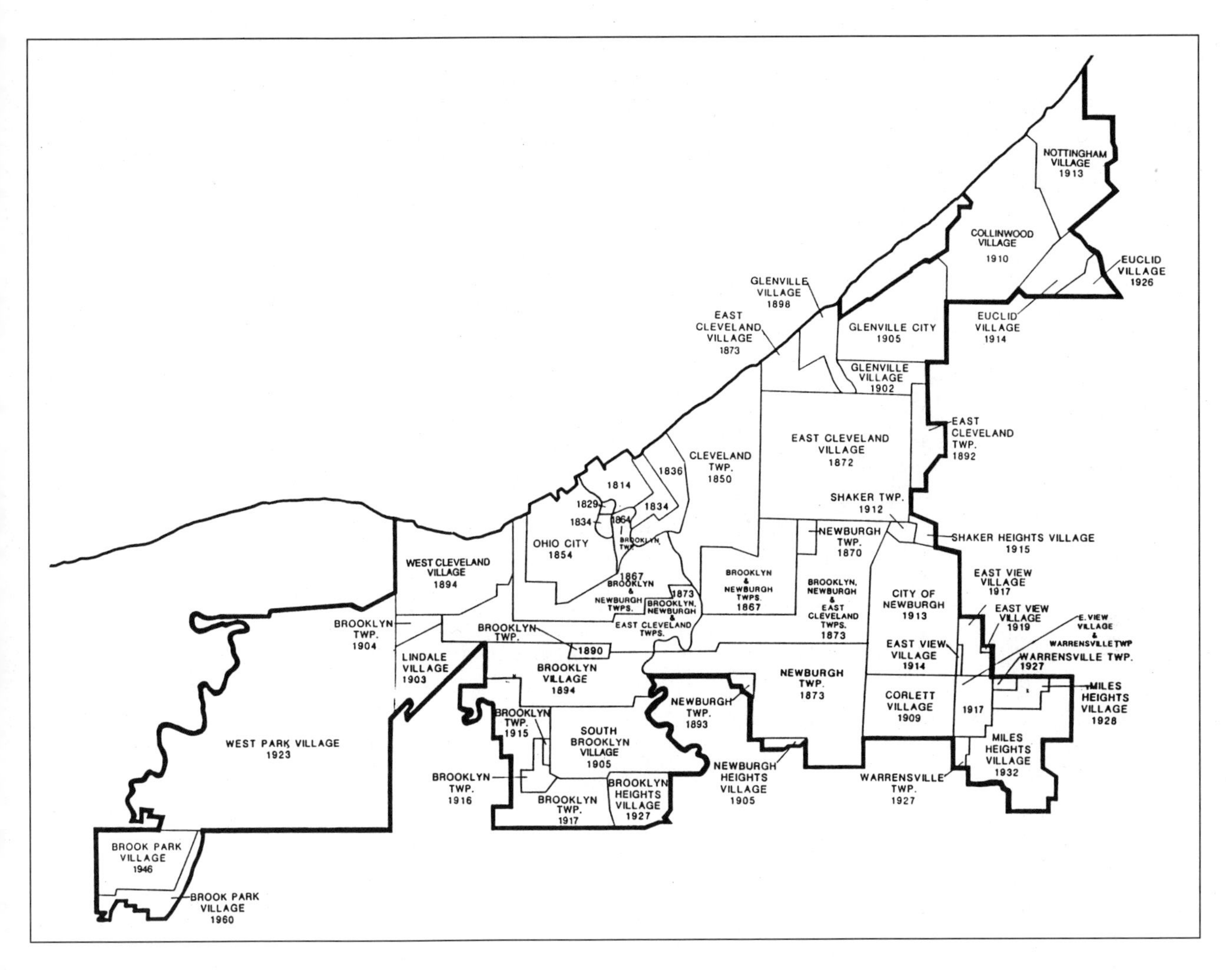

Fig. 9.2.1. The Development of Cleveland and Its Neighboring Communities

mined by the employment growth of its finance, insurance, real estate (FIRE), business and consumer service sectors, and its technical, scientific, and knowledge-based industries. Other industries—including construction, manufacturing, transportation, communications, and public utilities—are declining (Table 9.2.3).

Employment growth in the advanced service sector industries is occurring mainly in Cleveland's downtown, but these same industries are now downsizing, showing strong decentralization and suburbanization tendencies. Midlevel projections (Table 9.2.4) indicate that employment growth is projected in advanced service sector industries for the remainder of the century, while employment in traditional industry sectors continues to shrink.

Somewhat obscured in the industry employment statistics are a number of other important trends. For example, in Cuyahoga County, outside the Cleveland downtown area, employment declines are projected for the wholesaling and retailing trades, as sectoral employment growth quickly shifts to the suburbs. While a modest increase in total employment is expected in the downtown area, overall employment in both Cuyahoga County and the city should stagnate, eventually declining by the twenty-first century.

Economic restructuring and industry decentralization impacts are being felt most ponderously in the city. Some of the city's out-migration and high unemployment are directly linked to the emergence of an employment structure incapable of absorbing a redundant, deskilled, and unadaptable labor force. The mismatch between worker skills and emerging job opportunities can be observed in Table 9.2.5, which shows the distribution of total employment by occupation for 1970 and

Table 9.2.1

Population of Cleveland, Suburban Cuyahoga County, and PMSA Counties, 1870–1990 and Year 2000 Projection

YEAR	CLEVELAND		SUBURBAN CUYAHOGA		CUYAHOGA COUNTY		FOUR-COUNTY PMSA	
	Population	Change	Population	Change	Population	Change	Total	Change
1870	92,829		39,181		132,010		182,227	
1880	160,146	72.5%	36,796	–6.1%	196,943	49.2%	248,973	36.6%
1890	261,353	63.2%	48,617	32.1%	309,970	57.4%	363,436	46.0%
1900	381,768	46.1%	57,352	18.0%	439,120	41.7%	497,502	36.9%
1910	560,663	46.9%	76,762	33.8%	637,425	45.2%	698,620	40.4%
1920	796,841	42.1%	146,654	91.1%	943,495	48.0%	1,013,265	45.0%
1930	900,429	13.0%	301,016	105.3%	1,201,455	27.3%	1,288,220	27.1%
1940	878,336	–2.5%	338,914	12.6%	1,217,250	1.3%	1,319,734	2.5%
1950	914,808	4.2%	474,724	40.1%	1,389,532	14.2%	1,532,574	16.1%
1960	876,050	–4.2%	771,845	62.6%	1,647,895	18.6%	1,909,483	24.6%
1970	750,903	–14.3%	970,397	25.7%	1,721,300	4.5%	2,064,194	8.1%
1980	573,822	–23.6%	924,578	–4.7%	1,498,400	–13.0%	1,898,825	–8.0%
1990[a]	501,348[b]	–12.6%	910,792	–1.5%	1,412,140	–5.8%	1,831,122	–3.6%

Sources: T.A. Bausch, L. R. Cima and J. T. Bombelles, Regional Economic and Demographic Analysis for Cleveland, Ohio, 1974; The Cleveland Bureau of Government Research, A Compendium of Population Study Resources, 1960; Cleveland State University, Population and Household Projections: Cleveland Metropolitan Area, 1985–2020, Cleveland, Ohio: College of Urban Affairs, 1988, pp. 62–63.

[a] Department of Commerce, Bureau of the Census, 1990 Census of Population and Housing, STF1A.

[b] Review of preliminary census findings suggest a final population estimate of 505,616.

Table 9.2.2

Households in Cleveland, Suburban Cuyahoga County, and the PMSA, 1940–2005

YEAR	CITY OF CLEVELAND			SUBURBAN CUYAHOGA COUNTY			FOUR-COUNTY PMSA	
	Households	Change	Share	Households	Change	Share	Households	Change
1940	262,267		72.0%	94,252		28.0%	364,793	
1950	265,973	9.8%	65.5%	139,956	48.5%	34.5%	447,092	22.6%
1960	269,891	1.5%	54.3%	227,035	62.2%	45.7%	588,066	31.5%
1970	248,280	–8.0%	44.8%	305,959	34.8%	55.2%	650,138	10.5%
1980	218,297	–12.1%	38.7%	345,181	12.8%	61.3%	694,401	6.8%
1990	198,365	–9.1%	35.2%	364,878	5.7%	64.8%	712,362	2.6%
Projected								
2000	156,845	–20.9%	28.8%	387,259	6.1%	71.2%	730,349	2.5%
2005	134,701	–14.1%	25.5%	392,441	1.3%	74.5%	727,423	–0.4%

Sources: Census of Population and Housing, 1940, 1950, 1960, 1970, 1980, 1990; Cleveland State University, Population and Household Projections: Cleveland Metropolitan Area, 1985–2020, Cleveland, Ohio: College of Urban Affairs.

1980, along with midlevel projections at five-year intervals from 1985 to 2000.

In the decade from 1970 to 1980, with the exception of a modest employment increase in the downtown area, employment declined in manufacturing-related categories—such as precision production–crafts and operatives-laborers. On the other hand, employment increased in job categories associated with advanced

Table 9.2.3
Total Employment by Industry (in Thousands)

YEAR	1965	1970	1975	1980	CHANGE 1965–70	CHANGE 1970–75	CHANGE 1975–80
Geographical Area							
Cleveland PMSA							
Construction	32.8	33.4	30.5	29.8	1.8%	–8.7%	–2.3%
Manufacturing	296.8	296.8	260.3	254.9	0.0%	–12.3%	–2.1%
TCPU[1]	47.3	50.6	48.0	45.9	7.0%	–5.1%	–4.4%
Wholesale Trade	48.3	55.1	58.7	63.9	14.1%	6.5%	8.9%
Retail Trade	105.8	126.1	134.2	141.4	19.2%	6.4%	5.4%
FIRE[2]	35.3	41.4	43.2	48.2	17.3%	4.4%	11.6%
Services	107.1	127.5	158.0	187.5	28.4%	14.9%	18.7%
Government	90.8	112.2	116.2	122.6	23.6%	3.6%	5.5%
Total	764.2	853.1	849.1	894.2	11.6%	–0.5%	5.3%
City of Cleveland							
Construction	N/A	11.0	9.0	8.1	N/A	–18.2%	–10.0%
Manufacturing	171.3	131.0	120.8	92.5	–23.5%	–7.8%	–23.4%
TCPU	N/A	32.9	30.5	29.5	N/A	–7.3%	–3.3%
Wholesale Trade	32.6	27.8	25.8	24.0	–14.7%	–7.2%	–7.0%
Retail Trade	47.2	41.4	36.9	30.2	–12.3%	–10.1%	–18.2%
FIRE	N/A	20.7	28.4	21.7	N/A	37.2%	–23.6%
Services	32.1	36.8	33.4	42.1	14.6%	–9.2%	26.0%
Government	N/A	39.1	N/A	35.9	N/A	N/A	N/A
Total		340.7		284.0			

Source: Center for Regional Economic Issues, Cleveland Economic Analysis and Projections, Cleveland, Ohio: Weatherhead School of Management, Case Western Reserve University, 1987.

[1] Includes transportation, communications, and public utilities.

[2] Includes finance, insurance, and real estate.

service sector industries—such as professional-technical and executive-management. Nonetheless, the city showed a marked tendency for job disappearance. Approximately 15.3 percent of all jobs found outside the downtown vanished in the decade from 1970 to 1980. Concurrently, Cuyahoga County lost 4.7 percent of its jobs, including many in the administrative support occupations.

Both downtown Cleveland and the PMSA gained jobs from 1970 to 1980. Downtown Cleveland realized a 27.6 percent increase in jobs, mainly in occupations associated with professional-technical (72.5 percent), executive-management (49.1 percent), administrative support (23.9 percent), precision production–craft (53.3 percent), and service worker (14.7 percent) sectors. Sales and operatives-laborers jobs declined in importance. With the exception of precision production–crafts jobs, the PMSA has also experienced an increase in these same occupational categories and in addition, has gained jobs in sales occupations.

REI midlevel projections of total employment by occupation from 1985 to 2000 suggest that precision production–crafts and operatives-laborers jobs should continue to decline in the region. The recent recession may have accelerated this trend; the Council for Economic Opportunities in Greater Cleveland, on examining the Office of Business Employment Service data

Table 9.2.4
Mid-Level Projections of Total Employment by Industry (in Thousands)

YEAR	1985	1990	1995	2000	CHANGE 1980–85	CHANGE 1985–90	CHANGE 1990–95	CHANGE 1995–2000
Geographical Area								
Cleveland PMSA								
Construction	27.8	26.3	24.8	23.4	–6.7%	–5.4%	–5.7%	–5.6%
Manufacturing	210.7	189.2	170.4	154.1	–17.3%	–10.2%	–9.9%	–9.6%
TCPU[1]	41.3	39.0	36.6	34.0	–11.1%	–5.6%	–6.1%	–7.1%
Wholesale Trade	63.0	63.6	63.9	64.1	1.4%	0.9%	0.5%	0.3%
Retail Trade	145.9	147.1	148.0	148.3	3.2%	0.8%	0.6%	0.2%
FIRE[2]	49.4	52.3	55.2	58.3	2.5%	5.9%	5.5%	5.6%
Services	214.3	238.0	258.5	274.4	14.3%	11.1%	8.6%	6.1%
Government	115.9	116.3	116.5	116.7	–5.5%	0.3%	0.2%	0.2%
Total	868.3	871.7	873.9	873.3	–2.9%	0.4%	0.2%	–0.1%
City of Cleveland								
Construction	9.1	8.6	8.1	7.6	12.3%	–5.5%	–5.8%	–6.2%
Manufacturing	81.5	73.9	67.0	60.9	–11.9%	–9.3%	–9.3%	–9.1%
TCPU	24.0	23.0	22.0	20.9	–18.6%	–4.2%	–4.3%	–5.0%
Wholesale Trade	23.5	21.9	20.4	19.0	–2.1%	–6.8%	–6.8%	–6.9%
Retail Trade	31.1	28.4	25.9	23.6	3.0%	–8.9%	–8.8%	–8.9%
FIRE	24.2	24.4	24.5	24.7	11.5%	0.8%	0.4%	0.8%
Services	95.6	95.0	93.4	91.0	127.1%	–0.6%	–1.7%	–2.6%
Government	53.2	52.2	51.1	50.0	48.2%	–1.9%	–0.2%	–2.1%
Total	342.2	327.3	312.5	297.7	20.5%	–4.3%	–4.5%	–4.7%

Source: Center for Regional Economic Issues, Cleveland Economic Analysis and Projections, Cleveland, Ohio: Weatherhead School of Management, Case Western Reserve University, Executive Summary, Table 2, 1987.

[1] Includes transportation, communications, and public utilities.

[2] Includes finance, insurance, and real estate.

from June 1990 through February 1992, noted that Cleveland had lost 42,000 jobs, including 14,000 jobs in manufacturing, 6,000 jobs in construction, and 13,000 jobs in wholesaling and retailing. The PMSA suffered a total loss of 127,000 jobs, primarily in manufacturing (62,000), retail trade (31,000), construction (29,000), and wholesale trade (5,000).

More critical to the economic health of Cleveland is projected job losses in all occupational categories in the city's neighborhoods. Both relative and absolute job losses are expected. Moreover, neither downtown Cleveland, Cuyahoga County, nor the PMSA can be counted on to provide replacement jobs. If regional outmigration does not transpire, a job shortfall is imminent, with serious consequences for displaced, unskilled, and functionally unemployed workers.

Housing Markets in Disarray

Economic restructuring, interregional population shifts, and intraurban population movements are significantly affecting the economic viability of the city's housing. Structural shifts and cyclical fluctuations in the economy are causing household losses, increasing mortgage default, and concomitant increases in poverty. Moreover, public sector efforts to achieve diversity and increase the vitality of neighborhoods by reducing the isolation of low-income and minority groups have not promoted

Table 9.2.5
Total Employment by Occupation (in Thousands)

YEAR	1970	1980	PROJECTED 1985	PROJECTED 1990	PROJECTED 1995	PROJECTED 2000
GEOGRAPHICAL AREA						
Cleveland PMSA						
Professional/Technical	82.9	107.2	153.8	159.1	163.5	166.8
Executive/Management	60.9	70.3	76.3	77.4	78.3	78.9
Sales	54.1	61.2	62.8	63.5	64.0	64.3
Administrative Support	135.6	140.0	166.3	170.3	173.8	176.5
Precision Production/Crafts	87.7	84.7	101.3	96.9	93.0	89.3
Service Workers	73.4	81.2	134.3	140.9	146.4	150.6
Operatives/Laborers	161.7	132.8	173.5	163.7	154.9	146.9
Total	656.3	677.4	868.3	871.7	873.9	873.3

Source: Center for Regional Economic Issues, Cleveland Economic Analysis and Projections, Cleveland, Ohio: Weatherhead School of Management, Case Western Reserve University, 1987, pp. 74, 75.

racial and economic integration, spatial deconcentration, or interjurisdictional mobility (Margulis 1982). Efforts to secure neighborhood stability through the geographic and economic targeting of grants and loans for housing rehabilitation in the 1970s and 1980s have not altered the real estate market's tendency toward disinvestment and property tax delinquency (Margulis 1996, 1988, 1987; Wilson, Margulis, and Ketchum 1994; Margulis and Sheets 1985).

A recent study of elderly homeowners in Cleveland and its suburbs to determine how the processes of recomposition and congregation are affecting the housing choices of younger households and how the effects of cumulative inertia and frozen occupancy are influencing the condition of housing occupied by the elderly reveals that in aging neighborhoods (communities with large numbers of older homeowners), housing turnover has slowed. Younger, first-time buyers, forced to compete for affordable housing outside these aging neighborhoods, find a more limited space offering fewer housing amenities. In addition, as a consequence of cumulative inertia, elderly-owned properties are diminishing in value, and many aging homes nearing the end of their depreciation cycle are at risk of falling out of the housing inventory in the near future (Margulis 1993). These turn-of-the-century 1-, 1½-, 2-, and 2½-story properties are no longer fashionable, costly to rehabilitate, and largely unsalable. Ravaged by age, the cost of rehabilitating these architecturally obsolete structures exceeds their resale value.

As the number of households contracts, the city has been left with an excess of housing units. Real property transfers for 1991 show 3,750 vacant residential parcels that are land banked, 10,686 parcels that contain occupied buildings, and 4,840 unoccupied buildings that are tax delinquent. At the same time, decreasing housing demand has caused median sales prices to fall. In contrast, by the end of the 1980s median sales prices in the suburbs were outpacing gains in the city. However, in the inner suburban ring of older pre–World War II and postwar housing, median housing values have increased modestly, as younger homebuyers priced out of newly constructed peripheral suburban housing seek starter homes in these older communities.

Slackening housing demand threatens to reproduce some of the conditions that have devasted inner-city neighborhoods. A deficit of new residential construction in the city, continued construction of housing in the outermost suburbs, and relative price depreciation in the inner-ring suburban housing stock are hastening the outward movement of minority and marginal working-class families from the inner city. Since demographic projections predict a decline in the region's population and households, the prime source of suburban-bound migrants appears to be the central city. Since the city's minority population is also declining, intraurban suburbanization should slacken, as the pool of upwardly mobile home-seekers shrinks. Relative depreciation in suburban housing markets appears as

Table 9.2.6

Residential Segregation in Cleveland, 1990

COMPARISON GROUP	DISSIMILARITY INDICE[1]	
	CLEVELAND	CUYAHOGA COUNTY
White NH[2] vs. Afro-American NH	87.6	84.3
White NH vs. Asian[3]	44.9	34.0
Afro-American NH vs. Asian	86.9	80.9
Afro-American NH vs. Hispanic[4]	87.0	82.3
White NH vs. Hispanic	46.9	55.4
Hispanic vs. Asian	50.8	56.0

Source: 1990 Census of Population and Housing, Cleveland Metropolitan Area, STF1A

[1] The index of dissimilarity has the equality $ID=\frac{1}{2}ABS / (H_i/H) - (A_i/A) /$ where H_i and A_i are the population subtotals of two ethnic categories (Hispanics and Afro-Americans) in each areal unit, and H and A are population totals of each group. The absolute total is divided by one-half and result is multiplied by 100.00. The ID can be interpreted as the percentage of either group that would have to move in order to eliminate segregation between the groups and therefore produce a score of zero.

[2] NH equals non-Hispanic.

[3] Includes Chinese, Filipino, Japanese, Asian Indian, Korean, Vietnamese, Cambodian, Hmong, Laotian, Thai, and other Asians.

[4] Includes Mexicans, Puerto Ricans, Cubans, and other Hispanics.

an endemic and implacable feature of the suburban landscape.

The Fraying Social Fabric

Race relationships remain divisive in Cleveland. While the total number of African Americans in the city diminished from 1970 to 1980, overall segregation of minorities did not decline. In fact, the 1990 census shows that the vast majority of census tracts on Cleveland's east side were over 75 percent African American, with many close to 100 percent (Reshotko 1991). Table 9.2.6 shows the index of dissimilarity for subsets of the populations of Cleveland and Cuyahoga County. In comparing the indices for whites, African Americans, Hispanics, and Asians, no index involving African Americans is lower than 80.9, the index for African Americans versus Asians in Cuyahoga County. At the same time, no index not involving African Americans is higher than 56.0, that for Hispanics versus Asians in Cuyahoga County. The highest index is 87.6 for whites versus African Americans in Cleveland. The lowest is 34.0 for whites versus Asians in Cuyahoga County.

The high concentration of African Americans on Cleveland's east side is the obvious cause of the high dissimilarity index for African Americans in the city. Yet the county index is only slightly lower at 84.3. To some degree, central city segregation has given way to suburban resegregation, since the population of suburbs—such as East Cleveland and Warrensville Heights—are almost entirely African American and substantial minority populations are now found in Bedord Heights, Cleveland Heights, Maple Heights, North Randall, and Shaker Heights.

In comparison, Cleveland's whites, Hispanics, and Asians are relatively well integrated, with dissimilarity indices that range from 44.9 to 50.8 (Margulis 1991). Whites and Asians are especially well integrated in Cuyahoga County, with an index of 34.0. The dissimilarity indices for whites versus Hispanics and Hispanics versus Asians are higher in the county, reflective of the strong Hispanic base on the west side of Cleveland.

Darden (1987) suggests that African Americans remain segregated because higher social and economic status cannot easily be converted into residential and neighborhood quality due to discrimination in housing. Perhaps more insidious is the chronic unemployment caused by losses in manufacturing and governmental jobs. According to Hill (1992), the 1990 unemployment rate for African American men was 16.8 percent, and for African American women, 13.3 percent. The nonemployment rate was even higher. In effect, while the city has made remarkable progress in reversing downtown decline, African Americans and their neighborhoods have not felt the effects of these changes. Cleveland thus remains one of the most segregated American cities.

Cleveland Looks to the Future

For Cleveland to survive as a visible economic entity, a set of alternative policy models and scenarios must be developed. One such model is the city's Civic Vision 2000 Plan (Cleveland City Planning Commission 1988, 1989), essentially an urban development strategy package that outlines specific policy directions and provides benchmarks against which development can be assessed.

The Downtown Plan envisions redirecting the path of development by accommodating change and upgrading institutions and environmental conditions. A number of goals are identified, all designed to strengthen the downtown by creating a high-quality commercial center and a core of surrounding activities, especially amenity clusters—public open spaces and parks—which capitalize on the downtown's physical location. To make the city more people oriented and thus attractive to the middle class, the plan proposes to create new downtown housing and to develop convention and tourist activities. The need to retain light industry and service industries in the downtown area is also stressed.

Sustained efforts are being made to develop additional office, retail, and hotel space. All government entities are being encouraged to build new office space downtown and consolidate dispersed governmental facilities in downtown centers. A key objective is to establish the downtown as a regional sports and entertainment center, hence the city's emphasis on the Gateway Project and the Inner Harbor developments. To rejuvenate downtown retailing, reinvestment in compact, pedestrian-oriented retailing districts around Public Square (Figure 9.2.2) is being fostered. Other retail districts—the Euclid-Prospect District, the Warehouse District, East 9th Street–Erieview, Playhouse Square, and Cleveland State University—are to be strengthened, with programs directed at upgrading each area's physical appearance. These programs are to include improvement of major gateways to the downtown, streetscape beautification, facade rehabilitation, development of multiuse public open space (particularly along Lake Erie and the Cuyahoga River), street-level retail space, pedestrian connectors between major retail complexes, and well-designed, dual-use parking garages. Where necessary, incentives will be provided to developers to ensure that the downtown continues to absorb office development and to function as the region's chief office market.

To retain downtown industrial districts—such as the port of Cleveland, the Lakeside Industry District, Flats Oxbow North, and Flats Oxbow South—infrastructure improvements, strict code enforcement, design review, and financial incentive programs are proposed. Furthermore, these areas are to be promoted as unique, centrally located industrial districts linked to and benefiting from the office and retailing complexes developed in the downtown core.

Improving the transportation network would enhance regional accessibility and solidify the downtown area as a regional hub. The most ambitious transportation project, the Dual Hub Corridor Project, proposes developing a five-mile light rail system connecting Tower City Center in the downtown with the businesses, institutions and neighborhoods located along the eastbound Euclid Avenue corridor. More than 160,000 people are employed along the corridor, and buses operating there carry 62 percent of the region's public transit ridership (about 128,000 people per day).

Rebuilding a city for the twenty-first century essentially involves regulatory reform—to revise the city's zoning code and zoning map—resource development, and a reconfiguration of existing downtown land use patterns to reinforce agglomeration. The Downtown Plan consists of a set of goals and policies for guiding as well as refining public programs. Since the city of the future should be a place of high productivity in knowledge-based industries, the plan focuses on creating a concentration of central office administrative functions in an attractive downtown environment.

Judging development proposals essentially involves determining whether the proposed plans are compatible with surrounding land uses and consistent with the plan's goals and policies. Policy implementation and land use changes require that the city allocate public funds for capital improvements, and the plan provides a means for prioritizing, monitoring, and evaluating these capital improvements. In essence, the plan broadly describes appropriate land use types for each site, general patterns of development, and directions for future development. Because the plan will likely take years to achieve, it is intended to be flexible, thus accommodating alternative uses and changing conditions.

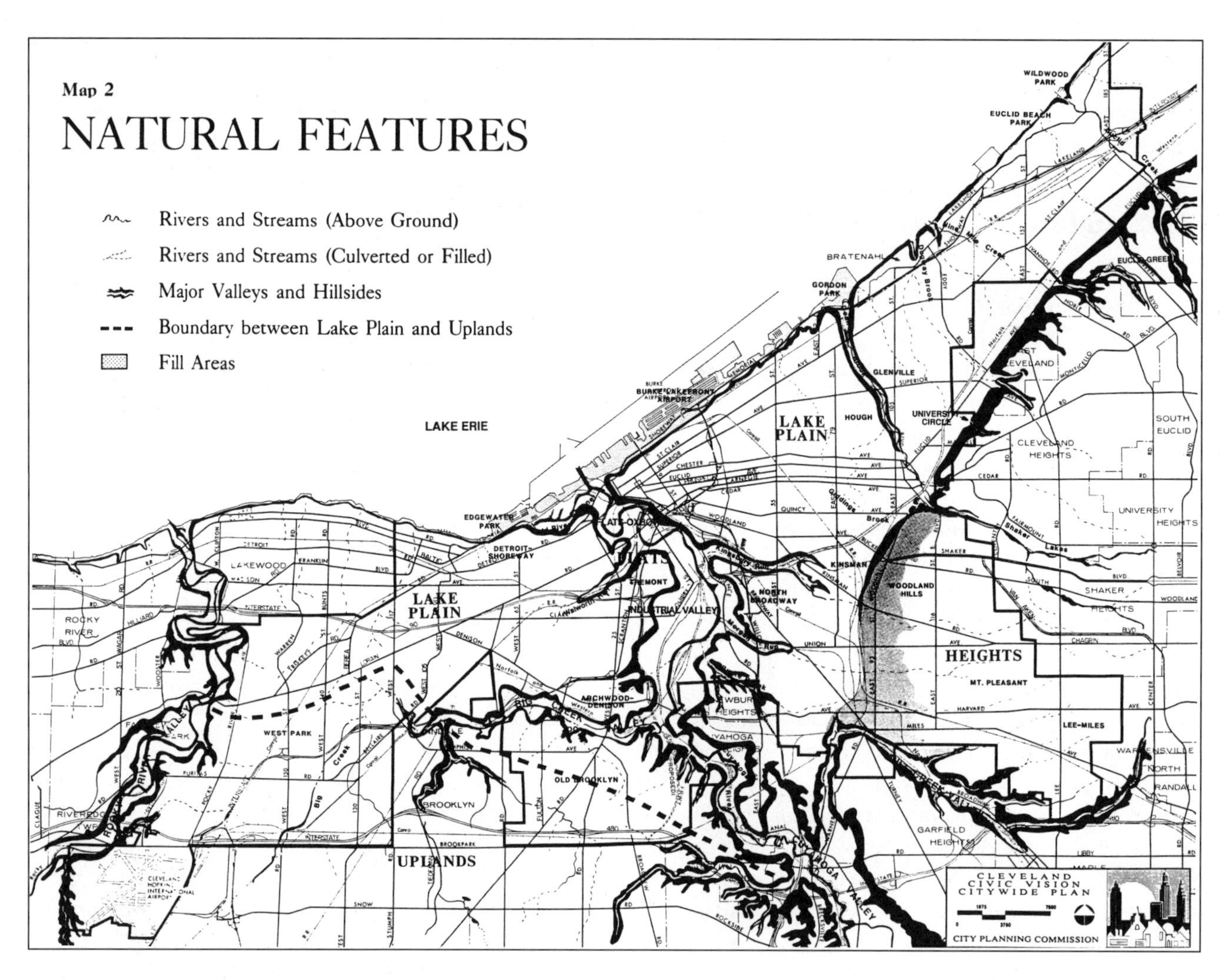

Fig. 9.2.2. Cleveland's Natural Features (Map Courtesy of Cleveland City Planning Commission)

As an illustration of how the plan could operate, one goal proposes creating neighborhood conditions sufficient to meet the needs of all residents, no matter their age or income. A number of alternative policy directions can be used to achieve this goal. For example, the city could facilitate adaptive reuse of marginal retailing facilities for low-density housing or conversion of buildings to multifamily residential uses. Alternatively, public assistance could continue to be directed to housing rehabilitation or new construction. Large vacant sites could potentially be reused to construct comprehensively designed residential developments or affordable, quality-designed manufactured housing. Most important, the report identifies potential sites where public investment could stimulate housing development or where existing housing could become more strongly anchored by new, compatible development.

In stabilizing neighborhoods, other choices might involve consolidation of scattered retail businesses, commercial renovation, financial assistance for business relocation, promotion of neighborhood entertainment centers, or encouragement of state legislation permitting creation of commercial assessment districts or community-based receivership programs. Even consolidation of multiuse recreational facilities at accessible locations and expansion of recreational facilities on the lakefront and riverfront are possible options. Other possibilities involve relocating industrial and office park developments near sites with freeway access, providing assistance in relocating incompatible industries, or improving infrastructure to support industrial retention or expansion.

Reversing decline is clearly a long-term process that must follow the ebb and flow of opportunities. In contrast, doing nothing can only result in further economic and physical decay, and inevitably, the demise of Cleveland. The Civic Vision 2000 Plan permits the city's private and public leadership to advance agendas for rebuilding. Though the process is imperfect and fraught with pitfalls, it nevertheless permits the city to face its future and influence its own fate. In other words, the zoning map enables the city's leadership to be proactive rather than reactive in confronting environmental changes. Population shrinkage has created opportunities to rebuild neighborhoods, diversify employment, and rationalize land uses. What is merging is a vision of a new Great Lakes city, a city able to meet the challenges of the twenty-first century.

REFERENCES

Cleveland City Planning Commission. 1988. *Cleveland Civic Vision 2000 Downtown Plan: Civic Vision 2000 Citywide Plan.* Pt. 1: Citywide Subject Chapters; Pt. 2: Sub-Area Analysis Chapters. Cleveland, Ohio.

———. 1989. *Civic Vision 2000 Citywide Plan,* Pt. 1: Citywide Subject Chapters. Cleveland, Ohio.

Darden, Joe T. 1987. Choosing Neighbors and Neighborhoods: The Role of Race in Housing Preference. In Gary A. Tobin, ed. *Changing Patterns of Racial Segregation.* Newbury Park, Calif.: Sage Publications.

Gleisser, Brian S., and Harry L. Margulis. 1988. Housing Rehabilitation and Enterprise Zones: Cleveland's Target Area Investment Program. *The East Lakes Geographer* 23:96–110.

Hill, Edward. 1992. Prospective: Contested Cleveland. *Urban Affairs Association Newsletter* (winter): 2–6.

Margulis, Harry L. 1982. Housing Mobility in Cleveland and Its Suburbs. *The Geographical Review* 72 (January): 36–49.

———. 1987. Neighborhood Perception and Housing Maintenance in Older Suburban Communities. *Urban Geography* 8:232–50.

———. 1988. Homebuyer Choices and Search Behavior in a Distressed Urban Setting. *Housing Studies* 3 (April): 112–33.

———. 1991. Creating an Asian Village in Cleveland Ohio: A Case Study of Planned Urban Morphogenesis and Urban Managerialism. *The East Lakes Geographer* 26:15–25.

———. 1993. Neighborhood Aging and Housing Deterioration: Predicting Elderly Owner Housing Distress in Cleveland and Its Suburbs. *Urban Geography* 14 (1): 30–47.

———. 1995. Housing Credit Lending and Housing Markets: A Canonical Analysis of Pooled Longitudinal Data. *Urban Affairs Review* 3 (September): 77–103

Margulis, Harry L., and Catherine Sheets. 1985. Housing Rehabilitation Impacts on Neighborhood Stability in a Declining Industrial City. *Journal of Urban Affairs* 7 (summer): 19–35.

Miller, Carol Poh, and Robert Wheeler. 1990. *Cleveland: A Concise History, 1796–1990.* Bloomington: Indiana University Press.

Reshotko, Adam. 1991. The Racial Makeup of Cleveland: White, Black, Hispanic, and Asian. First College, Cleveland State University, Cleveland, Ohio. Unpublished manuscript.

Wilson, David, Harry L. Margulis, and James Ketchum. 1994. Spatial Aspects of Housing Abandonment in the 1990s: The Cleveland Experience. *Housing Studies* 9 (4): 493–510.

CINCINNATI
Diversified and Stable

Howard A. Stafford

Cincinnati (population 1,744,124), the twenty-third largest metropolitan area in the United States (U.S. Bureau of Census 1990), was first settled by Europeans a little over 200 years ago. Since then the city has spread up the tributary valleys of the Ohio River and across the border into Kentucky. However, it was approximately 7,000 years ago that the first human inhabitants came to the lands along the Ohio River, between the Great and Little Miami Rivers. The Folsom peoples, the Mound Building Indians, and the Hopewell Indians first hunted and fished the abundant game of the region and later farmed the fertile and well-watered land. The first European settlement was on an Ohio River floodplain, across from the northward flowing Licking River. Here a small village was established by settlers floating down the river on flatboats (Silberstein 1982).

Almost from its beginning in 1788, Cincinnati's main business was commerce. Originally a loading and unloading center on the Ohio River, the city evolved into the major transportation route in the region. Soon, local manufacturing developed. These early characteristics foreshadowed Cincinnati's future. Being approximately 100 miles from Columbus, Ohio, Indianapolis, Indiana, and Louisville, Kentucky, and 85 miles north of Lexington, Kentucky, Cincinnati is now the commercial center for southwestern Ohio, northern Kentucky, and southeastern Indiana, and today its large and diverse manufacturing output serves markets worldwide.

By 1826 the population was over 16,000, of whom an estimated 800 were employed in trade and mercantile pursuits, 500 in navigation, and 3,000 in manufactures. In 1826 it was predicted that because of "its geographical features [which] indicate Cincinnati as possessing greater local advantages than any other site within this region . . . the rapid growth of Cincinnati may be safely predicted" (Drake and Mansfield 1827). The prediction was accurate. In the 1840s Cincinnati was known as "Porkopolis," a reflection of its role as the largest pork-packing center in the world. Today the meatpacking industry is almost gone, but its legacy continues; over the entrance to a new city park along the reviving waterfront are two "flying pig" sculptures.

Manufacturing remains strong, but in different forms, for the Cincinnati region's manufacturing sector has diversified. Located in the eight-county Cincinnati Consolidated Metropolitan Statistical Area—Hamilton, Butler, Warren, and Clermont Counties in Ohio; Boone, Campbell, and Kenton Counties in Kentucky; and Dearborn County in Indiana—are over 3,000 manufacturing plants. Major companies with more than 1,000 employees include General Electric (aircraft engines), Procter and Gamble (soaps, food, toiletries), Kroger (food stores and processing), Armco (steel), Cincinnati Milicron (machine tools), Ford Motor Company (automatic transmissions), Kenner Products (toys), Avon Products (cosmetics), U.S. Shoe Corporation (apparel and related products), Merrell Dow Pharmaceutical (pharmaceuticals), Gibson Greetings (greeting cards), Monsanto (plastics), and Steelcraft (metal doors). Manufacturing, however, no longer reigns as the dominant sector of the region's economy. Today, in line with statewide and nationwide trends, 25 percent of Cincinnati's labor force is in services, 20 percent in retail trade, 20 percent in manufacturing, 12 percent in government occupations (including education), and the remaining 23 percent divided among construction, transportation, public utilities, wholesale trade, finance, insurance, and real estate. Overall, the economy is healthy and expanding, supporting a total regional population of some 1,700,000.

The cultural attributes of the region's people are very typically American. The area is predominantly white and Christian, but there are significant numbers of African American and Jewish residents, with other races and religions represented in lesser numbers. The earliest settlers were of British heritage. Then in the mid 1800s there was a large influx of German migrants, giving Cincinnati a German flavor that remains today. The majority of these Germans were Roman Catholic, but many were Jewish; in fact Reform Judaism was born in Cincinnati. Excepting concentrations of white Appalachians or African Americans in some inner-city areas, the Cincinnati region does not have readily identifiable ethnic or racial subareas. Generally, unlike other large United States cities, there is a fairly even cultural mix across the region.

The Economic Base

Hamilton County is the core of the Cincinnati region, and the city of Cincinnati the heart within the county. This initially compact city—set in a basin formed by surrounding hills and, to the south, the Ohio River—began to expand up the tributary valleys, especially the Mill Creek. When all available flat land was occupied, expansion continued up the hillsides and onto the surrounding uplands. Though the earlier hilltop suburbs were incorporated into the political city of Cincinnati, the more distant, newer suburban areas have remained independent. Early commercial activity was likewise concentrated in the basin but has steadily expanded outward, predominantly along the valleys and especially, in the past three decades, along the routes of the controlled access highways. The central city grew until the mid 1950s but since then has declined by approximately 30 percent, while the metropolitan area has continued to grow. There are now more people living in suburban Cincinnati than in the central city. The central city still has more manufacturing plants than the suburbs, but the rate of change favors the suburbs; it appears only a matter of time before the majority of manufacturing is located in the suburbs.

The key to Cincinnati's changing urban spatial structure is the locational pattern of controlled-access highways. These are I-75, which runs generally north-south along the historically important Mill Creek Valley and on into Kentucky; I-71, which comes in from the northeast, intersects I-75 downtown, and continues southwestward into Kentucky; I-74, which runs from the central city northwest to Indiana; two east-west interconnectors, which link I-75 and I-71, known as the Norwood Lateral and Cross County Highway; and I-275, which circles the region approximately 15 miles from downtown Cincinnati (Figure 9.3.1). The modern importance of these highways can be illustrated by looking at the office buildings in the region bounded by I-275.

Thirty years ago virtually all major office buildings were located in Cincinnati, with most of these in the central business district (CBD). Now the pattern is dramatically different (Figure 9.3.2). Though the central city still dominates and major new office buildings continue to be built in this core area, major office centers are springing up in the suburbs, including concentrations in Kenwood, Blue Ash, and the Tri-County area in the north; and Florence, Kentucky, and the Greater Cincinnati International Airport region to the southwest. It appears that the commercial future of the CBD will be mainly to provide office space for large corporations and the many smaller support firms, such as lawyers and accountants, together with other services, such as restaurants and print shops.

Clearly, there is a correlation between the pattern of office location and the controlled-access highways: the central city lies at the convergence of I-75, I-71, and I-74; Florence is served by I-75 and I-71; and the airport is accessible via I-275. Likewise, the most important concentration of suburban offices, that to the north, lies astride I-75, I-71, and the connecting link section of I-275. The suburban office concentrations are generally in the same areas that have experienced the most suburban population expansion. However, though major suburban growth has occurred near controlled-access interchanges, the existence of such an interchange does not guarantee explosive growth. The western portions of Hamilton County, for instance, now have controlled-access highways, but intensive commercial usages have not developed to the same degree as in the triangle formed by Florence, the central city, and the Tri-County–Blue Ash area.

Manufacturing will continue as a critical part of Cincinnati's economy, with services and other forms of commerce steadily assuming even greater importance. The regional economy—historically resilient, not prone to periods of either boom or bust—is diverse and healthy and is expected to continue its current steady

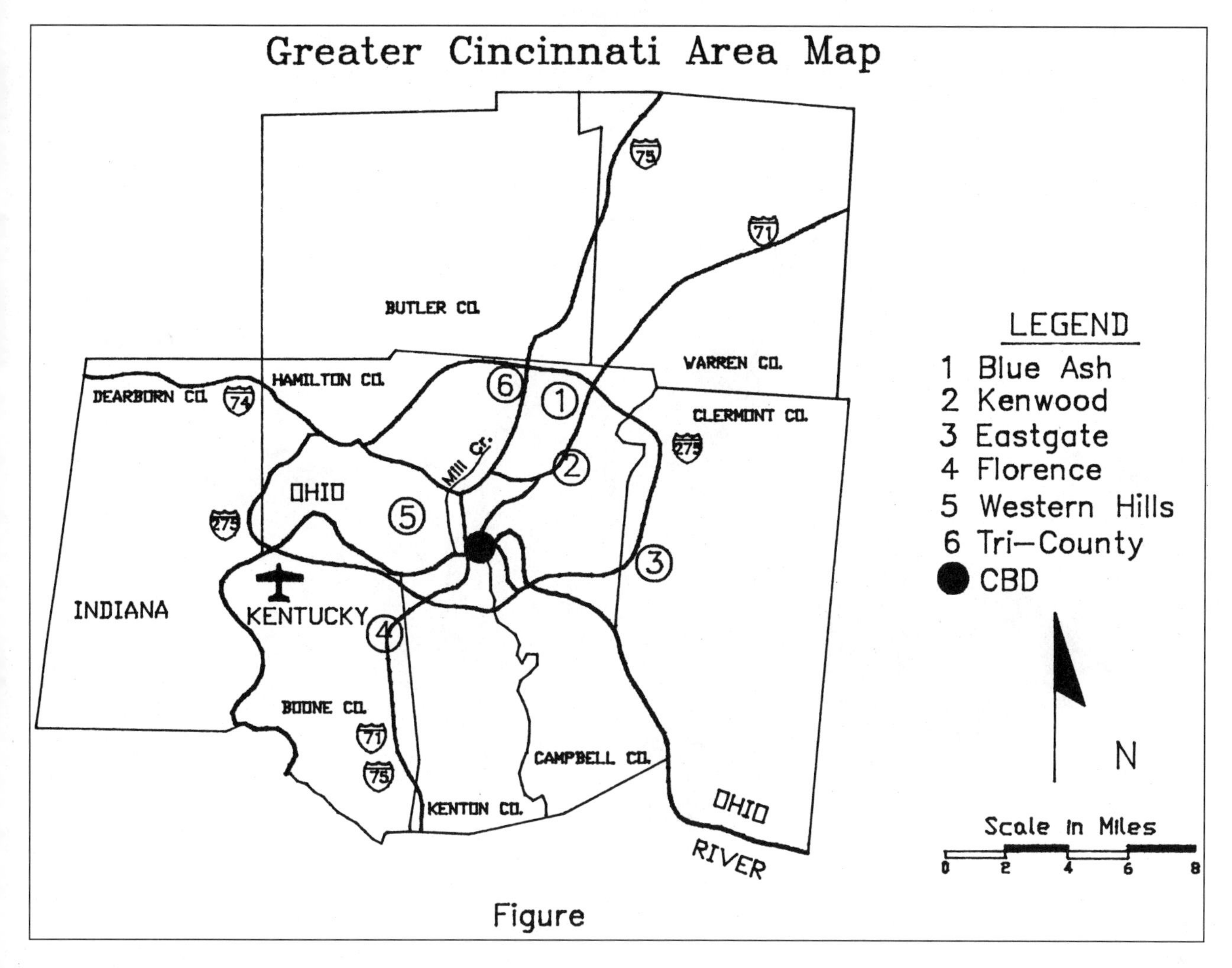

Fig. 9.3.1. Greater Cincinnati Area Map

growth. Future suburban expansion seems assured. For the next few years, the northern and southwestern suburban nodes will likely expand most rapidly. These areas are already experiencing the problems of growth, especially increasing land and building costs and traffic congestion. Local governments will be forced to provide more intermediate roads and sewers and expand other utilities, all of which will further increase costs. At some point in the near future, these communities will lose their suburban advantages, and expansion will accelerate to other, less crowded suburban nodes. This process has already begun in the Eastgate area northeast of downtown and in western Clermont County. Major developments have not yet been built to the northwest, out along I-74, but it seems likely that planning and construction will begin soon.

Development of Population

Population changes within the region provide a general view of growth and spatial redistribution (Tables 9.3.1 and 9.3.2). According to figures for the eight-county Consolidated Metropolitan Statistical Area (CMSA) and the political city of Cincinnati, during the 100-year period from 1850 to 1950, all areas grew in resident population, with the regional total increasing from 305,000 to 1,170,000, an almost fourfold increase in 100 years. The years 1950–90 saw continued growth to a population of 1,744,000, a 49 percent increase in 40 years. Again, all counties showed an increase, but the city of Cincinnati, for the first time in its history, lost population, a loss of approximately 25 percent.

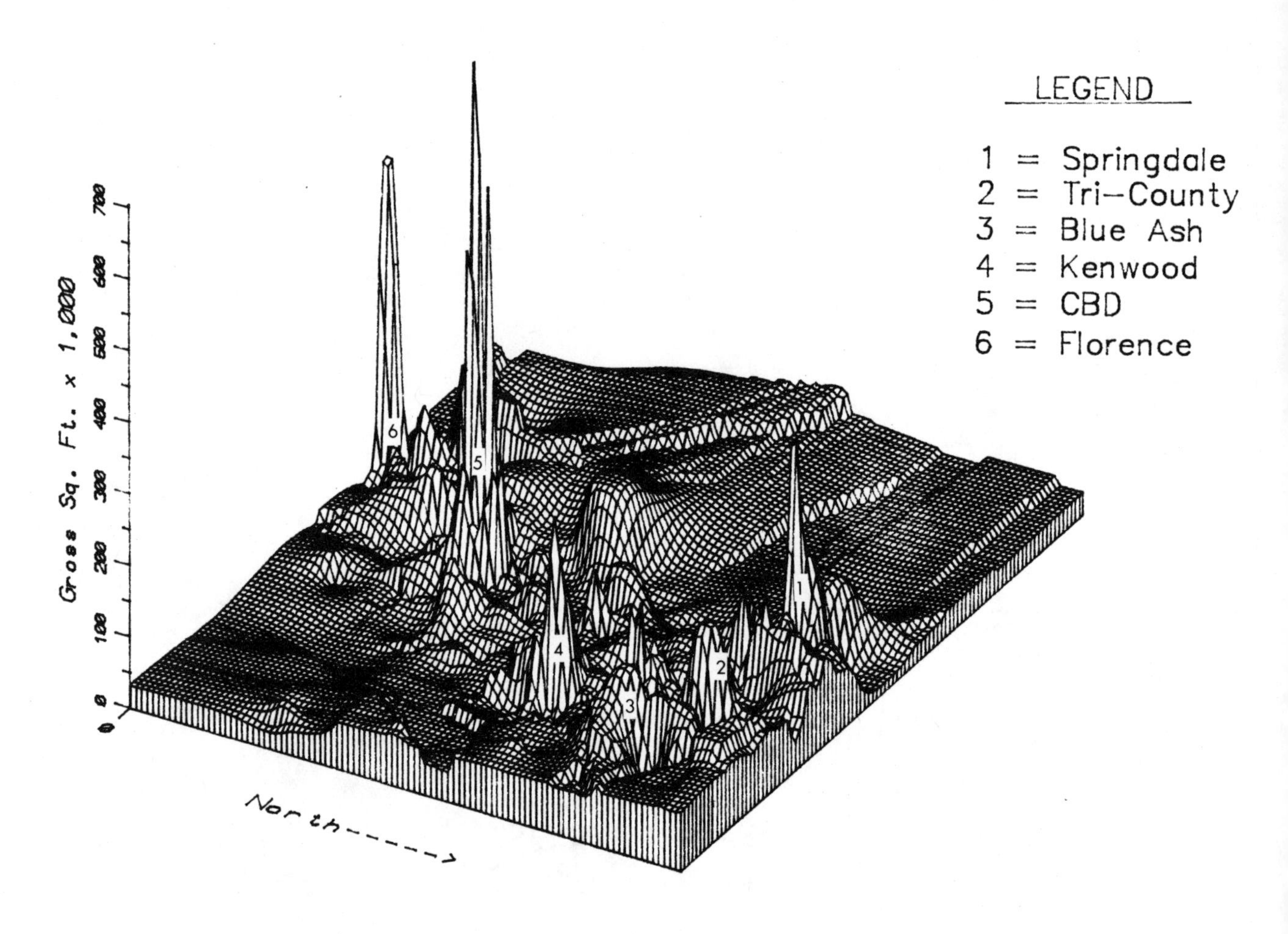

Fig. 9.3.2. Cincinnati Office Landscape

It is clear from Table 9.3.1 that population is not evenly distributed over the region. In 1850, Hamilton County had fourteen times more people than the smallest county, Boone in northern Kentucky. In 1985, Hamilton County had twenty-four times more people than Dearborn County in southeastern Indiana. Likewise, the rates of change have been very uneven, as shown in Table 9.3.2. During the period 1850–1900, some areas grew little or not at all, while other areas saw large increases. Most of the growth was in the central counties of Hamilton, Campbell, and Kenton; the city of Cincinnati alone gained more people than the rest of the subareas combined, an almost threefold increase over its 1850 population. The growth patterns were different in the 1950–90 period. Growth slowed in the central counties of Hamilton and Campbell, while rates of increase exploded in the more peripheral counties, especially Warren, Clermont, and Boone.

The last column of Table 9.3.1 shows the projected populations in absolute numbers for the year 2010, and the last column of Table 9.3.2 shows the projected rates of change from 1990 to 2010. The total regional population is expected to remain relatively stable, reflecting continued, but not excessive, vitality. As tables 9.3.1 and 9.3.2 show, the core counties of Hamilton and Campbell are expected for the first time to decrease in population. Within Hamilton County, the city of Cincinnati probably will decline somewhat. In contrast, the peripheral counties in Ohio and Indiana and the closer northern Kentucky counties of Boone and Kenton are expected to continue growing. The greatest relative increases are expected in Boone County, Kentucky, and in Clermont County, Ohio.

Clermont County illustrates the suburbanization trends which continue throughout the region. Just a few years ago Clermont was predominantly rural, but in the past decade or so it has been rapidly transformed into a more urban or suburban center. Some of Clermont's change has been the result of new manufactur-

Table 9.3.1
Population

COUNTY	1850	1900	1950	1990	2010
Butler	30,789	56,870	147,203	291,479	291,713
Hamilton	156,844	409,479	723,952	866,228	801,113
Warren	25,560	25,584	38,505	113,909	121,187
Clermont	30,455	31,610	42,182	150,187	176,960
Boone	11,185	11,170	13,015	57,589	71,577
Campbell	13,127	54,223	76,196	83,866	78,913
Kenton	17,038	63,591	104,254	142,031	152,050
Dearborn	20,166	22,194	25,141	38,835	42,320
Regional Total	305,164	674,721	1,170,448	1,744,124	1,735,813
City of Cincinnati	115,435	325,902	503,998	364,040	352,000

Table 9.3.2
Percentage Change of Population

COUNTY	1850–1900	1900–1950	1950–1990	1990–2010
Butler	84%	158%	85%	7%
Hamilton	161%	77%	8%	–6%
Warren	0	51%	178%	13%
Clermont	4%	32%	250%	20%
Boone	0	17%	349%	23%
Campbell	313%	41%	6%	–2%
Kenton	273%	64%	41%	3%
Dearborn	10%	13%	52%	11%
City of Cincinnati	182%	55%	–28%	–2%

ing plants, but most growth has occurred in residential and retailing areas. Some of Hamilton County's most desirable residential areas lie on the east side, and suburban housing has spilled over into Clermont County. Perhaps even more significant has been the completion of the I-275 beltway to the east, which has brought Clermont County, especially its western portion, closer to the rest of the metropolitan area. Boone County, Kentucky, likewise owes its growth to improved access. Located astride I-75, the single most important highway in the region, and housing the Greater Cincinnati International Airport, Boone County is now a choice area for new manufacturing plants and suburban office parks. Its northern counterpart, in terms of manufacturing and office growth, is the area running through northern Hamilton County and southern Butler and Warren Counties, where a section of I-275 is anchored by I-75 and I-71.

Conclusion

The Cincinnati metropolitan area always has been dependent on commerce and will remain so. Its patterns of growth have been both a contributor to and a consequence of transportation and access. The first important transportation routes were the Indian trails and the Ohio River; the Miami and Erie Canal and the railroads were added in the 1800s. Then truck and automobile transportation became dominant, and roads were built where needed; in turn, the patterns of these roads further influenced growth patterns. The best predictors of internal spatial changes in the region over the next two decades are the present locations of controlled-access highways. In the year 2010, Hamilton County will continue as the largest county by far, and Cincinnati will still be the county's largest city, but neither will be as overwhelmingly dominant as they once were.

REFERENCES

Drake, B., and E. D. Mansfield. 1827. *Cincinnati in 1826.* Cincinnati: Morgan, Lodge, and Fisher.

Silberstein, Iola. 1982. *Cincinnati Then and Now.* Cincinnati: League of Women Voters.

Editor's Note: Population figures prior to 1985 are from *Population Abstracts of the U.S.,* 1983, U.S. Bureau of the Census. Population figures for 1990 are from the *State and Metropolitan Area Data Book,* 1991, U.S. Bureau of the Census. The county projections for 2010 were published in the *Cincinnati Enquirer,* November 26, 1989, and are from the Ohio Department of Development, the University of Louisville Urban Studies Center, and the Indiana Board of Health. The projection to 2010 for the city of Cincinnati was made by the author.

TOLEDO
A City in Transition

William A. Muraco

Metropolitan Toledo provides an excellent illustration of a traditional manufacturing city confronted by several dynamic geographic challenges, including restructuring related to interregional and international relocations of its traditional economic base. The city has enthusiastically confronted the challenges of downtown revitalization and regional development, within the context of these major economic issues.

Geographic Situation

Located along the estuary of the Maumee River and stretching for 5 miles inland from the western end of Lake Erie (see Figure 9.4.1), Toledo has historically been perceived as a gateway city with outstanding water and land access to a productive agricultural hinterland and to major local and international markets. With respect to total tonnage shipped, the city is the largest Ohio port and the second largest Great Lakes port. In 1988 the port of Toledo shipped 14.4 million tons of cargo, including exports of grains and coal and imports of iron ore, steel, potash, and other general cargo (Harvey 1989). In addition to providing access to the Great Lakes and the St. Lawrence Seaway, the city serves as a major railroad hub. It has the largest number of high-tonnage railroad lines of Ohio's major cities: four Class A lines (20 million tons or more annually) and one Class B line (5–20 million tons annually).

Complementing the excellent rail system are interstate highway links that place Toledo at the intersection of major east-west (I-80/I-90) and north-south (I-75) routes. The excellent accessibility of the city to the interstate highway system has attracted more than 100 motor freight lines that maintain terminals or offices in the Toledo area. Two new interchanges that provide direct connections between I-80/I-90 (the Ohio Turnpike), I-75, and Toledo Express Airport were completed in 1990. It is anticipated that these new interchanges will encourage even greater growth in transportation-related and air cargo industries.

The Population

The Standard Metropolitan Area of the city contains 614,128 persons residing in a three-county area (Bureau of the Census 1992). The city of Toledo, with a population of 332,943, is located in Lucas County. Suburban development in Toledo is relatively strong, with approximately 44 percent of the Standard Metropolitan Area's population residing outside the central city. This strong suburban orientation has been a primary factor in the loss of retailing and employment activity in the downtown. The city is ringed with four major enclosed malls and numerous suburban strip developments.

The urban density gradient is characterized by a very uneven distribution across the metropolitan area. The Maumee River has acted as a major perceptual barrier, with much of the development east and southeast of the river associated with heavy industries—such as oil refining and glass production—and rail and water port activities. The northeastern quadrant of the city has also seen considerable industrial development, primarily auto parts production and transportation service industries. The industrial character of the eastern portions of the city has generally deterred residential development, encouraging a very strong sectorial growth pattern generating from the central business district to the northwest, the west, and the southwest. The development of the northwestern and southwestern

Fig. 9.4.1. Power Plants Along Maumee River

high-rent corridors is consistent with the Hoyt sector model of urban development. The completion of the I-475 outer belt in the late 1960s helped to intensify high-value residential, planned industrial, and commercial developments in the western sectors of the city.

The social geography of the Toledo area is consistent with the findings of the classical and factorial models of urban structure. The strong sectorial density of the residential areas demonstrate considerable variations in social status and ethnicity. High social status neighborhoods exist upstream from the CBD on both sides of the Maumee River. The largest high rent sector is located to the northwest of the downtown, originating in a historically rich Victorian area, the "Old West-End," and moving outward through the village of Ottawa Hills to the suburban city of Sylvania. Cultural features prominent in this sector include the prestigious Toledo Museum of Art, the University of Toledo, four major hospitals, and several large shopping districts. An emerging growth corridor is developing due west of the downtown and out toward the Toledo Express Airport, which is located in the far western section of Lucas County.

The greatest social status and demographic spatial disparities occur between the inner-city zones and the outer city and suburban areas. A concentric zone pattern emerges with respect to housing activity and rehabilitation. Toledo has never experienced the "boom town" growth cycles that have recently characterized many Sun Belt cities. Growth has generally been slow, resulting in development waves generated primarily outward from the CBD in a western direction. These systematic growth waves result in housing zones that are defined chronologically by their development periods. These outward shifts have resulted in a central city largely devoid of major housing concentrations. For example, it is estimated that less than 6 percent of the city's population are residing in the nine census tracts

Fig. 9.4.2. Waterfront Development, Toledo

located adjacent to the CBD (Bureau of the Census 1980). This lack of a sizable downtown residential population, coupled with the strong suburban orientation of the city, has been a serious barrier to revitalization of the central business district.

Downtown Redevelopment

Though the early 1980s brought Toledo more than $210 million in downtown development projects, serious economic and population distribution problems challenged revitalization efforts in the downtown. Investment efforts have focused on the development of a Festival Marketplace, associated riverfront developments (Figure 9.4.2), and substantial investments in downtown hotels, parking facilities, office buildings, and a major 150,000-square-foot convention center (Moon and Muraco 1989). While these projects have significantly improved the general appearance of the downtown, the degree to which they represent an economic success remains very unclear at this time.

The Economy

The primary economic base of the community has historically concentrated on specialized manufacturing activities (Table 9.4.1) that complement the economies of adjacent metropolitan areas.

These situational linkages are most obvious with respect to the Detroit metropolitan area and the city's dependence on automobile parts and accessories manufacturing. The centrality of the Toledo location has placed it within 500 miles of 77 percent of North America's automobile assembly plants, with 46 percent of the plants located 250 miles or less from the city (Toledo Chamber of Commerce 1988). The city's accessibility—via the interstates—to auto assembly operations is important with respect to recent transportation innovations such as just-in-time deliveries, which permit

Table 9.4.1
Toledo's Major Manufacturing Firms

FIRM	MAJOR PRODUCTS
Major Automobile Products Manufacturing Firms	
AP Parts Co.	Auto Exhaust Products
Acustar, Inc. (Chrysler)	Automotive Truck Components
Champion Spark Plug Co.	Spark Plugs
City Auto Stamping Co.	Metal Stamping for Autos
Dana Corporation	Component Parts Vehicular
Doehler-Jarvis Castings	Auto Die Casts Parts
Dura Corporation	Automotive Hardware
E.I. duPont de Nemours Co.	Teflon Automotive Finishes
Ford Motor Co.	Body Parts for Cars & Trucks
Gen Corp Polymer Products	Vinyl Coated Fabrics
Glasstech, Inc.	Glass Tempering Auto Glass
Hydra-matic Division (GM)	Auto & Truck Transmissions
Jeep Eagle (Chrysler)	Jeep Vehicle Assembly
Johnson Controls Inc.	Batteries
Libbey-Owens Ford Co.	Glass
Manville Corp.	Fiberglass Products
TRINOVA Corp.	Auto Parts
Other Major Manufacturing Firms	
The Andersons	Agribusiness
Art Iron, Inc.	Fabricated Metal
Beatrice Hunt-Wesson Foods	Food Products
Brush Wellman Inc.	Beryllium Metal Alloy
The DeVilbiss Co.	Coating & Health Products
General Mills, Inc.	Packaged Food Products
Jobst Institute, Inc.	Elastic Support Hosiery
Owens Corning Fiberglass	Building Products
Ownes Illinois Inc.	Glass, Plastic & Paper
Sharon Manufacturing Co.	Brazed Stamped Assemblies
Teledyne CAE	Gas Turbine Engines
Tempglass, Inc.	Customized Glass

Source: Toledo Area Chamber of Commerce, December 1988.

manufacturers to avoid warehousing of parts and components. Closely interrelated with this concentration of automobile and related industries is the glass industry, and Toledo's heavy participation in this industry accounts for its nickname, the "Glass City." Road, water, and rail access to the agricultural hinterland of the Midwest has also fostered a concentration of several major food and agribusiness industries.

Unfortunately, the dominant role the manufacturing sector has played in the city has recently experienced serious erosion. The changing economic character of Toledo is illustrated in Table 9.4.2, which provides comparative employment and payroll profiles for 1964, 1986, and 1992.

The decreased employment and payroll levels of 1992 associated with the manufacturing segment are largely due to interregional and international relocations. The implications of these changes suggest that a new functional role may be developing for Toledo. While manufacturing remains a major income and em-

Table 9.4.2

Changes in the Toledo Economic Base, 1964–1992

SECTOR	PERCENTAGE TOTAL EMPLOYMENT			PERCENTAGE TOTAL PAYROLL		
	1964	1986	1992	1964	1986	1992
Agricultural Services	.2%	.3%	.4%	.1%	.2%	.3%
Mining	.1%	.1%	.1%	.1%	.1%	.1%
Construction	3.7%	3.7%	4.0%	4.1%	5.9%	5.7%
Manufacturing	42.1%	28.7%	18.6%	54.6%	42.4%	30.2%
Transportation/Public Utilities	6.1%	4.5%	5.6%	6.8%	5.7%	6.8%
Wholesale	8.4%	6.2%	5.7%	8.8%	7.2%	7.2%
Retail Trade	19.0%	21.5%	22.0%	11.8%	11.0%	10.9%
Finance/Insurance/Real Estate	4.5%	5.4%	5.0%	4.1%	5.5%	5.7%
Service	15.8%	28.8%	38.7%	9.5%	21.4%	33.0%
Unclassified	.1%	.1%	.1%	.1%	.1%	.1%

Source: Computed from 1964, 1986, and 1994 U.S. Department of Commerce, Ohio County Labor Market Statistics

Table 9.4.3

Payroll Per Capita by Major Economic Sector in the Toledo Metropolitan Area, 1964–1992

SECTOR	PAYROLL PER CAPITA		
	1964	1986	1992
Agricultural Services	$ 795	$14,420	$16,157
Mining	1,139	25,367	28,593
Construction	1,588	32,680	34,254
Manufacturing	1,831	30,456	38,762
Transportation/Public Utilities	1,574	26,259	29,077
Wholesale	1,472	23,834	30,464
Retail Trade	875	10,532	11,816
Finance/Insurance/Real Estate	1,291	21,204	27,456
Service	850	15,238	20,357
Unclassified	965	15,348	35,905

Source: Computed from 1964, 1986, and 1994 U.S. Department of Commerce, Ohio County Labor Market Statistics

ployment activity, its relative share is decreasing; meanwhile significant increases are taking place in the service sector. Primary service activity appears to be associated with sizable employment clusters in restaurants, businesses, and health services.

In any city, the shift from a manufacturing to a service base carries several negative economic consequences. Perhaps the most serious impact of this shift is the lower wage and salary levels that characterize the tertiary sector, for manufacturing-oriented activities have traditionally been associated with higher wages than tertiary activities (Table 9.4.3).

By 1992, the increased proportion of service sector employment had brought about a sizable redistribution of employees to wage levels below the mean sectorial level. The impact of these downshifts in relative wage levels is illustrated in Table 9.4.4. This table also shows that the proportion of employees in sectors above the mean have decreased from 60.3 percent to 34 percent in the 22-year period.

Table 9.4.4

Proportions of Toledo Employees Above and Below the Sectoral Mean Payroll, 1964–1992

CATEGORIES	1964		1986		1992	
	TOTAL EMPLOYEES	%	TOTAL EMPLOYEES	%	TOTAL EMPLOYEES	%
Below the Mean	57,837	39.7%	119,493	51.4%	128,059	66.0%
Above the Mean	87,689	60.3%	112,772	48.6%	65,899	34.0%
Sectoral Mean	$1,411		$20,596		$27,284	

Conclusion

The city of Toledo represents a dynamic environment in which internal and external geographic forces are shaping its postindustrial future. Like many cities in Ohio, Toledo faces the serious challenges of central city revitalization and economic restructuring. A major geographic attribute, its accessibility, places the city in a unique situation relative to larger, nearby industrial markets. As a consequence of its location, it is in a unique position to attract the benefits of these larger markets (i.e. access to major employment clusters, shopping and cultural activities) without experiencing the extensive social and environmental liabilities facing very large urban places.

REFERENCES

Harvey, Hank. 1989. Air Shipping Complex Will Push to Toledo to Ohio's Premier Transportation Center. *Toledo Blade,* 30 August.

Moon, Henry E., and W. A. Muraco. 1989. Waterfront Revitalization, Conventions and Tourism: The Case of Toledo's Portside. Paper presented at the 10th Annual Travel and Tourism Research Association, Censtate Chapter Meetings, 20–22 September, at Indianapolis, Indiana.

Toledo Area Chamber of Commerce. 1988. *Toledo Area Automotive Profile.* Toledo, Ohio: Toledo–Lucas County Port Authority, Toledo Area Private Industry Council, and Toledo Area Chamber of Commerce.

U.S. Bureau of the Census. 1980. *Census of Housing 1980: Toledo Census Tracts.* Washington, D.C.: U.S. Government Printing Office.

———. 1988. *County and City Data Book.* Table C. Washington, D.C.: U.S. Government Printing Office.

———. 1992. *County and City Data Book.* Washington, D.C.: U.S. Government Printing Office.

U.S. Department of Commerce—Bureau of the Census. 1964. *Counties Employees Payroll and Establishments by Industry; County Business Patterns 1964: Ohio.* Washington, D.C.: U.S. Government Printing Office.

———. 1986. *Counties Employees Payroll and Establishments by Industry; County Business Patterns 1986: Ohio.* Washington, D.C.: U.S. Government Printing Office.

———. 1994. *Counties Employees Payroll and Establishments by Industry; County Business Patterns 1992: Ohio.* Washington, D.C.: U.S. Government Printing Office.

AKRON
The City between Three Rivers

Warren Colligan, Thomas L. Nash, and Frank Costa

As the Cuyahoga River completes its twenty-mile bend north toward Lake Erie, it is met by a small tributary, the Little Cuyahoga River. Both of these rivers are unremarkable. Little Cuyahoga lies mostly hidden under and behind industrial and residential areas of the city of Akron. Why did the city of Akron develop just south of the confluence of these two rivers? The answer lies in part in the fact that a third river, the Tuscarawas, runs just 10 miles to the south of the intersection.

Only shallow draft craft, such as canoes and log barges, were able to travel the upper parts of the Tuscarawas and the lower Cuyahoga. The Indians knew that they could travel and transport goods from Lake Erie to the Ohio River if they carried, or *portaged,* their goods through the area between the Cuyahoga and Tuscarawas Rivers. Fur traders operating trading posts in this area called the route *Portage Path.* The route was incorporated into early plans for a transportation artery linking the Cuyahoga and Tuscarawas Rivers. These plans called for an improved road between the two rivers, but politics and settlement patterns in Ohio delayed construction until the mid-1820s.

Another artery of communication instrumental in the development of the city was the Ohio canal system, approved by the legislature in 1825. At the time, Ohioans were concerned with opening up the interior of the state to trade with the Atlantic coast and the South. Ohio's population was concentrated in the southern part of the state close to the Ohio River, the water highway to the Mississippi and the South's trading center at New Orleans. A second canal, the Pennsylvania and Ohio, which connected Akron and Beaver, Pennsylvania, was opened in 1840. The canal era, which lasted from 1825 to 1913, allowed migration to inland areas and facilitated trade and industrial development in Ohio. New settlements and communities sprang up along the canals, and the state's population increased 350 percent from 1820 to 1850. This was reflected in the growth of Akron (Figure 9.5.2).

The site of the city of Akron marks the highest point on the Ohio and Erie Canal. Many locks were needed to lift barges to an elevation of 964 feet above sea level (Figure 9.5.3). Progress through the area was slow and delays on the canal created Akron's first business ventures: inns, taverns, boat repair yards, and warehouses. By the mid-nineteenth century, Akron's major industries involved the processing of local raw materials: iron foundries used local bog iron, woolen mills refined raw wool from local sheep, and a distillery used local grain.

The first major industry in the city was the production of clay products. Good quality clay was found in the area, as was wood for charcoal and later, coal and natural gas. The inventiveness of Akron's early pottery manufacturers was the greatest contributor to the growth of the industry; they developed new machinery to produce superior molded clay products and closely followed new trends in the industry. The development of two early machine-made ceramic products—tobacco pipes and beer bottles—led to the manufacture of ceramic pipes for other uses. In addition, there was a growing demand for municipal sewers, and the development of an improved vitrified ceramic sewer pipe helped make pipe production a major industry. However, changes in the costs of fuel, transportation, labor, and raw materials caused the clay pipe industry to gradually leave Akron during the years 1910–40.

Matches were also produced in Akron. A small match factory opened in the early 1840s and grew into

Fig. 9.5.1. Akron Skyline

the largest match company in the country. The availability of coal, abundant water, and cheap child labor were major factors in the growth of Akron's match industry. These companies gradually moved out of the area by the mid-twentieth century. Another industry for which Akron was known, cereal production, was partly a logical extension of the flour mills that sprang up along the Ohio Canal. Industries located near one of the canals for two reasons: water power and shipping facilities. Akron's cereal industry began in 1854 when a German immigrant grocer, Ferdinand Schumacher, who had perfected a process for making oatmeal, began selling his new breakfast food. By 1870 his company was America's largest cereal producer. The local flour mills also grew, and Akron became a major grain processing center. Large secondary industries developed in Akron as a result of the flour and cereal production; Akron had large paper and cloth bag factories, and by 1880, coopers were making 50,000 barrels a year for the flour and cereal industry. When breakfast food and flour companies became part of the Quaker Oats Company in 1901, flour milling and some other production was moved out of the city.

Industrial Development

For 30 years, 1865–95, Akron was a major producer of horse-drawn mechanical farm implements. Lewis Miller, Thomas Edison's father-in-law, was credited with starting the city's farm implement industry. During the Civil War, Akron's location near the major western farm markets, its ample skilled labor, and its good transportation network for heavy machinery, made the city an important producer of wartime materials, particularly farming tools. And as the wartime demand for food rose, orders for new farm machinery poured in. This industry continued to thrive for 30 years and then rapidly declined. The causes of this decline are threefold: first, the market for heavy farm machinery moved further west, due to the expansion of the upper Midwest Wheat Belt; second, there was a consolidation of the industry, with larger companies in Chicago buying out

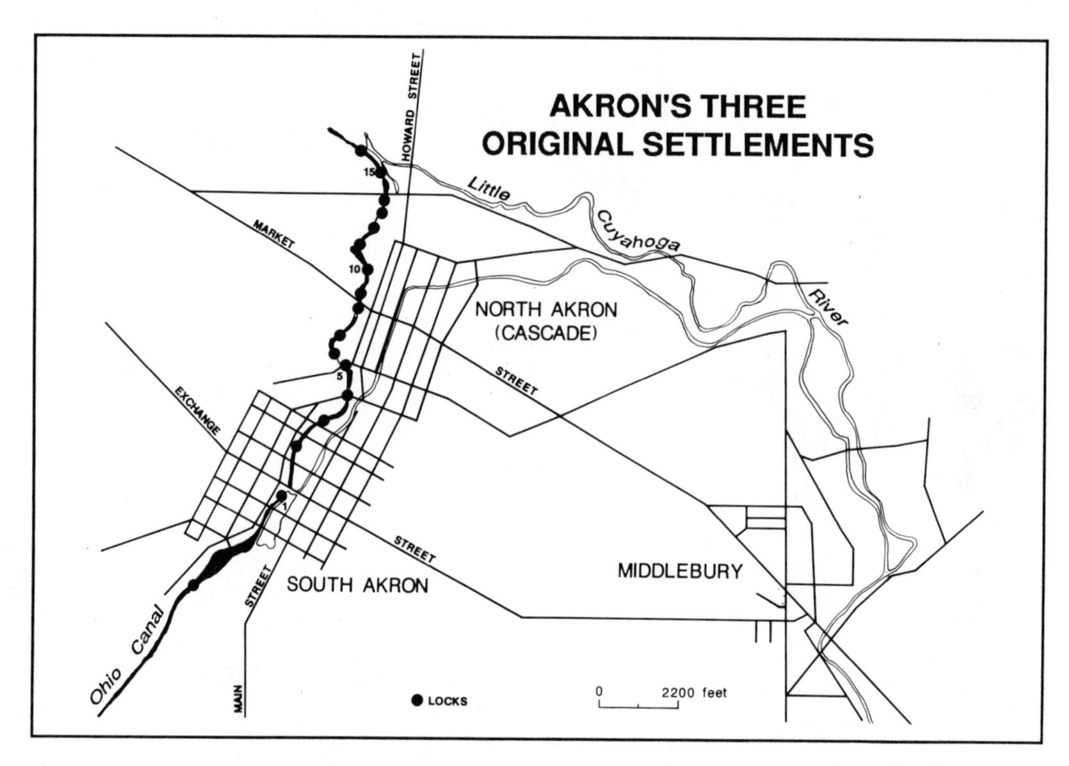

Fig. 9.5.2. Akron's Three Original Settlements

and underselling the smaller Akron companies; and third, the panic of 1893, which forced many farmers into bankruptcy, halted purchases of large, expensive farm machinery. The last farm implement company in Akron was purchased by International Harvester in 1907 and converted into a truck plant; trucks were built there until 1925.

In 1888 Akron's largest industries were farm machinery and related products, matches, clay products, and the milling and cereal industries. One of the largest companies, however, was the B. F. Goodrich Company, which had been producing rubber products in Akron for 17 years. Goodrich began by producing fire hoses, but by 1880 the company was making many other rubber products. American rubber companies were initially small concerns. Then in the early 1890s safety bicycles with pneumatic tires were perfected, and the bicycle became a major mode of American transportation. Goodrich made its first bicycle tires in 1891, and by 1893 pneumatic bicycle tires were a nationwide standard. America's first pneumatic automobile tires were built by Goodrich in 1896.

At the turn of the century, other industrialists tried to capitalize on the new demand for rubber tires. The Goodyear Tire and Rubber Company was founded in 1898 by the sons of one of Akron's farm implement manufacturers. The Firestone Tire and Rubber Company was formed in 1900 by Harvey Firestone, three local investors, and an inventor who held a valuable rubber carriage tire patent. Early automobile and truck tires were basically advanced bicycle and carriage tire designs. Automobile tires were a major product line for the large Akron rubber companies. They were also expensive; a set of original equipment tires often cost the car manufacturer 11 to 14 percent of the retail cost of the car.

With the introduction of the Model-T Ford in 1908, cars became more common in the United States, and three of the major automobile tire manufacturers—Goodrich, Goodyear, and Firestone—were located in Akron. Prior to World War I, the major companies had established large engineering and technical departments in Akron. By the mid 1920s, four of the five largest rubber companies in the country had their headquarters, and most of their manufacturing capacity, in the city. With the increasing demand for car and truck tires plus the wartime demand for all types of rubber products came a concomitant increase in the number of jobs. Consequently, the city's population jumped from 69,067 in 1910 to 208,435 in 1920. This era in

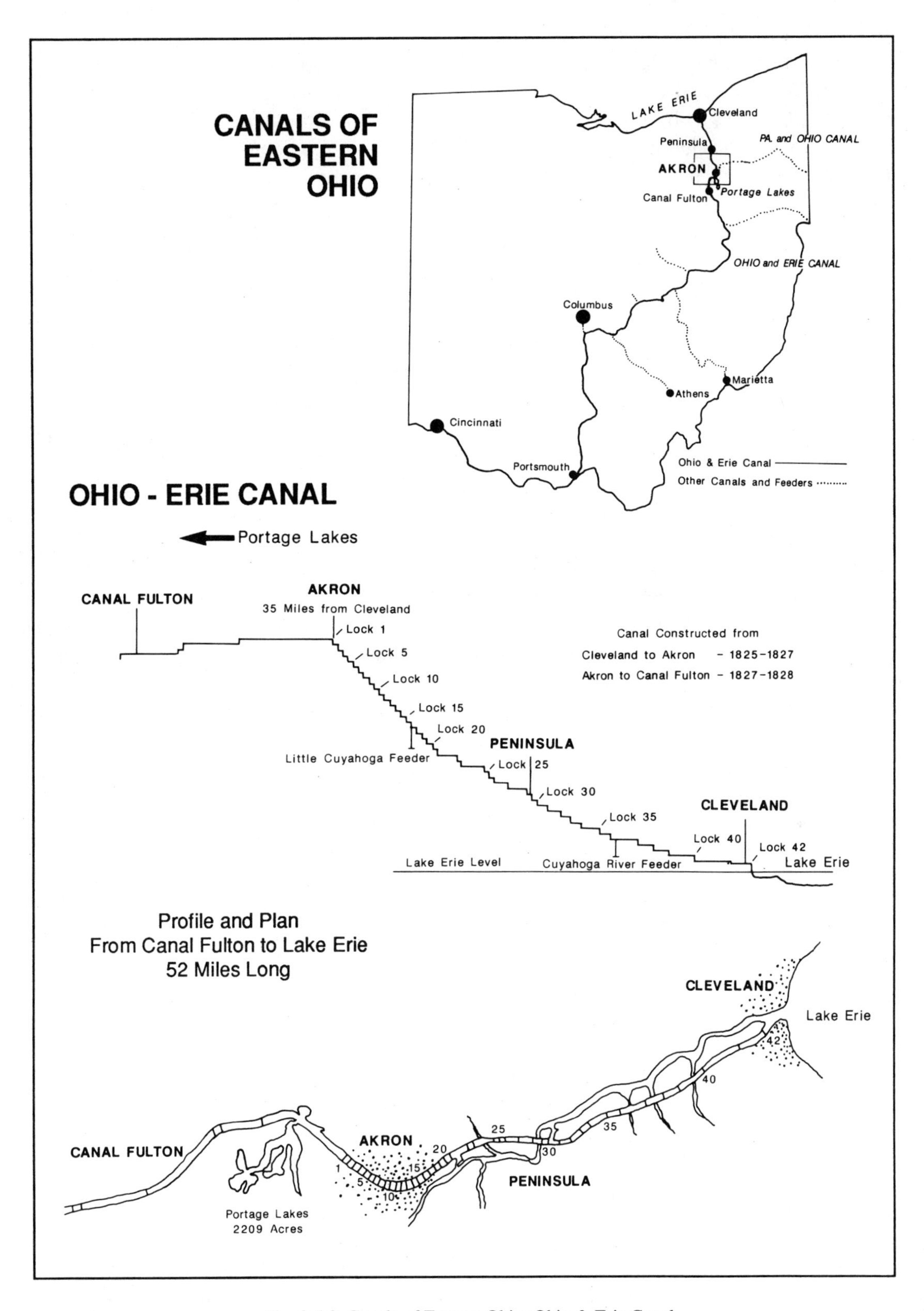

Fig. 9.5.3. Canals of Eastern Ohio; Ohio & Erie Canal

Akron's development is known as the "rubber boom." During the rubber boom housing was in short supply, rubber workers slept in shifts at boarding houses, the public transit system was overloaded, and the Akron water and sewer system could not keep up with the additional demand. The tire and rubber industry was Akron's primary industry from about 1910 until the early 1970s. These large companies were so predominant that they developed whole sections of the city to house their workers: Goodyear Heights still dominates eastern Akron, and Firestone Park is a large section southeast of the Firestone offices.

Akron's aerospace industry, primarily an outgrowth of the rubber industry, began as early as 1909 when Goodyear's factory superintendent brought two experienced workers to Akron from a factory in Scotland to make rubberized fabric for airplane wings. Goodyear was soon the major supplier of aircraft fabric for America's infant aircraft industry. During World War II, Akron's aerospace industry greatly expanded, producing planes, blimps, and air frames. All of the rubber companies were involved in war work, much of it aircraft production. At the end of the war, the rubber companies continued to be large military contractors, and Goodyear Aircraft remained a major aircraft parts supplier. With the coming of the space age during the 1960s, the rubber companies began producing component parts for space exploration vehicles: Goodyear built the tires for the first manned vehicle on the moon, and Goodrich built the first space suits and the tires and brakes for the space shuttle *Columbia.*

Spurred by the need to ship their products, the rubber companies helped create another major Akron industry: long-distance trucking. First Goodyear in 1917 and then Firestone in 1919 began extensive advertising campaigns promoting the use of trucks with pneumatic tires for shipping freight long distances. Before this time, most large trucks used solid rubber tires. These tires gave such a hard ride that they quickly wore out both the truck and the driver, limiting the ability of both to travel long distances. The rubber industry was much more eager to ship goods by truck, as opposed to by railroad, than other industries for three reasons. First, they realized that it was good for business; a long-distance trucking company was potentially a good customer, whereas a railroad was not. Second, since the independent trucking companies were competing with the large railroads, and railroad shipping rates were regulated by the Interstate Commerce Commission (ICC), they expected the unregulated truck lines to offer them lower shipping rates. Third, they realized that with so many small tire dealers scattered throughout the country, the truck lines could make faster deliveries to out-of-the-way customers than could the railroads. Akron became an ideal place to start up a truck line; the largest local customers were friendly to the new business, and the major routes involved shipping tires and inner tubes to Detroit and hauling raw materials from the large seaport cities, places which offered the potential of good return traffic. Akron's rubber companies started using truck lines to ship tires to distant warehouses, tire dealers, and automobile manufacturers during the 1920s, and American roads improved enough during these years that long-distance trucking became practical.

Industrial Decline

There is little evidence today of Akron's seven historic major industries. No match, farm machinery, or cereal companies remain, although one small industrial ceramic company still exists. In 1990 only a very limited number of racing, high speed, and experimental automobile tires were being built in Akron, with the rest of America's car and truck tires made elsewhere. Of all the large Akron rubber companies, only Goodyear remains as a wholly owned American company. Firestone's tire production facilities are now owned by Bridgestone, a major Japanese rubber company. In the 1980s Goodrich merged with Uniroyal and has since been purchased by Michelin, the large French rubber company. The General Tire and Rubber Company, renamed Gencorp, sold its tire manufacturing to Continental, a West German rubber company.

Several small and medium-size rubber companies in the Akron area still produce specialty products such as stair treads and printing rubber. A few industrial rubber products—conveyor belts, for instance—are still built by the larger of these companies, but most employees are involved in administration, with lesser numbers in sales, engineering, and marketing. Few are directly involved in production. Aircraft parts and special products for space exploration continue to be produced. Greater Akron remains the home of many of the nation's truck lines, although a number of these companies have moved to Richfield, a suburb halfway between Akron

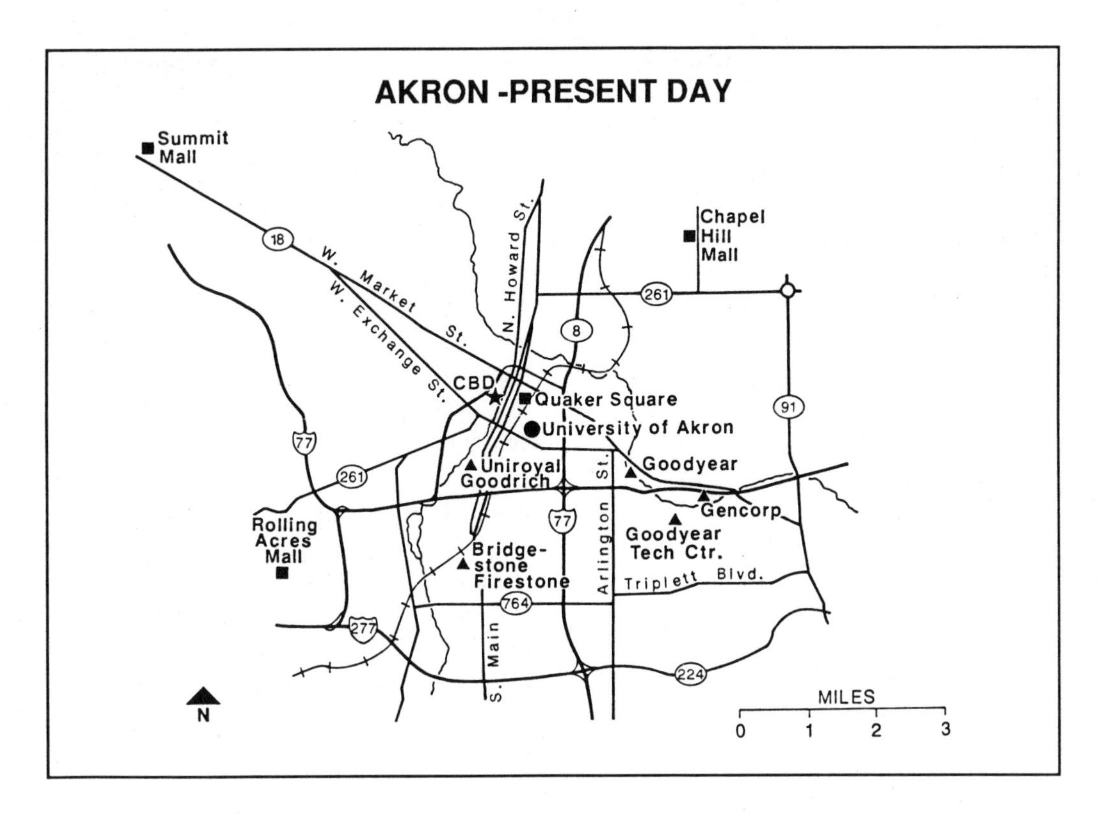

Fig. 9.5.4. Present-Day Akron

and Cleveland. This move allows them to be close to both markets and also to the three major interstate highways that intersect near Richfield.

City Redevelopment

As Akron has moved toward a service economy, several plans for the redevelopment of the downtown area have been put forward, and some have been implemented. One of these focused on Quaker Oats's large empty grain silos and factory, which had become symbols of one of Akron's lost industries. In 1975 the factory was converted into a downtown shopping center and in 1980 the silos became a hotel (see Figure 9.5.4). The two have combined into the thriving Quaker Hilton hotel and shopping center. A number of other plans for the redevelopment of the central city have focused on the old Ohio and Erie Canal. One of these was the American City Plan for downtown Akron. This plan, developed by the American City Corporation, seeks to link together a number of important downtown redevelopment initiatives of recent years, including the Cascade Plaza, Quaker Square, and O'Neil's Department Store. Planners seized on the old Ohio and Erie Canal as the ribbon that would tie together the various parts of the proposed redevelopment scheme. Perpendicular connectors to the canal in the form of attractive pedestrian walkways and covered skyways would link development throughout the CBD into a network of interconnected spaces. Another important feature of the plan was its goal of linking adjacent areas to the downtown through use of the canal both north and south of the CBD. A direct connection north through the new Cascade Park toward the Cuyahoga Valley National Recreation Area and a connection south to Opportunity Park and beyond was envisaged as a means to tie these two areas together into a lineal development that would increase the appeal of the downtown area for many more potential users. Serious fault can be found with this plan. First, it was unveiled as the local economy plunged into its deepest crisis since the Great Depression. Its ambitious timetable was simply unrealistic given the economic conditions of the early 1980s. Second, it failed to recognize the economic and physical redevelopment importance of a linkage eastward toward the University of Akron and the Goodyear Tech Center in east Akron. Though the plan justly stresses the importance of the canal as a link to the north and the south, the resulting connections are primarily to recreational and residential areas. These, of course, are very important linkages that have a decidedly favorable impact on overall quality of life in the central city. But economically important links are those that tie together government, industry, and research activities with product development and product marketing ventures.

A far more innovative design plan, and one with far-reaching economic development implications, is the "Span-the-Tracks" linkage being undertaken jointly by the University of Akron and the city. It had long been a goal of both university and municipal planners to bridge the physical distance between the campus and the city's downtown with a connector that would engender mutual attraction between the two districts, but past ideas had not considered functional interrelationships along the path of the connector. The attractive pedestrian concourses that were proposed were believed to be sufficient in themselves to create a successful link.

"Span-the-Tracks" builds on the functional linkage between the university and the Akron CBD. Although the plan is far from complete, important elements are in place including the university's Polymer Science Center and the city's new convention center. Thus, at one end of the proposed link is the University of Akron's polymer research complex, while at the other end is the hotel–convention center–exhibition hall complex where new polymer products will be exhibited to potential buyers. Included at midpoints along the linkage are conference and continuing education facilities and the university's Performing Arts Hall. This design plan is in line with the original intent to use existing facilities, augmenting these so that participants in polymer research and development activities are provided with attractive, linked facilities.

What do the 1990s hold for this city between three rivers? One trend is the city's rapid movement toward a service economy. Akron's growing employment sectors are in education, medical care, and wholesale and retail sales. Akron has four large hospitals, and there are two more in the adjacent suburbs. The medical care facilities are also becoming very specialized, serving a much larger region. The Children's Hospital Burn Center, for example, serves the area between Akron and Columbus. Thus Akron has come full circle. Its origin was as a service center for the canal; the city then developed into a major manufacturing city. Today the manufacturing has almost disappeared and Akron is once more a service center.

REFERENCES

Allen, Hugh. 1949. *Rubber's Home Town*. New York: Stratford House.

Blower, Arthur. 1955. *Akron at the Turn of the Century*. Akron, Ohio: Summit County Historical Society.

Grismer, Karl H. 1952(?) and updated. *Akron and Summit County*. Akron, Ohio: Summit County Historical Society.

Kenfield, Scott Dix. 1928. *Akron and Summit County Ohio: 1825–1928*. Chicago: S. J. Clarke.

Knepper, George W. 1981. *Akron: City at the Summit*. Tulsa, Okla.: Continental Heritage Press.

Nichols, Kenneth. 1975. *Yesterday's Akron*. Miami, Fla.: E. A. Seeman Press.

YOUNGSTOWN

The Decline of a Steelmaking Giant

David T. Stephens

Youngstown's early growth and development were greatly influenced by its site and situation (see Figure 9.6.1). Today this situation helps explain some of its economic woes. When John Young of Whitestown, New York, platted his town, he selected a point along the Mahoning River where the stream emerges from a steep-sided valley and the floodplain begins to widen. This location was astride an important path connecting Lake Erie and the Ohio River. It was a route initially established by animals and later utilized as a portage by Indians. In subsequent years this path would be followed by a canal, four major railroads, and some of the more important transcontinental highways in the United States. Two miles from the site of Young's town on Mill Creek, a Mahoning tributary, was a waterfall that provided an ideal location for a mill. Also nearby was a well-known source of salt. Unknown to its founder, this salt source was a precursor to the development of important local mineral wealth that would help catapult the town into the industrial age as one of the premier iron and steel centers in the world.

Platted in 1798, Youngstown was laid out on the north bank of the Mahoning River in a grid form with a central square. This feature of town design, found in New England, was imitated in many communities in the Western Reserve. Unlike most town founders in the Reserve, Young laid his community out in a valley and not on a hill. No doubt this contributed to some of Youngstown's early problems with diseases, but it later proved to be a blessing when routes were selected for a canal and railroads. In subsequent years the river valley site proved an ideal location for the space-consuming manufacturing plants requiring easy access to water or rail transportation and large volumes of water for cooling and waste disposal.

It has been suggested that Youngstown, which was both the Reserve's first settlement and the first along the Mahoning River, was the initial destination of three-fourths of the newcomers to the newly opened Western Reserve (Butler 1921). Some of the new arrivals stayed but most went on to other locations in the Reserve or further west. One of Youngstown's initial roles was as an outfitting point for settlers, but it was not long until the city began to develop other businesses. In 1802 bog iron ore was discovered in the area, and a year later a blast furnace—which utilized local bog ore, limestone, and charcoal—was in operation along Yellow Creek, a Mahoning River tributary about five miles downstream from Youngstown. This would be the first of the many iron facilities that would fuel the region's growth, and indeed the growth of the city parallels the development of the iron and steel industries in the surrounding area. The city, indeed the region, is a classic example of an area built around a single industry. This heavy dependence on a single type of manufacturing helps to explain some of the problems found there today.

Before 1835, three furnaces and a blooming mill were constructed in the vicinity of Youngstown. The city's growth had been modest, reaching about 1,000 residents by 1830. At that time, additional expansion seemed questionable, for the region's fledgling iron industries were facing two problems: limited access to markets and depletion of local iron and fuel resources. In the following decade, 1840–50, both problems were solved. In 1841, the completion of the Pennsylvania and Ohio Canal linked Youngstown to markets in Cleveland and Pittsburgh. Then the shortage of fuel for the blast furnaces was remedied when it was discovered that local coals were a satisfactory substitute for charcoal, and the shortage of iron ore was rectified by

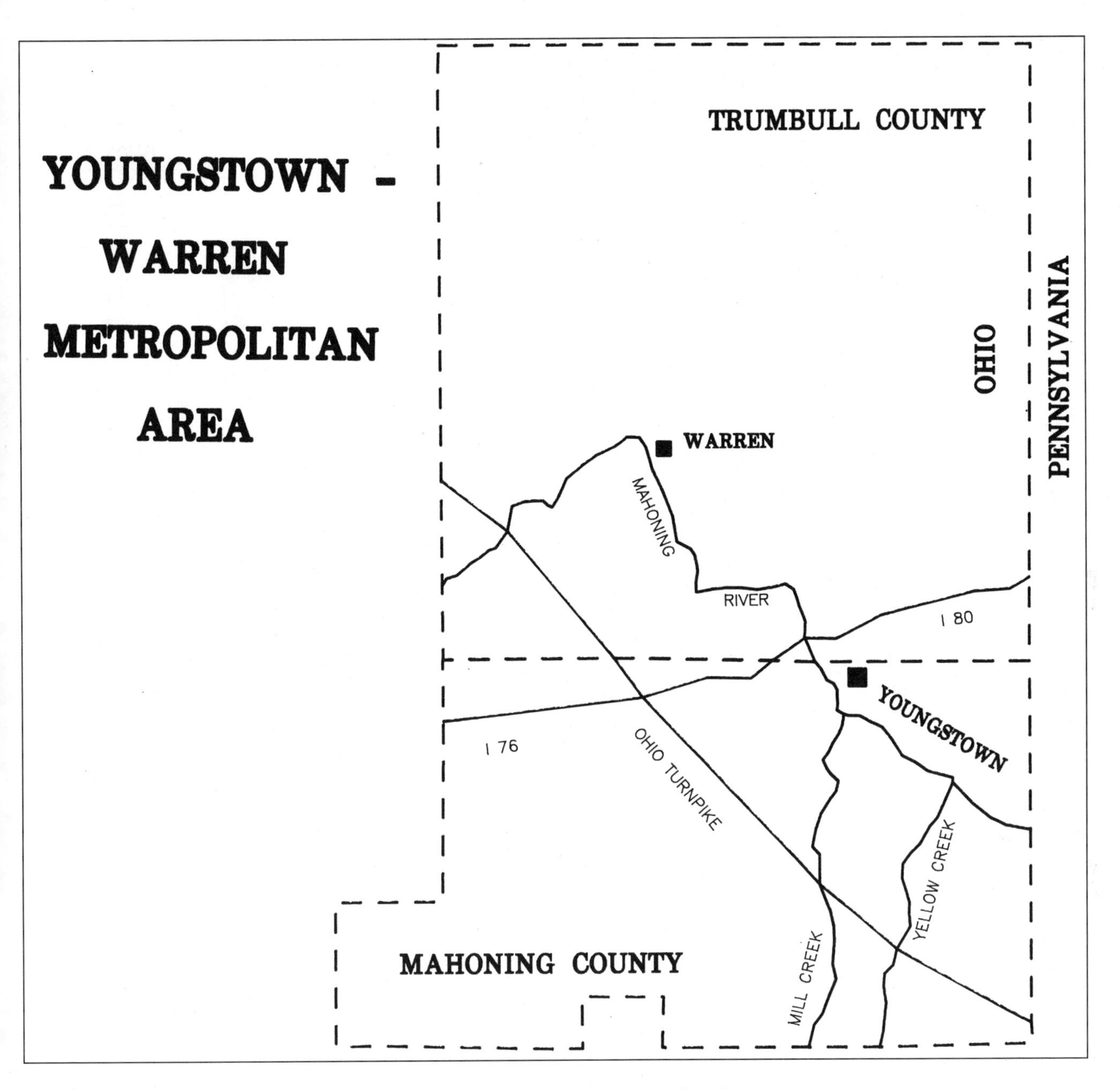

Fig. 9.6.1. Youngstown-Warren Metropolitan Area

using imported ores and a band of iron ore newly discovered beneath a local coal seam. These events ushered in an era of rapid growth that saw the city increase its population by more than 25 percent in every decade through 1930.

The Pennsylvania and Ohio Canal, like most Ohio canals, had a short life: 1841–72. Among the benefits of the canal's improved access were the development of additional local iron and iron-related industries and a rapid increase in population. However, it soon became evident that an even faster and more efficient form of transport was needed. In response to this need, a railroad link between Cleveland and Youngstown was completed in 1856. There followed a rail building binge that eventually saw Youngstown served by four major trunk railroads: the Baltimore and Ohio, the New York Central, the Pennsylvania, and the Erie. These lines and their various spurs were constructed to move immense volumes of raw materials—coal, iron ore and limestone—to the continually expanding iron

Table 9.6.1
Youngstown, Ohio, Population Characteristics, 1890–1990

YEAR	TOTAL POPULATION	% CHANGE	FOREIGN BORN	% FOREIGN BORN	BLACK	% BLACK
1890	33,220					
1900	44,885	35.1	12,207	27.2	915	2.0
1910	79,066	76.2	24,869	31.4	1,936	2.4
1920	132,358	67.4	33,938	25.6	6,662	5.0
1930	170,002	28.4	32,938	19.4	14,552	8.5
1940	167,720	–1.3	26,671	15.9	14,615	8.7
1950	168,330	0.4	21,410	12.7	21,459	12.7
1960	166,689	–1.0	16,851	10.1	31,677	19.0
1970	140,909	–15.5	9,003	6.4	35,285	25.2
1980	115,436	–18.1	5,557	4.8	38,478	33.3
1990	95,732	–17.1	2,879	3.0	36,478	38.1

Source: U.S. Bureau of Census, 1890–1990

industry of the Mahoning Valley and to transport Youngstown's products to other parts of the country. Commenting on rail traffic in its 1920s heyday, Howard Aley noted that "more train cars per day pass beneath the Center Street Bridge than anywhere else in the nation" (Aley 1975).

Improved transportation, inertia from an early start in the iron industry, local supplies of limestone, and a location between the coking coals of Pennsylvania and West Virginia and the iron ores from the upper Great Lakes all combined to make Youngstown one of the most important iron centers in the world. The first steel was manufactured in the Mahoning Valley in 1895 and steel quickly became the area's most important industrial product. Butler, writing in 1921, suggested that the Youngstown district might surpass Pittsburgh as the world's leader in iron and steel production. This was not an idle boast, as Butler offered credible statistical evidence to buttress his prediction (Butler 1921).

As noted above, the city's population grew rapidly, from 3,000 in 1850 to over 130,000 in 1920. (Table 9.6.1 documents this growth and subsequent population trends.) To accommodate this influx, the physical city expanded from the floodplain, spreading northward toward higher ground. As was typically the case in American cities of this generation, the move away from the central business district toward higher ground was led by the affluent, who sought to distance themselves from both "undesirable" land uses—manufacturing—and "undesirables"—immigrants.

Some industrial expansion took place on the opposite side of the river, west of the CBD, but this was modest growth. Until 1899, expansion to the south was blocked by the steep escarpment that formed the southern bank of the Mahoning River. In that year, a viaduct was constructed that connected the city's CBD with the south bank of the Mahoning, and residential and commercial development moved rapidly to the south. Similarly, construction of a bridge in 1903 opened the west to rapid residential and commercial growth. During the first quarter of the twentieth century, the move south and west was aided by the construction of streetcar and interurban lines that focused on the Youngstown CBD. These transportation corridors gave the city a starlike pattern of development and provided the impetus for the growth and development of the satellite industrial communities of Campbell, Struthers, and Lowellville, among others, and the residential communities of Boardman and Austintown. The improvements in personal transportation permitted the American and western European settlers who had gained an early start in the iron and steel industry and climbed several rungs on the socioeconomic ladder to move to these outlying subdivisions. The new arrivals to the area took up the residences left behind.

As the iron and steel industry continued to expand, John Young's decision to select a river valley location to plat the town proved providential. The floodplain of the Mahoning River provided ideal sites for manufacturing plants and railroad yards and sufficient space for storing vast quantities of raw materials and finished products; in addition, the river provided the requisite water for industrial processing and cooling as well as a convenient place for the discharge of industrial wastes. By 1920 there was a nearly continuous string of iron and steel works and related industries spanning a 25-mile stretch along the Mahoning River, from Warren through Youngstown to the Pennsylvania border. An unusual legacy from this period of expansion was the creation of one of the outstanding urban parks in the country along the gorge of Mill Creek. This is not the type of development one generally associates with rapidly industrializing communities. The park remains today as a legacy of its forward-looking architect, Volney Rogers.

Youngstown's initial settlers came from New York and New England, and those areas continued to contribute new residents. As time passed, Pennsylvania and Virginia also became important sources for new settlers. According to Aley these early settlers had ancestries that could be traced to England, Scotland, Wales, and Ireland (1975). One could logically add Germanic peoples to his list, as many Pennsylvanians had roots that could be traced to the German-speaking areas of Europe.

In the years before 1880, the Youngstown area was the recipient of new settlers from Eastern states and from several areas in Europe. Many of these migrants were drawn by specific economic opportunities. The building of the Pennsylvania and Ohio Canal and later the railroads created a demand for construction workers, a demand filled by Irish immigrants who arrived in considerable numbers throughout construction. In addition, Welsh miners were attracted to the valley's coalfields in the 1830s and 1840s, and Scotch immigrants came to the Mahoning Valley around 1860 to work in the coal mining and iron industries. As the city's industrial base grew in the last quarter of the nineteenth century, the need for labor expanded even more rapidly, and the source of immigration shifted from western to southern and eastern Europe. The changing nature of the immigrant stream is documented in Table 9.6.2, which also shows contemporaneous ancestry patterns. Census data for 1910 indicated that nearly one-third of the city's population was foreign born. By 1930, 54 percent of Youngstown's population was of foreign birth or extraction (Buss and Redburn 1983). An additional 8.5 percent were African Americans who had been recruited during World War I to fill the void created by the cutoff of European immigrants.

The influx of immigrants created a mosaic of ethnic neighborhoods usually anchored by a church and a social club. A large number of these structures are still evident in the city's landscape. To many residents of the Youngstown area today, ethnicity is still a matter of considerable importance, and for some groups, social clubs are a significant focus of community life.

In many ways 1930 marked a high point for the city of Youngstown. Its population stood at more than 170,000, and the CBD reached its maximum areal extent (City Planning Associates, Inc. 1968). The Youngstown district ranked as the third leading steel-producing center in the world (City Planning Associates, Inc. 1968). The Depression affected Youngstown as elsewhere: population growth ceased; immigration slowed to a trickle; and unemployment rose. World War II brought a return to full employment, and the Mahoning Valley again enjoyed prosperity. In the postwar period, the area's fate was very much tied to the ups and downs of the iron and steel industry. There were good times and times that were not so good. As the market for metal products changed and the sources of raw materials shifted, Youngstown's situation, once its strength, now became its Achilles' heel. All the materials needed to make iron and steel reached Youngstown via rail or truck, and the area's output of metal products began their trip to market via these same two modes of transportation. Tidewater locations and newer, more efficient plants placed Youngstown's metals industries at a comparative disadvantage. This handicap was evident with each economic downturn, as Youngstown's mills were the first to be shut down when the market contracted and the last to be returned to production when things improved.

Aggravating the marginal nature of production in Youngstown were high labor costs and a lack of capital investment in local production facilities. In 1983, Buss and Redburn cited an additional contributing problem: "A major change affecting many older cities is the trend of ownership of the communities' major industries by corporate conglomerations headquartered elsewhere." Given all these factors, by the mid 1970s the iron and steel industry that had been Youngstown's economic base for nearly 170 years was in deep trouble.

On September 19, 1977, a date destined to be known locally as "Black Monday," the Lykes Corporation, one of those conglomerates referred to by Buss and Redburn that were located outside the area, announced the closing of its Campbell works, in one

Table 9.6.2

Youngstown's Foreign-Born Population, 1900–1930, and Youngstown's Ancestry, 1990

COUNTRY	1900	1910	1920	1930	1990
Arab					818
Austria	492	4,005	3,160	848	179
Belgian					24
Canadian	302	341	509	546	103
Czechoslovakia			2,096	4,454	
Czech					91
Danish					26
Dutch					1,026
England	2,278	2,239	2,536	2,284	5,350
Finnish					25
French	32	48	131	74	1,100
Germany	1,632	2,100	1,469	1,333	13,963
Greek	11	134	1,297	782	716
Hungarian	1,031	5,490	2,684	2,052	2,305
Irish	2,124	1,842	1,578	1,230	11,246
Italian	1,331	3,604	5,538	6,977	13,253
Lithuanian					217
Norwegian					102
Polish	193		2,601	2,238	3,600
Romanian	1	158	1,375	1,367	384
Russian	166	1,691	2,214	1,872	507
Scotch-Irish					1,058
Scottish	675	819	1,024	1,377	809
Slovak					8,498
Subsaharan Africa					149
Swedish	343	567	769	771	682
Swiss	74	109	120	111	221
Ukrainian					1,550
U.S. or American					1,429
Welsh	1,351	1,181	1,103	1,097	1,553
West Indian					210
Yugoslavia			2,579	2,195	35
Other	171	532	1,051	1,330	117,047

Source: U.S. Censuses of Population.

Note: Data are based on a sample and subject to sampling error.

stroke idling 4,100 steel workers. Unfortunately, the worst was yet to come, for in January of 1980, two other outsiders—U.S. Steel and Jones and Laughlin—announced the closing of Youngstown-area facilities that employed a total of 4,900 steel workers. The ripple effect this had on other economic activities was soon evident: in 1982 the unemployment rate for the Mahoning Valley climbed to 19.7 percent (*Youngstown Vindicator* 1990, E-1).

This ripple effect has continued in metals and other industrial sectors. An indication of the impact of the decline of local steelmaking can be seen by examining employment in primary metals. Data in Table 9.6.3 from the Ohio Bureau of Employment Services show that during the third quarter of 1968 nearly 50,000 persons were employed in the primary metals industries in the Youngstown-Warren SMSA, whereas in December of 1993 the Youngstown-Warren MSA em-

Table 9.6.3

Selected Employment Data, Youngstown-Warren SMSA, MSA, 1965 to 1993 (1,000s)

INDUSTRY	1965	1970	1975	1980	1985	1993[1]
Construction	8.7	9.1	6.5	7.8	7.4	7.4
Transportation & Utilities	9.3	10.3	10.3	9.4	7.4	6.9
Wholesale	5.7	6.6	7.1	7.8	9.0	10.9
Nondurables	6.6	7.0	5.8	5.7	5.5	5.2
Stone, Glass & Clay	2.7	2.5	2.2	1.7	1.7	N/A
Primary Metals	48.8	46.5	38.1	28.2	20.5	17.7
Fabricated Metals	7.7	8.6	9.4	8.9	7.7	6.3
Non-Electric Machines	7.1	6.6	7.5	6.7	3.9	N/A
Electric Machines	3.9	4.5	3.6	3.0	2.5	N/A
Transportation Equipment	5.2	13.2	9.9	11.1	8.4	7.5
Durable Goods	75.4	81.9	70.7	59.5	44.7	39.9
Retail Trade	25.3	31.3	34.8	37.2	37.4	43.1
Finance, Real Estate & Insurance	4.6	5.8	6.6	7.2	7.5	9.0
Services	22.7	27.4	34.3	38.2	41.7	51.4
Government	15.5	18.8	21.9	24.0	21.1	25.9
Totals	173.8	198.2	198.0	196.9	181.7	200.1

Sources: Ohio Bureau of Employment Services and Stocks

[1] Column will not total 200.1 because some industrial categories were not available.

ployed less than 18,000 such workers. Predictions by a local economic forecaster suggest that the number of workers in this industry will drop to around 17,000 by 1994 (Stocks 1992, 4). The industrial corridor that once belched smoke and blazed with the fire of furnaces from Warren through Youngstown all the way to the Pennsylvania border is only a shadow of its former presence. It is characterized by abandoned facilities and in some cases by the complete disappearance of industrial complexes. Fortunately some of this industrial legacy has been preserved in the recently opened Center of Industry and Labor.

Helping to offset the rapid decline in primary metals employment was the growth of the local automobile industry. In 1966, General Motors began operating at Lordstown what was then the largest automotive assembly plant in the world. GM's decision to locate at Lordstown was, in part, related to the community's situation. Lordstown is well located to gather component parts from various suppliers throughout the western section of the manufacturing belt and the finished products could be shipped expeditiously to a variety of eastern markets. Interstate 80 is the shortest highway link to northern New Jersey and New York City. The Ohio and Pennsylvania Turnpikes provide a direct connection to the Philadelphia–southern New Jersey market. A branch from the Pennsylvania Turnpike provides access to the Baltimore and Washington, D.C., markets.

In the third quarter of 1971, more than 16,000 workers were employed in the transportation equipment industry in the SMSA (Ohio Bureau of Employment Services). Unfortunately, this industry has proven as vulnerable to economic fluctuations as the iron and steel industries; December 1993 data shows 7,500 workers remaining in the transportation equipment industry (OBES); and the local economic forecast projects no growth for this sector (Stocks 1992). The production of vans at the Lordstown facility was phased out at the end of the 1991 model year, and there is some possibility that the entire plant will close, as General Motors seeks to become more competitive in the world market.

Table 9.6.3 provides information on the nature of changes that have occurred in the employment structure of the Youngstown area. Several trends are evident from the table. First, as discussed, the Youngstown area, like other older industrial districts, has experienced an erosion in the number of job opportunities in manufacturing. This is especially true for the sectors involved with machinery and metals. The ripple effects of Black Monday and subsequent steel plant closings are very evident in the data for 1985. It has taken the area over 10 years to return to near pre-plant-closing employment levels. On the positive side, the number of wholesale, retail, service, and government jobs expanded. Unfortunately, jobs in these sectors do not have the same high wage rates and generous fringe benefits that are traditionally associated with employment in the industrial sector.

In the early post–World War II period, population grew very rapidly in suburban communities around Youngstown. Austintown and Boardman Townships had just over 10,000 residents in 1950; by 1980 they counted more than 33,000 and 41,000, respectively. Reflecting the erosion of the local employment base, in the 1980s both townships lost population, as did most of the other surrounding suburban communities. In the years immediately following World War II, Youngstown's population remained rather constant, at about 160,000. Then in 1960 it began to decline, hitting less than 96,000 by 1990. As in many other large cities, the population of Youngstown has grown older, poorer, and more ethnic in recent decades, while the suburban population is younger, more affluent, and predominantly white.

In looking to the future, the Youngstown area has some lingering liabilities. The region's economic malaise has translated into problems in supporting the provisioning of public services. Many local governments and school systems find themselves facing financial woes. Unfortunately, one of the ways they are meeting their shortfalls is to postpone improvements to infrastructure, buildings, roads, bridges, and utilities. Such a short-range perspective leads to additional problems in the future. The area must find a way to cope with the changing employment structure that is replacing high-paying industrial jobs with lower-paying positions in the retail and service sector. One apparent answer to the employment problems of the 1980s was the meteoric growth of a national discount drug chain headquartered in Youngstown. Unfortunately, in August of 1992, just after the opening of their 300th store, Phar-Mor was forced into bankruptcy, and there are serious doubts as to its ability to restructure. The downsizing of local operations had resulted in the layoff of more than 1,300 employees by September of 1992, and it is not known whether the headquarters will remain in downtown Youngstown.

A phaseout or additional downsizing of operations at the GM Lordstown complex looms as another economic threat. Amid these difficulties, an especially vexing problem is the lack of interregional cooperation among local governments. There is a history of noncooperation on the part of officials from Trumbull and Mahoning Counties and the cities of Youngstown and Warren. Without cooperation across political boundaries, little can be done to solve regional ills. The problem is compounded by the ebbing of the region's political power in Washington and Columbus, a result of declining population. The area also has a long tradition of organized labor, a factor translated by some to mean that the local work force has been overpaid and underproductive. Unions are still well entrenched in the area, but that does not necessarily suggest a work force with low productivity.

On the asset side of the ledger, the region is very well connected to the highway network of the Eastern United States. This network appears to be the major factor in the rapid growth of the Youngstown area as a center for distribution, a trend evidenced by the growth of employment in the wholesale sector. In the last 10 years, several major retailers have opened distribution centers to capitalize on the region's nodality in the highway network. Youngstown is home to two of the nation's largest mall developers: the DeBartolo Corporation and the Cafaro Company. Other assets include the area's ethnic diversity, which provides a cultural variety matched by few places. Additionally, the former industrial barons, before they moved elsewhere, left the area with a rich cultural legacy. The Butler Museum of Art has a renowned collection of American art. Youngstown also supports a symphony. And Mill Creek Park is one of the nation's outstanding urban parks. For a region that is in a state of economic decline, supposedly down on its luck, Youngstown has a surprising array of cultural and leisure amenities.

REFERENCES

Aley, Howard C. 1975. *A Heritage to Share: The Bicentennial History of Youngstown and Mahoning County, Ohio.* Youngstown, Ohio: The Bicentennial Commission of Youngstown and Mahoning County, Ohio.

Buss, Terry F., and F. Steven Redburn. 1983. *Shutdown at Youngstown.* Albany, N.Y.: State University of New York Press.

Butler, Joseph G. 1921. *History of Youngstown and the Mahoning Valley, Ohio.* Chicago: American Historical Society.

City Planning Associates, Inc. 1968. *Preliminary Report 1 for Youngstown, Ohio Community Renewal Program: Historical Development of Youngstown and the Mahoning Valley.* Youngstown, Ohio: City Planning Associates.

Ohio Bureau of Employment Services. February issues, 1966, 1971, 1976, 1981, 1986, 1993. *Labor Market Review.*

Stocks, Anthony. 1992. Review of the Economic Performance of the Youngstown-Warren Metropolitan Area (MSA) for 1991 and Forecast for 1992. January. Mimeographed.

Youngstown Vindicator. 1990. Business Outlook (11 February).

DAYTON-SPRINGFIELD METROPOLITAN AREA:
The Miami Valley Region

Mary-Ellen Mazey

The Dayton-Springfield Metropolitan Region, also known as the Miami Valley Region, is composed of four counties—Montgomery, Greene, Miami, and Clark—and according to the 1990 census had a total population of 951,270. The central city of the region, Dayton, has less than 20 percent of the region's population with a 1990 population of 182,044. As with other midwestern cities, Dayton's population peaked in 1960 and has been in decline ever since. Because of the small percentage of the region's population that resides in the central city, Dayton ranks eighty-ninth in terms of the one hundred largest cities in the country but is eighth in terms of decentralized metropolitan areas which is determined by the percentage of the population that resides in the central city as compared to the suburbs. Today, the region's economy is continuing not only to build on its traditional manufacturing base, since it is the second largest concentration of General Motors workers in the United States, but also to seize the aerospace opportunities associated with Wright-Patterson Air Force Base, the major research installation of the United States Air Force.

Historical Perspective on the Region

On April 1, 1796, the region's first settlers—a group of sixty people—arrived in Dayton from Cincinnati, and in that same year the first settlers were reported in Clark and Greene Counties, with the first permanent residents arriving in Miami County the following year. Therefore, Dayton in 1996 is celebrating its bicentennial, and the two hundred years of population growth and development can be attributed to the initial prime advantage of the region—its central location and excellent transportation system, connected to the early major waterways and today associated with the location of the intersection of two major interstate systems, I-70 and I-75.

At the time of the region's earliest settlement, the transportation advantage needed to establish thriving agricultural communities was provided by the Great and Little Miami Rivers. Later, in 1825, Ohio's General Assembly authorized the construction of the Dayton-Cincinnati Canal. The canal boosted commerce and the export of material from the region such as flour, whiskey, and pork. In addition, the canal system, and subsequently the railroads, facilitated the growth and development of the Miami Valley as a major industrial center of the Midwest.

Historically, Dayton has been recognized as a center of creativity and progressivism. For example John Patterson, founder of the National Cash Register Corporation (owned under another name by AT&T 1990–96, now once again NCR), perfected the concept of the modern factory as well as the techniques of modern sales training. At Patterson's death in 1922, an estimated one-sixth of the heads of the nation's major corporations were former NCR men who spread Patterson's techniques throughout the business world. Just as importantly, Patterson was known not only for his business leadership but also as a community leader. He introduced employee welfare programs, health clinics, and libraries, demonstrating his concern for the overall well-being of the labor force. His civic contribution was the introduction of the city-manager form of government in Dayton, making it the first major city in the United States to adopt this model form of local government.

Dayton also is famous as the home of Orville and Wilbur Wright, aviation pioneers. Additionally, the work of Edward Deeds and Charles Kettering, inven-

Table 9.7.1
Nonagricultural Employment

1990 EMPLOYMENT AREAS	DAYTON-SPRINGFIELD MSA	DAYTON CITY
Manufacturing	22.78%	18.62%
Government	15.55%	19.04%
Retail Trade	17.71%	17.64%
Health Services	9.57%	10.73%
Total Employment	438,828	70,730
% Labor Force Employed in Central City	16.12%	

Source: Center for Urban and Public Affairs, Wright State University, 1993

tors of the self-starter for automobiles and the rural electric generator, illustrates the Miami Valley's diversity with links to the rural farm areas and the industrial giants, Delco and General Motors.

Economic and Population Base

In 1990 the Dayton-Springfield MSA was the fifty-third largest MSA in the country. Montgomery County is the core of the region with approximately half of the region's inhabitants. However, Dayton has declined in the role of the central city. In 1930 Dayton composed 40 percent of the region's population while in 1990 this percentage had dropped to 19. Just in the decade of the 1970s, Dayton's population declined by 20 percent. Moreover, at the time of the central city's population decline, the region's population increased with rapid growth and development of the suburbs and exurbs. In part, this growth can be attributed to the region's transportation assets including Interstates 75, 70, and 675, the latter completed in 1985. This excellent highway accessibility makes Dayton, with over 4.2 million people residing within ninety minutes of the city by highway travel, the nation's eighth largest ninety-minute car/truck market. An additional transportation asset is the Dayton International Airport which is the national "superhub" of Emery Air Freight Corporation.

Like many other regions of the Midwest, the Dayton-Springfield area has undergone rapid economic changes in recent years. When compared with other MSAs and the U.S. average, this metropolitan area has a higher percentage of the labor force employed in manufacturing, service, and the governmental sectors of the economy than in wholesale trade, retail trade, and the financial, insurance, and real estate sectors. This is illustrated by examining the four major employers in the region—Wright-Patterson Air Force Base, General Motors with five divisions, Navistar Corporation, and National Cash Register.

With an economy oriented toward the industrial, governmental, and service sectors, the slowdown in manufacturing industries and the downsizing of government has made the region's economy vulnerable to national trends. As Table 9.7.1 indicates the central city and the region maintain a strong industrial base, but nearly all of the region's recent economic growth has been in the nonmanufacturing sector. In 1970, manufacturing provided nearly 40 percent of the MSA's employment, but by 1990 only 22.8 percent of the jobs in the MSA were in manufacturing. This decline in manufacturing employment has been offset by an increase in service sector employment.

With Wright-Patterson Air Force Base as the major employer in the region, it is not surprising the Dayton-Springfield metropolitan area has the highest per capita concentration of engineers in Ohio. This research-oriented base employed 16,007 civil service employees and 9,578 military employees in fiscal year 1992; however, this was a 13 percent decline from 1988.

As a region the Dayton area is well endowed with institutions of higher education. These include the public universities—Wright State University and Central State University; the private universities—University of Dayton, Wittenberg University, Wilberforce University, Cedarville College, and Antioch College; and the two-year colleges—Sinclair Community College, Clark State Community College, and Edison State Community College. Even though Dayton area residents have a high degree of accessibility to higher education (see Table 9.7.2), there is a considerable difference in educational

Table 9.7.2
Dayton-Springfield-Fairborn MSA Demographics

	DAYTON	MONTGOMERY COUNTY	GREENE COUNTY	MIAMI COUNTY	CLARK COUNTY	MSA
Population	182,044	573,809	136,731	93,182	147,548	951,270
Households	72,670	226,192	48,351	34,559	55,198	364,300
% Renter Occupied Housing	49.00%	37.10%	30.60%	27.30%	30.90%	34.30%*
% Owner Occupied Housing	51.00%	62.90%	69.40%	72.70%	69.10%	65.70%*
Median Family Income	$24,819	$36,069	$39,776	$35,898	$32,597	$35,999
Median Contract Rent	$253	$316	$352	$291	$256	
Median Age	31.00	33.30	32.40	34.30	34.30	
Median Housing Value	$43,200	$65,000	$78,200	$65,000	$54,900	$64,200**
% Over 65 Years	13.10%	12.50%	9.7	12.60%	13.90%	12.40%
% Below Poverty	26.50%	12.60%	9.5	8.40%	13.40%	11.90%
Persons Per Household	2.41	2.49	2.70	2.67	2.60	
Persons Per Family	3.10	3.04	3.11	3.11	3.06	
Race						
White	58.40%	80.80%	90.70%	97.10%	90.30%	85.30%
Black	40.40%	17.70%	7.00%	1.90%	8.80%	13.30%
American Indian	0.20%	0.20%	0.30%	0.20%	0.20%	0.20%
Asian	0.60%	1.00%	1.60%	0.70%	0.40%	1.00%
Other	0.00%	0.00%	0.00%	0.00%	0.00%	0.00%***
Education (25+)						
< 9th grade	10.70%	7.60%	5.90%	6.90%	8.10%	7.40%
9th to 12th, no degree	20.90%	14.60%	11.80%	16.50%	18.50%	15.00%
High School Grad.	30.80%	30.80%	31.50%	39.80%	38.50%	33.00%
Some College, no degree	19.90%	20.40%	18.90%	17.50%	17.10%	19.40%
Associate's Degree	5.30%	6.50%	5.90%	5.20%	5.60%	6.20%
Bachelor's Degree	8.10%	13.00%	15.00%	9.70%	7.60%	12.10%
Graduate/Professional Degree	4.20%	7.00%	11.10%	4.40%	4.60%	6.90%
% High School or Higher	68.30%	77.70%	82.40%	76.60%	73.40%	77.60%

Source: U.S. Bureau of Census, 1990. Calculations: Center for Urban and Public Affairs, Wright State University
* calculated from census data
** calculated from real estate data
*** percentages are in the thousandths and rounded to 0%

attainment between the central city and its entire home county of Montgomery and between the central city and the more suburban and exurban counties of Greene, Miami, and Clark. The difference is somewhat less between Dayton and Clark County because of Clark's central city of Springfield.

The disparity between the central city and outlying areas is further exemplified by examining median family income and poverty levels. The city of Dayton has the sixth highest poverty level of the one hundred largest cities in the United States; its poverty rate is over double that of the surrounding suburbs and more rural areas (Table 9.7.2). The city's median family income is about two-thirds of that of the region as a whole, as is the city's housing value. The Dayton region is also highly segregated. Over 40 percent of the city's residents are African American while the region's percentage is 13.

Conclusion

The Dayton-Springfield Metropolitan Area, as with other Ohio metropolitan areas, is moving toward an increasing emphasis on the global economy. In fact, a number of Japanese plants have located along the Interstate 75 corridor in recent years. As the Miami Valley positions itself for this next century, the emphasis will be upon applied science and technology in aviation/aerospace, information systems, and the traditional automotive industry and its associated parts, in addition to building on the higher educational system and the preservation of the region's environmental assets. The major problem the region faces is the decentralization of the population and the accompanying issue of the educational and economic disparity that has been created between the central city and its surrounding area.

REFERENCES

State of the Region. 1993. Center for Urban and Public Affairs, Wright State University.

U.S. Department of Commerce, Bureau of the Census. 1990.

CHAPTER TEN

Small Towns

Allen G. Noble

Ohio's numerous small towns are a reflection of the original settlement process, which ended around the mid-nineteenth century. Few towns were created after that time, although some earlier settlements disappeared, a trend that continues today. Those that have disappeared have been mostly agricultural service centers, whose limited functions were usurped by neighboring locations with better centrality or access.

Although the number of Ohio small towns may appear at first glance to be increasing, the opposite is true. The confusion results from the U.S. Census Bureau's practice of changing its definitions from census to census. Even more important than a decline in number of towns is the stagnation, or even decrease, in the size of small towns.

According to the *1990 Census of Population,* Ohio had just over 1,000 incorporated or census-designated places (Bureau of the Census 1993). Out of the total of 1,008 places, 162 had a population of over 10,000 each, 213 had populations between 2,500 and 10,000, and 633 were small settlements of less than 2,499. During the preceding decade there was a clear trend toward larger settlements, with an accompanying decline in and actual disappearance of the smallest settlements, called *hamlets.* Some of the independent settlements enumerated in the census were so close to major metropolitan areas that they functioned largely as suburban dormitory communities, but most were truly independent small towns.

Cultural Imprint of Ohio's Small Towns

The Small Town Persona

The significance of Ohio's small towns lies not in numbers but in the range of ideas, practices, and values associated with them. In a very real sense the outlook of Ohio is that of the small town, incorporating all of the deficiencies and shortcomings of that perspective, together with the considerable stability, self-reliance, and abundant inner strength of basic and traditional viewpoints and approaches. Ohio small towns are considered by their inhabitants to be "friendly places, typified by a widespread community pride and spirit of involvement" (Noble and Korsok 1975). Other frequently noted advantages include lower costs of living and real estate, low taxes, and less government. Conservatism, a small town ambiance, and an orientation toward the past often go hand in hand. Many Ohioans yearn for a time when life was simpler, easier to comprehend and deal with, a time when problems were smaller in scope. Thus, they resist governmental activities, even when such activity could benefit them in the long run. The relatively low expenditures for public higher education in Ohio are an excellent example of such limiting attitudes.

Ohioans generally possess a strongly held view of the virtues of small towns. Life in small towns proceeds at a more leisurely pace than in cities, and the surroundings are quieter and more spacious. Small towns are thought to be good places to raise a family, places where crime and juvenile delinquency are minimal. The truth, however, is that modern-day mobility permits crimes to be committed over a larger venue, and crime rates are steadily increasing in small towns, especially as drugs become more widely available.

The Small Town Landscape

Far more than large cities, it is small towns that provide the distinctive looks that characterize the differ-

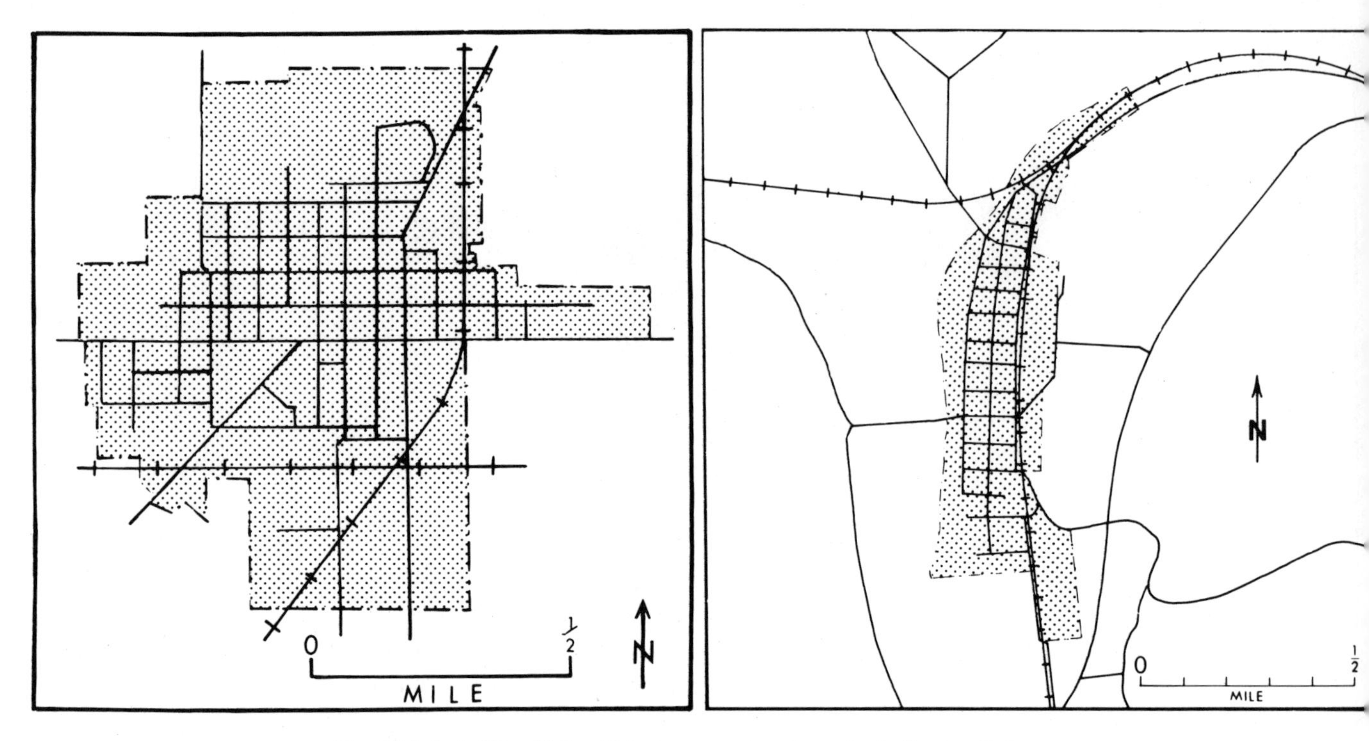

Fig. 10.1. *Left:* West Unity, Ohio, is compact and regular indicating little topographic control or restriction. *Right:* Killbuck, Ohio. The elongated form of the town is a response to the topography of a river valley.

ent sections of the state. Such looks are a combined product of topographic conditions, economic orientation, and ethnic roots. In the valleys of hilly southern Ohio, communities are often elongated and sinuous, whereas those on the essentially flat, glacial till and lacastrine plains of northwestern Ohio tend to be compact and rectangular in layout (Figure 10.1). Coal mining and river transportation impart a flavor which is quite unlike that created by the arrow-straight railroad lines, grain elevators, and hog-feeding lots of central Ohio, and the towns of each section respond to these different attributes through their layout and form.

Everywhere, the small town provides a different cultural aspect. The presence of a central town green, simple Classical Revival houses, and white New England churches give a distinctive character to northwestern Ohio quite unlike either the hardscrabble, helter-skelter arrangement of the many originally Scotch-Irish settlements of southern Ohio or the tight precision of the red brick and timber frame German communities of western Ohio. In these latter towns, the tall, slim spires of German churches dominate, especially when seen at long distances across the open farmland. And throughout the state, small county seats cluster around Victorian courthouses.

The Town Patterns

The physical layout of Ohio's small towns shows considerable variation, although most are quite compact, and all are governed to some extent by transportation facilities. The basic orientation of a town's streets often is controlled by the location of early cross-country roads or by the position of later transportation routes, such as railroads or canals. The overwhelming popularity of the rectangular grid street pattern in Ohio's small towns is not difficult to explain: rectangular blocks are easy to lay out in uniform building lots, which in turn accept rectangular buildings with little wasted space.

In rare instances, survey district boundaries divided a settlement, resulting in different orientations. Bellefontaine, in Logan County, is one example. West of Ludlow Street the street orientation is north-south and east-west and is determined by the Between-the-Miamis original land survey. East of Ludlow Street, in the Virginia Military District survey area, the orientation is irregular (Figure 10.2).

Another disparity occurs in Kenton (Hardin County), where streets to the north of the Scioto River are oriented to the General Land Office Survey—a system that, although common throughout much

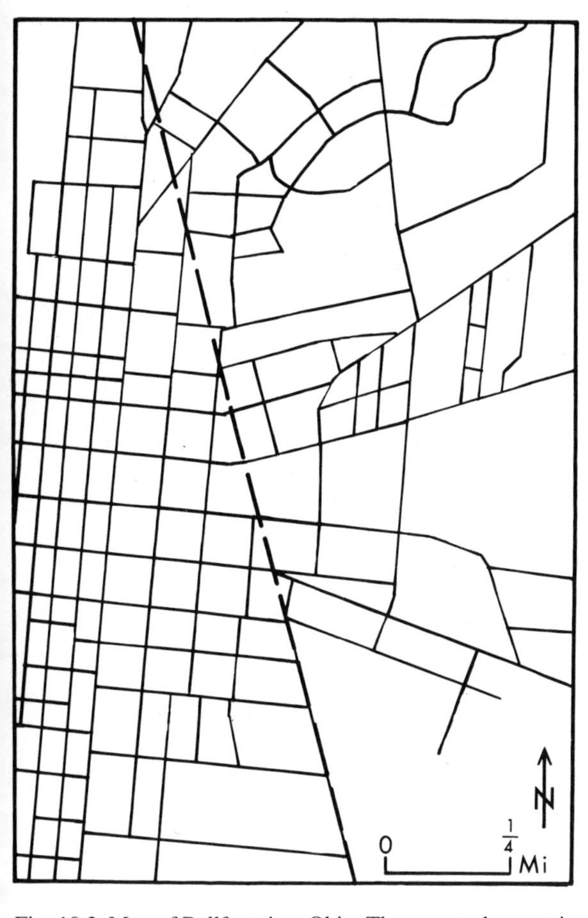

Fig. 10.2. Map of Bellfontaine, Ohio. The area to the west is part of the Between-the-Miamis land survey and that to the east lies in the Virginia Military District survey.

of the United States, was first employed in Ohio. Land south of the river originally was part of the Virginia Military District, and two basic orientations are in evidence, neither matching that north of the river. Over the course of time, as the river meandered, land which was originally north of the river and surveyed as such has ended up south of the river, producing an incongruous north-south, east-west street pattern in a small area.

Even in areas of difficult or limiting terrain, the grid pattern of streets predominates. A town with a nonrectangular street pattern generally indicates a community settled and platted before government survey. Rock Creek in Ashtabula County is one example.

The Influence of Transportation on Small Towns

The transportation system affects the layout of many small towns. Most Ohio towns are focal points that, at least originally, provided services to a surrounding hinterland. This history explains the widespread phenomenon in which roads radiate outward, connecting a town with all parts of its hinterland. A map of one such town, West Union in Adams County, reveals quite clearly the distinction between the town's grid pattern streets and its radiating roads (Figure 10.3).

Not all Ohio towns are products of road development. A few communities—Deshler in Henry County is perhaps the best example—have street patterns determined by the location of railroads (Figure 10.4). The boundaries of Deshler, as well as its connecting roads, are orientated to the General Land Office Survey, but the community's streets are oriented to the northeast-southwest position of the Baltimore and Ohio Railroad, which passes through the center of town. Curiously, the other end east-west line of the Baltimore and Ohio Railroad had little effect on the orientation of the streets, although it functions as a significant barrier to north-south road traffic.

In at least one village the influence of both road and railway is seen. The northern section of Hamler, in Henry County, has its streets oriented to the Baltimore and Ohio Railroad, whereas the streets of the southern section fit the north-south, east-west road orientation of the General Land Office Survey. But however important railroads may be in determining the structure in some urban places, in others they appear to have had no impact. A case in point is Ohio City in Van Wert County (Figure 10.5): though three rail lines converge on this community, the town's street pattern shows absolutely no effect.

Another group of villages were largely products of the canal period. Generally, canal towns are elongated in line with the axis of the canal, and with only a few bridges across the former canal bed. In Spencerville (Allen County), the Miami and Erie Canal, whose line can be traced easily today, determined the position of the town's streets (Figure 10.6). The building of canals promoted the establishment of such urban centers as Canal Winchester, Spencerville, Canal Lewisburg, and Roscoe; later, the decline of canals saw the corresponding decline of these centers, except in such cases as Canal Fulton, where additional urban functions had been developed (Figure 10.7). A similar response occurred a bit later, in the case of communities tied to railroads.

The canal-created settlement of Roscoe was able to avoid extinction only because it had declined so precipitously and been abandoned so completely that its

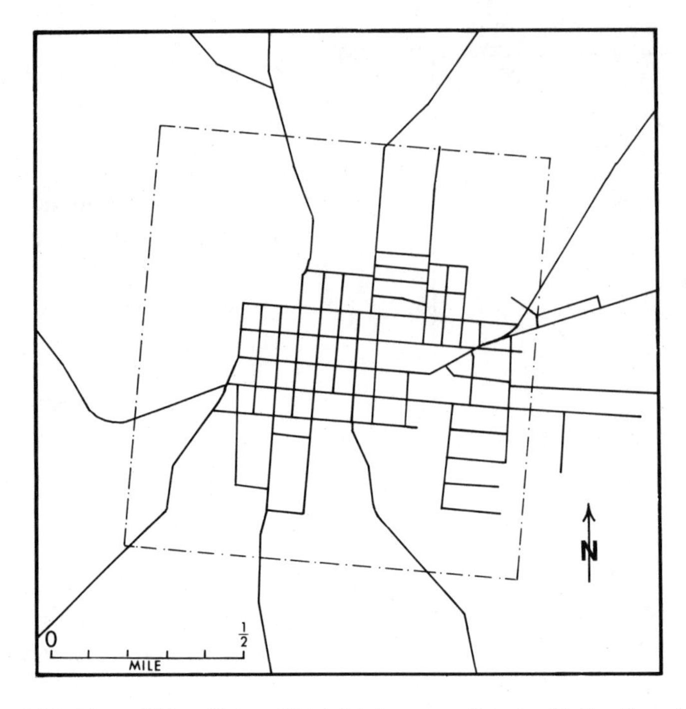

Fig. 10.3. Map of West Union, Ohio. Highways radiate in all directions from the town.

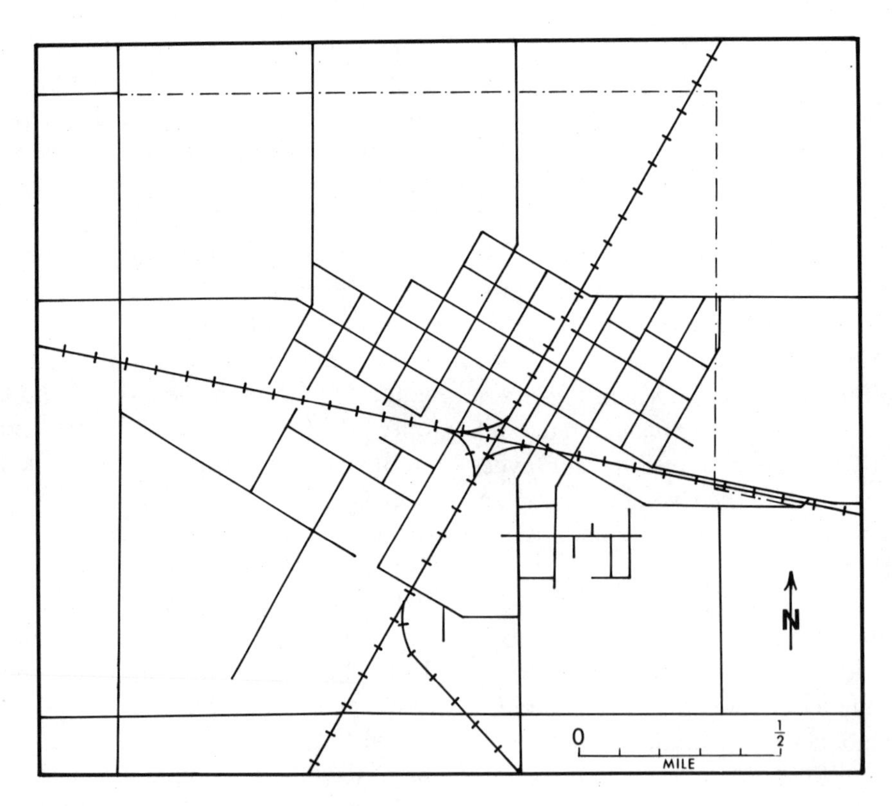

Fig. 10.4. Map of Deshler, Ohio. The street pattern is oriented to the north-south railroad, but not the east-west one.

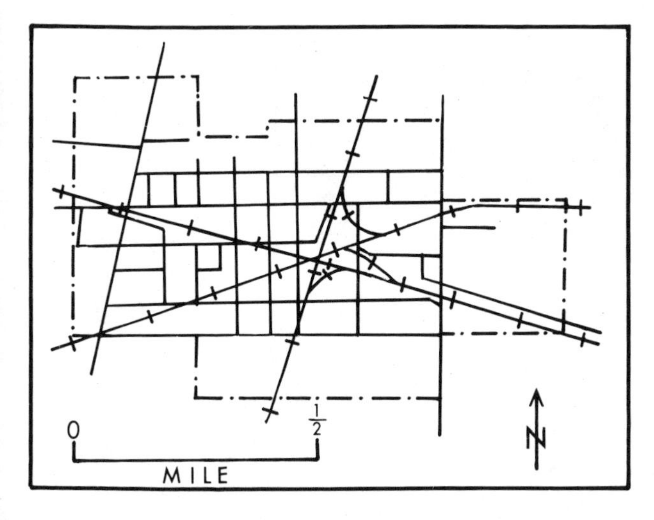

Fig. 10.5. Map of Ohio City. The railroads and roads present a chaotic pattern with numerous oblique crossings.

canal-era buildings were preserved intact, making it possible for preservationists of the 1960s to restore the community as a tourist site. Thus it rose a second time, but with an entirely different urban function: tourism had replaced transportation.

Because of the improvement of roads and the ever-wider ownership of automobiles, vans, and pickup trucks, rural residents are no longer oriented to small, local, service center communities. A journey to work of 20 miles is not uncommon, and longer trips for specialized services such as medical consultation, college education, and governmental services has come to be expected. The consequence for the small rural community has been a gradual but steady decline in facilities and population.

Only in close proximity to urban areas is this decrease balanced by an influx of urbanites in search of a less costly, more spacious, more bucolic atmosphere. Such influx, however, does not usually bring commercial, industrial, or other revenue-producing land uses, thus the tax base of these rural communities steadily erodes. Many local facilities have had to be closed or consolidated. The centralized school system has largely replaced the local schools, common as late as midcentury. Medical facilities, too, are tending toward consolidation, often in a rural setting equidistant from two or more small towns, each of whose residents drive a few miles to the medical center.

The Western Reserve Town

Much of northeastern Ohio lies within the Connecticut Western Reserve and its western extension, the Firelands. The first of these areas was originally unsettled land set aside in 1786—when that state renounced further claims to western American territory—for the expansion of Connecticut settlement. The Firelands were set aside in 1792 as compensation to Connecticut victims of Revolutionary War burnings and other devastation.

Throughout both of these areas, settlers from New England, especially Connecticut, established a cultural landscape reminiscent of their area of origin. Houses, barns, and other buildings, as well as the forms of the villages themselves, mimicked those of New England. From its very beginnings, New England settlement revolved around villages surrounded by agricultural fields in a medieval pattern referred to as the three-field system. When settlement expanded, new villages—sometimes called "daughter" villages—were formed by calving-off some inhabitants from the original village. Always, however, the village was the dominant focus of settlement. When New Englanders entered northeastern Ohio, they introduced the same settlement system. Although a pattern of dispersed farmsteads quickly replaced one of scattered nuclear villages, the distinctive form of the New England village was retained.

A distinguishing feature of the Connecticut Western Reserve village that derived from its earliest, medieval forerunners in England was a central open space, termed the *green* in New England and the *commons* in the Connecticut Western Reserve. Originally designed as a more or less enclosed grazing area for the village's larger domestic animals, it has come over time to function as a focal point for the village. The form of a commons may vary, from irregular (Twinsburg in Summit County), to rectangular (Chardon in Geauga County), square (Medina in Medina County), oval (Burton in Geauga County), or circular (La Grange in Lorain County).

The relationship of village streets to the central commons also varies. In some instances the commons conveniently occupies a block surrounded by bordering streets. In other cases the streets converge on the commons, radiating outward from it (Tallmadge in Summit County). Commercial properties

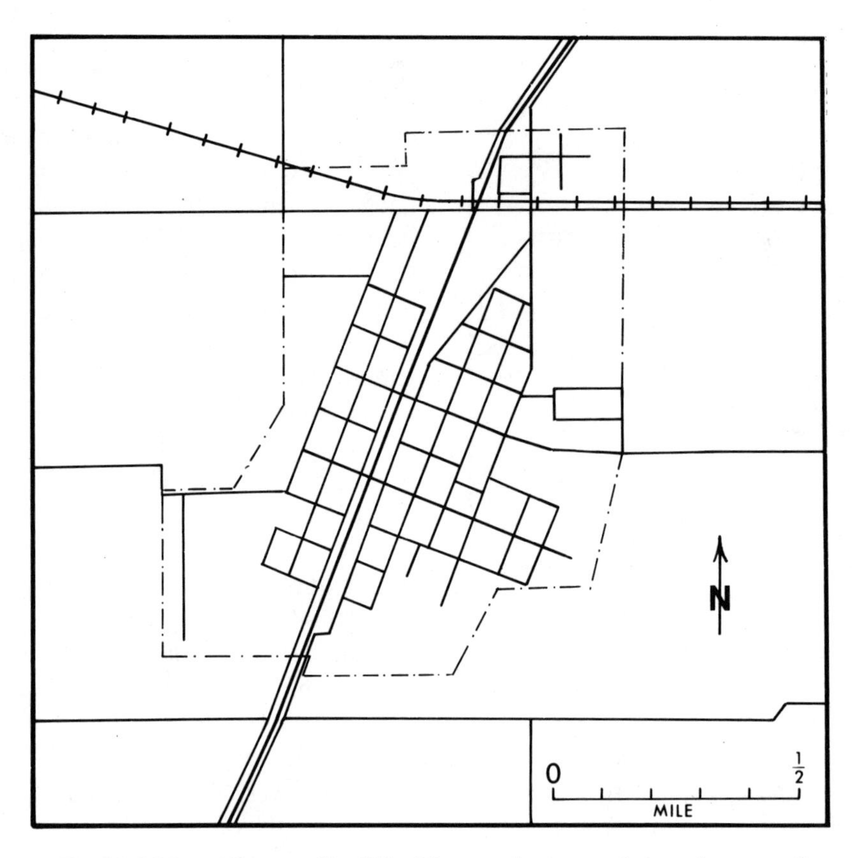

Fig. 10.6. Map of Spencerville, Ohio. The town is elongated along the route of the canal and few streets cross the canal.

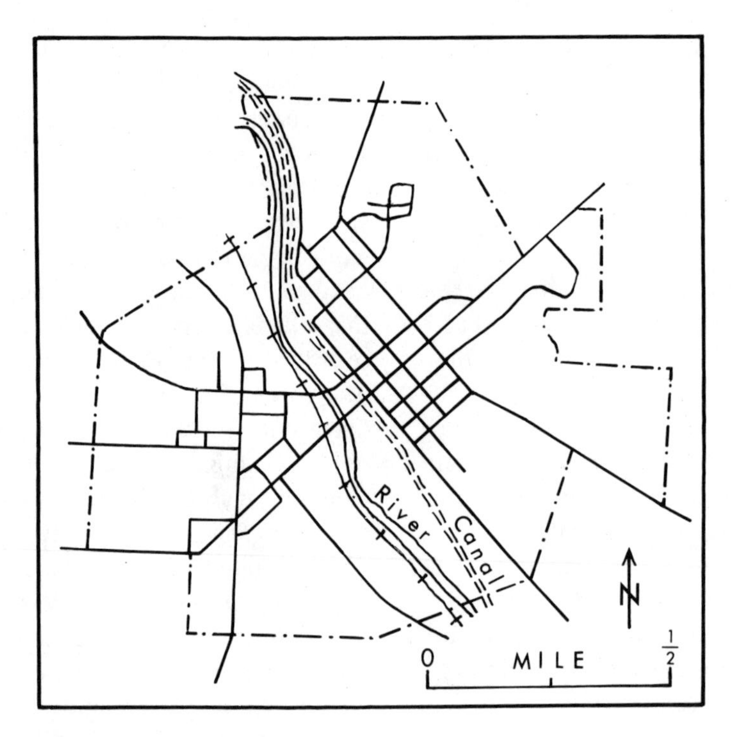

Fig. 10.7. Map of Canal Fulton, Ohio. The river and canal have determined the orientation of the streets.

and public facilities—such as a village hall, library, and churches—line the streets around the commons and, if the village has grown large enough or if the commons is small, extend along a main street leading away (Burton in Geauga County). In a few instances public buildings are actually situated on the commons itself; in others, the commons has all but disappeared, although some public buildings—like the Portage County courthouse in Ravenna—may remain.

Another feature of Western Reserve towns—Chardon in Geauga County, for instance—is the irregular size and shape of town blocks, with the smaller blocks generally toward the center and the larger blocks on the periphery (Figure 10.8). There is by no means a regular order of increasing block size, however. A third, additional, characteristic element of Western Reserve towns is the railroad, which normally occupies a position halfway between the commons and the edge of town (Figure 10.8). The tracks of these railroads most often formed a broad curve and thus cut into a high percentage of the main roads that focused on the green.

More than any of these features, except the presence of the commons, it is the overall appearance of the village itself which distinguishes these Western Reserve communities. In many cases the villages have not grown very much, or at least, where new residences have been added, they have been limited to peripheral locations, away from the core of the village. Thus, the flavor of a nineteenth-century small town is preserved. Tree-lined streets, spacious front lawns, the regular spacing of ubiquitous white houses, and the homogeneous, Classical Revival architecture (clearly derived from a Yankee–New England tradition) complete the strong visual unity of those towns.

Ohio Place-Names

The definitive study of Ohio place-names has yet to be written. The following discussion is based upon William Overman's study, the best of existing works (1959).

For the most part, Ohio's place-names follow patterns commonly found elsewhere. A wide scattering of Indian names remains: Osceola (Crawford County), Ottokee (Fulton), Moxahala (Perry), Tymochtee (Wyandot), Piqua (Miami), Wapakoneta (Auglaize), and Catawba (Clark), for instance. Most small towns, however, were named after first settlers, surveyors who initially laid out the community, or wealthy early landowners, merchants, or industrialists. Thus explains Ohio town names like Kinsman (Trumbull), for an early large landowner; Fredrickstown (Columbiana), for the first mill owner; Russell (Geauga), for the first settler; Sabrina (Clinton) and Botkins (Shelby), for each town's surveyor; Farmer (Defiance), after Nathaniel Farmer, the first prominent settler; Fletcher (Miami) for an early merchant; and Buchtel (Athens), for a prominent businessman and landowner. These are only a few examples; there are hundreds of others.

Some community names are more intrinsically interesting and often more geographically instructive. A few represent geographical relationships at the most basic level. Thus we find Middlebourne (Guernsey County), so named because it lies halfway between Wheeling, West Virginia, and Zanesville; Middleburg (Van Wert) halfway between Van Wert and Decatur, Indiana; Middlefield (Geauga) halfway between Warren and Painesville; Middlepoint (Van Wert) halfway between Delphos and Van Wert; Middleport (Meigs) on the Ohio River halfway between Pittsburgh and Cincinnati; Middletown (Butler) halfway between Cincinnati and Dayton; Midvale (Tuscarawas) halfway between New Philadelphia and Uhrichsville; and Midway (Madison) halfway between Philadelphia and Chicago. In the seemingly great distances of nineteenth-century Ohio, halfway points were viewed as important locations, reached after considerable effort and with evident relief.

Another type of distance relationship applies to a few communities. Seventeen (Tuscarawas) was the home of the seventeenth lock on the Ohio Canal, and Twenty Mile Stand (Warren) was a stagecoach stop 20 miles from downtown Cincinnati. Some community names—Big Plain (Madison), Bay View (Erie), Stony Ridge (Wood), Big Springs (Logan), and Crestline (Crawford)—derive from obvious geographical features, the latter from its placement on the Lake Erie–Ohio River drainage divide. Other place-names are clearly associated with local resources: Coalton (Jackson), Ironton (Lawrence), Gypsum (Ottawa), and Limestone (Ottawa). Cannelville (Muskingum) is less so, its name derived from a local deposit of cannel coal. Celeryville (Huron), which suggests a different sort of resource, began as a community of Dutch farmers who specialized in celery and other vegetables.

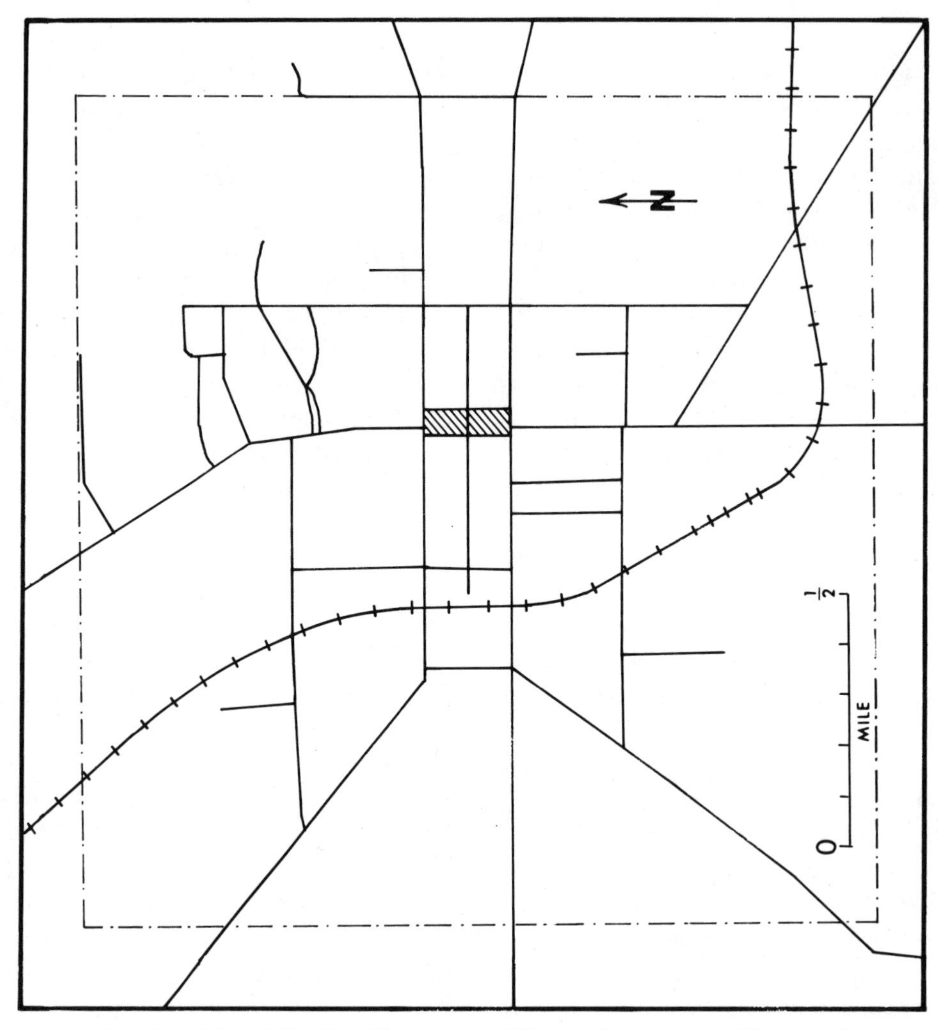

Fig. 10.8. Map of Chardon, Ohio, a typical Western Reserve town. The central commons is shaded.

Place-names often designated the ethnic and geographic origins of early settlers. Because most early settlers came from the East or Southeast, names reflect these areas, filling out the details of migration. Community names in the Western Reserve of northeastern Ohio, for example, have their origins in the New York–New England region, especially Massachusetts and Connecticut. The smaller Firelands area of north-central Ohio reflects more precisely its Connecticut connection. Thus in Huron County, Connecticut place-names—including Greenwich, New Haven, New London, North Fairfield, and Norwalk—abound. Similarly in Highland County, which is part of the Virginia Military District—an area originally designed for Virginia Revolutionary War veterans—names such as Danville, Fairfax, Leesburg, New Market, New Petersburg, and Lynchburg appear.

Place-names also reveal European connections. Some examples include Antrim (Guernsey), established by Irish settlers; Belfast (Highland), by Scots-Irish; Belfort (Stark), by French; Brandon (Knox), by Welsh; Flushing (Belmont), by Dutch; and Glasgow (Columbiana), by Scottish. One of the more interesting names is Russia (Shelby), which does not identify the origin of the town's early inhabitants but was chosen because its rolling plains reminded the original French settlers of the country they had campaigned in as members of Napoléon's army. German place-names are clearly the most common of those of foreign origin. Though for the most part they are widely scattered, names such as Geneva, Breman, North Berne, Basil, New Strasburg, and Hamburg in Fairfield County attest to an early concentration of German and German-speaking Swiss settlers.

Not only do place-names reveal something of the ethnic and geographic origins of early settlers, they also disclose much about the culture and history of the times in which the land was settled. One note-

Fig. 10.9. The "Salems" of Ohio.

worthy example is Mount Healthy (Hamilton County), so named to commemorate deliverance from a cholera outbreak in 1850.

In early nineteenth-century Ohio, education, when available at all, was often classically oriented, and the society's underpinnings were strongly religious. Hence the plethora of classical names—Alpha (Greene), Adelphi (Ross), Arcadia (Hancock), Athens (Athens), Castalia (Erie), Delphi (Huron), Delphos (Van Wert), Homer (licking), Lithopolis (Fairfield), Sparta (Morrow), Xenia (Greene), and Omega (Pike)—as well as those with a biblical origin—Bethel (Clermont), Bethesda (Belmont), and Bethlehem (Richland, and also, Marion). A few biblical names have negative connotations: witness Sodom (Trumbull) and River Styx (Medina), the latter so named because much of the area was originally swampy.

The study of place-names also provides insight into the history of the state at the time of initial settlement. In the late 1840s, a time when agricultural settlement was filling in the countryside, the impact of the Mexican War was being felt. Communities such as Monterey (Clermont), Buena Vista (Hocking and Scioto), Montezuma (Mercer), Mexico (Wyandot), New Matamoras (Washington), Vera Cruz (Brown), Santa Fe (Auglaize), and Rio Grande (Gallia) attest to this influence. Somewhat later, the construction of railroads added a new layer of town names. Gano (Butler), Cecil (Paulding), Willard (Huron), Patterson (Hardin), and Beach City (Stark) are among the names that memorialize railroad figures.

Geography is not always accurately portrayed in place-names. As the *Cleveland Plain Dealer* once observed (July 10, 1951), starting from Salem (Columbiana), a person would have to travel south to reach North Salem (Guernsey). West Salem (Wayne) is further north than North Salem, and South Salem (Ross) is farther west than West Salem. Salem Center (Meigs) is not in the middle of the other Salems but rather on the southern fringe of the state, farther south than South Salem, Lower Salem (Washington), and New Salem (Fairfield) (Figure 10.9).

Geographers have a continuing interest in the study of place-names, also called *toponyms,* because of what such study can reveal about both the cultural and physical landscape of an area. The most comprehensive geographical study of toponyms is that compiled by the Australian geographer M. Aurousseau (1957). For eastern North America, the work of Wilbur Zelinsky (1955) is well known. Perhaps as Ohio geographers become more proficient in toponymic research they will be able to explain the origins of such fascinating Ohio place-names as Charm (Holmes), Mudsock (Franklin), Getaway (Jackson), Businessburg (Belmont), Bangs (Knox), Funk (Wayne), Worstville (Paulding), Steam Corners (Morrow), Zone (Fulton), Revenge (Fairfield), and Knockemstiff (Ross).

REFERENCES

Aurousseau, M. 1957. *The Rendering of Geographic Names.* London: Hutchinson University Library.

Cleveland Plain Dealer. July 10, 1951.

Noble, Allen G., and Albert J. Korsok. 1975. *Ohio: An American Heartland.* Columbus: Ohio Department of Natural Resources, Division of Geological Survey.

Overman, William D. 1959. *Ohio Town Names.* Akron, Ohio: Atlantic Press.

U.S. Bureau of the Census. 1993. *Census of Population and Housing.* Washington, D.C.: U.S. Government Printing Office.

Zelinsky, Wilbur D. 1955. Some Problems in the Distribution of Generic Terms in the Place Names of the Northeastern United States. *Annals of the Association of American Geographers,* vol. 45 (December).

CHAPTER ELEVEN

Energy Production and Consumption

Richard W. Janson and Leonard Peacefull

The economic development of Ohio from the very beginning is synonymous with the scientific application of inanimate energy to production. Ohio's locational advantage—between Lake Erie, the southernmost of the lower lakes, and the Appalachian Mountains of West Virginia—assured a pivotal and decisive influence in shaping the destiny of the nation, first as a gateway to the West and later as the heart of the manufacturing belt.

Tools of production are designed to reflect current state-of-art technology. In the realm of energy applications in Ohio the flow of innovations ranges from technologies for millwheels on running streams to technologies for nuclear power generation and rocket engine propulsion. From the opening of the Northwest Territory in 1787, the kinetic energy of running water assisted the pioneers in their conquest of nature. After 1835, steam power was quickly adopted to drive the engines that powered Ohio River boats; to drive the locomotive engines of railroad trains that linked hundreds of Ohio communities in a network system of factory production; and to power the fixed engines in factories that brought, in turn, the use of assembly lines, interchangeable parts, and mass production, resulting in high standards of living. Ohio's second century was shaped by two other fundamental innovations in power systems: the use of electric energy by motors, and the use of liquid energy sources for internal combustion engines. The electric motor and the internal combustion engine have shaped our culture in a thousand ways, both at the macro level of the landscape and at the micro level of everyday life. This chapter examines Ohio's energy production and consumption by both domestic and industrial consumers at the end of the twentieth century.

Energy in Historical Context

Western civilization has advanced in population, prosperity, and health in a magnificent way since the Industrial Revolution commenced in England only 250 years ago and diffused to Ohio along with the earliest settlers after the Revolutionary War. The essence of the Industrial Revolution is the application of inanimate energy to the production systems of human society. The invention of the steam engine and its improvements, and its application to coal mine pumps, stationary engines in factories, locomotive engines, and steamboats, brought rapid increases in productivity and efficiency in almost all industrial activities. Every mill wheel on a stream transformed the kinetic energy of the moving water into grinding, sawing, carding, spinning, or weaving. These millstream applications preceded steam engine applications for the same purposes. The first steamboat on the Ohio River began operating in 1835, 10 years after Robert Fulton introduced regular passenger service on the Hudson River. The steamboat and railroad engine, using energy in the form of coal for fuel, opened the North American continent in the last two-thirds of the nineteenth century to massive, rapid immigration.

The spatial diffusion of steam machinery brought in its wake the factory system, interchangeable parts, production lines, mass production, mass assembly, mass consumption, and immense energy inputs into production. Electric motors, which convert the energy of an electric current into a rotating shaft, came into use late in the nineteenth century, shortly before the rise of the internal combustion engine. Both gasoline engines and diesel engines burn liquid petroleum products in the presence of air to expand gases, causing the movement of a piston. The chemical energy that binds the atoms of the fuel into molecules is transformed into heat energy in the burning process and the heat is transformed into mechanical work as the expanding steam drives the piston. The combination of the internal combustion engine with the widespread availability of electricity resulted in another dispersion of factories. Continuous investment in the highway network and the general use of automobiles and truck transport has allowed degrees of freedom in both plant location and residential location during the last half of the twentieth century that were not reasonable options during the steam age.

Each of the major advances in new energy applications resulted in corresponding advances in standards of living. The key to prosperity has been the successive application of more and more energy to production systems. Ohio has been a leader in the successive waves of energy use innovations, and the economic growth of Ohio has been based on these innovations from the very beginning.

Energy Production

Chapter 4 describes the distribution of fuel resources throughout the state and the various methods of production. The value of coal shipments from Ohio mines exceeds the combined value of natural gas and petroleum from Ohio wells. Most Ohio coal is converted into electrical energy which is then used for lighting, heating, and driving machinery.

Besides electricity, other forms of energy are in everyday use, including gasoline, diesel fuel, and aviation petrol. These liquid fuels, all refined from petroleum, are used to drive internal combustion engines, such as those that power our automobiles, trucks, locomotive engines, and airplanes. All of the major contemporary transportation prime movers use liquid fuel for ease of fuel delivery within the engine. Contemporary society is utterly dependent on the technologies of liquid fuel. In addition to these liquid fuels, our homes and factories use natural gas for space heating, and gas is also used to generate process heat in industry and to operate gas turbines to generate electricity. The energy source taken most for granted is electricity, because cross-country distribution grids make electric power available almost everywhere.

Electrical Energy

Electrical energy is a secondary energy source derived from primary energy sources, which include the fossil fuels—coal, petroleum, and natural gas. When electricity is produced in a typical Ohio fossil fuel plant, the latent chemical energy within the fuel is ultimately transformed by the generator into a moving stream of electrons—the electrical current. The utility company generates the electric current, transmits power at high voltage cross-country, and distributes the electricity locally to the consumers of energy. Some states by law require public utilities to buy electrical power produced by nonconventional plants, such as those using solar radiation, wind, the burning of refuse, waste heat, and geothermal sources of energy. Although these alternative energy sources are only trivial contributors to Ohio's energy requirements, such requirements constitute one way of supporting research into alternative solutions.

Most privately owned electrical utility plants in Ohio are large, probably because average costs normally decline as plant size increases. One reason electricity is preferred is because the environmental problems are spatially concentrated at very large plant sites, and after an initial substantial investment in pollution control equipment, these plants create little pollution at the site of power generation. There is no pollution at the site of power use. Also, control devices, sensors, motors, starters, and other apparatus have widespread availability.

Electricity in normal use requires a completely closed circuit that connects the appliance being used to the generating plant. This means that electricity must be generated simultaneously with its use. Hence the capacity of an electrical plant is defined by its maximum power output. Plants are usually designed for maximum use in the summer, when air conditioning units are intensively used.

The distribution of Ohio's electricity generating plants (shown in Figure 11.1) indicates that coal is by far the most important of the three indigenous fossil fuel resources used for electricity production. There are a few natural gas stations powered by generators and some small oil-fired generators. The map shows the major fuel use at each facility. Where two fuels have similar inputs both are noted. Many other stations use a secondary fuel source mainly for stand-by generation in times of emergency. There are also some

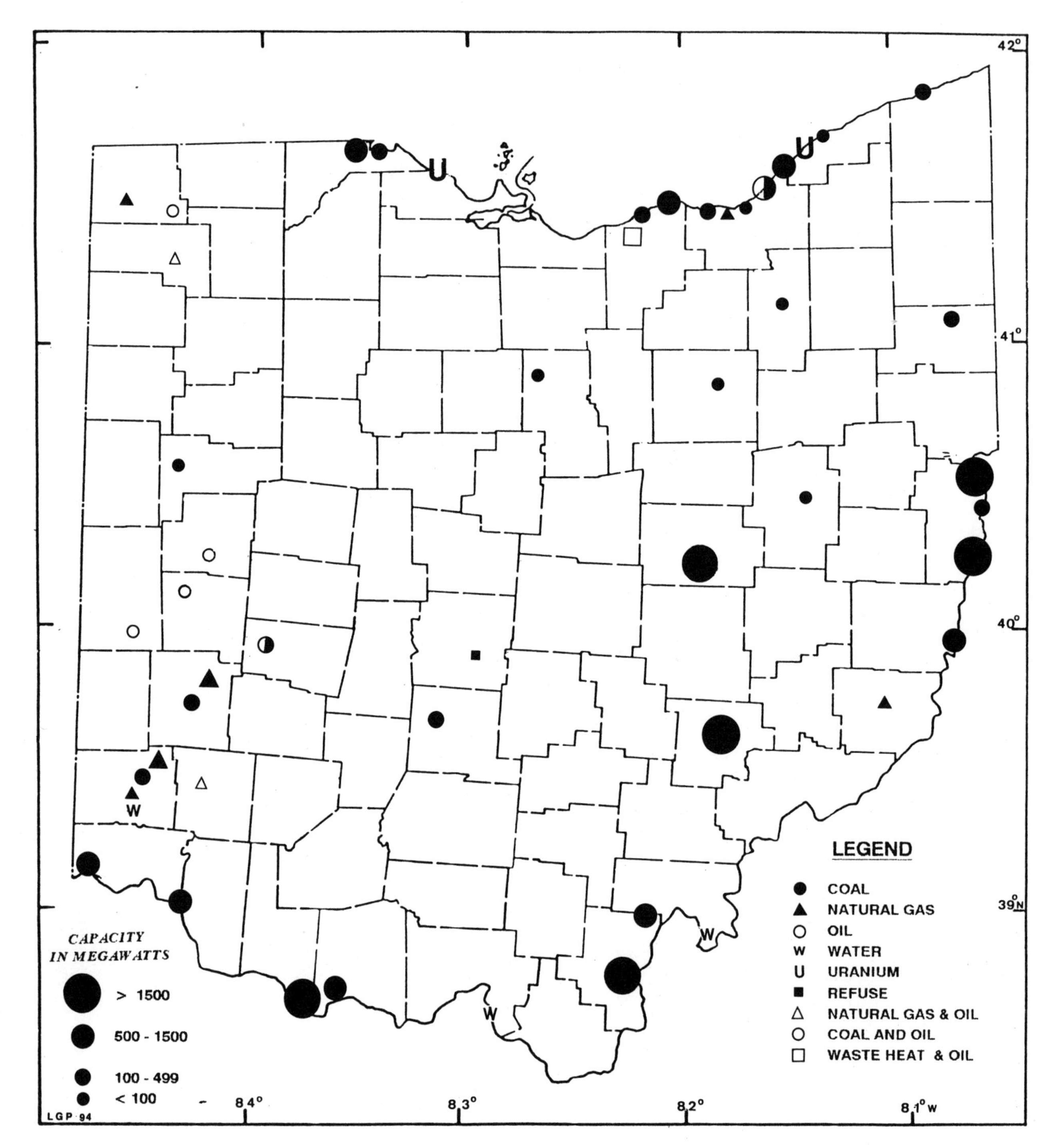

Fig. 11.1. The Location of Power Stations Showing Output Based on Predominant Fuel Source

plants that do not use any of the three fuel minerals, including two plants that generate electricity by burning garbage (Figure 11.2), a process sometimes confused with waste heat. Waste heat is steam heat that is surplus to the needs of an industrial plant that produces it. Through a process called cogeneration this excess heat is then used to generate electricity. Also on the list of Ohio fuel requirements is uranium.

Fig. 11.2. The Waste Burning Plant, Akron (Photo by Leonard Peacefull)

Ohio's two nuclear-powered generating sites, Davis-Besse and Perry, are the largest noncoal generating stations in the state. Both are situated near Lake Erie, far from the coal deposits.

Ohio's electric generating power plants are distributed in three distinct clusters, together with a fourth and smaller group. The three clusters are the Ohio River Valley, the shore of Lake Erie, and the southwestern Miami Valley. The fourth area is suggested by two large plants in the Muskingum Valley. There are also several small plants scattered around the state that serve their immediate area without exporting any significant energy elsewhere.

The distribution of generating plants throughout the state is governed by two factors: The first is proximity to water, essential for cooling in electricity production. Of the four discernable producing areas, three are in river valleys and one is along a large lake. The second factor is proximity to the coal fields. The largest capacity stations are situated on a coalfield or nearby. There are also negative correlations between the size of the plant and the distances from a coal source and from a sufficient source of water. These considerations explain why many west-central counties tend to have no power generation, and why western counties rely on noncoal fuels for their energy production.

Energy from Coal

Coal is a heavy, dirty, difficult, heterogeneous material containing impurities that are hazardous to health if allowed to escape into the atmosphere when the fuel is combusted. Yet coal represents the largest energy reserve in both the United States and Ohio. Modern utility plants are so efficient that one kilowatt-hour of electric power can be produced from a single pound of coal. Ongoing industry research continues toward the design of combustion systems that are reasonable in

Table 11.1

Estimates of Energy Input at Electric Utilities, 1960–1990, Ohio (in Trillion Btu)

YEAR	COAL	NATURAL GAS	PETROLEUM	NUCLEAR	HYDRO	OTHER	TOTAL (*)	INTERSTATE (**)	AGGREGATE
1960	512.5	3.1	1.2	0.0	0.1	0.1	516.9	169.8	686.7
1965	587.3	3.0	1.3	0.3	0.1	0.1	592.1	178.9	771.0
1970	794.7	21.9	9.0	0.0	0.1	0.1	825.7	169.7	995.4
1975	1037.2	5.3	23.2	0.0	0.1	<0.05	1065.8	140.1	1205.9
1980	1110.5	4.7	13.4	23.1	0.1	<0.05	1151.8	160.9	1312.7
1985	1103.3	0.7	3.8	21.0	1.8	2.7	1133.4	285.0	1418.4
1990	1160.8	1.3	3.5	113.9	1.8	2.8	1284.1	264.0	1548.1

Source: State Energy Data Report 1960–1990, Energy Information Administration, Table 233 and Table 228.
(*) Energy Input at Electric Utilities in Ohio
(**) Net Interstate Inflow of Electricity

cost, efficient in production, and clean in operation, degrading the environment as little as possible during the combustion process.

Ohio is a prodigious user of coal, exceeded only by Texas. As Table 11.1 shows, 75 percent of the fuel used to produce Ohio's electricity comes from coal. In total, the electric utilities consume over 80 percent of Ohio's coal output. About 9 percent is destined for the coke plants that are vital to Ohio's steel industry. Coke is essentially baked coal from which volatile gases have been driven off. Coal (except anthracite) will not burn at a sufficiently high temperature to melt iron out of ore, whereas coke, which is almost entirely carbon, will burn at the temperature required. Approximately 7.5 percent of coal produced is used in other industrial processes, with the remainder consumed by residential and commercial sectors.

Despite recent improvements in combustion, unwanted wastes remain a problem. These unwanted products include: (1) solid particles such as fly ash that escape through the smokestack; (2) sulfur dioxide gases; (3) nitrogen oxide products; (4) carbon oxide gases; and (5) hydrocarbons. All of these unwanted by-products are detrimental to the environment and noxious to anyone with bronchial disease. Congress established the Environmental Protection Agency (EPA) in large part to monitor coal users (MacAvoy 1987).

Turning coal into electricity can be done in several ways. Coal can be burned in a crushed form on a moving grate (stoker furnace), pulverized into a very fine powder that is blown into suspension and then burned at a high temperature (pulverized coal boiler), or pulverized by crushing only and then blown very rapidly around the perimeter of a cylinder tilted in relation to the horizon (cyclone furnace). In this last case, the temperature at which the crushed coal is burned is lower than that used in those furnaces burning powdered coal that has been blasted vertically upward in the combustion chamber. The lower temperatures of the cyclone are important because less noxious nitrogen oxides will be produced. If limestone is introduced into a cyclone furnace, most of the troublesome sulfur in the coal combines with the limestone to form slag, a by-product used in the highway industry (EIA Coal Distribution 1987).

The most recent advance in coal-burning technology is the fluidized bed furnace. This technology has been advanced by a $100 million bond issue for research in coal technologies passed by Ohio voters in the early 1980s. Using this method, crushed coal is literally supported on a bed of sand held suspended by rapid upward airflow. When the coal is burned, the resulting energy heats boilers that produce steam in the conventional way to turn steam turbines. New fluidized bed designs use the gaseous by-products of combustion to simultaneously turn gas turbines. The fluidized bed design operates at about 1500 degrees Fahrenheit, rather than 3000 degrees that is typical of pulverized coal vertical furnaces. This lower temperature produces about half the amount of nitrogen oxides. An array of other coal research projects are also being financed by Ohio.

The environmental problems related to the use of coal from mining, through transport, preparation, burning, waste collection, and waste disposal are of such enormous scale that an obvious energy strategy has been to use coal at fixed sites in large plants where intensive pollution control equipment can be installed and maintained. Thus most coal is combusted at electric utility generating plants. One effect of this energy strategy is to conserve liquid fuel resources (petroleum) for our transportation system, which is almost totally based on liquid fuel technologies and for feedstocks for the polymer industries.

Nuclear Power

Among countries using nuclear power the United States produces the largest amount, numerically speaking, of electricity from nuclear plants. In 1988 the nuclear power industry produced about 19 percent of total U.S. electricity, amounting to some 550 megawatt hours of energy. In the same year, France produced about 300 megawatt hours of electric energy using nuclear fuel sources, but the French production represented about 70 percent of total French electric production. In France nuclear plants are still being added to the power plant inventory, but the policy is being strenuously challenged. In the United States no new nuclear power plants have been ordered since the late 1970s. The difference in the spatial patterns clearly reflects the availability of alternative power sources in the United States, but there are other relevant reasons with implication for future energy policy for both Ohio and the nation.

Civilian nuclear power in the United States began as an offspring of Admiral Hyman Rickover's successful development of nuclear propulsion for U.S. Navy submarines. The civilian nuclear power industry quickly adopted a strategy for large plants, but because performance parameters for these large plants were reestablished with each new facility, each civilian plant was of a different design. A burst of construction activity between 1965 and 1975 shaped the industry. In Ohio, Davis-Besse 1 near Port Clinton became commercially operable in 1977, and Perry 1 at North Perry, east of Cleveland, became operational in 1987. Perry 2 is still not on line and may never be commissioned. These are the only nuclear plants in Ohio (U.S. Council for Energy Awareness 1989a).

By statute, Congress limited the liability of nuclear plants to $560 million, with the government providing indemnity for $500 million. This was later amended to require that the industry be liable through a pooled arrangement and to change this limitation to approximately $635 million. Nuclear plants in the U.S. are closer to population centers than they otherwise would be without the limitation to liability, a situation that has intensified citizen concerns regarding nuclear safety. Citizen protests at proposed new plant sites and simultaneous lawsuits claiming environmental degradation have delayed plant construction causing costs to escalate. In Ohio, Perry was the site of many demonstrations, both during and after completion. Another utility consortium—Dayton Power and Light, Cincinnati Gas and Electric, and Columbus Southern—cancelled the Zimmer nuclear plant on the Ohio River, after spending more than $1 billion, citing the inability to control the costs of construction and regulation. The consortium has converted the facility into a coal-fired plant. This negative experience was typical for the rest of the United States; consequently not a single nuclear plant has been ordered since 1979. The present environmental imperatives are tantamount to the requirement that almost all the social costs of a nuclear plant be paid for by the utility through its consumer base. Considering today's uncertainties and contingencies, capacity expansion by nuclear power is simply not cost-effective anywhere in the United States. There is a discernible trend by utility commissions to require electricity prices to more closely follow total social costs. This has changed relative fuel prices, making natural gas a preferred fuel for power plants.

The eventual reemergence of nuclear power as an alternative energy source in the U.S. and in Ohio will

Table 11.2

Energy Consumption by End-Use Sectors (Trillion Btu) Ohio, 1990

SECTOR	NATURAL GAS	MOTOR GASOLINE	ELECTRICITY	OTHER PETROLEUM	COAL	DISTILLATE FUEL	NET ENERGY	ELECTRICAL SYSTEM ENERGY LOSSES	END-USE SECTOR TOTAL
Transport	10.4	566.4	0.1	71.4	0.0	147.6	795.9	0.3	796.2
Industrial	278.0	5.1	237.8	228.1	248.2	29.9	1027.0	519.4	1546.4
Commercial	166.7	5.5	118.9	4.0	10.3	9.6	314.9	259.8	574.7
Residential	321.0	N/A	129.3	18.7	5.5	23.8	498.3	282.4	780.8
Total	776.1	577.0	486.1	322.2	264.0	210.9	2636.1	1061.9	3698.1

Source: State Energy Data Report 1960–1990, Energy Information Administration, Tables 7, 8, 9, and 10.

Note 1: Wood energy is not included in the figures. The residential sector for the United States consumed an estimated 786 trillion Btu of wood for energy, and the industrial sector for the United States consumed an estimated 1562 trillion Btu of wood energy (primarily for pulp and paper industry).

Note 2: Very substantial electrical system energy losses are incurred in the generation, transmission, and distribution of electricity, and use in electrical generating plants and in unaccounted electrical system energy losses.

probably require all of the following: (1) new technology for nuclear plant design; (2) general acceptance that nuclear power plants are less damaging to the environment than fossil fuel plants; (3) rising costs for fossil fuels, especially petroleum—the energy source price leader that will make the nuclear alternative cheaper by comparison; (4) provable assurances that new nuclear plants will be safe from any possibility of meltdown and that radioactive materials, both inputs and outputs, can be safely transported and stored; and (5) development of acceptable disposal methods and acceptable storage sites for radioactive waste.

Energy Consumption

The consumption of energy is governed by several factors: the availability of the fuel source; the market price; and especially, the geographic factors of time of year, location of the consumer, and the distance that the energy has to travel. The major energy sources are listed in Table 11.2, and the apportionment of consumption is illustrated in Figure 11.3.

In the modern world, there is great reliance on electricity as an energy source. Almost everyone takes electricity for granted. At the flick of a switch power is at our disposal. Americans are inveterate consumers of electrical energy. In 1990, Ohioans consumed 486.1 trillion Btu (British thermal units, the quantity of heat required to raise the temperature of one pound of water one degree Fahrenheit at 39.2 degrees F.).

Aggregated data for the nation indicate that sales of electricity are closely divided among three major consuming sectors: residential (34.6 percent), commercial (27.1 percent), and industrial (34.8 percent), with another 3.5 percent going to other sectors, such as streetlighting. However, this does not hold true for Ohio, where the corresponding percentages are 26.5, 24.5 and 49.0 (Figure 11.4). By evaluating the considerable percentage of electricity sales to Ohio's industrial customers, we can see the significance to Ohio industries of dependable coal resources to fuel the state's electric utilities.

Yet Ohio consumers use more energy in the form of natural gas and gasoline than electricity (Table 11.2). Natural gas is a very efficient heating fuel; it is also used to air condition homes and factories and in manufacturing processes that require heat. However, the residential consumption of energy during the years

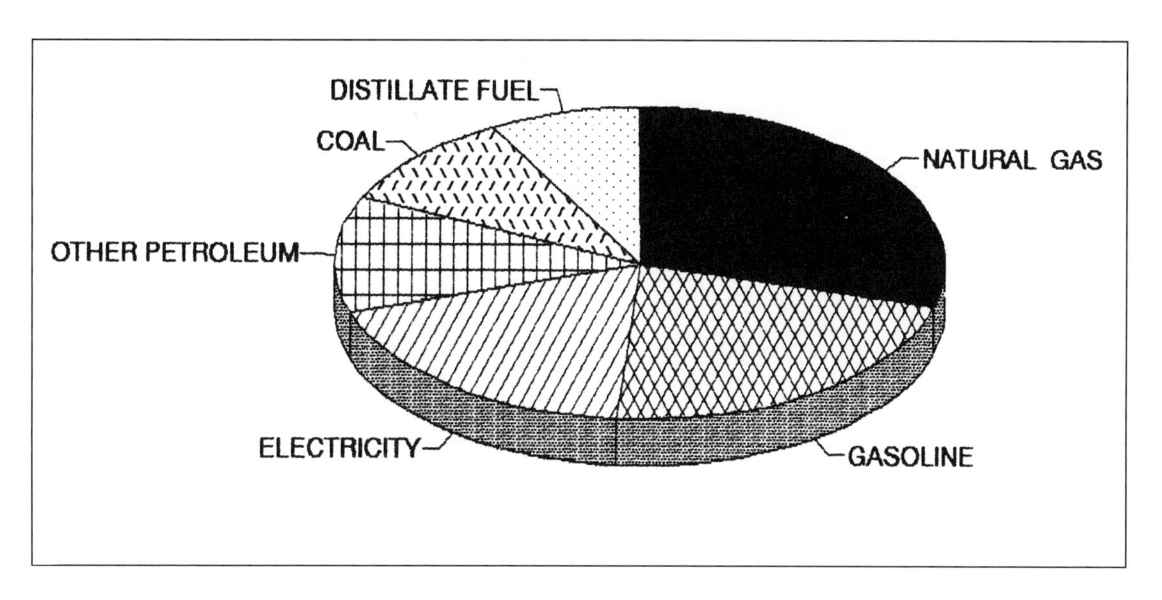

Fig. 11.3. Energy Consumption by Source, 1990

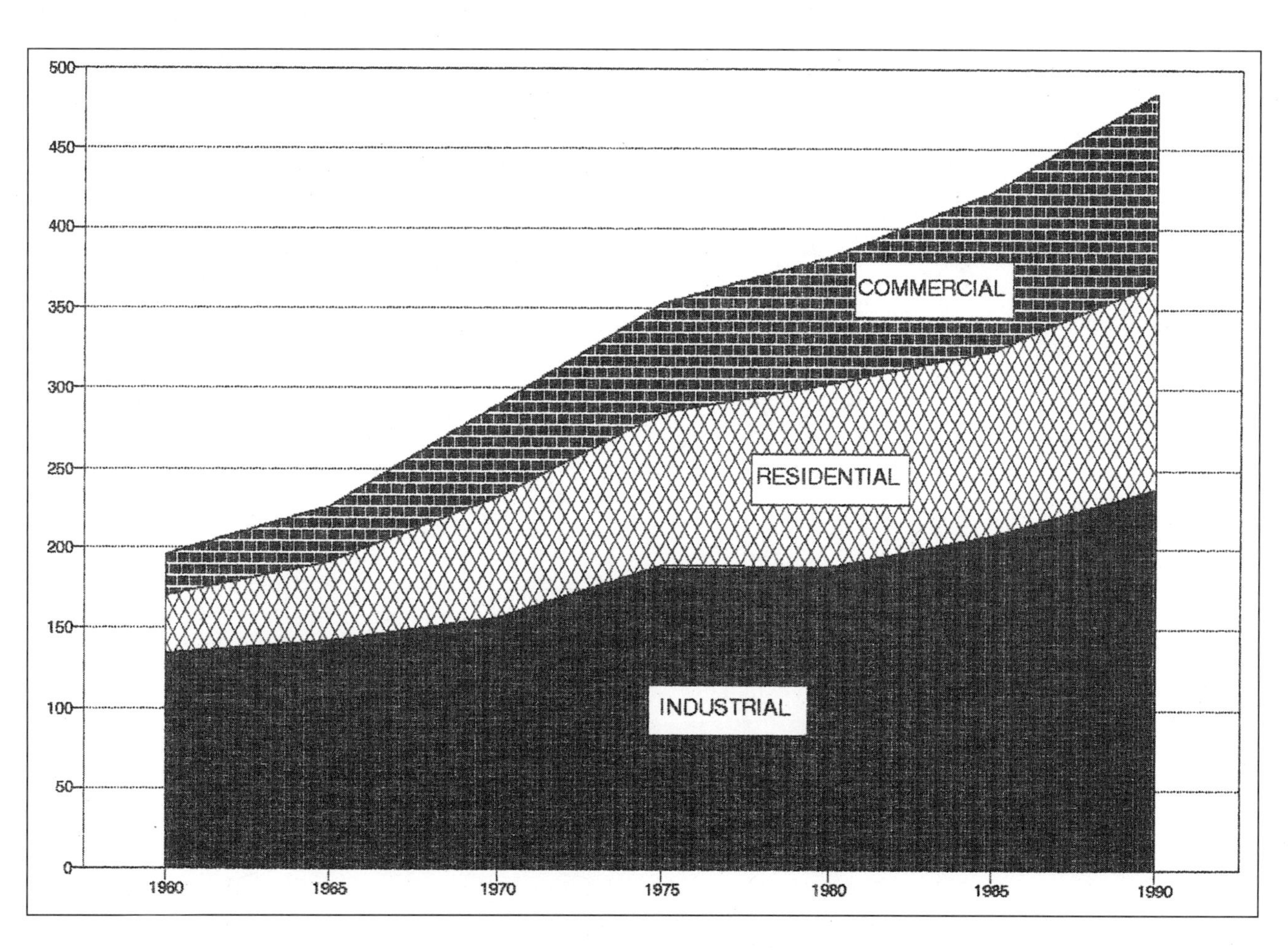

Fig. 11.4. Electrical Energy Consumption by Sector

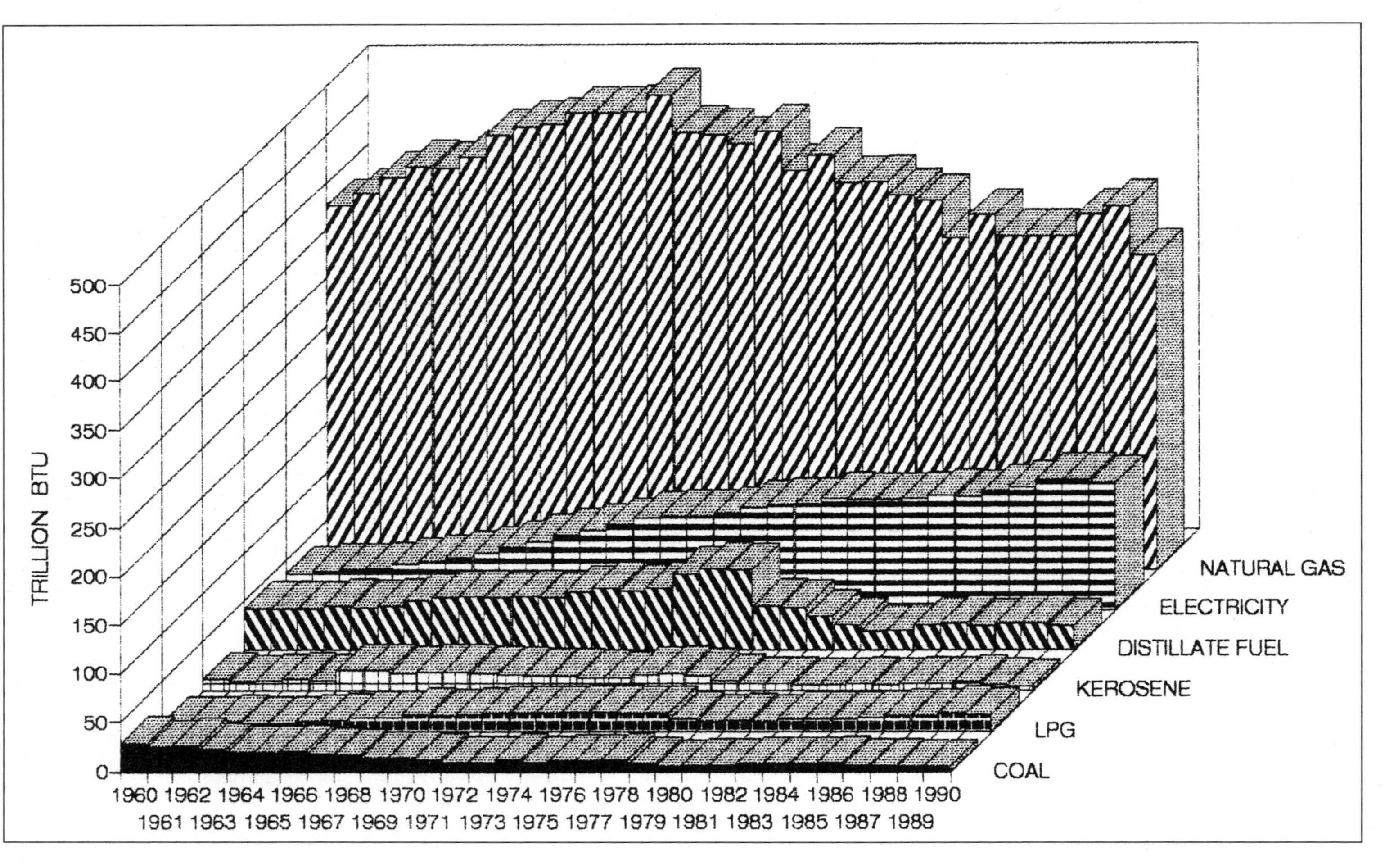

Fig. 11.5. Residential Energy Consumption

1960–90 (shown in Figure 11.5) illustrates that natural gas consumption in the home has declined from a peak in the mid 1970s, while the use of electricity has steadily risen. Several reasons may explain these trends. Consumers are now more energy conscious, are taking measures that will conserve heat—insulation, draught exclusion, and so forth—and thus reduce the amount of natural gas consumed. As a further conservation measure, some are turning away from oil-fired heating, which has become expensive, and are installing electric heating. Other exogenous market forces also play a part in the choice of fuel, such as the unstable price of oil on the world market in the 1980s. Adding to the electric burden, more and more electrical appliances are finding their way into our homes, thus increasing the consumption of electricity.

Ohio's reliance on automobiles and trucks is illustrated in the amount of gasoline and distillate fuel consumed. Nearly all gasoline consumed in the state is for road transport, with a very small amount going to commerce and industry. Examples of this latter are stationary generators and power tools used by the lawn care industry.

Energy for Ohio Manufacturing

Manufacturing has been the basis for Ohio's prosperity since the 1870s. First agricultural machinery, then steel, vehicle assembly, components manufacturing, engine manufacture, metal fabrication, and a thousand other product lines formed the economic basis for Ohio's real income. These industries represent traditional strengths in Ohio's economy, and are the foundation for future prosperity.

The manufacturing sector is a division of the industrial sector and includes the industries that produce goods by specified activities. The production activities of manufacturing include change of form of tangible material inputs, change in chemical composition, assembly, and blending.

The other divisions of the industrial sector, in addition to the manufacturing division, are the following: utility industries, mining industries, agriculture, fishing, and forestry industries. The entire industrial sector of the United States consumes 36.3 percent of total national energy consumption by all end-use sectors. The oil price shocks of the 1970s clearly impacted the growth of energy use in the industrial sector, which has been more or less stable since 1975.

The Energy Information Administration (EIA) defines several interrelated measures of energy consumption by the manufacturing sector as an end use of energy. The purpose of specifying end use is to eliminate the counting of energy used to produce fuel as primary energy consumption. For example, the oil used in a refinery to produce gasoline is not included as primary consumption of energy by the manufacturing sector, but the gasoline fuel used by the manufacturing sector is recorded as primary energy consumption. By the same definition, consumption of energy for the production of nonfuel goods is defined as primary energy consumption by the manufacturing sector. By this definition, electricity is not a fuel.

There is clear evidence that the amount of purchased fuels and electricity has steadily declined relative to industrial production since 1975 (EIA, Manufacturing Energy Consumption Survey 1985). Indices for consumption and production record the drop in purchased fuels in absolute figures. In 1981, 11.6 quads of fuel and electricity were purchased by the manufacturing sector. By 1985, this figure decreased to 9.7 quads, a 16.4 percent decrease, while real output of U.S. manufacturing increased by 14.4 percent. This represents a conservation of about 37 percent in 4 years.

A fundamental principle of economics is that of substitutability of factors of production. When relative prices change, then changes in input structure quickly follow. The Arab oil embargo was one event that greatly influenced rapid adaptation to new methods. Cogeneration became more widely adopted, as industries producing large quantities of surplus heat channeled heat into electricity generation. Additionally, insulation was installed in many factory buildings; thermostats were more carefully monitored; and control systems to minimize energy inputs were put in place, both for process control and for environmental control of buildings. The motivation to reduce energy inputs will continue through the remainder of the century and beyond. This is because relative prices for petroleum will increase as population growth and economic development imply rapid expansion of energy needs worldwide, while at the same time, petroleum becomes relatively scarcer and more valuable.

Energy for Transport

The internal combustion engine was introduced to the world in 1879 at the Paris World's Fair by Professor Rudolf Diesel of Munich. In a little more than a single century all of the important modes of transportation—automobiles, trucks, trains, ships, and airplanes—have become dependent on liquid fuel, and these technologies have profoundly transformed the patterns of human activity on the surface of the earth. Many great trends of change in this century—industrialization of agriculture, mass factory production, total war mobilization, spatial agglomeration, suburbanization, and globalization of markets—are utterly dependent on the technologies of liquid fuel.

The fundamental reason for liquid fuel's preeminence is its ease of delivery into and within the engine where the transformation into mechanical energy occurs. All of our contemporary transport systems are rendered obsolete if liquid fuel is not available.

The transportation technologies that characterize all modern economies have caused successive waves of cost reduction in the movements of factors of production. Globalization of markets has assured competitive adjustments among producers and has caused declines in real prices. Prosperity is passed on around the world to larger and larger consumer constituencies. Transport technologies underlie prosperity through trade. Petroleum is the feedstock for gasoline, diesel fuel, and aviation petrol. The convenience of liquid fuel and the enormous infrastructure of the oil industry provide inertia to change. Even if new superior technology becomes economically viable, existing plants and equipment will be used as long as marginal cost is less than marginal revenue.

Price, in a free market for oil, is determined by the costs of production. In the long run, each successive discovery of oil-bearing strata is likely to be more costly to exploit, with escalating costs for environmental safety magnifying the economic burden. On the demand side, the market for energy will continue to expand as more nations industrialize. For the future, the most important reasons for phasing out liquid fuel technologies based on petroleum are: (1) the increasing expense of the discovery and exploitation of finite petroleum sources; and (2) the increasing disapproval by the public of contingent pollutants during the whole chain of fuel discovery, production, and utilization, especially the noxious exhaust pollutants from engines.

Energy Usage and Material Welfare

The correlation between energy usage and advancement in material welfare is manifest at the largest scale and in the long run. Nevertheless there are wide differences in energy utilization per capita that do not correlate well with gross domestic product per capita. There are several reasons for this dilemma.

First, projections of energy usage should be based on sectoral analysis, rather than on macro variables. The use of energy varies widely among industries. For example, minimills use scrap metal for inputs and electric furnaces for melting. These processes require substantial power requirements, especially when compared, for example, with a leisure industry based on an aggregation of golfing resorts. Within manufacturing sectors, basic metal and chemical industries require substantially more energy per dollar value of output than assembly operations.

Second, projections of energy usage should be regionally based to capture the sectoral differences among regions that result from the comparative advantage of spatial differentiation. Ohio's comparative advantage in metal fabrication is based primarily on relative location among strip steel producers and customers in the interindustry webs of manufacturers that comprise the manufacturing belt. For heavy industry the comparative advantage is based on highly capitalized, fixed plants that required massive amounts of energy inputs. An analysis of comparative advantage implies sectoral projections that will differ among regions.

Third, as a region matures, there is a relative shift from manufacturing employment to service employment. This shift implies less energy usage per capita within the region, as population increases over time.

Fourth, as incomes increase, the demand for some products diminishes. This is because the marginal income is not used to purchase the same proportion of commodities—such as automobiles and refrigerators—that require high energy inputs. Higher demands for leisure, including travel and education, can be expected as average incomes increase within a region.

For all of these reasons the correlation of the rate of energy usage within a region with its gross domestic product per capita is weak in the short run. From a longer perspective, the relationship of energy usage to real income per capita is undeniable, and this correlation is the essence of the material progress that has occurred throughout the world since the dawn of the Industrial Revolution.

The Greenhouse Effect and Ohio Energy

Most of Ohio's energy needs are satisfied directly or indirectly by the combustion of fossil fuels. An inevitable by-product of this consumption is CO_2, or carbon dioxide, a "greenhouse effect" gas. The term "greenhouse" gas indirectly refers to a property of energy radiation. Radiation of energy is proportional to the fourth power of absolute temperatures and radiation is a characteristic of every real object of the universe.

$$R = kT^4$$

In the equation for radiation, R denotes radiation, k denotes a constant that corresponds to the units of measurements used for the variables, and T denotes absolute temperature. Because the temperature of the sun is extremely high compared to objects on earth, the radiated photons are of extremely high energy. These high energy photons can pass through glass, as in a greenhouse, or through atmosphere that contains the greenhouse gases, especially CO_2. On the other hand when energy is radiated from earth, the lower absolute temperature of earth-bound objects results in photon radiation of far lower frequency. Glass and atmosphere are not so transparent to these low energy frequencies, resulting in a buildup of heat within the system, whether the greenhouse or the whole planetary system.

The incremental changes of the greenhouse effect are not trivial on a human scale. If fossil fuel use continues to increase at the present marginal rate, our civilization—based on present technologies, fundamentally dependent on burning fossil fuels—would be extinct within 200 years. By that time the heat generated by combustion per unit of time would equal the insolation of the sun per unit of time. It is clear that the accelerating worldwide energy usage of fossil fuels, especially for electric power generation, makes the future problematic.

Conclusion

Our technologies of transport are based on liquid fuel, and therefore petroleum and liquified coal will continue in wide use for several decades, even though sulfur and nitrogen by-products are major health considerations. Producing electricity from coal or petroleum creates the same environmental problems, but these problems will be confined to the sites of large utility plants where major capital equipment will be installed to remove the noxious products of combustion. Thus coal will continue to be used extensively for many decades. Natural gas, when burned cleanly, produces only water and carbon dioxide derivatives and so will continue to be a preferred energy source for space conditioning, electricity generation, and process heat, for as long as it is available. This is especially so because of current policy trends to internalize the costs of environmental sustainability.

The combustion of fossil fuels, on which our material civilization is based, adds heat and carbon dioxide to the global environmental system. When this factor is considered along with the global population explosion, nuclear power becomes a likely alternative for new generating plants during the next 10 to 15 years. Society must learn to use nuclear power safely, because the alternative scenarios are even less desirable. Unconventional energy sources—geothermal power, solar power, wind and tide power, biomass, trash burning, waste heat utilization, and other alternatives—will contribute in a very small way to the accelerating demands for power that come with burgeoning world populations.

Editor's Note:
Fuels vary in their energy content. Coal beds in particular have different histories in terms of compression, temperature, uplift, bending, and the like, while being formed. Ohio coal on the average has 23,790.4 Btu per short ton, compared with an average value for the United States of 20,901.8. The figures for petroleum and gas are closer; for petroleum, 6,284.3 Btu per barrel nationally versus 6,063.5 in Ohio; and for gas, 1,028.3 per 1,000 cu. ft. nationally, versus 1,011.6 in Ohio.

REFERENCES

Aubrecht, Gordon. 1989. *Energy.* Columbus, Ohio: Merrill Publishing.

Energy Information Administration. Office of Energy Markets and End Use. 1985. *Manufacturing Energy Consumption Survey: Energy Intensity in the Manufacturing Sector 1980–1988.* DOE/EIA-0552(85). Washington, D.C.: Supt. of Documents.

———. Office of Coal, Nuclear, Electric and Alternate Fuels. 1989. *Commercial Nuclear Power 1989: Prospects for the United States and the World.* DOE/EIA-0438(89) (UC-98). Washington, D.C.: Supt. of Documents.

———. Office of Coal, Nuclear, Electric and Alternate Fuels. 1989. *Electric Power Annual 1989.* DOE/EIA-0348(89). Washington, D.C.: Supt. of Documents.

———. Office of Oil and Gas. 1990. *Natural Gas Annual 1990.* Vols. 1 and 2. DOE/EIA-0131(90/1 and 90/2). Washington, D.C.: Supt. of Documents.

———. Office of Coal, Nuclear, Electric and Alternative Fuels. 1991. *Commercial Nuclear Power 1991: Prospects for the United States and the World.* DOE/EIA-0438(91). Washington, D.C.: Supt. of Documents.

———. Office of Energy Markets and End Use. 1992. *Monthly Energy Review: September 1992.* DOE/EIA-0035(92/09). Washington, D.C.: Supt. of Documents.

Gordon, Richard L. 1987. Coal Policy in Perspective. In *Energy: Markets and Regulation.* Ed. Richard L. Gordon, Henry D. Jacoby, and Martin B. Zimmerman. Cambridge, Mass.: MIT Press.

MacAvoy, Paul W. 1987. The Record of the Environmental Protection Agency in Controlling Industrial Air Pollution. In *Energy: Markets and Regulation.* Ed. Richard L. Gordon, Henry D. Jacoby, and Martin B. Zimmerman. Cambridge, Mass.: MIT Press.

Management Information Services, Inc. Management Analysis Company. 1989. *Economic Growth and the Requirements for Electric Power During the 1990s in Illinois, Indiana and Ohio.* Prepared for U.S. Council for Energy Awareness (USCEA). June.

U.S. Council for Energy Awareness (USCEA). 1988. *Nuclear Energy Facts and Figures.* Washington, D.C.

———. 1989a. *Electricity from Nuclear Energy.* Washington, D.C.

———. 1989b. *INFO 243.* Washington, D.C.

Zimmerman, Martin B. 1987. The Evolution of Civilian Nuclear Power. In *Energy: Markets and Regulation.* Ed. Richard L. Gordon, Henry D. Jacoby, and Martin Zimmerman. Cambridge, Mass.: MIT Press.

CHAPTER TWELVE

Agriculture

David T. Stephens

Agriculture has played a very important role in shaping Ohio's geography. The state's early pioneer settlers were primarily agriculturalists, and they immediately began to modify the landscape to make it more conducive to farming. The cultural baggage that accompanied the settlers included a variety of agricultural commodities and techniques. Their different agrarian experiences coupled with the state's physical diversity led to a mid-nineteenth century Ohio with five distinctive agricultural regions (Jones 1983). Regional differentiation of Ohio's agriculture began at an early date and continues today.

Gaining better access to Eastern markets for agricultural products was a driving force behind many of the state's first transportation improvements. During the early days of settlement, much of the produce from Ohio farms reached market via a long and arduous route involving the Ohio and Mississippi Rivers. Ohio's initial road-building efforts, its canal boom, and much of the early railroad construction were driven by the desire to move the bounties of Ohio's agriculture expeditiously to the Eastern markets. As these markets became more accessible and demands for food rose, vast areas of forests were removed to create more cropland. At statehood, 1803, it is estimated that Ohio was 97 percent forested. At the peak period of land clearance in the mid 1940s, less than 20 percent of the state remained forested. The impact on vegetation was not limited to forests, for extensive drainage projects turned massive swamps into some of the best agricultural land in the eastern United States. In the first 100 years of statehood, residents greatly modified Ohio's landscape to meet the growing needs for additional agricultural land. By the middle of the nineteenth century, Ohio was among the national leaders in many categories of agricultural production. Agriculture, as the driving force behind much of the early settlement of the state, has left an indelible imprint on Ohio's physical and cultural landscapes.

Today, agriculture's role in the state's economy appears rather minor. It is not a major contributor to the gross state product; during the period 1971–86 it accounted for less than 2 percent of the state's total (Ohio Data Users Center 1989), while manufacturing contributed nearly one-third of the state's gross product during the same period. Based on these data, agriculture seems to be unimportant in present-day Ohio. However, other yardsticks present a different picture. Fifteen percent of Ohio's workforce is employed in agriculture or food processing (Vonda 1992), with most of these in food processing. Nearly 57 percent of the state's total land area is devoted to agriculture. If forested land is counted as agricultural land—much of Ohio's woodland is used for pasture—the total rises to over 80 percent of the state's area. Thus, in terms of land use, agriculture is by far the most significant activity in the state. On a national basis, in 1988 Ohio ranked ninth in the value of farm crops marketed, eighteenth in livestock sales, and tenth in farm value (Shkurti and Bartle 1989).

The Physical Base

Agricultural practices are a function of cultural, economic, and physical factors. To understand Ohio's agricultural geography one needs to consider the physical attributes—such as topography, soils, drainage, vegetation, and climate—that impact on the distribution of agriculture practices. One of the most significant aspects of Ohio's physical environment for agriculture is the fact that two-thirds of the state was subjected to glaciation (see chapter 1). The glaciers impacted two aspects that relate to agriculture: terrain and soil formation. South of the glacial boundary, the land is characterized by numerous hills, ridges, and narrow valleys. In the unglaciated section there is so little flat land that a typical 200-acre farm may have only 12 acres of tractor land (Jones 1983).

On the opposite side of the boundary, the glacier acted like a giant bulldozer, leaving behind a nearly level or gently rolling surface. Not all of the glaciated section is flat. The Appalachian Plateau in northeastern Ohio has some rugged areas, as do the Till Plains along the lower course of the Miami River. However, most of the glaciated area poses no major topographic barriers for agriculture. Of particular interest in the glaciated section are extensive areas of nearly flat land once occupied by vast swamps. One of these swamps, the Black Swamp of northwest Ohio, represents one of the more extensive land drainage areas in the United States, covering some 1,500 square miles (Kaatz 1955). Many of Ohio's former swamps constitute the best agricultural land today.

Soils are often an important factor in explaining the distribution of agricultural activities (see chapter 3), and glaciers played an important role by influencing the parent materials for many of Ohio's soils. In the southeast, where glaciers did not reach, the parent materials are mainly sandstone and shales. Soils derived from these rocks are low in fertility, acidic, and unable to produce sustained crop yields without liberal applications of fertilizer and lime. Given the local relief, the soils of southeast Ohio are prone to erosion. The best soils in this region are those derived from alluvial material associated with the floodplains. Unfortunately, floodplains are not well developed in this dissected region. Another area of good soil is found on the section of the Lexington Plain that extends into Adams and Brown Counties. Here the parent materials are limestone and shale. Although these moderately fertile soils are subject to severe erosion because of the hilly terrain, they remain important to Ohio's tobacco production.

In the glaciated section, bedrock and glaciation combined to produce several distinct soil types. During the Wisconsin Glaciation, clays and silt were laid down in ancient lake beds over much of northwestern Ohio's Lake Plain. This clay and silt became the parent materials for soils with a high lime content, thus a high inherent fertility but poor drainage. The region known as the Black Swamp possesses this type of soil. These soils, when properly drained, have supported much of the state's soybean and grain production. On the eastern portion of the Lake Plain, the water-laid materials have eroded, leaving parent materials that are mainly shale and glacial till. Though the soils that have developed from these parent materials are acidic and poorly drained, with heavy applications of fertilizer and improved drainage, these soils can support specialized agricultural endeavors.

On the Till Plains of western Ohio, the parent materials are primarily glacial till with occasional pockets of outwash or lacustrine materials. In west-central Ohio the till is of Wisconsin age and contains a large amount of lime. This has produced soils that are deep and fertile but that require extensive drainage. They support much of Ohio's corn and soybean crop. Further south, near the glacial margins, the soils have developed from the older and more weathered Illinoian till. These soils tend to be acidic and silty, with moderate fertility. Not as inherently productive as those derived from the Wisconsin till, they are used primarily for grain crops.

The soils of the Glaciated Appalachian Plateau are derived mainly from till of Wisconsinan and Illinoian age containing fragmented sandstone and shale. A few areas were once lake beds or outwash plains. The tills of the plateau tend to be low in lime, low in fertility, acidic, and not well drained. An exception to this generalization is the soils in an area that extends from northeastern Columbiana County west as far as Richland County. While not as inherently fertile as western Ohio, this eastern zone supports a large variety of grain and forage crops. The soils over the balance of the plateau are well suited to most small grains and forage crops. Dairying tends to be the dominant agricultural activity in areas with these soils.

Organic soils occur in the Till Plains and the Glaciated Appalachian Plateau. These soils are composed mainly of vegetative materials and have a limited areal extent, rarely covering more than 1,000 acres. Generally they extend only 3–4 acres. Organic soils are highly productive when properly drained but are subject to a multitude of problems, including compaction, wind erosion, and fire. These soils form the basis for much of the state's vegetable production.

Based on the characteristics of many of the soils discussed above, one would expect drainage to play a significant role in Ohio agriculture, and in fact it has been important in two ways: the first was the use of streams to move agricultural products to market during the early stages of settlement; the second has been the role of drainage in the creation of additional cropland.

It is not surprising that the state's first settlements concentrated along the Ohio River and its navigable tributaries, particularly the Muskingum, Scioto, and Miami. Until canals were constructed, most of Ohio's agricultural products journeyed to market via the Ohio and Mississippi Rivers to New Orleans and thence on to the East Coast. Many early agricultural areas and practices in Ohio were geared to this market situation. As Ohio's canals were completed and the route to market changed, so did the nature of the state's agricultural output.

As the inherent fertility of some of Ohio's soils became known, efforts were made to expand the area that could be cultivated. This expansion was achieved partly through deforestation. More important was the drainage of swamps or bogs, for with the advent around 1850 of machine-made clay drainage tiles, an assault began on Ohio's wetlands. Although plastic pipes have long since replaced clay tile, wetland drainage has continued. By 1985 Ohio ranked forth in the nation in the amount of land that had been drained. Even more telling is the fact that 50 percent of the state's cropland is drained land (Thompson 1989).

As noted previously, most of Ohio was once forested, and clearance of the forest was a primary objective of early agriculturalists, beginning with Ohio's aboriginal population. Often white settlers took advantage of lands cleared by their predecessors. With 97 percent of the state in forest, these Indian oldfields were very attractive to the new arrivals who, if they were lucky, could clear only an acre or two of land annually.

Settlers to Ohio had little prior experience with grasslands and found only a few grassy areas on arrival. The best known was the Pickaway Prairie near Circleville. There were also "barrens" in Madison and Clark Counties and another area of grassland in the oak savannas of Fulton and Lucas Counties. The grasses in these prairies served as the basis for the development of Ohio's early cattle industry. Eventually most of these grasslands were converted to crops, because cropping offered a greater return.

Climate affects Ohio's agriculture in three important ways: precipitation, temperature, and length of growing season. Ohio's precipitation is concentrated in the growing season—spring through summer—and is generally adequate for agriculture. An exception occurred in the late 1980s and early 1990s, when crop yields were significantly reduced by drought. Heavy spring rains can hinder planting, especially of corn and soybeans, and fall corn harvest occasionally is delayed by early snowfalls. An indication of the adequacy of precipitation comes from examining irrigation in the state. Only 2 percent of the state's farms are irrigated, and irrigated land accounts for less than 4 percent of total cropland. Much irrigated land is used for specialty crops such as vegetables. Specialty crop farms use land intensively and are small in size. It is not unexpected, then, to find that 73 percent of Ohio farms with irrigation have less than 10 acres under irrigation.

Average annual temperatures vary only 6–7 degrees across the state. Such differences have a minimal impact on agriculture. Far more important is the length of time between the last and the first frosts—the growing season. One would expect the southern portion of the state to have the longest growing season. But in fact the longest growing season is found in northern Ohio along the shores of Lake Erie, where the different cooling and heating rates of land versus water operate. Their effects are especially evident too in the fall of the year, when the land cools more rapidly than the water. Winds blowing across the lake keep temperatures warmer than would be expected, extending the frost-free season. This longer growing season, as well as the proximity to a large urban market, helps explain the presence of specialized agricultural activities along the shore of Lake Erie. The lake's modification of temperatures is less significant in the spring. The other area having a long growing season is southwestern Ohio along the Ohio River. One result of this long growing season has been the concentration of Ohio's tobacco production in the region bordering the river from Gallia County to Clermont County. Ohio's shortest growing seasons occur on the Appalachian Plateau, in the northeastern part of the state, though a growing season of at least 130 days—found in some part of the plateau—is adequate to allow the production of most mid-latitude crops.

An indication of the physical qualities of farmland can be gleaned from its value and the value of the agricultural commodities it produces. High land values and returns from sales suggest land with superior physical attributes, whereas lower revenues and values suggest lands with lesser physical qualities. This method of land evaluation is subject to some qualification because of the tendency of urban land values to drive up the prices of nearby agricultural land. To compensate for the higher land costs, farmers must use their land more intensively, thus producing higher agricultural revenues. Figure 12.1 indicates the market value per acre for agricultural products sold in 1987. The map shows that land in the counties of southeast Ohio produced less than $150 per acre in agricultural sales in 1987. Conversely, the state's most urbanized counties, the lake shore counties, and two clusters of counties—one in the east-central and the other in the west-central—all yielded over $250 per acre. These patterns generally mirror the physical attributes of Ohio's agricultural land.

Recent Changes in Ohio Agriculture

Ohio's agriculture has undergone significant changes in recent years. Table 12.1 provides some insights into a few of these changes. Like the rest of the nation, Ohio has experienced a substantial decrease in the number of farms. Since 1954 the total number of farms has been more than halved. Among the factors contributing to this decline are the cost-price squeeze, the growth of nonfarm occupations, and the increasing marginality of small farms. The reality of the cost-price squeeze is evident from the following: With farm prices (revenues) indexed at 100 in 1977, in May of 1982 farmers received for their labors a price index of 143; meanwhile their expenses came to an index value of 153 (Hurt 1988). Since that date things have improved little. If it costs farmers more to produce a product than they receive for that product at market, it will not be long before they are forced to seek other sources of income. In 1987, more than half of Ohio's farm operators reported primary occupations other than farming. Either Ohio farmers are finding other economic pursuits more lucrative or it has become necessary to supplement farm income with wages from other types of employment. One suspects that both factors have operated to reduce the number of farmers.

Part of the decline in the number of farms can be attributed to the decrease in total farmland. The data show the loss of 5 million acres, or 25 percent, of farmland since 1954. Some of these lost acres have succumbed to urban encroachment, while others have been consumed by transportation or mining. Over 1 million acres have been diverted from production to federal acreage reduction programs or conservation reserve programs (Bureau of the Census 1989). Figure 12.2 shows that in the period 1954–87, the greatest percentage losses in agricultural land occurred in the urban counties of Cuyahoga, Lake, Summit, and Hamilton, where agricultural land was lost to urban expansion. Particularly in Lawrence, Jackson, and Vinton Counties, but also throughout most of the eastern and southeastern parts of the state, the loss can be attributed to the marginal nature of the land for agricultural uses, the ravages of strip-mining, and reforestation. Fayette, Mercer, Van Wert, and Henry Counties lost less than 10 percent of their farmland, and a band of counties in the west and northwest experienced only modest decline. These counties tend to be among the most productive in the state, and therefore have retained more of their agricultural land.

In the period 1982–87 the counties experiencing the greatest percentage decrease in number of farms were those found in the state's urban axis—stretching from Cincinnati to Cleveland—and in the marginally productive southeast. Yet even with the loss of agricultural land, the size of Ohio's farms has continued to increase, a reflection perhaps of the necessity of economies of scale in modern farming operations. In discussing this trend toward larger units in the Corn Belt, John F. Hart notes farmers had to "get bigger or go under" (Hart 1986, 55). Farm size in the state seems related to several factors. Figure 12.3 indicates that the smallest farms occur in the highly urbanized counties around Akron, Cleveland, Canton, and Youngstown in the northeast, with a similar zone of urbanization in the southwest around Cincinnati. In these urban counties, the high cost of land caused by competition from other land uses and the intensive and specialized nature of their agriculture combine to keep farms small. Lawrence County in the extreme south also has small farms, mainly because of the hilly terrain and extensive forests. Holmes County in east-central Ohio has an average farm size of less than 125 acres. Here the labor intensive, nonmechanized farming of the Amish is the explanatory factor. The largest farms are found in west-central Ohio, where favorable terrain and economies of scale are common.

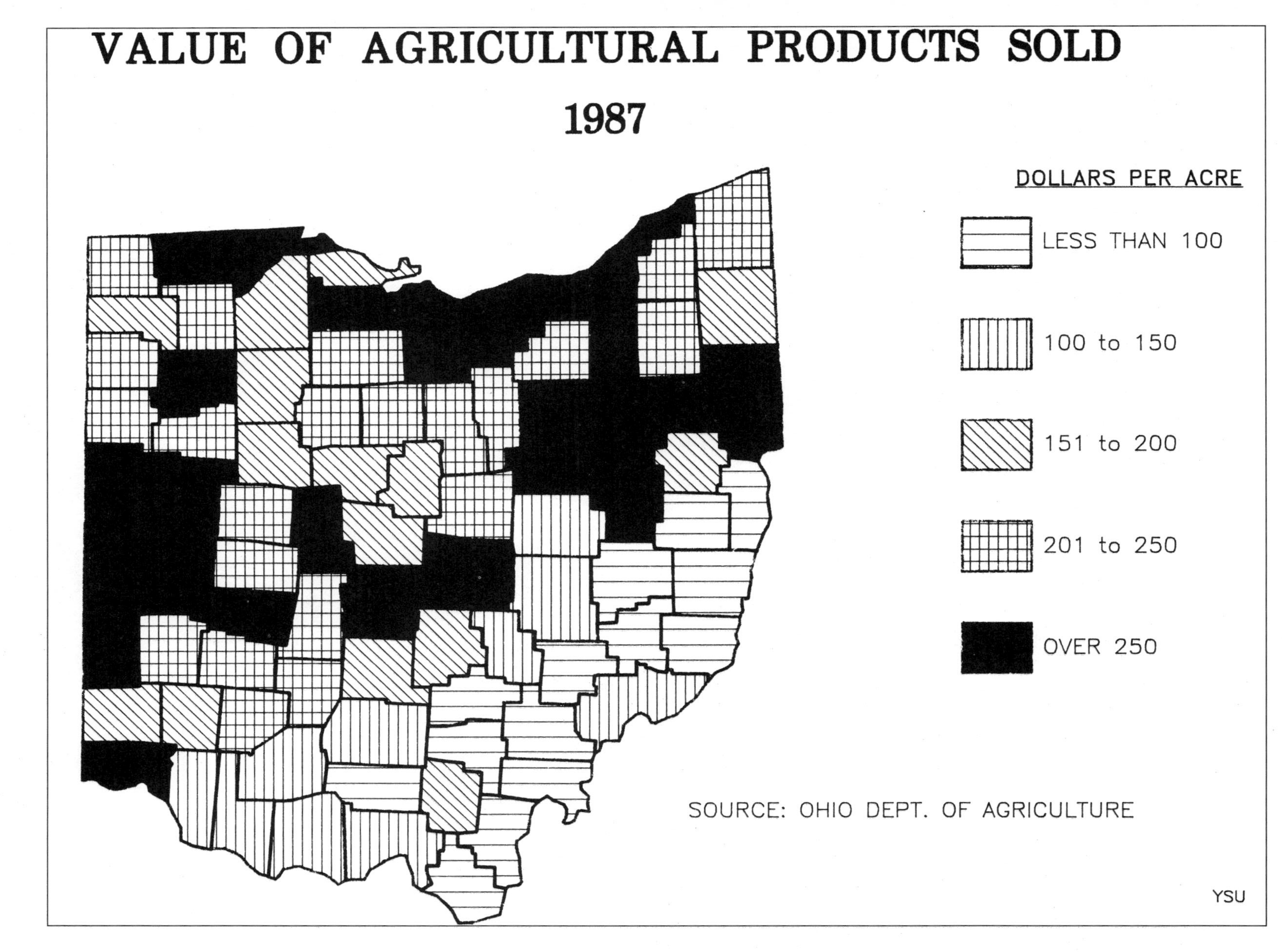

Fig. 12.1. Value of Agricultural Products Sold, 1987

Table 12.1

Changes in Ohio Agriculture, Selected Years, 1954–1987

CATEGORY	1954	1964	1974	1982	1987
Farms	177,074	120,381	92,158	86,934	79,277
Land in Farms (mil. acres)	20	18	16	15	15
Avg. Farm Size (acres)	113	146	170	177	189
Value of Land and Buildings					
Per Farm $	20,973	43,373	119,964	267,899	227,341
Per Acre $	185	295	706	1,504	1,199
Market Value of Goods Sold					
Avg. per Farm $	4,766	8,411	24,551	38,966	43,317
% from Crops	44	42	59	55	51
% from Livestock	56	58	41	45	49

Sources: Census of Agriculture, 1954, 1964, 1974, 1982, and 1987.

Table 12.1 also indicates a rapid rise in the value of farmland and buildings through the late 1970s and early 1980s. Farmland reached an all-time high of $1,831 per acre in 1981 (Hurt 1988). Farmers used this increase in value as collateral to borrow more money to finance expansion of their farming operations. As a result of a worldwide recession in the early 1980s, agricultural prices dropped dramatically, and the value of farmland began to fall. By 1987 it was down to $1,097 per acre, and many farmers found their credit greatly overextended. In the spring of 1986, 30 percent of Ohio's farmers were experiencing extreme financial difficulties or serious cash flow problems (Hurt 1988). With revenues declining, some farmers, unable to meet payments on their loans, were forced into foreclosure and bankruptcy. By 1991 the situation had improved slightly. Farm real estate was up to $1,217 per acre, and only 2 percent of the state's farmers were reported to be in serious financial difficulty (Keck 1991).

Figure 12.4 shows the value of farmland and buildings throughout the state. As would be expected, high land values are associated with the highly urbanized counties. These values reflect competition by other users that are willing to pay a higher price than agriculturalists for the land. There is also a spillover effect of above-average land values evident in counties on the periphery of these urban areas, as well as a band of higher than average values in the western third of the state. This western section benefits from terrain well suited to agriculture and inherently fertile soils. The lowest values are in the southeast, where the land is less productive.

Farm income has risen steadily during the years shown in Table 12.1. In 1954 and 1964, the majority of agricultural income was attributable to livestock, but in 1974 and 1982, crops were predominant. In the most recent period listed, 1987, the two sources of income are about equally divided. Given these developments it is evident that the state's farmsteads and land use patterns have changed in the last 35 years.

Crops

In 1987 slightly more than one half of Ohio's farm income derived from the sale of crops, the bulk of that revenue coming from soybeans and corn. Table 12.2 indicates the contributions of various commodities to total crops revenue. During 1987, soybeans eclipsed corn in importance in both revenue and acreage. This is consistent with the increasing consideration farmers in the Midwest are giving to soybeans (Rumney 1988). Because of this change in emphasis, it has been suggested that the region's well-known title, the "Corn Belt," should be expanded to: the "Corn-Soybean Belt" (Rumney 1988). In Ohio, soybean acreage has doubled since 1965, while during the same period, corn acreage has increased by less than 25 percent (Bureau of the Census 1989).

The third leading source of income from crops comes from nursery and greenhouse products, which, when combined with vegetables and fruits, account

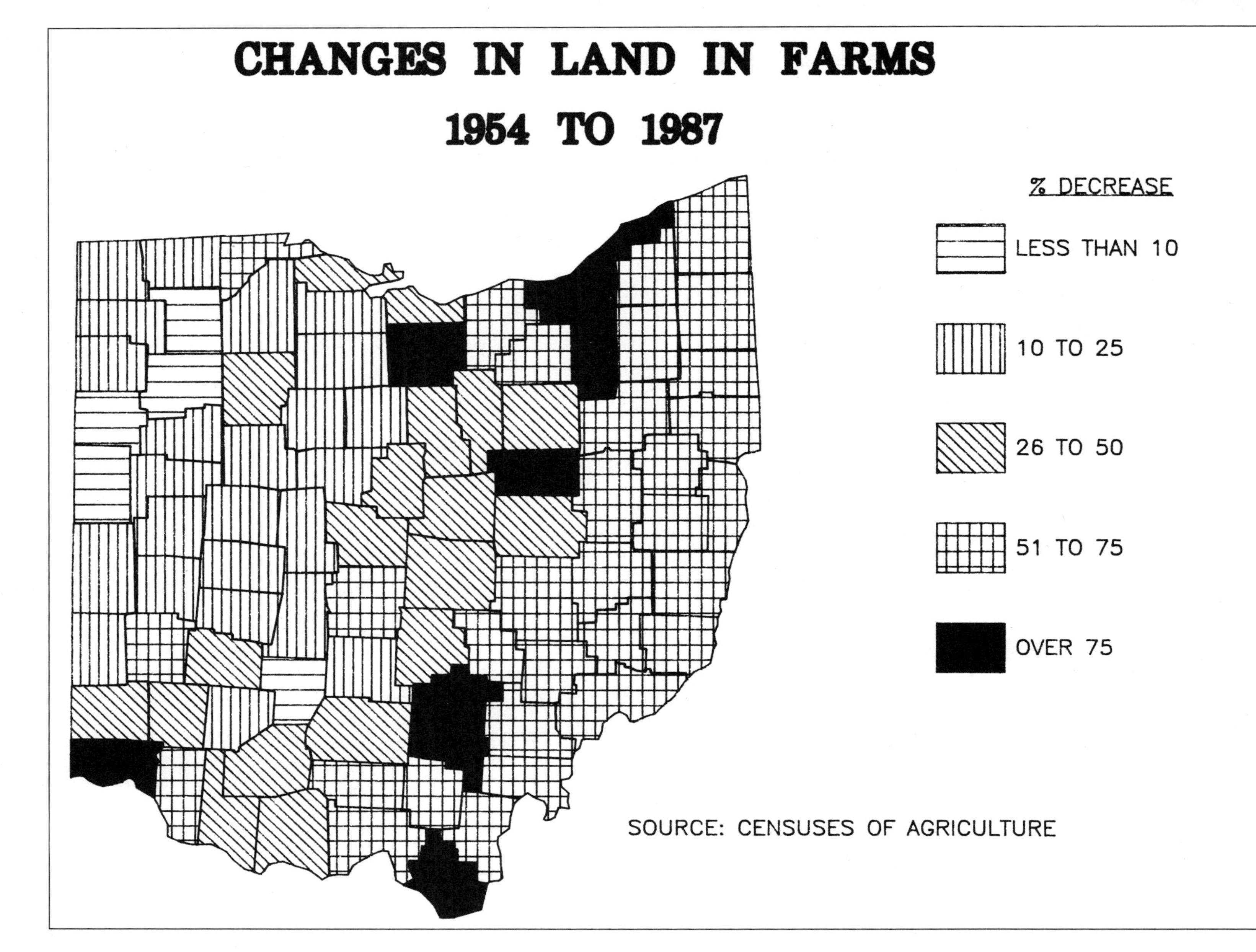

Fig. 12.2. Changes in Land in Farms, 1954 to 1987

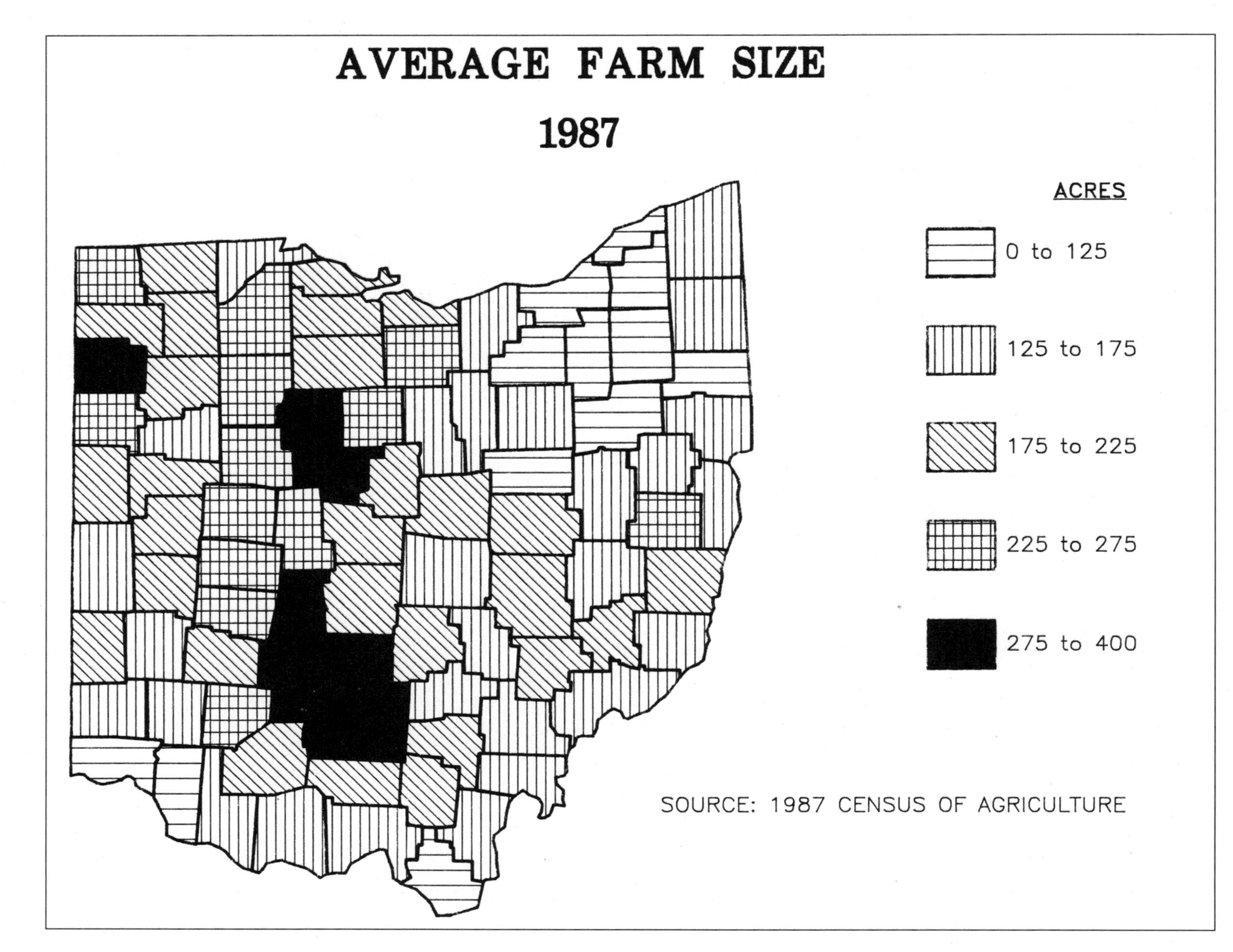

Fig. 12.3. Average Farm Size, 1987

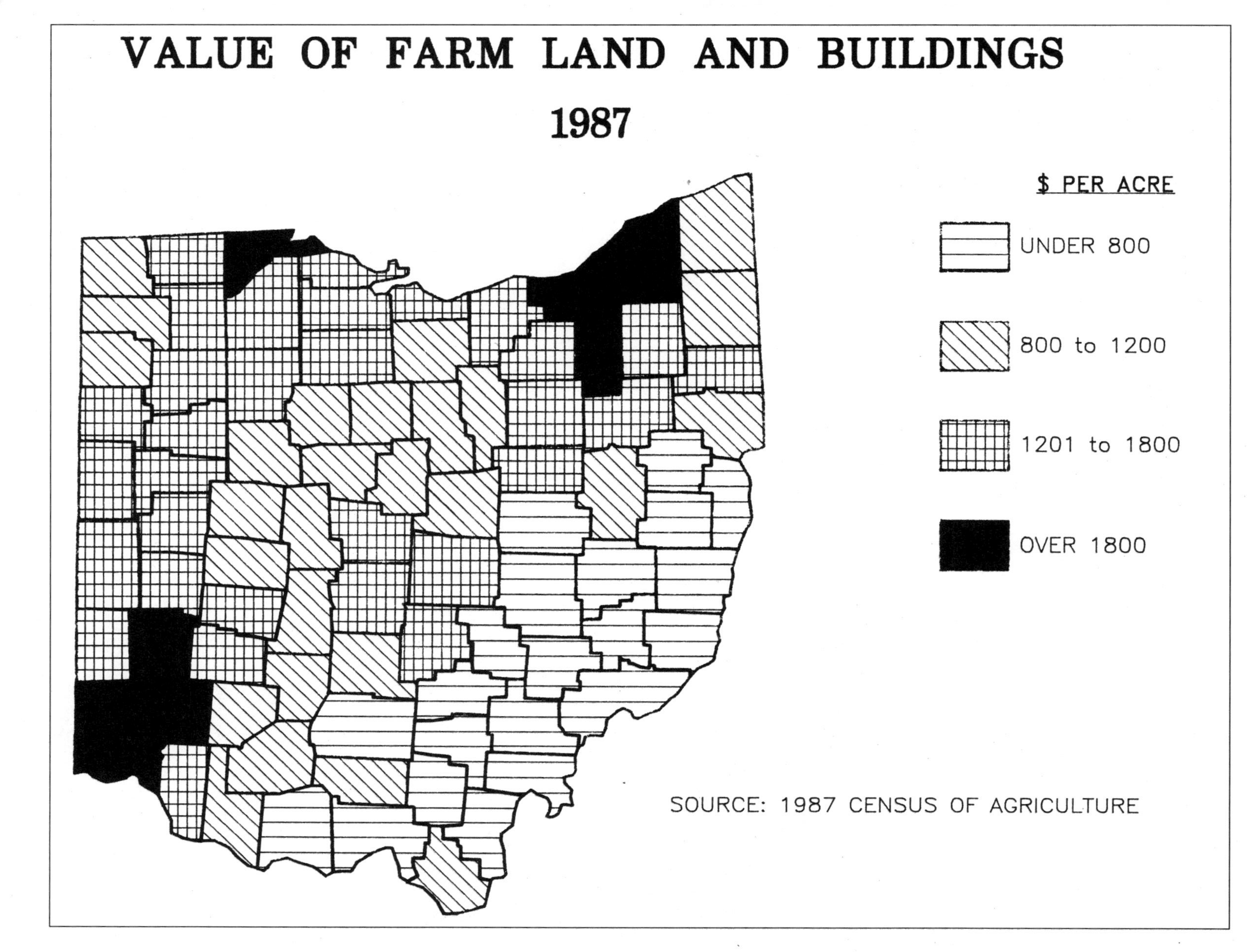

Fig. 12.4. Value of Farm Land and Buildings, 1987

Table 12.2
Crop Revenues, 1987

CROP	SALES ($1,000)	% OF TOTAL SALES
Soybeans	667,540	38.1
Corn	555,237	31.7
Wheat	105,660	6.0
Sorghum for grain	103	0.1
Barley	94	0.1
Oats	9,371	0.5
Other grains	8,920	0.5
Tobacco	17,462	1.0
Hay, silage, and field seeds	47,773	2.7
Vegetables, sweet corn, & melons	84,072	4.8
Fruits, nuts, and berries	28,354	1.6
Nursery and greenhouse crops	209,031	11.9
Other crops	17,206	1.0
Total	1,750,823	100.0

Source: 1987 Census of Agriculture

for more than one-sixth of the state's crops income. Wheat is the only other category that contributes more than 5 percent to crop revenues. Just as each crop or group of crops vary in their importance to Ohio's farm income, they also vary in their spatial pattern within the state. The explanations for these differences lie in variations in the physical requirements of various crops, historical patterns, and economic considerations.

Soybeans

Soybeans are a cash crop, not used directly as feed for livestock. They are a clean till crop, and as such are a considerable erosion risk. To reduce soil losses, beans are generally confined to fields with little relief. Soybeans, a legume, have an extra added value: as a nitrogen fixer they are valuable in a rotation scheme, fitting especially well into a rotation with corn, which has a very high nitrogen requirement. Soybeans and corn are also compatible crops in terms of equipment and labor, and frequently both crops are grown on the same farm. In spite of this compatibility the two differ somewhat in the areas where their production is concentrated. Table 12.3 lists the ten leading producer counties for both crops. Only four counties, Darke, Madison, Van Wert, and Wood, appear on both the lists for soybeans and corn.

Figure 12.5 shows that soybean production is concentrated in the northwest part of the state. Outside this area, bean production tends to occur on level land, such as that found in Madison County or on the floodplains of the state's major streams. John Fraser Hart notes that this affinity between soybeans and level land is common throughout the Corn Belt (1986). Also evident from the map is the lack of soybean production on the rolling terrain of southeastern Ohio. After studying this crop, Thomas Rumney noted four trends in Ohio's soybean production: (1) concentration of production in the western part of the state; (2) expansion of production in the east-central region; (3) increasing concentration of the original area of production in western Ohio; and (4) increasing occurrences of production in suburbanized areas (1986). Some of these patterns are reflected in Figure 12.5 which shows soybean production in 1987.

Corn

Historically, corn has been one of the most important crops in terms of value and acreage. It was almost the only crop ascribed to squatters in the Ohio Country prior to 1788 (Jones 1983). During the early 1800s the Miami Valley of southwest Ohio evolved into the seedbed for the Corn Belt. By 1850, corn had become the preeminent crop in the Scioto and Miami Valleys.

Table 12.3

Ohio's Agricultural Leaders Top Ten Crop Producers for Selected Years

Soybean Production 1990 (1000 Bushels)		Corn Production 1990 (1000 Bushels)		Wheat Production 1990 (1000 Bushels)	
Darke	4,969	Darke	12,273	Wood	3,770
Wood	4,686	Wood	12,179	Paulding	3,473
Van Wert	4,490	Pickaway	11,905	Hancock	3,168
Hancock	4,363	Fulton	11,760	Putnam	3,125
Seneca	4,253	Van Wert	11,905	Henry	2,968
Henry	4,035	Champaign	11,032	Darke	2,705
Putnam	3,821	Preble	10,793	Van Wert	2,650
Hardin	3,560	Clinton	10,749	Mercer	2,519
Marion	3,530	Madison	10,229	Hardin	2,513
Madison	3,428	Fayette	10,074	Seneca	2,499
Oats Production 1990 (1000 Bushels)		**Hay Production 1990 (1000 Tons)**		**Corn Silage 1987 (1000 Tons)**	
Wayne	933	Wayne	279	Wayne	242
Stark	776	Holmes	178	Mercer	151
Seneca	750	Carroll	143	Darke	97
Holmes	712	Columbiana	138	Holmes	90
Columbiana	688	Ashtabula	124	Shelby	73
Richland	613	Muskingum	112	Ashland	68
Mercer	585	Ashland	109	Columbiana	67
Ashland	484	Mercer	109	Tuscarawas	65
Ashtabula	454	Coshocton	106	Ashtabula	62
Paulding	421	Tuscarawas	104	Stark	60
Tobacco Production 1987 (1000 Pounds)		**Vegetable Sales 1987 (1000 Dollars)**		**Horticulture Sales 1987 (1000 Dollars)**	
Brown	4,226	Huron	19,503	Lake	30,097
Adams	3,501	Sandusky	9,891	Lorain	28,743
Clermont	1,458	Stark	5,772	Cuyahoga	13,351
Gallia	1,188	Putnam	3,823	Lucas	12,711
Highland	786	Wood	3,737	Clark	12,133
Scioto	501	Seneca	3,158	Hamilton	7,562
Lawrence	433	Henry	2,703	Miami	6,100
Pike	150	Lucas	2,439	Montgomery	5,411
Darke	75	Fulton	2,289	Columbiana	4,813
Warren	48	Ottawa	1,981	Stark	4,773

Sources: Ohio Department of Agriculture 1990 Annual Report and 1987 Census of Agriculture

In early Ohio, aside from its use on the farm as animal feed, corn was an important contributor to the development of the distillery industry. Cincinnati was reported to have been the greatest whiskey market in the world in 1840 (Jones 1983), possibly because corn in a liquid form (whiskey) was much easier and more profitable to transport and market than were sacks of grain.

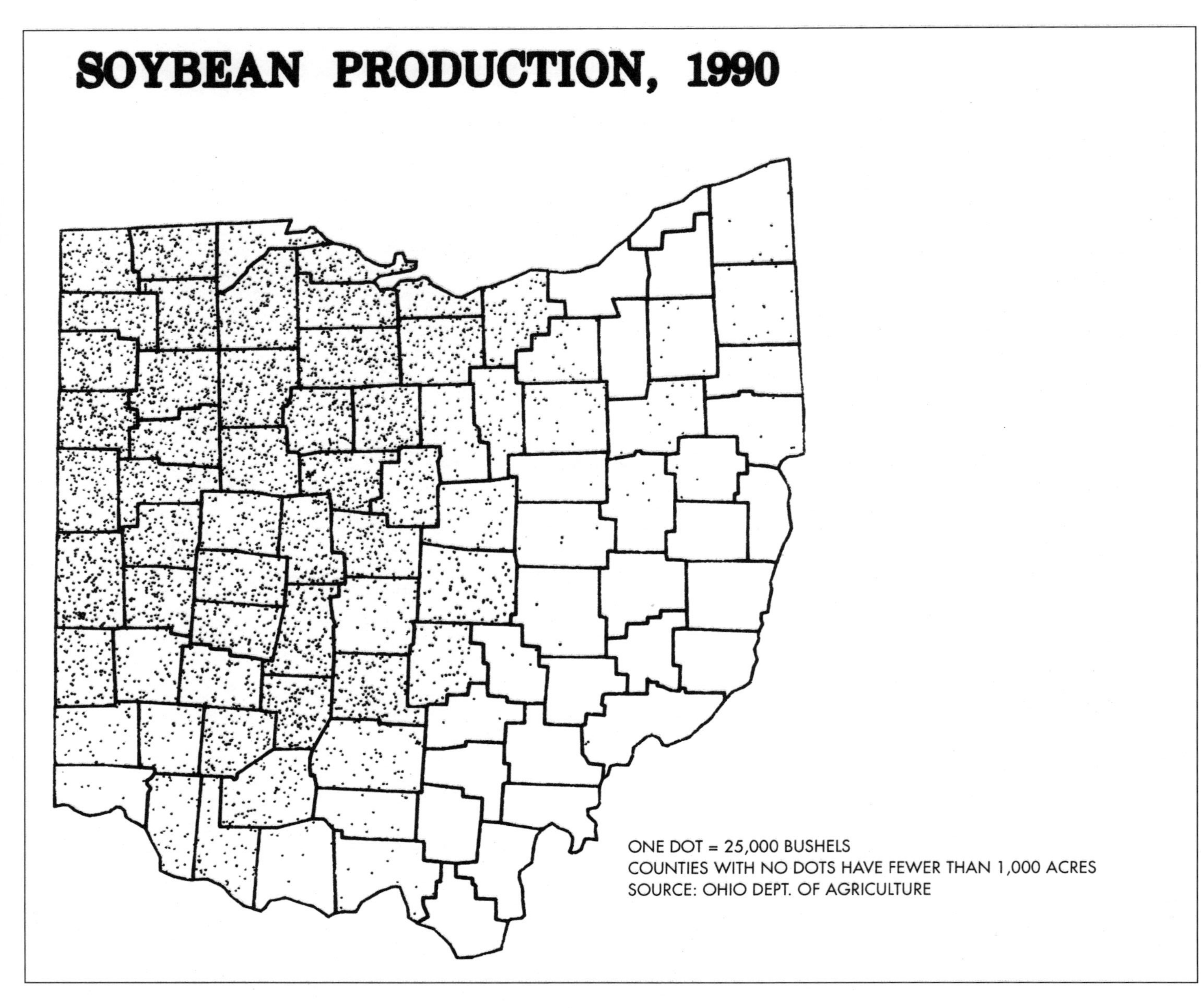

Fig. 12.5. Soybean Production, 1990

As noted above, in 1987 corn was surpassed in revenue production and acreage by soybeans. The two crops, while having many complementary attributes, do not have the same pattern of distribution. Corn, like soybeans, can be a cash crop, but a significant amount of corn is used directly on the farm to feed livestock. In that sense, the corn "walks to market." Figure 12.6 shows the distribution of corn production for 1990. The pattern of 1850, with concentrations in the Scioto and Miami Valleys, has changed relatively little in the last 150 years. There is also a very strong correlation between the production of corn and the raising of livestock, especially hogs. This relationship will be explored more fully in the section on livestock. Corn production is more widely dispersed than soybeans. The production in the northeastern counties is often harvested as silage to provide forage for dairy cattle. Corn production is not very important in the southeast.

Wheat

Ohio was once the center of wheat growing in the United States, leading the nation in wheat production in 1839. The changing geography of wheat production underscores the importance of transportation in influencing agricultural location. Ohio's first wheat fields were planted along the Ohio River and its navigable tributaries. The market was New Orleans, via the Ohio and Mississippi Rivers. With the opening of the Ohio and Erie Canal in the 1830s, the market shifted east to Buffalo, and the center of wheat growing moved to the backbone counties of Ashland, Belmont, Carroll, Columbiana, Coshcoton, Harrison, Knox, Licking, Muskingum, Richland, Stark, Tuscarawas, and Wayne (Jones 1983). In the 1840s after the completion of the Miami and Ohio Canal, both the Miami Valley and the northwestern region joined the ranks of important wheat producers. Gradually the Wheat Belt migrated westward, while in Ohio wheat was displaced by more profitable crops—first corn and, more recently, soybeans. The peak in Ohio's wheat acreage occurred in 1899 (Ohio Department of Agriculture 1990). Today wheat is grown for cash sale and as part of a crop rotation scheme to provide relief from the taxing demands of corn. It also can be used a winter pasture for livestock, although this use is not common in Ohio. As Figure 12.7 shows, the leading wheat-producing counties are in the northwest region. There is a high degree of spatial association between wheat and soybean production; the lists of the ten leading counties for each crop contain many of the same names.

Oats

In the past, oats were grown for three reason: as a rotation crop, as feed for horses, and as feed for dairy cattle. The peak year for oat acreage was 1927, a time when Ohio agriculture began increasingly to use tractors (although horses remained an important power source until World War II). Historically, there has been a strong spatial association between oats and wheat. As Robert Jones notes, "The burgeoning of the wheat industry created the need for horses, and horses required oats; therefore, the leading wheat-producing counties tended to be foremost in oats" (1983). That association has not continued, and today oats production is concentrated outside the traditional cash grain area of northwest Ohio. Nearly 15 percent of the state's production is centered in two counties: Wayne and Holmes. It is no accident that oats production is there, for these two counties have the highest Amish population in the state. Both counties are home to more horses than any other county, and both derive a significant amount of income from dairying. Most of the other important oats-producing counties are major milk producers.

Other Grains

Ohio's other grain crops include barley, emmer, spelt, rye, popcorn, sorghum, and sunflower seeds. Barley tends to have the same distribution as oats. Nearly one-third of the crop comes from Holmes and Wayne Counties. Emmer and spelt, used primarily as livestock feed, are produced almost exclusively in the northeast; again, Holmes and Wayne are the leading producers. Rye has a somewhat similar distribution but also is produced in some of the cash grain counties of northwest Ohio. Popcorn production is concentrated in the northwest counties, where Van Wert and Hardin are responsible for more than one-third of production. Because of their limited acreage, the data are insufficient to gain a true picture of the production patterns of sunflower seeds and sorghum. Collectively this category of crops, labeled "other grain crops" by the Ohio Department of Agriculture, is rather

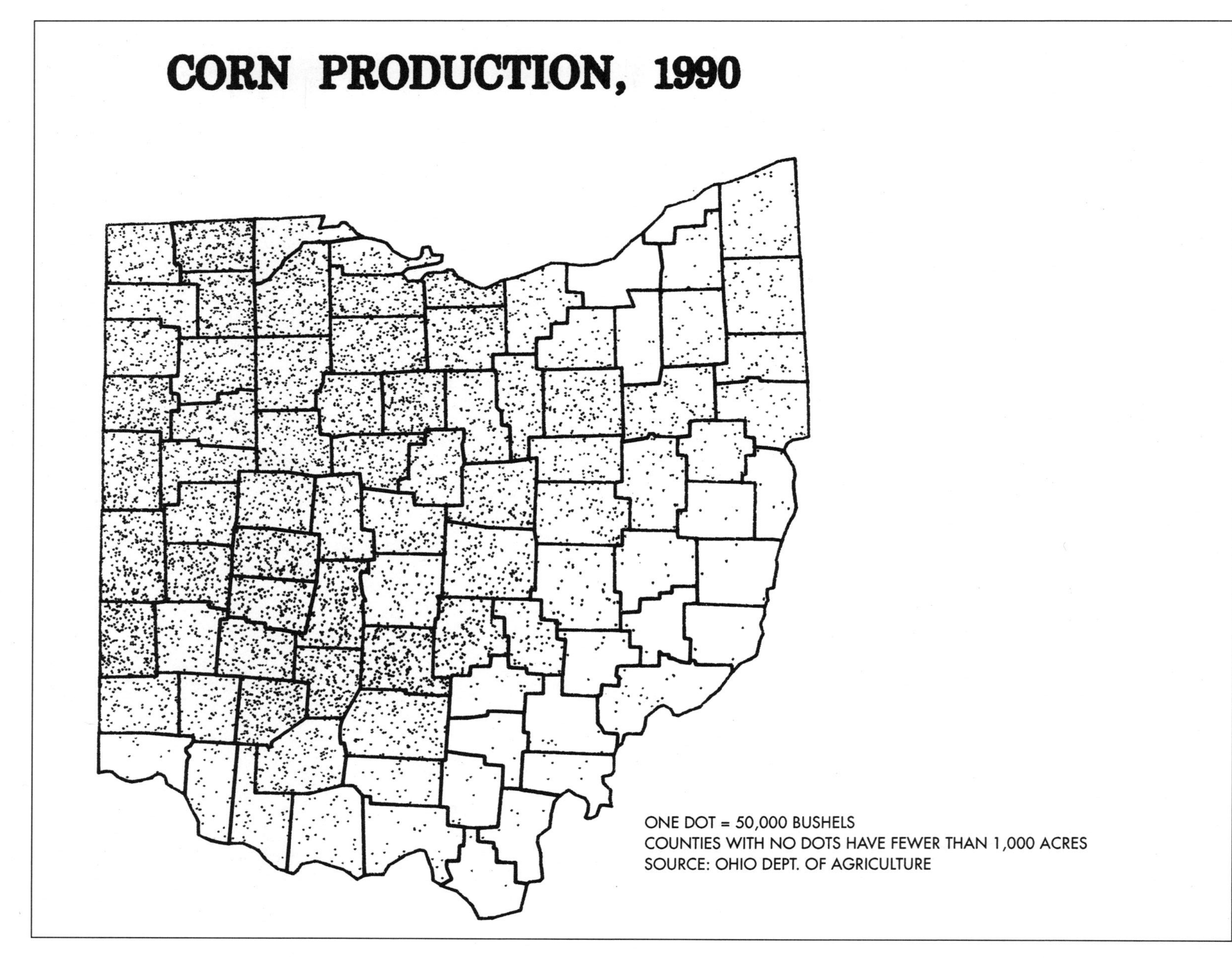

Fig. 12.6. Corn Production, 1990

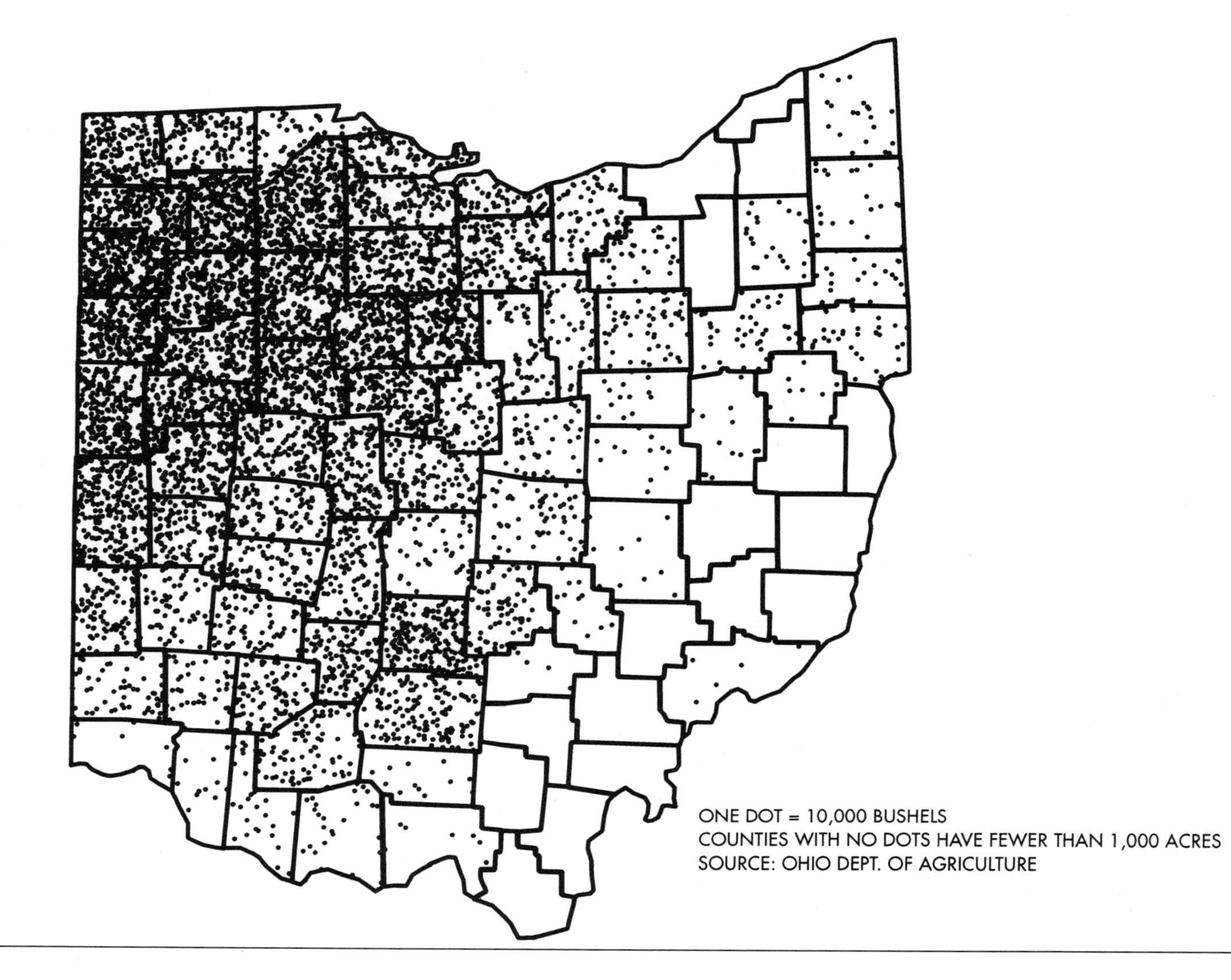

Fig. 12.7. Wheat Production, 1990

insignificant, accounting for less than 1 percent of crop revenues. They have some limited regional significance, but statewide they are unimportant.

Tobacco

A minor crop in terms of revenues—but one of considerable regional importance—is tobacco. Although today tobacco accounts for only 1 percent of crop revenues, it was once far more important. The state's first tobacco district was in the eastern hill counties. That district's rise to prominence began about 1825. Jones cites an 1825 report from Washington County stating that tobacco was ten times more profitable than any other type of farm produce (1983, 250). Prior to the Civil War, this district alone was producing more tobacco than was produced in the entire state in 1987; statewide production of tobacco in 1863 was more than 37,000,000 pounds, nearly three times the current output. A second district, an expansion of tobacco production in Kentucky, developed in the "ABC Counties—Adams, Brown, and Clermont—in the 1840s. A third tobacco-producing district appeared a bit later in the Miami River Valley.

In 1987 production was centered in two areas: the counties along the Ohio River from Gallia to Clermont, and the Miami River Valley. The Ohio River counties are the most significant tobacco producers, especially Adams and Brown; in 1987, 61 percent of tobacco revenues came from these two counties. In 1982 it was reported that 40 percent of farm income in these two counties came from tobacco leaves: cut the counties in half with an imaginary east-west line, and in the southern half that figure exceeds 60 percent (Bernstein 1982).

Vegetables

Ohio farmers produce a variety of vegetable crops, primarily for two distinct markets: canneries and fresh produce for urban centers. In 1990 the state ranked sixth nationally in the production of processing vegetables (Vonda 1992). Vegetable production is greatly influenced by soil conditions, and the state's peat and muck soils have made possible some very intensive vegetable farming. The importance of vegetables in some communities is even reflected through place-names, as evidenced by Celeryville in southern Huron County. In addition, festivals commemorating vegetables abound in the state; they include the Millersport Corn Fest, the Circleville Pumpkin Festival, and festivities at Reynoldsberg honoring the tomato.

Figure 12.8 indicates the pattern of vegetable sales in the state. Unfortunately a complete picture is not available because the Bureau of the Census does not disclose data for counties with only a few producers. Still, some conclusions can be drawn from the map. Very apparent is the concentration of sales in the northwest, where soils are favorable for the production of many vegetable crops. The organic soils of Huron and Sandusky Counties produce well over one-third of total sales. The map suggests that some production is destined for sale in urban markets, as counties in highly urbanized northern regions have significant vegetable sales. Sales in nonurban counties, such as those in the northwest, are often tied to a local cannery. Though the data do not permit a detailed analysis of all vegetable crops, some assessment can be made of regional specializations in a few crops (see Table 12.4).

The significant Mexican-American populations in many northwestern Ohio counties are legacies of vegetable production. Attracted to the area initially as migrant laborers, many of these immigrants have taken up permanent residence. Migrant labor still is important in the production of many of Ohio's vegetable crops.

Fruits and Berries

Although data problems similar to those for vegetables hinder a detailed analysis of Ohio's fruit crop, enough data are available to permit a limited assessment. Based on a map of sales (Figure 12.9), climate and proximity to urban markets appear to be the critical factors influencing production of these crops. The extended growing season along the shore of Lake Erie is the primary reason for the high sales in that area. More than one-third of fruit crop revenues comes from counties that border the lake. Counties near the state's major urban centers tend to have high sales. "Pick your own" fields and roadside markets abound in the urban rural fringe of major cities. Some regional specializations and absences are evident. Fruits and berries are notably unimportant in both the southeast and the northwest regions. Apples are produced in most parts of the state, but Columbiana, Ashtabula, Licking, and Sandusky Counties are the major pro-

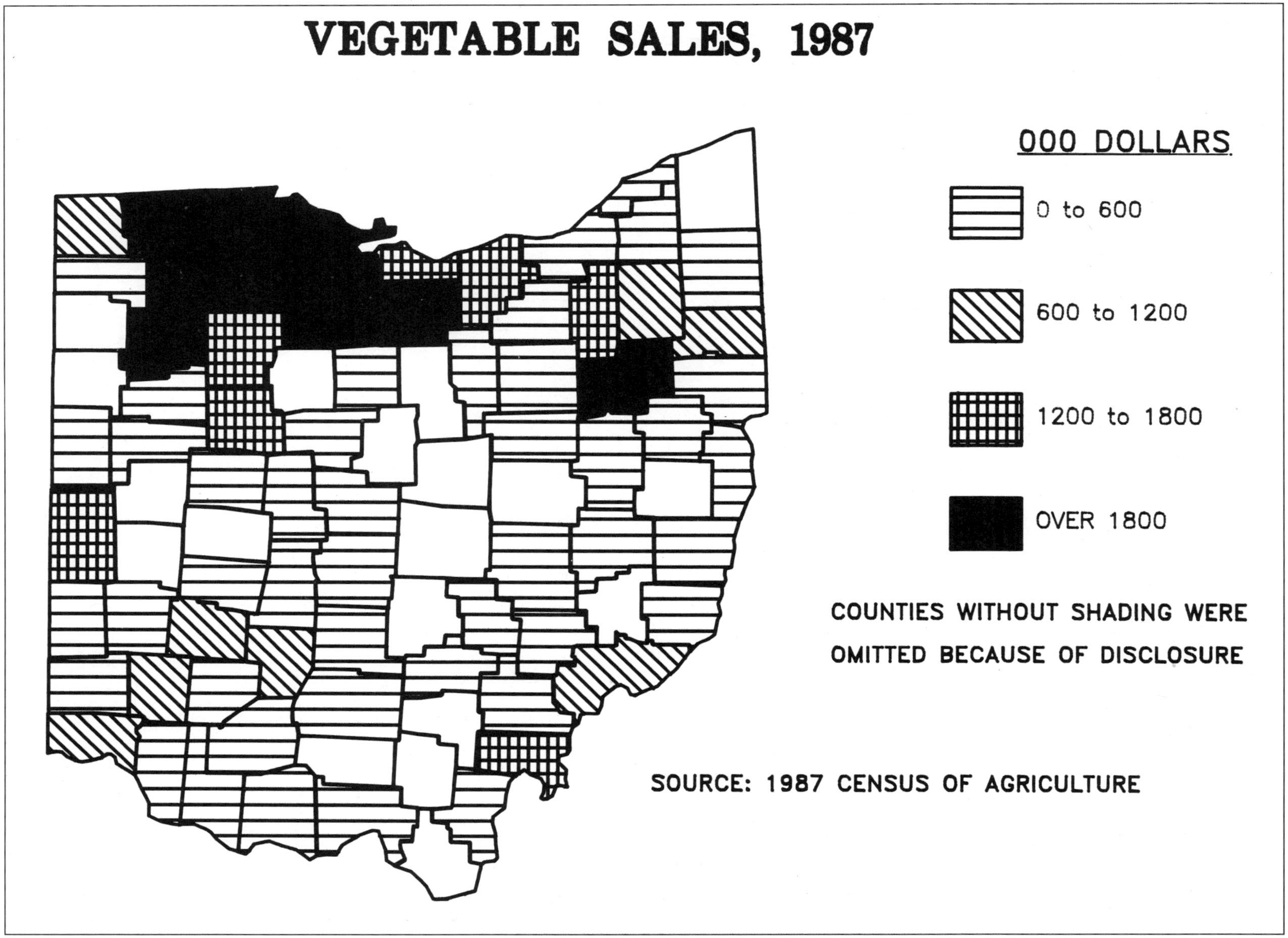

VEGETABLE SALES, 1987

000 DOLLARS

0 to 600

600 to 1200

1200 to 1800

OVER 1800

COUNTIES WITHOUT SHADING WERE
OMITTED BECAUSE OF DISCLOSURE

SOURCE: 1987 CENSUS OF AGRICULTURE

Fig. 12.8. Vegetable Sales, 1987

Table 12.4
Regional Specialization in Vegetables

CROP	COUNTY
Cabbage	Erie and Sandusky
Cantaloupes	Mahoning and Sandusky
Carrots	Henry
Cucumbers	Sandusky
Lettuce	Huron
Pumpkins	Sandusky
Snap Beans	Hamilton
Squash	Lorain
Sweet Corn	Lorain and Lucas
Tomatoes	Fulton, Sandusky, Seneca, and Woods

Source: 1987 Census of Agriculture

ducers. The grape crop is concentrated along the eastern shore of Lake Erie, with over half the crop coming from Ashtabula, Lake, and Lorain Counties. Columbiana and Ross Counties are the major centers of peach production. Strawberry production occurs in most counties, with Columbiana, Mahoning, and Tuscarawas the leading producers.

Nursery and Greenhouse Crops

These crops account for a surprising amount of the state's crop revenue—nearly 12 percent. Again, climate and proximity to urban markets are the most important locational factors affecting production. One-third of the revenue for these crops comes from sales in Lake, Cuyahoga, and Lorain Counties. Much of the production in Cuyahoga and Lorain is under glass or similar protection. The urban counties of Cuyahoga, Lucas, Lorain, and Hamilton are important centers for bedding plants, cut flowers, and various other potted plants. Lake and Clark Counties are both major producers of nursery crops.

Livestock

Cows, horses, and pigs were among the agricultural assemblages that early white settlers brought to Ohio. These animals provided the basis for the state's livestock industry—one of the two major sources of agricultural revenue. The proportional contributions of various animals and animal products, as of 1987, are indicated in Table 12.5.

Dairying

Owing to the poor milking quality of early cattle and the high labor demands of dairying, Ohio's dairy industry was relatively slow to develop. Jones suggests that New Englanders introduced dairying to Ohio around 1796, most likely in Washington County (1983). By 1803 Trumbull County cheese was being sold in Pittsburgh, and by 1815 the Western Reserve had become the "cheesedom" of Ohio. In 1820 cheese from the Western Reserve was marketed along the Ohio River (Lloyd, Falconer, and Thorne 1918). Early Ohio cheese was not known for its quality. One exception was the Swiss (Switzer) cheese manufactured by the German-speaking Swiss immigrants that settled in Holmes and Tuscarawas Counties. Today this area continues to be an important center for dairying and still enjoys a reputation for its excellent Swiss cheese.

Today dairying has risen in importance. Nationally Ohio ranks eighth in milk production. In fact, more counties report dairying as their leading source of income than any other type of agriculture. Dairying ranks first as an agricultural income generator in twenty-five counties. Historically the industry has been concentrated in the Western Reserve, and the leading dairy counties currently cluster along the southern margin of the Reserve and in areas immediately to the south (Figure 12.10), forming a dairy zone ideally located to supply milk to the Cleveland and Pittsburgh markets. This region, at one time glaciated, tends to have rolling terrain and some of the shortest growing seasons in the state. Given this short growing season, much of the region's corn crop is used for silage to feed dairy cows. A second concentration of dairying is found in west-central Ohio, primarily on the rougher moraine lands not well suited to cropping. It, too, is well situated to serve urban markets, in this case those of Cincinnati and Indianapolis. Dairying is also important in several of the counties bordering the Ohio River in the southeast part of the state. Here again, the restraints terrain has placed on cropping seem to be an explanatory factor. Dairying is of minor importance in most of the nonriver counties of southeastern Ohio, and in the counties of the northwest where the land is well suited to cropping.

Cattle and Calves

In early Ohio cattle served three purposes: as draft animals, as a meat supply, and as a source of milk. Ini-

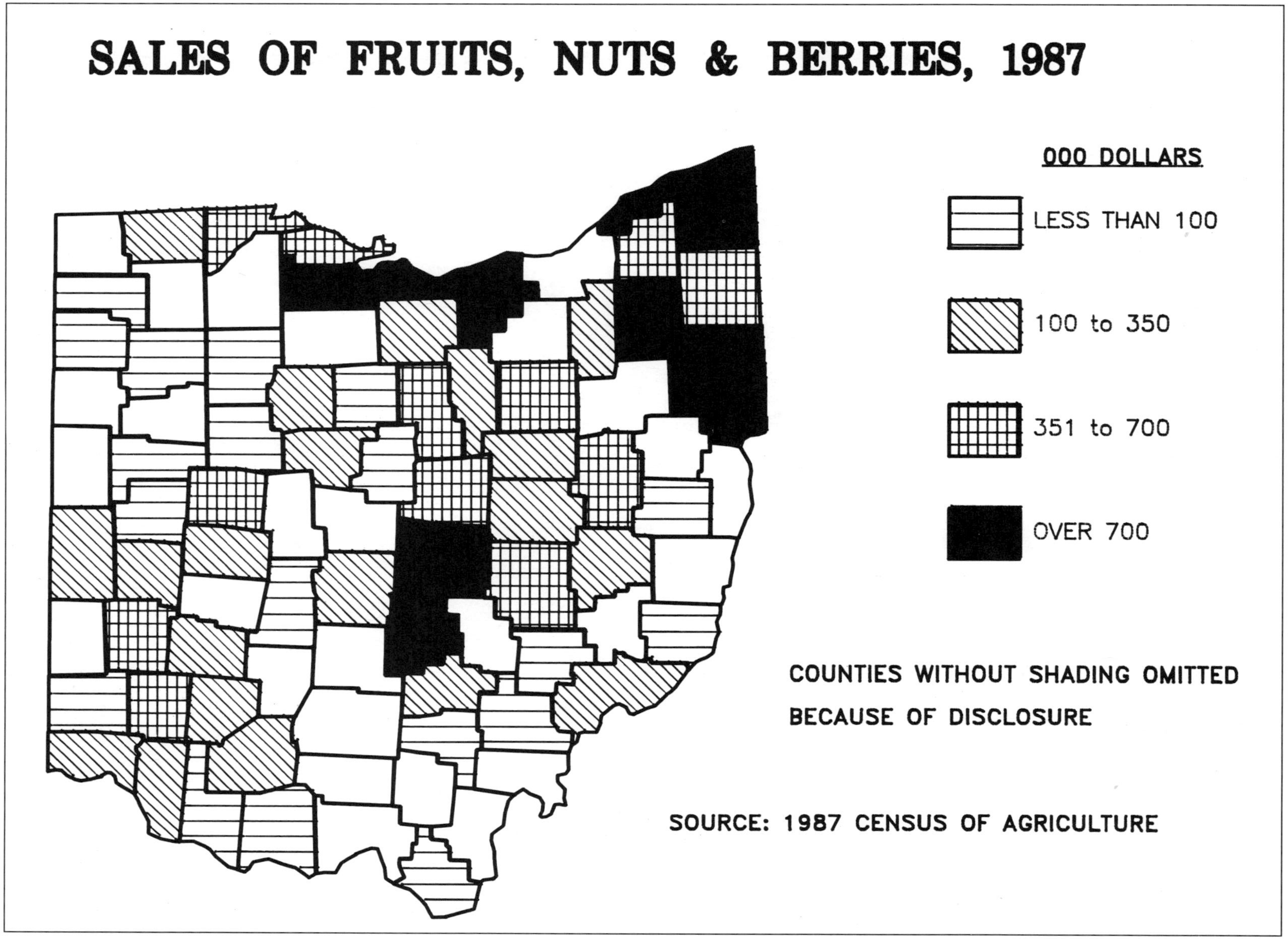

Fig. 12.9. Sales of Fruits, Nuts, and Berries, 1987

Table 12.5

Sales of Animals and Animal Products, 1987

ANIMALS OR PRODUCT	SALES ($1,000)	% OF TOTAL SALES
Dairy Products	543,691	31.8
Cattle & Calves	426,114	25.3
Hogs	383,227	22.8
Poultry & Poultry Products	281,232	16.7
Sheep	15,201	0.9
Other Livestock & Products	42,816	2.5
	1,683,281	100.0

Source: 1987 Census of Agriculture

tially the latter two uses were less important. Most of the early cattle were driven to Ohio from Kentucky, Pennsylvania, and New York. However it was not long until cattle were being driven east, from Ohio to Philadelphia or Baltimore, often pausing in Virginia to be fattened along the south branch of the Potomac before moving on to market. By 1805 Zane's Trace was used to move corn-fattened cattle from the Scioto Valley to eastern markets (Ray 1970). Three areas emerged as important early centers for the cattle industry: the "barrens" of Madison and Clark Counties were used to graze (grass fatten) cattle from Kentucky or states to the west of Ohio; the Scioto Valley became the premier producer of corn-fattened cattle; and, as noted above, the Western Reserve was the center for dairying.

The sale of cattle and calves is the leading source of agricultural income in eight counties, most of which are in southeast Ohio. It should be noted that these are not the leading counties in terms of either the number of cattle or the sale of cattle and calves. In this area, where arable land is in short supply, cattle offer a means of reaping some return from the land. Such areas often emphasize cow-calf operations, which produce calves for sale to other places with large stocks of feed grains such as corn. Feeders in these corn-rich areas fatten the calves and then send them to packers. Figure 12.11 shows the distribution of cattle and calves in the state. The pattern is very similar to that of dairy cows. Outside the dairy concentrations there are large numbers of cattle in both the east and west-central parts of the state. The latter is an important corn-producing region and thus the center of feedlot operations. Cattle are not found in large numbers in the nonriver counties of the southeast and the extreme northwest and southwest parts of the state.

Hogs

The significance of hogs to Ohio agriculture is underscored by the name given to Cincinnati, "Porkopolis." By 1845 Cincinnati was the leading hog-packing center in the world. As Mark Bernstein quotes one observer, "Cincinnati originated and perfected the system that packs fifteen bushels of corn into a pig and packs that pig into a barrel and sends him over the rivers and oceans to feed mankind" (1986). An early account relates the streets of Cincinnati as often filled with hogs driven from the southwest. The by-products of the pork-packing industry gave rise to one of Cincinnati's most important employers: Procter and Gamble.

Apparently, early settlers found wild hogs on their arrival in Ohio. It is likely that these were escapees from areas further east or south. Initially these hogs were allowed to roam freely and feed upon the mast provided by the state's extensive forest; they were not considered fit for slaughter until they had been fattened on corn for five or six weeks (Jones 1983). Like cattle, hogs were driven to eastern markets. With the passage of time, two areas emerged as the centers of Ohio's swine production. One was the Miami River Valley, which is often cited as the heart of the Corn Belt—a system of agriculture that relies on fattened animals, particularly hogs, as a means of marketing corn. The second was the corn-rich valley of the Scioto River.

Although Ohio ranks sixth nationally in the number of hogs and pigs, the sale of these animals is the leading source of agricultural income in only two counties: Preble and Fulton. At the same time, they are one of the top four sources of income in many counties. Figure 12.12 shows the distribution of hogs and pigs

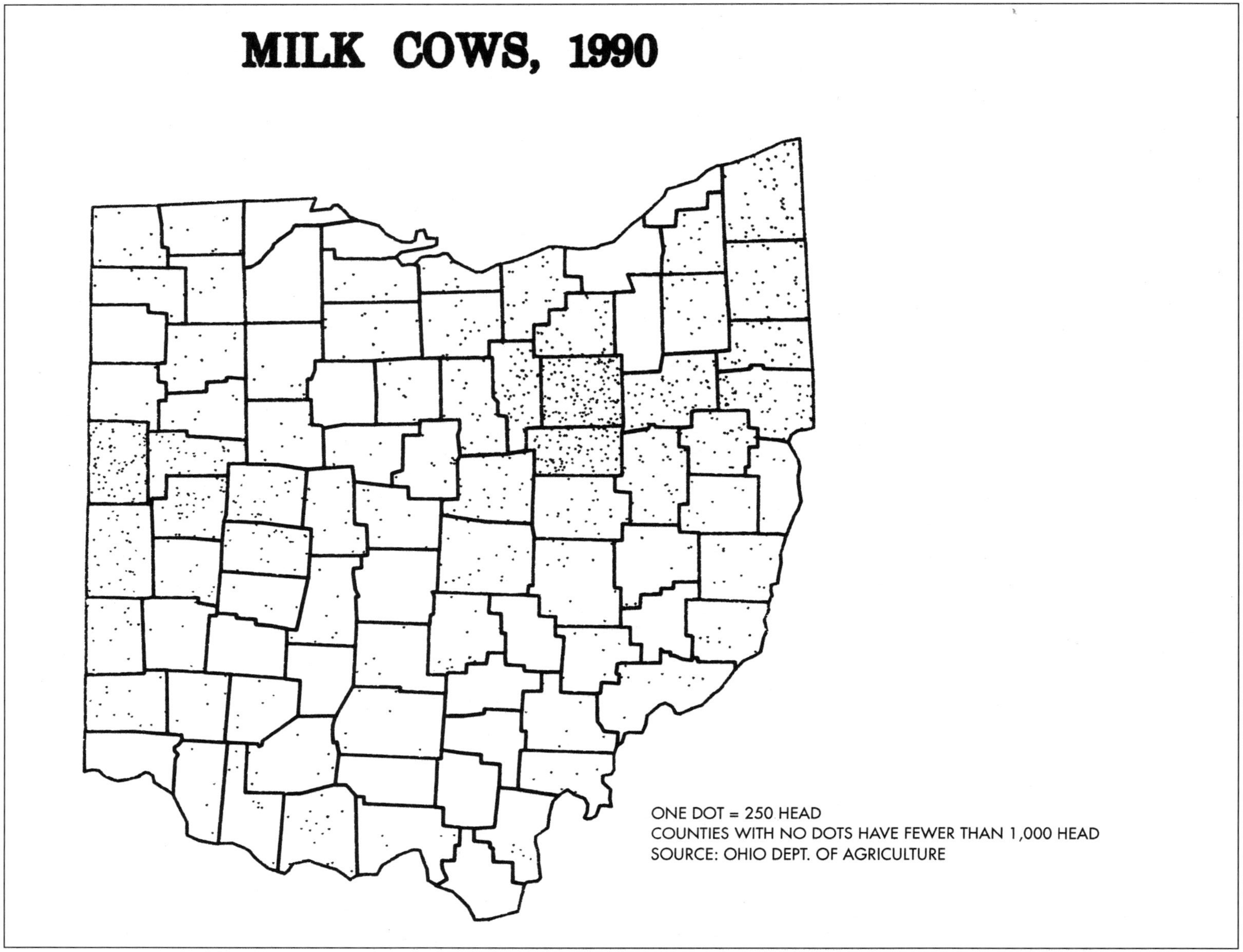

Fig. 12.10. Milk Cows, 1990

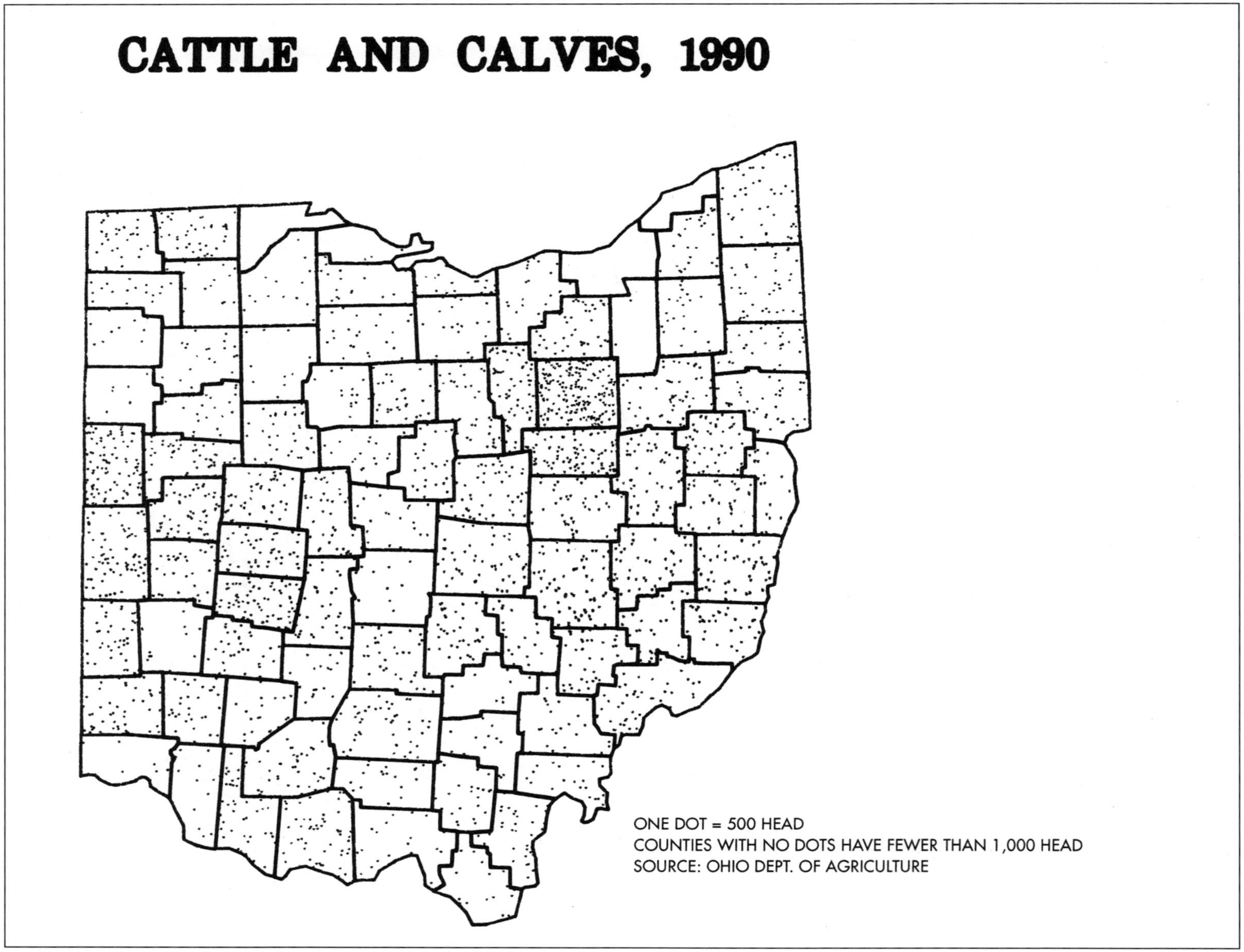

Fig. 12.11. Cattle and Calves, 1990

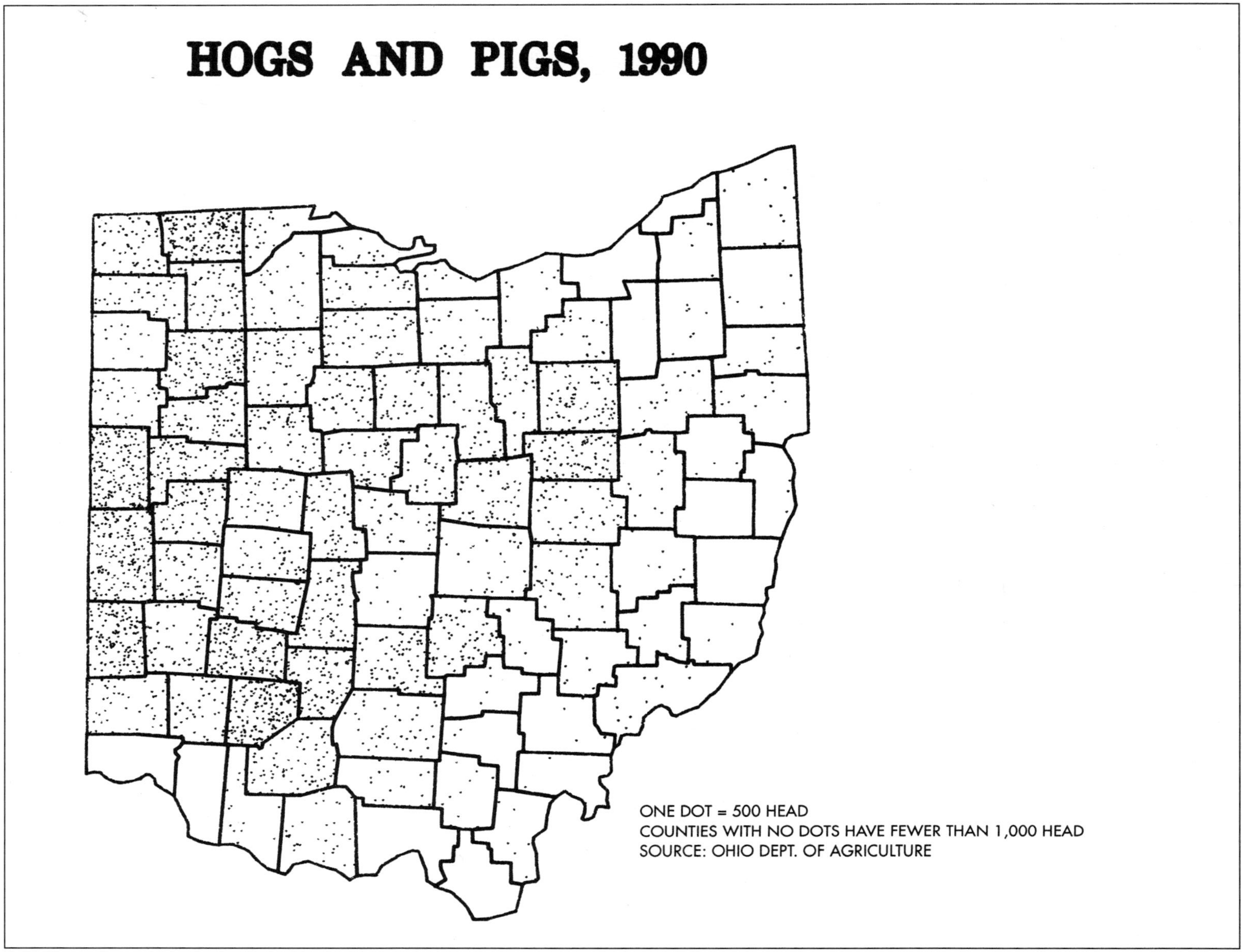

Fig. 12.12. Hogs and Pigs, 1990

in the state. With the exceptions of Wayne and Holmes Counties, hogs are not important in the eastern third of the state. The most important producing areas are found in three clusters: the southwest, including Highland, Fayette, and Clinton Counties; the west, with Preble, Darke, and Mercer Counties; and the northwest, where Fulton and Putnam are ranked among the top ten counties in the nation in their populations of hogs and pigs. The primary explanation for the emphasis on hogs in these counties is the availability of large amounts of feed, mainly corn. The system of marketing corn that developed in the Miami River Valley continues today.

Poultry

Poultry sales account for one-sixth of Ohio's livestock income. As Table 12.6 shows, poultry furnished significant revenues for several counties. As dietary habits change, one can expect that poultry sales will increase at the expense of red meats. Most of the important poultry-producing counties are in the northwest region where there are good supplies of feed grains.

Sheep

Although today sheep contribute only a minor amount to the state's agricultural revenues, they were once far more important. The primary use of sheep on the frontier was as a source of wool. Sheep did well in the hill country of eastern Ohio where other farming alternatives were rather limited. During several census periods in the last half of the nineteenth century, Ohio led the nation in its number of sheep. The peak year for sheep was 1870, when there were more than 6 million head in Ohio. Today, the number is less than 250,000. Production, now as in the past, centers in east-central Ohio.

Horses

In the early stages of agriculture, horses were less important than cattle and swine. Oxen were the preferred beast of burden for clearing land (Jones 1983), whereas horses were used for transportation and sporting functions. Like cattle and hogs, horses became an early export commodity, driven east from Ohio to markets along the seaboard. By 1850, Ohio was the nation's leading horse producer. That date also marks the era when horses began to replace oxen as the power source for Ohio farms. For the next 90 years, horses played an intricate role on Ohio farms. Following World War II their use diminished rapidly, as they were replaced on the farm by inanimate forms of power.

Horse-drawn agriculture persists today among Ohio's Amish, who shun the use of most mechanical forms of power. One can still find teams of horses tilling the land in Wayne, Holmes, Stark, and Geauga Counties, and to a lesser extent in other Ohio counties. Another concentration of horses exists in suburban counties, where "gentlemen farmers" maintain stables of Arabians and thoroughbreds. These horses are not for work but for show, racing, and play. Warren, Medina, and Delaware Counties are examples of counties with these so-called "horsey farms."

Agricultural Regions

In 1850, Ohio was characterized by five distinct agricultural regions: the backbone counties of the Upper Muskingum Valley (wheat growing); the Miami Valley (hog raising); the Scioto Valley (cattle raising); the Madison County (cattle grazing); and the Western Reserve (dairying) (Jones 1983). Roderick Peattie, writing in 1923, reported "no distinct agricultural provinces in Ohio." Apparently he was not prone to think in terms of regionalization, for more than 30 years later, in his *Economic Geography of Ohio,* Alfred Wright defined twenty-five distinct farming areas (1957). In 1975, Allen Noble reduced the number to twelve. The number has been reduced further for this analysis. The approach employed here differentiates between the agriculture of highly urbanized counties, where only a small portion of the land is used for agricultural purposes, and those where agriculture is the primary use. Among the latter several regions are identified, and within those regions some subregional specialization occurs. Figure 12.13 shows the regionalization of Ohio agriculture. These regions are discussed below. Readers are cautioned that regionalization cloaks much detailed variation in patterns of agriculture. Moreover, these regions are a reflection of conditions as they existed in 1987; agricultural regions are dynamic and subject to the changing nature of the marketplace.

Table 12.6

Ohio's Agricultural Leaders; Livestock Production for Selected Years

Milk Sold 1990 (Million Pounds)		Poultry Sales 1987 (1,000 Dollars)	
Wayne	515	Darke	53,754
Mercer	281	Licking	47,379
Holmes	248	Mercer	46,224
Summit	190	Knox	16,678
Tuscarawas	159	Holmes	9,854
Columbiana	157	Wayne	7,815
Ashland	151	Putnam	6,762
Ashtabula	144	Wood	5,627
Darke	136	Tuscarawas	4,981
Auglaize	133	Henry	3,333
Number of Cattle & Calves 1990 (Excludes Dairy Cows)		**Number of Sheep 1990**	
Wayne	61,800	Muskingam	19,600
Clark	36,700	Knox	12,200
Mercer	32,000	Coshocton	11,000
Muskingum	31,400	Licking	8,800
Darke	28,600	Harrison	8,400
Holmes	25,500	Union	7,400
Ashland	24,600	Seneca	6,400
Adams	24,400	Morrow	6,000
Ross	22,600	Ashland	5,500
Washington	22,000	Logan	5,300
Number of Hogs and Pigs 1990		**Number of Horses 1990**	
Darke	108,000	Wayne	6,831
Mercer	90,000	Holmes	3,260
Clinton	89,000	Geauga	2,680
Fulton	80,000	Medina	2,148
Preble	76,000	Delaware	1,745
Greene	75,000	Stark	1,720
Putnam	71,000	Trumbull	1,710
Fayette	55,000	Warren	1,699
Holmes	50,000	Portage	1,495
Auglaize	49,000	Lorain	1,476

Sources: Ohio Department of Agriculture 1990 Annual Report and 1987 Census of Agriculture.

Urban Agriculture Zones

The availability of large urban markets and the high cost of land in urban areas interact to produce some distinctive agricultural specializations in Ohio's principal urban areas, namely Cleveland-Lorain-Akron, Toledo, Youngstown, Canton, Columbus, Dayton-Springfield, and Cincinnati. Farms in these areas are small, but the land is used intensively. These areas usually have high labor, energy, and fertilizer costs. The emphasis is on producing products for the local market; often these products are highly perishable, making their expeditious movement to market vital. Roadside markets and pick-your-own operations are common in the suburban fringe of these communities. Most communities have a farmer's market in or near downtown. Commonly nursery, vegetable, and fruit crops are the leading income producers, with grain crops and most livestock unimportant. Two exceptions to this pattern are the presence of dairying and recreational horses.

Dairying Regions

Four types of dairying subregions are identifiable in the state:

Type A is found in the northeast. Here dairying is the leading source of agricultural income, followed by sales of cattle and calves. The two industries are interrelated. Crops are geared toward supplying forage and feed for livestock. Thus, improved pastures, corn for silage, oats, and hay occupy a considerable proportion of the cropland.

Type B is associated with the large Amish population centered in Wayne, Holmes, Tuscarawas, and Stark Counties. There is also a small enclave of Amish in Geauga county. The Amish have created a very distinctive landscape of small, intensively farmed units. While dairying provides their primary income, hogs, cattle, and poultry are important contributors to agricultural revenues. The diversity of crops in this region is not matched in any other location in the state. Here cultivated land is used to produce feed crops, such as corn and oats, and forage, like clover and hay. Most Amish farms have a large garden and an orchard, and some of the produce from these two sources finds its way to market.

Type C is found throughout much of southeast Ohio, where dairying ranks first or second in agriculture income and the sale of cattle and calves enjoys equal prominence. This area provides calves for fattening to some of the grain-producing counties to the west. Farms here tend to be small, and most operators have a principal occupation other than farming. Hay is frequently the most important revenue producer among crops. Pasture land is much more common than cultivated land.

Type D—centered in Shelby, Auglaize, and Mercer Counties—is physically separated from other dairying regions. These are among the most productive agricultural counties in the state, with poultry, dairying, hogs, and the production of corn and soybeans the most important revenue producers. Dairy operations tend to occur on moraines, while the crops are centered on the intermorainal areas.

Lake Erie Horticultural Region

The climatic and soil conditions along the shore of Lake Erie have fostered a region that now produces a host of specialty crops. Two types of subregions are identified. Type A stretches from Ashtabula County westward to Lorain County. Though the forte of this area is nursery and greenhouse products, vegetables and fruits are also major contributors to revenue. Animals, however, are noticeably absent. Type B begins in Lorain County and extends westward, through Lucas and into Fulton County. Here vegetables and grain crops, particularly soybeans and corn, dominate. Vegetables rank as the leading source of income for Huron County. Fruit crops and nursery products are of secondary import in this subregion. Livestock revenues are dominated by dairying, a reflection of the area's proximity to the large urban market found along the shore of the lake.

Southeast Cattle and General Farming Region

Here cattle are the major source of income, with hogs and poultry as other livestock contributors. Crops, limited because of the lack of arable land, tend to be used only to support cattle. Most common are hay and corn. Like the dairy subregion to the east, many operators in this area work at nonfarm occupations. The lowest farm incomes in the state are found in the counties that make up this region. It includes parts or all of Vinton, Hocking, Jackson, and Lawrence Counties.

Ohio River Cattle, Dairying, and Tobacco Region

Extending from Gallia County downstream through Adams, Brown, and Clermont Counties lies a region with a unique combination of agricultural revenues.

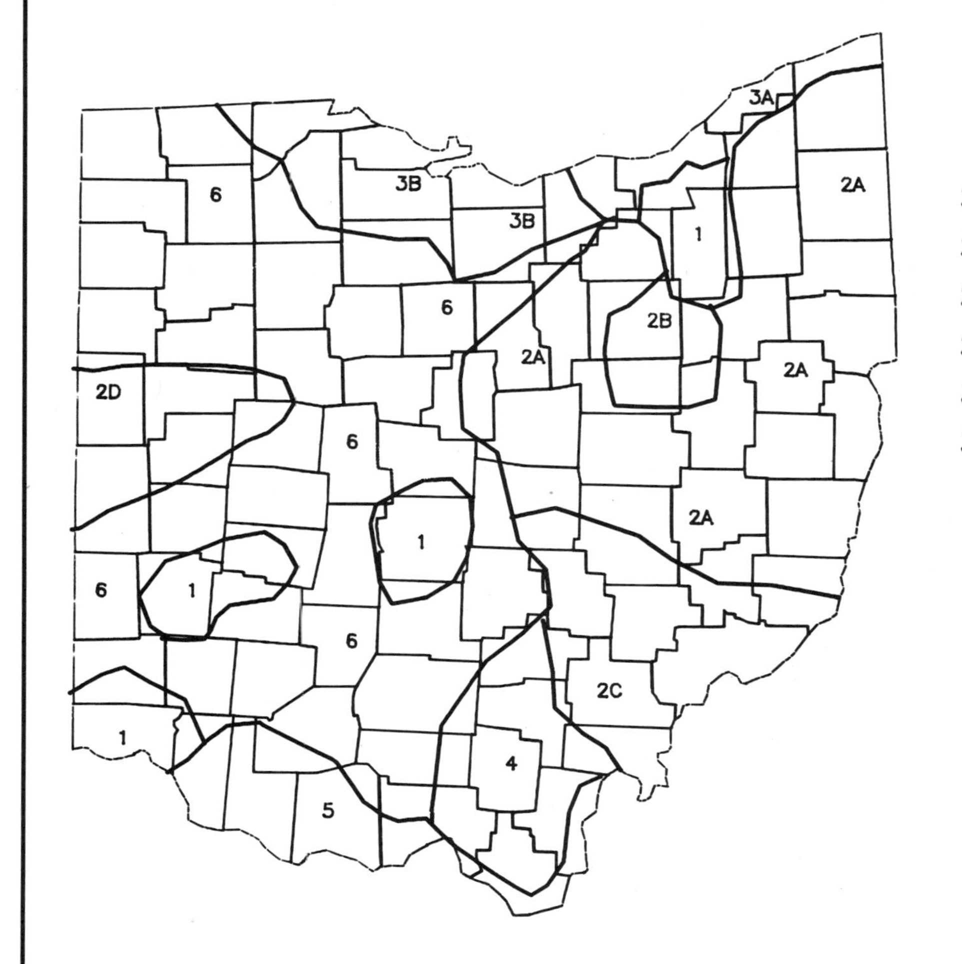

Fig. 12.13. Ohio's Agricultural Regions

Dairying and the sale of cattle and calves are the major income earners, and tobacco the leading income producer among crops. There is great diversity in this region when one compares the productive river bottoms to the marginal operations found in the uplands.

Cash Grain and Livestock

Over much of the western part of Ohio, corn and soybeans reign as the premier income producers. Soybeans rank first and corn, second; in both the north and the south the rankings are reversed. In most of this region either hogs or cattle rank as the third most important income source. This is an area traditionally viewed as Ohio's contribution to the Corn Belt. While corn remains important, soybeans are taking over as the leading income source. Changes are also evident in the livestock mix. An interesting change in American dietary habits seems to be reflected in this region. With a decrease in the consumption of red meats—beef and pork—has come an increased demand for poultry products. The availability of large amounts of feed grains and the proximity to large urban markets provided a base of support for the poultry industry. In the two largest counties in western Ohio, Mercer and Darke, poultry ranked as the leading source of income in 1987, surpassing both corn and soybeans.

Conclusion

Although agriculture will continue to be an important aspect of Ohio's geography, there is little prospect that employment in the industry will increase. It is likely that the number of farms and farmers will continue to decline as farms grow larger. But though the amount of land devoted to agriculture will continue to decline, agriculture will remain as the most important land use in the state. Its impact on Ohio's geography will continue in the form of modifications to the physical environment and a legacy of material culture in its agricultural settlements and farmsteads.

REFERENCES

Bernstein, Mark. 1982. Tobacco: The Green Gold of Southwest Ohio. *Ohio Magazine* 5, no. 8 (November): 43–51.

———. 1986. The Pork Papers. *Ohio Magazine* 9, no. 8 (November): 78–79.

Hart, John Fraser. 1986. Changes in the Corn Belt. *The Geographical Review* 76, 1:51–72.

Hurt, R. Douglas. 1988. Ohio Agriculture Since World War II. *Ohio History* 97 (winter-spring): 50–71.

Jones, Robert Leslie. 1983. *History of Agriculture in Ohio to 1880.* Kent, Ohio: Kent State University Press.

Kaatz, Martin R. 1955. The Black Swamp: A Study in Historical Geography. *Annals of the Association of American Geographers* 45 (March): 1–35.

Keck, Gail C. 1991. Ohio Farmers Speak Out at Farm Rally. *Ohio Farmer* 287, no. 14 (November): 32–33.

Lloyd, W. A., J. I. Falconer, and C. E. Thorne. 1918. The Agriculture of Ohio. *Bulletin of the Ohio Agricultural Station,* no. 326.

Noble, Allen G., and Albert J. Korsok. 1975. *Ohio: An American Heartland.* Bulletin 65. Columbus, Ohio: Division of Geologic Survey.

Ohio Data User Center. 1989. *The Final Market Value of Ohio Goods and Services: Gross State Product by Division: 1971–1986.* Columbus, Ohio: Department of Development.

Ohio Department of Agriculture. 1990. *Ohio Agricultural Statistics and the Ohio Department of Agriculture Annual Report.* Columbus, Ohio: Agricultural Statistic Service.

Peattie, Roderick. 1923. *Geography of Ohio.* Fourth Series, bulletin 27. Columbus, Ohio: Geological Survey.

Ray, John B. 1970. Trade Patterns along Zane's Trace, 1797–1812. *Professional Geographer* 22, no. 3 (May): 142–46.

Rumney, Thomas. 1988 Soybean Production: A Geographic Inquiry. *Ohio Geographer* 16:57–67.

Shkurti, William J., and John Bartle. 1989. *Benchmark Ohio 1989.* Columbus: Ohio State School of Policy and Management and Ohio State University Press.

Thompson, John. 1989. *The Introduction and Adoption of Steam Excavators in Land Drainage in the Midwest.* Paper Presented at 21st Annual Meeting, Pioneer America Society, 9 November, at St. Charles, Missouri. Mimeographed.

U.S. Bureau of the Census. 1956. *Census of Agriculture 1954.* Vol. 1, part 35. U.S. Dept. of Commerce. Washington, D.C.

———. 1967. *Census of Agriculture 1964.* Vol. 1, part 35. U.S. Dept. of Commerce. Washington, D.C.

———. 1977. *Census of Agriculture 1974.* Vol. 1, part 35. U.S. Dept. of Commerce. Washington, D.C.

———. 1983. *Census of the Population 1980.* Vol. 1, part 37. U.S. Dept. of Commerce. Washington, D.C.

———. 1984. *Census of Agriculture 1982.* Vol. 1, part 35. U.S. Dept. of Commerce. Washington, D.C.

———. 1989. *Census of Agriculture 1987.* Vol. 1, part 35. U.S. Dept. of Commerce. Washington, D.C.

Vonda, Damaine, ed. 1992. *The Ohio Almanac.* Wilmington, Ohio: Orange Frazer Press.

Wright, Alfred J. 1957. *Economic Geography of Ohio.* Bulletin 50, 2d ed. Columbus, Ohio: Division of the Geological Survey.

CHAPTER THIRTEEN

Manufacturing

Thomas A. Maraffa

Located in the heart of the American manufacturing belt, Ohio has long been thought of as an industrial state and, indeed, ranks among the leading states in various measures of manufacturing—third in employment, fifth in number of factories, and fourth in value added by manufacture. For decades manufacturing has been the state's largest employer; it currently employs nearly one out of every four workers.

Ohio has weathered the recent transition to a postindustrial economy, along with other states in the manufacturing belt. Manufacturing employment in the state peaked in 1967, but by 1987 nearly 300,000 manufacturing jobs had been lost. Ohio's position as a manufacturing state also eroded. In 1967 the state accounted for 7.8 percent of total U.S. value-added and 7.2 percent of manufacturing employment. By 1987 these figures had declined to 6 percent and 5.8 percent, respectively. Nearly every large Ohio city lost manufacturing jobs, some experiencing severe economic hardship when major employers laid off thousands of workers. The present geography of Ohio was largely shaped by the location of key industries. Manufacturing is a city-building activity, acting as a magnet for labor and providing the foundation for growth in the retail and service economies. Patterns of industrial location are influenced by access to both raw materials and markets; transportation costs; and production costs, including labor, energy, taxes, and land. Recently government incentives have become influential location factors. Many industries, however, simply took root and grew in the community where their founder resided.

Character of Ohio Manufacturing

Industries can be divided into twenty general sectors, based on type of product (see Table 13.1). Eighteen of those twenty categories are found in Ohio, in varying degrees of concentration. Two ways of evaluating the relative importance of different industrial sectors are, first, each sector's share of Ohio's manufacturing employment and, second, the shares of U.S. manufacturing employment found in Ohio.

Based on the share of Ohio's manufacturing labor force employed in each sector, Ohio's manufacturing is specialized. Over 38 percent of all manufacturing jobs are accounted for in the three largest sectors: nonelectrical machinery, transportation equipment, and fabricated metals. Sixty percent of manufacturing jobs are concentrated in the top six sectors, which include, additionally, electrical and electronic equipment, primary metals, and rubber and miscellaneous plastic products.

If the share of manufacturing jobs in an industrial sector located in Ohio is greater than Ohio's share of total U.S. manufacturing jobs, then that sector can be considered concentrated in Ohio. Ohio contains 5.8 percent of U.S. manufacturing employment. Six industrial groups have shares of employment in Ohio exceeding 5.8 percent: they are rubber and miscellaneous plastics, primary metals, stone, clay and glass products, nonelectrical machinery, transportation equipment, and fabricated metal products. These patterns of concentration are indicative of Ohio's industrial maturity (Noble and Korsok 1975). Most of these sectors produce or contribute to the production of durable goods, manufactured for the industrial rather than consumer market. Ohio's most important industries, except for primary metals, do not involve the processing of raw materials.

The distribution of manufacturing in Ohio reveals distinct patterns of concentration (Figure 13.1) that mirror urban distribution. Each county with more than

Table 13.1
Profile of Ohio Manufacturing, 1987

	SECTOR	EMPLOYMENT (000)	% OF OHIO TOTAL	% OF U.S. IN OHIO
20	Food and kindred products	54.5	4.9	3.8
21	Tobacco manufacturers	.1		.2
22	Textile mill products	4.9	.4	.7
23	Apparel and other textile products	13.2	1.1	1.2
24	Lumber and wood products	17.7	1.6	2.5
25	Furniture and fixtures	15.2	1.4	3.0
26	Paper and allied products	30.0	2.7	4.9
27	Printing and publishing	66.4	6.0	4.4
28	Chemicals and allied products	43.0	3.9	5.3
29	Petroleum and coal products	5.4	.5	4.6
30	Rubber and miscellaneous plastic products	86.2	7.8	10.4
31	Leather and leather products	1.5	.1	1.1
32	Stone, clay, and glass products	39.6	3.6	7.6
33	Primary metal industries	82.7	7.5	11.8
34	Fabricated metal products	140.2	12.7	9.6
35	Machinery, except electrical	140.5	12.7	7.6
36	Electrical and electronic equipment	77.1	7.0	4.9
37	Transportation equipment	145.3	13.2	8.0
38	Instruments and related products	26.9	2.4	2.7
39	Miscellaneous manufacturing	13.5	1.2	3.6

Source: U.S. Census of Manufactures, 1987

25,000 manufacturing jobs is a part of one of Ohio's metropolitan areas. Cuyahoga County alone contains nearly 164,000 manufacturing jobs, nearly 15 percent of the state's total. It is logical that manufacturing and urbanization should be geographically associated, because of the role manufacturing plays as a city builder and also because large cities are attractive locations for many manufacturers.

Industrial regions are evident from Figure 13.1. Northeastern Ohio developed as the dominant industrial region in the state; nearly 41 percent of the state's manufacturing jobs are concentrated in northeastern Ohio counties. Northwest and central Ohio are secondary concentrations. Although the Ohio River drew manufacturing first to southeastern Ohio, rugged topography, lack of large cities, and relative inaccessibility have made it the most industry-poor region of the state.

Descriptions of the geography of Ohio manufacturing from the early twentieth century to the late 1960s show that the state's industrial areas remained in place, attesting to the stability of once-established patterns (Wright 1953). For example, Sten DeGeer's 1920 description of the American manufacturing belt identified six Ohio industrial areas associated with major waterways, most notably Lake Erie, the Miami and Erie Canal, the Ohio and Erie Canal, and the Ohio Valley. The railroad and later the highway realigned the orientation of these regions, but the basic pattern has persisted.

Spatial Evolution of Ohio Manufacturing

Ohio manufacturing has evolved through three time periods, each with distinctive geographic patterns. The first period extends from the pioneer era to the Civil War and can be described as a quest for self-sufficiency. In the second period, which stretched from the second half of the nineteenth century to approximately 1970, Ohio manufacturing was influenced by the Industrial Revolution and the subsequent maturation of the American economy. Ohio manufacturing became integrated into the national and global

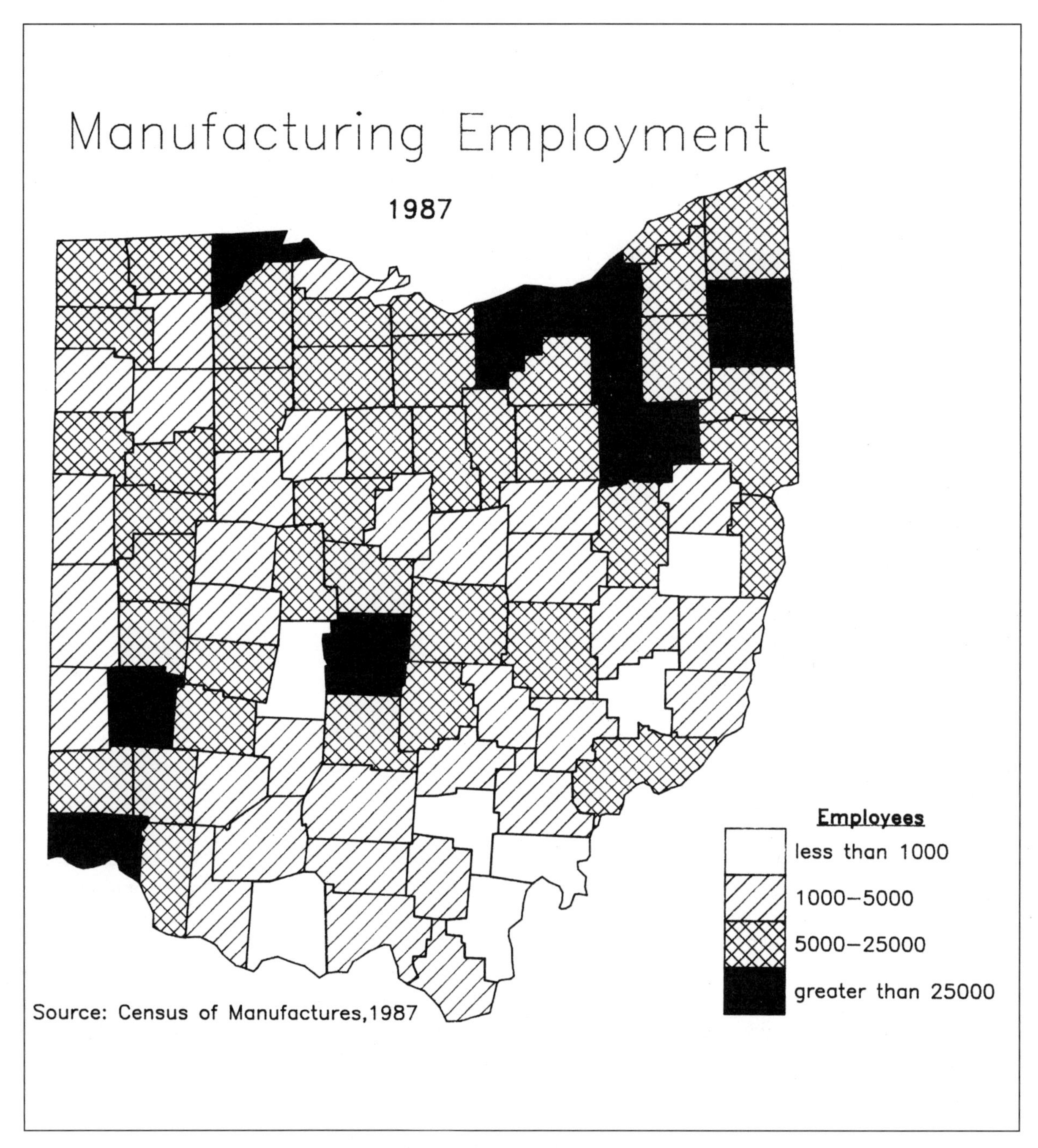

Fig. 13.1. Manufacturing Employment, 1987

economic systems. The third and most recent period is one of response to the postindustrial economy.

The Quest for Self-Sufficiency

Manufacturing came early to Ohio's economy, even though settlers were first attracted by Ohio's potential as an agricultural area. The historic record shows that a brick factory, a furniture factory, and several flour mills were in operation by 1800 in settlements along the Ohio River. By 1820 Ohio had 578 factories producing goods worth 3.1 million dollars (Mabry 1946). This total understates the amount of manufacturing that actually occurred in early Ohio. During the late eighteenth and early nineteenth centuries the lines between economic sectors were more blurred than they are today. Every farm family manufactured products—such as soap, cloth and clothing, tools, and processed foodstuffs—for their own use. Ohio's location on the frontier initially hindered the growth of manufacturing, for the state was too inaccessible for its industries to compete with Eastern manufacturers. As Ohio's population and cities grew and its farmers produced an ever-increasing agricultural surplus, this isolation from the East became a locational advantage. Demand exceeded supply for a broad spectrum of industrial goods, and delivered prices of available goods from the East were too expensive. Local entrepreneurs were quick to recognize this opportunity, and manufacturing thus spread throughout the state. Factories gravitated toward water locations, first along the Ohio River and its tributaries, later near the canals and Lake Erie.

During the first half of the nineteenth century Ohio's manufacturing served the needs of farmers, travelers, and residents of the early towns and cities. Sawmills, flour mills, and iron furnaces dotted the landscape, supplying the needs of local markets. Because of the importance of water transportation during this period, cities such as Cincinnati, Marietta, and Steubenville became early shipbuilding centers. Salt was an important manufactured product because supplies from the East were both too expensive and insufficiently available. In 1833, Morgan County was the location of thirty salt furnaces (Mabry 1946). By 1987 there were only fourteen manufacturing plants of any type remaining in Morgan County. As Ohio farmers became more productive, processing the agricultural surplus became an important manufacturing activity. Here Ohio manufacturers were able to penetrate Eastern markets. Cincinnati's nickname "Porkopolis" came from its importance as a meatpacking center. Hams from Ohio were considered a delicacy in the East.

Manufacturing played an economic role similar to that of central place activities during this initial period of industrial growth. As a rule, manufacturing served the local market once the demand for products was sufficient. The distribution of manufacturing was similar to the distribution of people and settlements; both were oriented to the major waterways, which for most of this period were the prime avenues of travel for people and products.

This is not to say that regional concentrations did not exist early in Ohio's industrial distribution. The Hanging Rock region—which includes the Ohio counties of Jackson, Gallia, Vinton, Scioto, and Lawrence, as well as several Kentucky counties—was the major iron-producing region in that era. Over 37 tons of iron were produced in this region in 1846, and the region was the source of several innovations in ironmaking (Wright 1953). Ironmaking concentrated in this area because of the quality of local iron ore and the availability of wood for charcoal and limestone. The ironmakers sold their iron for manufacture into a variety of tools and implements. By the twentieth century this region was supplanted by northeastern Ohio's iron and steel producers.

Seeds of future regional industrial prominence were also planted during this period. The pottery industry that began in East Liverpool in 1838 spread throughout eastern Ohio, which became a major national supplier of clay products. Early manufacturers emphasized tableware, but the industry diversified into construction-related products such as tiles and bricks. The industry was originally attracted to eastern Ohio by deposits of clay but remained after clays were imported; its locational pattern still shows a strong connection to the distribution of clay (Noble and Korsok 1975). Tuscarawas, Carroll, Columbiana, Jefferson, and Muskingum Counties as well as cities such as East Liverpool, East Palestine, Wellsville, Zanesville, and Salem are all Ohio centers of production. In recent decades the industry has declined in Ohio. Whereas in 1954 there were nearly 250 plants employing over 24,000 people in pottery and structural clay products industries, by 1987 the number of factories had declined to 142, and employment had fallen to 8,000. Particularly detrimental to the industry were the development of nonceramic substitutes and foreign competition.

The Industrial Revolution: Integration and Concentration

After the Civil War, the Industrial Revolution changed both the role of manufacturing in Ohio's economy and locational patterns of manufacturing in the state. Manufacturing became a basic activity. Products of Ohio manufacturers were increasingly sold in regional, national, and global markets, bringing money into the state and stimulating both the general economy and the growth of individual cities. Ohio became a prominent national center for production of iron and steel, machinery, clay products, glass, and tires. Industrial regions evolved within Ohio as areas capitalized on particular locational advantages.

Accessibility or an advantageous relative location was of critical importance to Ohio's emergence as an industrial state during the second half of the nineteenth century (Knepper 1989). Ohio's relative location made it a bridge between the major cities and markets of the East Coast and the rapidly growing West. Ohio manufacturers were well located to provide products to the Western markets at lower cost than their Eastern competitors. In fact several major Ohio industries began when Eastern business owners recognized the advantages of an Ohio location and relocated. B. F. Goodrich, originally located in New York, came to Akron in 1870; Libbey Glass moved from Massachusetts to Toledo in 1888.

As the transportation system evolved, Ohio's locational advantages were reinforced. The railroad further integrated Ohio with the national market and redirected the shipment of commodities from waterways. Major industrial centers such as Cleveland, Cincinnati, and Toledo developed partially because of their location as transhipment points between rail and water. Transportation itself provided a growing market for Ohio-manufactured products (Knepper 1989). The railroad boom stimulated Ohio's iron and steel industry beyond simply the need for the manufacture of railroad equipment and the rails themselves. During the twentieth century, highway construction created a market for cement, brick, and paving block—and of course the automobile and components industries.

It was during this second period that the regional patterns of manufacturing described earlier began to take shape. Many individual industries developed regional concentrations of their own, responding sometimes to geographic endowments and sometimes to the whims of entrepreneurship and chance. The development of the clay products industries in eastern Ohio has already been described in terms of the location of the raw material, clay. There are other examples of industry orientation to raw materials and energy. Sand and natural gas attracted the glass industry to the Toledo area; coal-generated electrical power drew aluminum manufacturing to the Ohio Valley (Wright 1953); the forests of southeast Ohio attracted wood products and furniture; the farms of northwest Ohio attracted food processing; the discovery of oil in northwest Ohio led to petroleum refining in Toledo, Findlay, and Cleveland; and the iron and steel industry grew in northeast Ohio because of the availability of coal, iron ore, water, and limestone in the same general area. For many of these industries, the local supplies of raw materials that gave them their start were eventually depleted; however, they continued to grow in the region where they began. This locational persistence can be explained by access to the transportation system, favorable location with respect to markets, and inertia.

The market and entrepreneurship help explain other regional concentrations in Ohio. The Miami Valley is a major center for the production of machinery and paper products, and Akron grew to world leadership in the production of tires and rubber products (Wright 1953). Both areas owe their existence to the ability of local companies to capitalize on their central location in the manufacturing belt. Their major markets tend to be other manufacturers; thus they contribute to the complex patterns of linkage between firms and industries.

The Postindustrial Era

The pattern of Ohio manufacturing established during the second period was disrupted by the events of the 1970s and 1980s. The postindustrial era, which began in the late 1960s, has been characterized by two shifts that have changed the role of manufacturing in the United States economy. First is the decreased importance of manufacturing employment; job growth in manufacturing has been stagnant, while thousands of new jobs have been created in nonmanufacturing sectors, such as services. Second is the type of manufacturing; new industries such as computers and semiconductors are growing, while job loss is the rule for traditional industries such as steel, machinery, and automobiles. These new industries have developed largely outside the traditional manufacturing belt and states such as Ohio.

As one of the early industrial states, Ohio was hit particularly hard by these changes. The numerical facts behind Ohio's loss of industrial standing were described at the beginning of the chapter. Ohio's major industries—steel, automobiles, machine tools, and rubber—lost competitive standing, not only from foreign manufacturers in Europe and Japan, but from domestic competitors in the South and the West. No single reason is sufficient to explain both the absolute loss of jobs and the erosion of industrial position (Knepper 1989). Ohio was perceived as a high cost state, particularly in terms of labor costs. Its traditional locational advantages in transportation and raw materials have lost significance because these factors are relatively unimportant to new industries. Much of Ohio's plant capacity was 50–60 years old. Not modernized during the prosperous years, these plants were therefore less efficient than newer facilities constructed elsewhere. Meanwhile, markets in Southern and Western states expanded as the population migrated to these regions. Thus the cycle of self-sustaining industrial and population growth that typified Ohio during the preceding period relocated to the South and West. Ohio became locked in a cycle of job loss and out-migration.

One outcome of this process was that areas of Ohio long associated with specific manufactured products ceased to be significant centers of what originally made them famous. As one observer wrote, "By 1980, the Mahoning Valley made little steel, Cleveland made few metal fasteners, Akron built just a few specialty tires, Dayton manufactured few business machines" (Knepper 1989).

The distribution of manufacturing employment within the state changed as a result. Figure 13.2 shows the change in manufacturing employment in Ohio between 1967–87. The similarity between this map and the map of 1987 manufacturing employment is striking. Ohio counties with the greatest industrial concentration lost the most manufacturing jobs. Over 286,000 manufacturing jobs were lost in Cuyahoga, Franklin, Hamilton, Lucas, Mahoning, Montgomery, Stark, and Summit Counties—the central counties of the state's largest metropolitan areas. The northeastern region experienced particularly heavy job loss; it is in this area that the state's declining sectors were most concentrated.

Numerous counties gained manufacturing jobs during this period, although the combined gains in these counties did not come close to offsetting the losses in the counties listed above. Thirteen counties—all either suburban or rural—gained more than 1,000 manufacturing jobs from 1967–82. The two largest gains were reported in Lake County, a suburb of Cleveland, and Clermont County, a suburb of Cincinnati. Suburban counties as a group nearly doubled their share of manufacturing jobs, from 6.6 percent in 1967 to 12.5 percent in 1987.

Several reasons explain why suburban counties were attractive locations for industrial expansion. First, it is difficult if not impossible to acquire the large tracts of land necessary for modern industrial layouts in large cities. Second, the cost of land is significantly lower in outlying counties. Taxes also tend to be lower, and labor is usually less expensive and less likely to clamor for unionization. Finally, companies and their employees are attracted to the perceived lifestyle of these counties (Knepper 1989).

Industry Profiles

Soft Drinks

Market-oriented industries are attracted to locations at or near the point of consumption. Several conditions lead to market orientation. For example, if the finished product is bulky or highly perishable, firms locate at the market to minimize transportation costs. Market orientation also results when frequent communication with customers is required, such as in the printing industry or specialized machine shops. Finally, industries in which the raw materials are readily available and in which these materials do not lose weight in the production process locate near their markets. Soft drinks are an example of this last case.

The geographic distributions of market-oriented industries are similar to those of population, urbanization, and market potential. The distribution of soft drink bottlers in Ohio is typical of this market-oriented pattern (Figure 13.3). Soft drink plants are found in metropolitan areas and certain nonmetropolitan counties in a pattern that generally conforms to Ohio's population trends.

During the past thirty years the soft drink industry has responded to an increasingly competitive environment by consolidating production at fewer plants. In 1963 there were 152 soft drink plants operating in over one-half of Ohio's counties and employing a total of over 5,500 people. The average plant had thirty-six employees. By 1987 there were only 40 soft drink plants operating in 24 counties; however, employment in the industry increased to 5,720 people, with the average plant employing over 132 workers.

Consolidation resulted in fewer plants in the more urban counties and the disappearance of production in rural counties. For example, from 1969 to 1986, the number of soft drink plants in Cuyahoga County

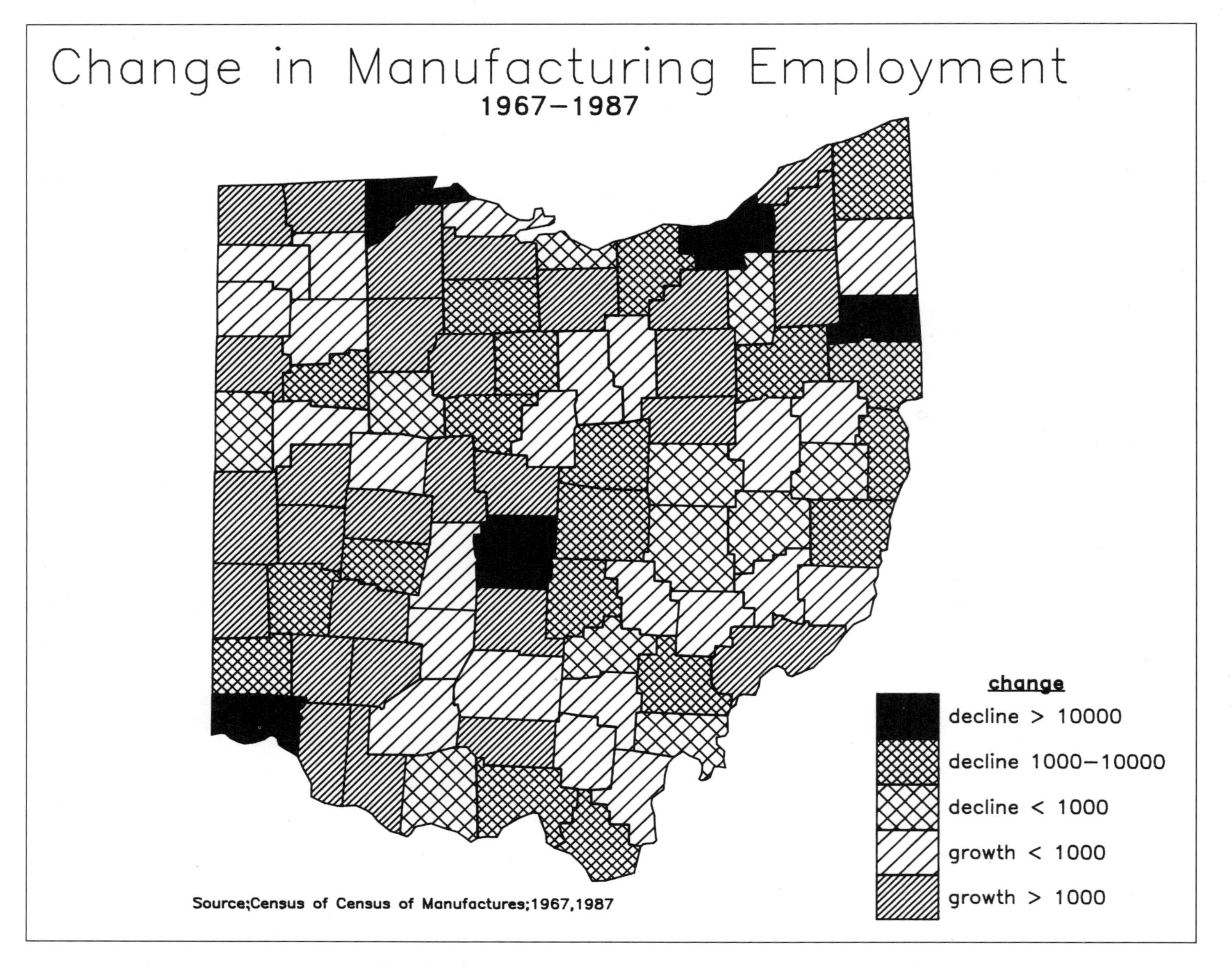

Fig. 13.2. Change in Manufacturing Employment, 1967–1987

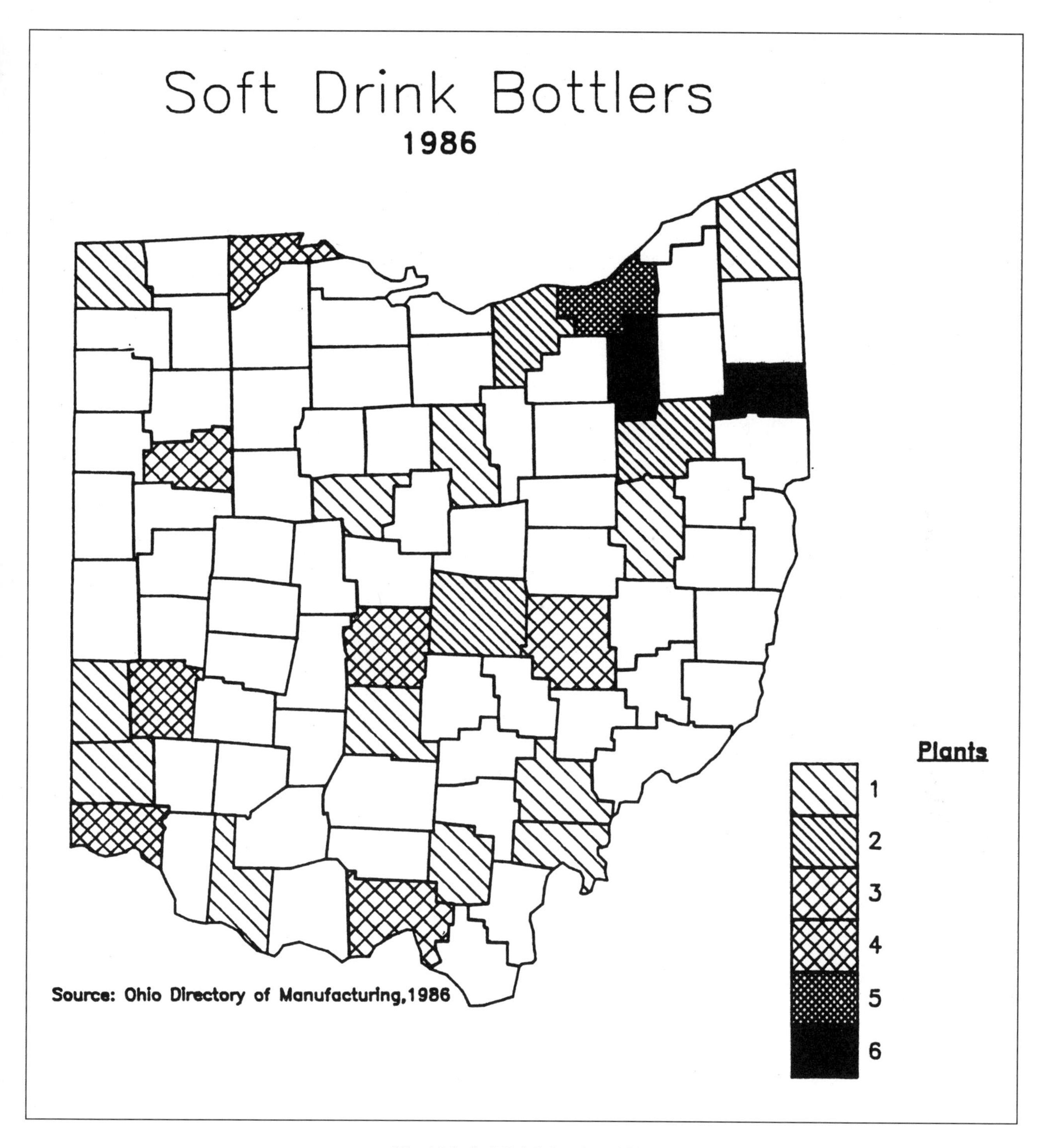

Fig. 13.3. Soft Drink Bottlers, 1986

decreased from eighteen to five, and in Hamilton and Montgomery County from nine to four, each. Rural counties such as Columbiana, Henry, Sandusky, Seneca, and Guernsey lost their soft drink bottling industry completely.

The competitive environment and the transportation system together have encouraged the recent pattern of consolidation. The market for soft drinks has stabilized, with young adults and teenagers as the largest consuming group. Because the American population is aging and its growth has slowed, soft drink bottlers are faced with growing competition for a nonexpanding market. The soft drink bottler is a franchise of the brand it manufactures, purchasing syrup from the company and distributing the finished product over an exclusive territory. Profitability for bottlers and their parent companies is increasingly tied to lower costs, and one way to lower costs has been to take advantage of the economies of scale possible at larger plants. The construction of the interstate highway system has enabled plants to distribute over larger territories.

Automobiles

Ohio is a national leader in the automobile industry. More cars are manufactured in Ohio than any other state except Michigan. The automobile industry is really two industries: the assembly of cars and the manufacture of the thousands of component parts. A strong case can be made that the automobile industry is Ohio's single most important industrial sector (*Cleveland Plain Dealer* 1989). Statewide, it employs over one out of ten manufacturing workers, more than any other sector. Automobile plants tend to dominate the communities in which they are located, because they are large employers. The GM assembly plant at Lordstown employs over 10,000 workers. The Lorain Ford plant employs 5,600; has a payroll of $250 million; purchases $63 million worth of local supplies and $9.1 million from area electric, natural gas, and water utilities; and pays $3.4 million in property taxes.

Parts suppliers employ fewer people but are the major employers in the smaller towns where they are increasingly located. For example, Bryan Custom Plastics, which manufactures molded plastics for automobiles, employs nearly 500 in a town of 8,000. The Goodyear tire plant in Logan, Ohio (pop. 6,900), employs 900.

The geographic pattern of the automobile industry in Ohio reflects the locational eras of the industry nationwide (Boas 1961). Until 1915, the automobile was in reality an experimental vehicle; production occurred in literally hundreds of small factories—which often produced fewer than twenty cars annually—located mainly in the Northeast and Midwest. Cleveland hosted a major concentration of these factories, which numbered at various times between twenty and twenty-five. Columbus, Dayton, Cincinnati, and Toledo were secondary clusters. In 1905 Ohio produced 2,521 cars; during that same year 2,800 cars were manufactured in Michigan.

After 1915 the industry consolidated into three major companies, and the number of assembly locations dramatically fell as plants expanded. The industry coalesced first on Detroit and later on locations outside the Midwest. By 1965 Ohio was the location of only two major automobile assembly plants: the Ford plant at Lorain and a GM plant at Norwood, a suburb of Cincinnati (Rubenstein 1986). Throughout 1965–86, Ohio benefited from the reinvestment of the industry in the Midwest. GM opened the Lordstown plant in 1966, and in the mid 1980s Honda opened its only U.S. assembly plant in Marysville, northwest of Columbus in Union County.

Since the automobile industry is dominated by large corporations, the jobs of Ohio autoworkers are subject to decisions made in corporate headquarters outside the state. The consequences of this arrangement were dramatically illustrated by GM's decision to close the Norwood assembly plant in 1986 and a subsequent decision to move van production out of the Lordstown facility in the early 1990s (Rubenstein 1987). Currently GM employs 63,000 Ohioans in its assembly and parts factories, Ford 30,000, Chrysler 13,000, and Honda 8,200 (*Cleveland Plain Dealer* 1989).

The auto parts industry traditionally has concentrated in the urban counties of northern Ohio along the Great Lakes (Rubenstein 1988). This concentration is due to the region's proximity to both the assembly plants in Michigan and to raw materials, primarily steel from Cleveland and the Youngstown area. During the past 20 years, this industry has shifted its orientation to rural counties and to central Ohio. One factor responsible for this shift is the growing demand for domestic components at the Marysville Honda plant and the need to be close to the plant because of the just-in-time inventory system favored by the Japanese. In addition, the substitution of plastics for steel components has freed firms from the need to be close to steel centers. Finally, firms have recently favored small-town locations because the company will likely be a major employer, and the rural labor force is perceived as more willing to resist unionization and hence more "cooperative."

The automobile industry also illustrates the degree of interdependence that is typical of an advanced industrial system such as Ohio's. To assemble a Ford Thunderbird in Lorain, Ohio, requires engines, instrument clusters, and sound systems from Canada; aluminum wheels from Italy; glass from Oklahoma; carpeting from Pennsylvania; and stampings, lamps, and instrument panels from Michigan and other locations in Ohio (*Cleveland Plain Dealer* 1989). The linkages that have developed between Honda and Ohio parts suppliers demonstrate the benefits of interdependence to Ohio industry. Interdependence, however, cuts both ways. Those Ohio parts manufacturers that are subsidiaries of GM and Ford were once guaranteed a market. But now these large auto companies, in order to reduce costs, are granting contracts competitively. Ohio companies are increasingly competing with foreign manufacturers in Mexico, Brazil, and elsewhere.

Steel Industry

The iron and steel industry is in many ways the classic industry of the manufacturing belt—and as such is an integral part of Ohio's industrial geography. Ohio has long been a leading state in both the production and consumption of steel products. Unfortunately, the decline of both the manufacturing belt and Ohio's position as an industrial state is synonymous with the decline of the steel industry. As recently as 1967, the iron and steel industry employed over 127,000 Ohioans. Fifteen years later that figure stood at 75,000. More recent estimates place employment even lower—below 50,000. Hartshorn and Alexander, in their textbook, *Economic Geography,* illustrate a chapter on the steel industry with a dramatic photograph showing the demolition of the abandoned blast furnaces at the U.S. Steel works in Youngstown (230). This photograph serves as a metaphor for much of Ohio's iron and steel industry.

Ohio's steel industry, which grew rapidly after the Civil War, developed separately from the pioneer iron forges discussed earlier (Wright 1953). Proximity to raw materials and accessibility to markets were major influences on the location of Ohio steelmaking. The manufacture of steel involves four raw materials: iron ore, coal, limestone, and water. Local sources of iron ore and coal initially supplied steel mills at the Mahoning Valley. When these sources became inadequate, coal was transported from Pennsylvania and Appalachia, iron ore from Michigan, Minnesota, and Canada. Steelmaking also concentrated in Cleveland, where iron ore and coal could easily be transported. Local supplies of limestone were more than adequate. The need for massive quantities of water drew steel mills to river valley locations within these regions. Cleveland and the Mahoning Valley were at one time two of the most important steel-producing regions in the world.

The market reinforced the position of Ohio steelmakers. Their major customers were the automobile, railroad, construction, and machinery industries, all of which were either geographically centered near Ohio or accessible via rail or water transportation. The relocation of the Armco steel mill to Middletown in the early twentieth century was partly a response to the market provided by the agricultural equipment industry. Today it is one of the largest employers in Middletown.

There are fewer large steel mills in Ohio today than at any time since the beginning of the industry. However, the contemporary distribution of large steelmaking facilities still reflects past patterns of development (Figure 13.4). Major concentrations of steelmaking remain in Cleveland, the Mahoning Valley, and north-central Ohio. Steelmaking is also present in larger urban counties such as Franklin, those counties along the Ohio River, and in Butler County, where Armco's Middletown works are located.

Many reasons have been offered for the decline of the steel industry in Ohio (Buss and Redburn 1983): the demand for steel products has fallen in the major consuming sectors due to the use of substitutes or due to slow growth; automobiles use more plastic and less steel; the bust in the domestic oil industry dried up demand for steel pipes, tubes, and machinery. In addition, steelmakers in Ohio, as in other parts of the manufacturing belt, were unable to compete with the foreign and domestic producers that emerged after World War II. This loss of competitiveness was attributed to high labor cost and antiquated production facilities. The newest blast furnace in the Mahoning Valley was constructed in 1921. In fact the Mahoning Valley was seen to be at a competitive disadvantage as far back as the 1930s but persisted as a major steel region by virtue of artificial transportation rate structures, the absence of competition, and inertia (Rodgers 1952).

Another contributing factor was the gradual takeover of local steel companies by out-of-state corporations. These companies, unwilling to invest in the older facilities, used their profits to finance other cor-

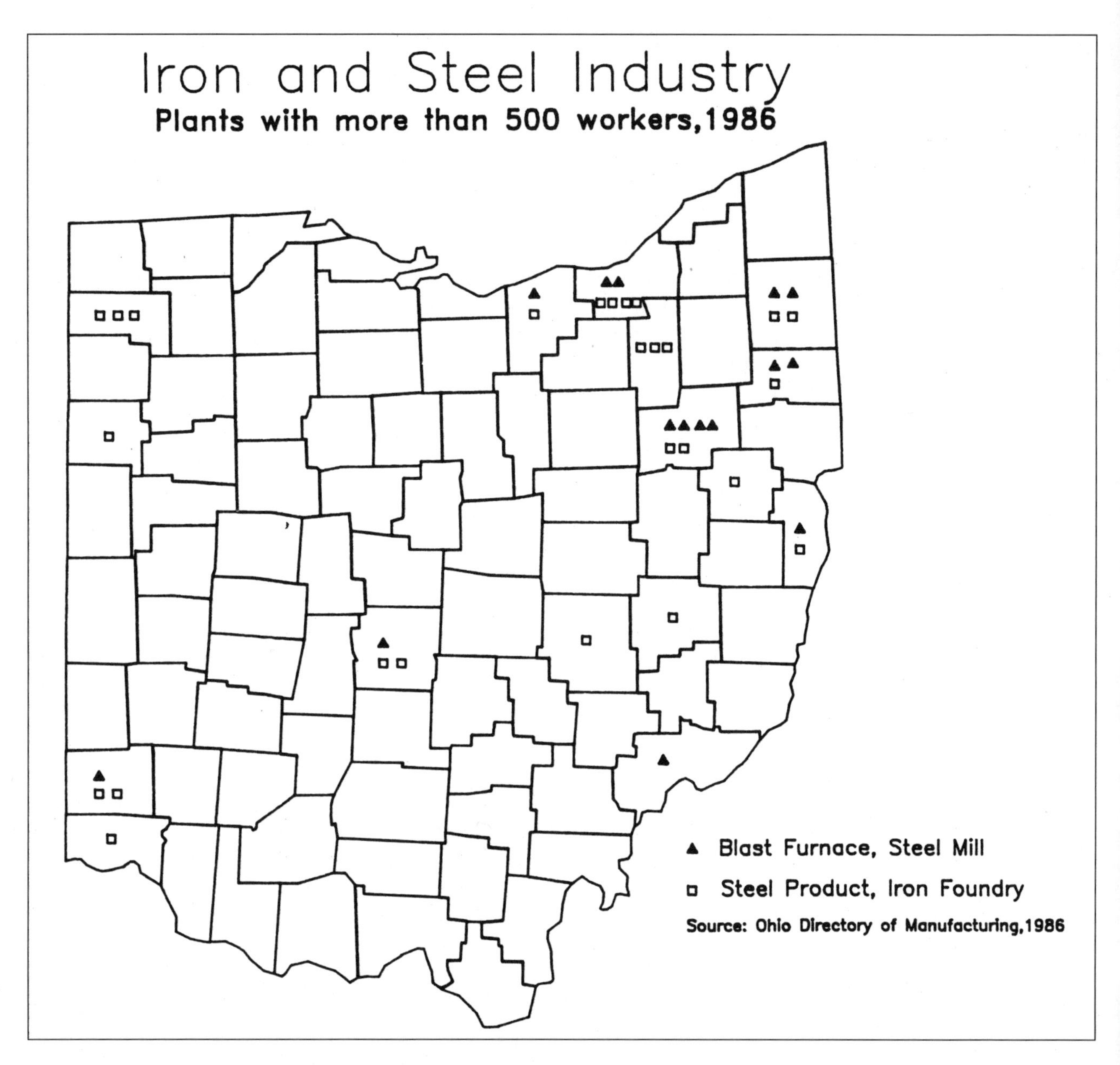

Fig. 13.4. Iron and Steel Industry (Plants with More than 500 Workers, 1986)

porate ventures, closing the mills when they became unprofitable. Lists of Ohio steel mill closures are dominated by major conglomerates, such as LTV and U.S. Steel.

It goes without saying that many Ohio communities were devastated by the collapse of steel. However, the Mahoning Valley was hit particularly hard, for its economy was even more dependent on this single industry than that of Cleveland, another hard-hit community. From 1973–83, the Youngstown-Warren area lost about 40,000 manufacturing jobs, as several major steel mills closed and the effects rippled through related manufacturing sectors.

Although the steel industry in Ohio will never regain its lost dominance, there are signs of revitalization and optimism. A recent report found that both employment and steel production have risen in Ohio since the early 1980s. With productivity substantially increased, steel executives are optimistic about the future and plan to continue investments in modernization for the rest of the decade (Peterson 1990). For example, Armco in Middletown and Timken in Canton have reinvested heavily in modern facilities, and Worthington Steel in Franklin County has won national acclaim as a model of a modern, efficient company able to compete globally. One trend to watch in the coming years is the shift toward smaller, specialty steel mills. These "minimills" concentrate on a few product lines and often use scrap as a raw material. One such minimill was recently opened within a portion of a formerly closed U.S. Steel facility in Youngstown.

Regional Patterns

Previous sections have alluded to regional patterns of manufacturing in Ohio. Efforts to develop a viable comprehensive regionalization of manufacturing are complicated by the fact that industrial location is based on factors of production such as markets, technology, transportation, and labor that are themselves not located in a continuous regional pattern. As a result, many Ohio counties contain a diverse mix of industrial groups instead of a single industry. For example, twenty-seven of Ohio's eighty-eight counties contain plants representing more than ten of the twenty two-digit Standard Industrial Classification (SIC) categories, while only nineteen counties had fewer than four SIC categories.

The following regionalization of Ohio industry should be interpreted in view of this diversity. It is based on two considerations. First, rather than focusing on single two-digit industrial groups, regional production complexes were identified that represent related or linked manufacturing processes. Second, the regionalization emphasizes those industrial sectors that are likely to produce for external rather than local markets. These basic industries form the foundation of the local economy, supporting other manufacturers as well as the retail and service sectors. Since large manufacturing plants are more likely to fulfill this role, the regionalization is based on plants that employ more than 100. Six manufacturing regions are identified and depicted in Figure 13.5.

Counties containing large metropolitan centers are considered separately. These counties include Cuyahoga (Cleveland), Franklin (Columbus), Hamilton (Cincinnati), Summit (Akron), Stark (Canton), and Montgomery (Dayton)—all counties with more than 100 manufacturing plants that employ at least 100 workers. Furthermore, nearly all two-digit SIC categories are represented in each county's industrial mix. The industrial structure in these counties is much more diverse than in Ohio's other metropolitan counties—Lucas and Mahoning, for example—and it is distinct from the counties that surround them.

A primary materials region, roughly corresponding to Appalachian Ohio, occupies much of southeast Ohio. Manufacturing in this region is closely linked to natural resources and includes such sectors as lumber and wood products, primary metals, chemicals, and stone clay and glass products. Counties in this region have relatively few large manufacturers. Only Tuscarawas County has more than 20 plants with over 100 employees, while Harrison and Meigs Counties have no plants of that size. Northeastern Ohio is dominated by industries that together comprise an equipment complex. Manufacturing in this region focuses on the linked sectors of fabricated metals and industrial, electronic, and transportation equipment. The large cities of Cleveland, Akron, Canton, Lorain, and Youngstown organize manufacturing in the surrounding counties. Automobiles and steel are major sectors.

Northwestern Ohio is a diversified industrial region, with Toledo as the major industrial center and small-town branch plants dominating the industrial landscape. This largely agricultural region contains industries ranging from food processing to equipment to stone, clay, and glass products.

West-central Ohio has evolved into a transportation equipment complex. Although equipment manufacturing has long been present in these counties, the Honda Marysville assembly plant has changed the industrial structure of this region by stimulating the location of automobile parts plants. Counties located along the I-75 corridor have experienced the greatest recent influx of automobile component manufacturing (Rubenstein 1988).

Southwestern Ohio is characterized by the prominence of metals and machinery manufacturing. Primary metals and industrial equipment are major sectors in this region. This old industrial area is most similar to northeastern Ohio.

Conclusion

Ohio's fortunes as an industrial state are similar to those of other manufacturing belt states. Until the 1960s manufacturing reinforced the growth of Ohio's large urban centers, and Ohio developed as a national

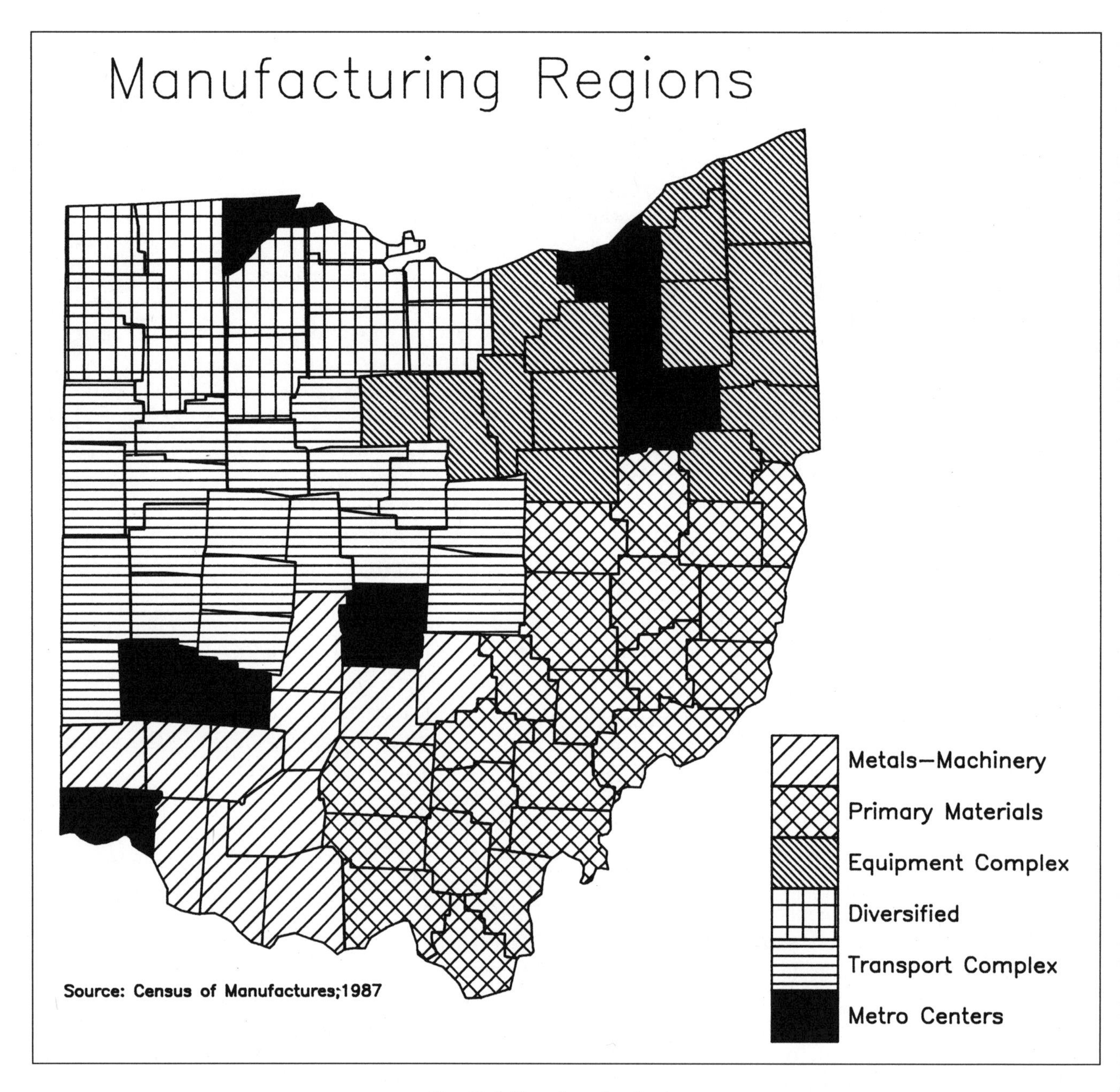

Fig. 13.5. Manufacturing Regions

and global leader in the manufacture of a variety of products. Regions within the state specialized to make the most of their comparative locational advantages. The postindustrial era abruptly ended decades of continuous manufacturing growth and was marked by the decline of key Ohio industries such as steel. The major geographic trend was deconcentration of manufacturing to suburban and rural locations, a trend that will likely continue for the foreseeable future. Despite recent instability in the state's industrial economy, the industrial patterns established earlier are still evident and will continue to persist. Future patterns of industrial location will be shaped by how well Ohio manufacturers are able to compete nationally and globally without their once-clear locational advantages.

REFERENCES

Boas, Charles W. 1961. Locational Patterns of American Automobile Assembly Plants, 1895–1958. *Economic Geography* 37:218–30.

Buss, Terry, and F. Steven Redburn. 1983. *Shutdown at Youngstown.* Albany: State University of New York Press.

Hartshorn, Truman A., and John W. Alexander. 1988. *Economic Geography.* Englewood Cliffs, N.J.: Prentice Hall.

Jensen, Christopher. 1989. Autos Drive Ohio's Economy *Cleveland Plain Dealer.* (3 October).

Knepper, George. 1989. *Ohio and Its People.* Kent, Ohio: Kent State University Press.

Mabry, William. 1946. Industrial Beginnings in Ohio. *Ohio State Archeological and Historical Quarterly* 55:242–53.

Noble, Allan G., and Albert J. Korsok. 1975. *Ohio: An American Heartland.* Bulletin 65. Columbus, Ohio: Ohio Geological Survey.

Peterson, Gil. 1990. *The Steel Industry in Ohio.* Center for Urban Studies. Youngstown State University.

Rodgers, Allan. 1952. The Iron and Steel Industry of the Mahoning and Shenango Valleys. *Economic Geography* 28:331–42.

Rubenstein, James M. 1986. Changing Distribution of the American Automobile Industry. *Geographical Review* 76:288–300.

———. 1987. Further Changes in the American Automobile Industry. *Geographical Review* 77:359–62.

———. 1988. Changing Distribution of American Motor Vehicle Parts Suppliers. *Geographical Review* 78: 288–98.

Wright, Alfred J. 1953. *Economic Geography of Ohio.* Bulletin 50. Columbus, Ohio: Ohio Geological Survey.

CHAPTER FOURTEEN

Transportation

Henry Moon

Transportation, the movement from one place to another, is one of the most important factors in our lives. In an economic sense transportation is the major factor in determining the cost of a good or service. Unfortunately it is also a fact of life that is taken for granted. Geographically speaking, transportation refers to transfer across space involving a center of origin (node), a path (link), and a destination (second node). Transportation is a process that indicates the relationships between places. As such it is spatially variable in most areas in the United States, particularly at the state and regional levels (Table 14.1). In an abstract sense this movement of phenomena from place to place involves both human and physical processes of spatial interaction. Human migration, commerce, and communication all occur via some form of transportation. Transportation exists in three realms: on the surface (roads, railroads, and bikeways, for instance); on water (canals, rivers, lakes); and in the air.

Surface Transportation

Bicycles and Bikeways

This first section will consider three major forms of surface transport—roads, railroads, and public transit—but will begin with the most basic of vehicles: the bicycle. The movement of people over land begins at the individual level en route to becoming a complex set of intermodal systems. This does not mean that Ohioans are failing to build or improve simple means of access or that individual transportation is becoming more complicated. Throughout the last decade bikeways have become required components of transportation improvement projects in many areas. It is estimated that there are 78 million bicyclists in the United States, 4 million of whom reside in Ohio (Ohio Department of Transportation 1989). Cycling provides transportation, recreation, competition, fitness, relaxation, and entertainment, while being both cost-effective and fuel efficient. Aside from walking, bicycling is the most socially equitable and environmentally sound means of human movement. The Ohio Bicycle Advisory Council was established by executive order to support the increasing role that bicycling plays across the state.

The Surface Transportation Assistance Act, passed in 1982 and renewed by Congress in 1987, provides Ohio with over $4 million per year for bicycle-related programs and facilities (Ohio Department of Transportation 1989). Bicycle-specific projects go through the same Ohio Department of Transportation (ODOT) review process as do highway projects. All of the state's urban areas and many of its rural counties feature both walkways and bikeways as integral parts of their transportation networks. Officially designated bikeways run alongside highways, through parks, and in an increasing number of cases, on abandoned railroad lines. Over fifty major bicycle tours attract thousands of bikers from across the Midwest each year to a system of trans-Ohio routes that crisscross the state. Six state departments and a host of local advocacy groups support bicycling, not only as a means of pleasure, but as an alternative to other modes of transportation. By the end of 1987, there were sixty-eight bikeways in various stages of planning or development being coordinated through the Ohio Bicycle Transportation Administration (ODOT 1989).

Currently there are twelve designated statewide bicycle routes open in Ohio featuring over 2,500

Table 14.1
Transportation Comparisons

VARIABLE	U.S. MEAN	OHIO	OHIO'S RANK	STATE TREND
Domestic travel expenditure in million $ (1990)	5,809	7,822	10	+
Federal highway trust funds in million $ (1993)	323	529	8	+
Federal UMTA funds in million $ (1986)	62	78	12	+
Transportation and public utility employment, thousands (1993)	114	214	8	+
Transportation energy consumed in trillion Btu (1991)	442	787	7	+
Energy expenditure on transportation in million $ (1991)	3,427	6,823	6	+
Gasoline tax rates in cents (1993)	18	22	13	+
Gasoline tax receipts in million $ (1992)	NA	1,142	5	+
Motor vehicle registrations in thousands (1992)	3,807	9,030	5	+
Motorcycle registrations in thousands (1992)	81	230	2	–
Number of autos per capita (1986)	.563	.648	7	+
Number of drivers licenses in thousands (1992)	3,463	9,169	5	+
Deaths from motor vehicle accidents (1992)	806	1,440	6	–
Number of licensed pilots (1986)	NA	23,954	6	+

Source: USDC, 1989, 1995; FAA, 1987

miles of multimodal roads. Fourteen counties and forty-four cities offer bikeways as integral transportation system components. Bikeways are designated as distinct paths (173.4 miles), lanes alongside highways (7.4 miles), and as routes shared with other vehicles (3,197 miles) (ODOT 1989). Interaction between bicycles and automobiles, trucks, or motorcycles, while never desirable, is often unavoidable. While data on bicycle accidents is sketchy at best, those figures available indicate a worsening problem. Detailed accident data is only available through ODOT for rural state highways from January 1979 to November 1986. The five counties with the greatest number of bicycle–motor vehicle collisions during that period were Stark (80), Hamilton (52), Franklin (51), Montgomery (50), and Trumbull (49) (ODOT 1989). The worst day for accidents is Friday (16.5 percent) while the best day is Sunday (10 percent). Of these accidents, 51 percent occurred between intersections, 22 percent at intersections, and 18.5 percent at driveways. In 1986, 3,469 injuries and 36 deaths were reported in the state as a result of bicycle accidents. With the number of cyclists, bicycle-related accidents, and rail abandonments all increasing across

Table 14.2
Vehicles Entering Ohio Daily on Primary Highways

HIGHWAY TYPE	PASSENGER CARS, VANS, AND PICKUP TRUCKS	OTHER TRUCKS	TOTAL
Interstate	141,600 24.4%	30,124 5.2%	171,724 29.5%
Ohio Turnpike	8,990 1.5%	5,420 .9%	14,410 2.5%
State	346,700 59.7%	48,330 8.3%	395,030 68.0%
Total	497,290 85.6%	83,874 14.4%	581,164 100.0%

Source: ODOT, 1989

Table 14.3
Mileage and Traffic Figures for Ohio's Classified Highways

HIGHWAY TYPE	MILEAGE	%	MILES OF TRAVEL	%
Rural Interstate	669.1	1.9	19,489	12.3
Rural Principal Arterial	1,679.9	4.9	10,189	6.4
Rural Minor Arterial	3,302.4	9.5	13,789	8.4
Rural Collectors	18,913.0	54.6	28,177	17.7
Total Rural	24,564.4	70.9	71,228	44.8
Urban Interstate	654.6	1.9	33,015	20.8
Urban Expressways	339.5	1.0	6,473	4.1
Other Urban Arterials	2,158.0	6.2	22,944	14.4
Urban Minor Arterials	2,802.4	8.1	16,296	10.3
Urban Collectors	4,107.2	11.9	8,897	5.6
Total Urban	10,061.7	29.1	87,625	55.2
Total	34,626.1	100.0	158,853	100.0

Source: ODOT, 1989
Note: Figures exclude 77,286 miles of local roads and streets

Ohio, the availability of bicycle paths and lanes is likely to increase.

Roads, Streets, and Highways

The overwhelming majority of mechanized transportation in Ohio occurs over the state's highway system. Ohio contains 111,912.04 miles of roads, streets, and highways, including 27,355.57 miles (24.4 percent) operating under the federal aid system (ODOT 1989). Each day, 158,853,000 miles are traveled in Ohio, with approximately 581,164 vehicles entering the state via its primary routes (Table 14.2), 29.5 percent on one of the state's interstate highways, 2.5 percent on the Ohio Turnpike, and 68 percent on state routes (Table 14.3 and Figures 14.1 and 14.2). Of the total entering vehicles, 85.6 percent are light vehicles—passenger cars, vans, pickup trucks—while 14.4

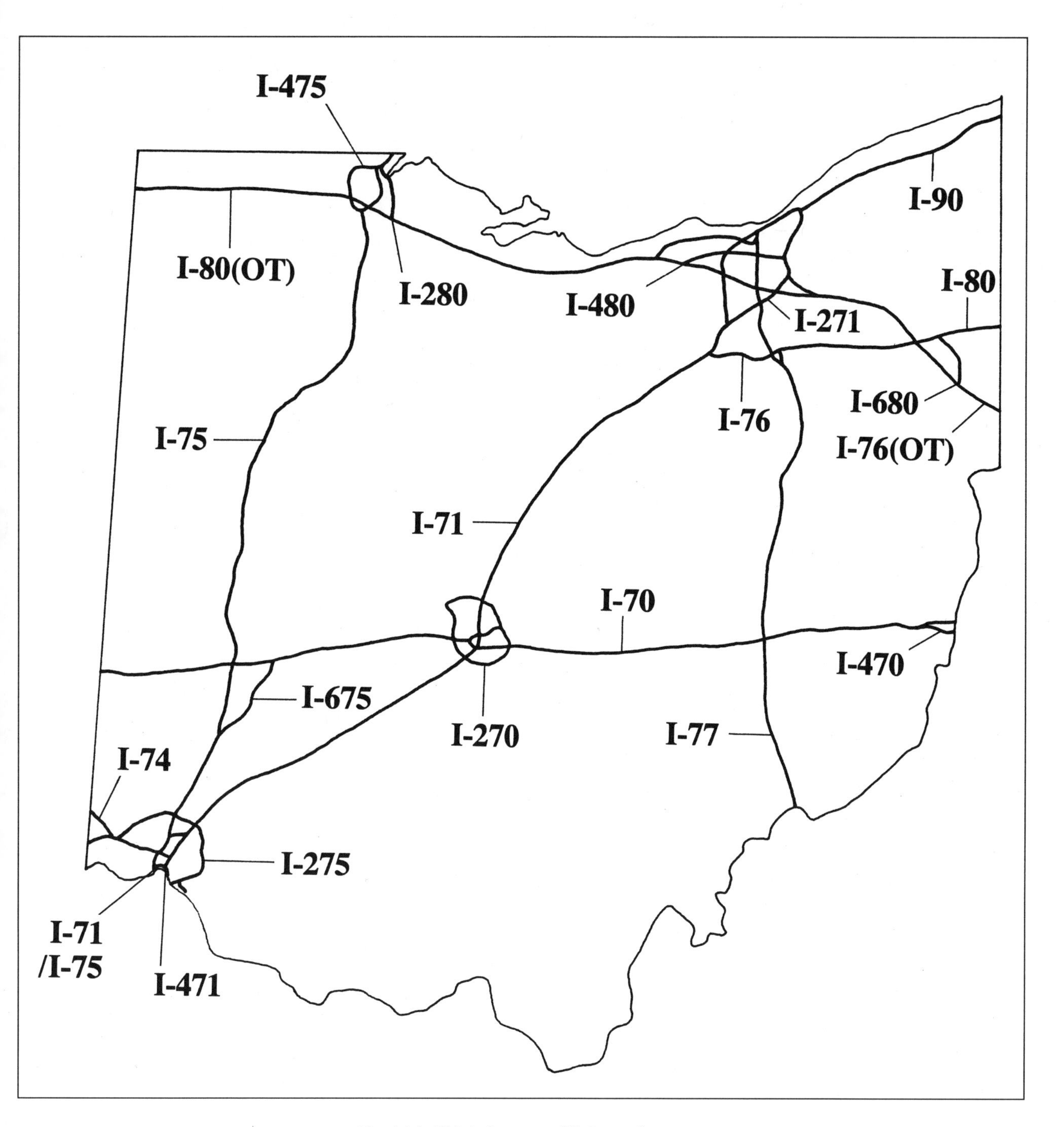

Fig. 14.1. Ohio's Interstate Highway System

percent are heavy trucks. Comprehensive data are not available regarding traffic volume or type on Ohio's many local streets and roads.

Over 110,000 miles of noninterstate roads, streets, and highways make Ohio one of the most locally accessible states. The majority of vehicles entering the state do so on these state and local routes, while an estimated 395,030 vehicles enter on primarily rural routes. Unlike the bulk of the traffic using interstate and toll roads, only 12.2 percent of that brought into Ohio on rural and municipal state routes consists of heavy trucks. It is generally thought that this figure could be lowered if the Ohio Turnpike ceased collecting tolls. With nearly 15 percent of all vehicles enter-

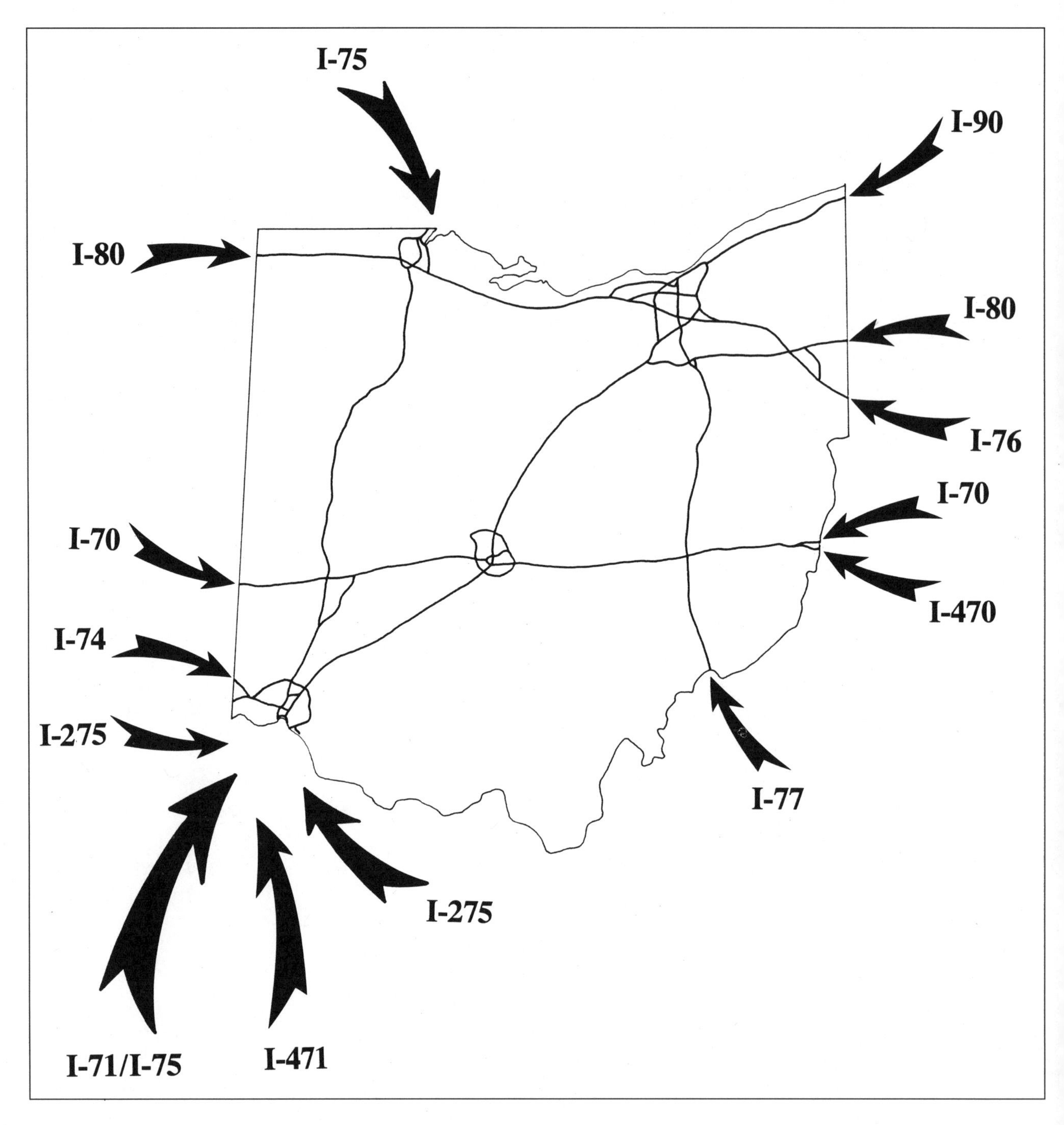

Fig. 14.2. Interstate Highway Traffic into Ohio

ing the state on interstate and state highways falling into the heavy truck category, and the average weight of these large trucks increasing, attention is constantly paid to car-truck relationships. Heavy truck traffic on state and local highways frequently invokes calls for bypasses, reduced speed limits, and the addition of more turn or traffic lanes on existing routes. Ironically, many of the adjustments made to reduce heavy truck traffic actually serve to increase it, for highway improvements tend to increase all forms of traffic.

Much of the traffic moving within and through the state is carried on a network of limited-access highways, or *interstates*. Three of these routes, I-71, I-75, and I-77, connect north to south, while only two interstates, I-80/I-90 and I-70, join east to west. In combination with I-76, which serves as a link between Pennsylvania and Akron, plus several bypasses and fender routes, these roads form an extensive limited-access highway system within the state.

I-75 is a major component of the national interstate highway system connecting Sault Sainte Marie and southern Florida. On a daily basis 18,595 vehicles, 29 percent of which are heavy trucks, enter the state from southeastern Michigan on I-75 (ODOT 1989). This main artery of the national interstate system features three supporting urban expressways and runs concurrent with I-71 in Cincinnati. I-475 is the expressway that veers off from I-75 to access western Toledo. I-675 is an expressway around Dayton that connects I-75 south of the city with I-70 near Wright-Patterson Air Force Base. I-71 extends from its intersection with I-64 in urban Louisville, runs through Cincinnati and Columbus, and ends at its intersection with I-90 near downtown Cleveland. The eastern part of Ohio is served by I-77, which connects Cleveland and Columbia, South Carolina. Daily, 9,930 vehicles, 14.0 percent of them heavy trucks, come into Ohio on this route from West Virginia (ODOT 1989). Around the city of Akron this highway splits in two: I-277 is an urban expressway connecting I-77 to I-76 south of the city, while I-77 runs concurrent with I-76 across the north of the city.

On average 47,440 vehicles enter the environs of Cincinnati daily on I-71/I-75 from northern Kentucky. An estimated 89.0 percent of these are light vehicles, while 11.0 percent are heavy trucks (ODOT 1989). This is not a surprising figure given the suburbanization of Kentucky's northern three counties and the Kentucky location of the Greater Cincinnati International Airport. Since the early 1980s, trucks without scheduled stops in Cincinnati have been directed onto I-275, a less direct and longer route. This beltway brings in an estimated 23,572 vehicles daily. I-471 is an appendage of I-71 that runs between Highland Heights, Kentucky, and downtown Cincinnati north of the I-71/I-75 split. An estimated 23,762 vehicles—of which 95.6 percent are light vehicles—enter Ohio daily via this short feeder. Northwest of Cincinnati, I-74 enters Ohio from Indiana, bringing 6,000 vehicles (75.3 percent light vehicles and 24.7 percent heavy trucks) each day (ODOT 1989). This interstate crosses I-275 and terminates at its intersection with I-75 near downtown Cincinnati.

One of the east to west cross-country interstates is I-70, which connects Baltimore, Maryland, with I-15 in central Utah. This highway brings in 9,315 vehicles daily from the East (20.4 percent of which are heavy trucks) and 8,685 vehicles from the West (41.9 percent of which are heavy trucks). I-470 extends into the western part of Ohio from Wheeling, West Virginia, as an urban expressway around that city. It brings 9,600 vehicles into Ohio each day; 18.0 percent of those vehicles are heavy trucks (ODOT 1989). Around the city of Columbus is I-270, which acts as an urban expressway-bypass for the capital city.

I-80 from New York City to San Francisco enters Ohio near Youngstown. An estimated 8,030 vehicles (57.2 percent light vehicles and 42.8 percent heavy trucks) use this route from Pennsylvania daily (ODOT 1989). This route operates as the Ohio Turnpike between its intersection with I-76, west of Youngstown, and Indiana. I-680 is an urban expressway that connects I-76 (the Ohio Turnpike), which runs south of Youngstown, with I-80, located north of the city. I-480 is a southerly bypass for the city of Cleveland. It connects I-80 at Streetsboro with I-271, the eastern relief route for Cleveland, and then travels through the southern and southwestern suburbs to rejoin I-80 in Elyria. This expressway provides important relief for the congestion that used to occur in downtown Cleveland, where I-90, I-71, and I-77 intersect. I-280 connects the Ohio Turnpike east of Toledo, in Wood County, with I-75 in the northern part of the city. It crosses the Maumee River via a drawbridge, one of the few on the national interstate system. Another notable east-west trans-America route is I-90, which crosses the state on its way from Boston to Seattle. Between its intersection with I-80 in Elyria and the Indiana border, I-90 becomes part of the Ohio Turnpike.

The Ohio Turnpike passes through the state as parts of I-76, I-80, and I-90 and joins the Indiana and Pennsylvania Turnpikes. An estimated 5,650 cars, vans, and pickup trucks plus 2,410 heavy vehicles travel west from Pennsylvania, while 6,350 vehicles, 47.4 percent of which are heavy trucks, enter from Indiana (ODOT 1989). The toll road is operated by the Ohio Turnpike Commission, an organization that operates outside of ODOT through an agreement between the Federal Highway Administration (FHWA) and the state of Ohio. Recently, both the commission and the turnpike have come under fire because of the tolls, the enormous budgets, high operating costs, and a lack of interchanges. Economic development officials complain that the northern tier is discriminated against by the toll-gathering operation. They cite higher levels of development along freeways and a greater number of interchanges thereon. Partially in response to these charges, the Ohio Turnpike Commission is planning a number of interchanges near Archbold, Delta, and Genoa in northwest Ohio, a region that remains relatively unconnected to the nation's main highway system, at least in terms of east-west movement.

Railroads

Like highways, railroads make Ohio one of the most internally accessible states (Figure 14.3). The state currently features 6,530.5 miles of track carrying twenty-one railroads, all operating as Class I, II, or III freight carriers, a classification based on gross annual revenues (ODOT 1989). Of this mileage, 93.5 percent is controlled by Class I carriers, 2.3 percent by Class II carriers, and 4.2 percent by Class III carriers. An additional 183.4 miles of track operate under the auspices of seven switching and terminal companies. Sixty-seven railroad yards serve as first-order nodes connecting the major and minor carriers with local switching operations (Figure 14.4). Intermodal (truck-train) loading facilities supplement Ohio's rail network at Bellevue, Cincinnati, Cleveland, Columbus, Montpelier, and Toledo. All of the state's river and lake ports are also fed by railroads.

Class I railroads are the largest operations and are usually national, with over $50 million in annual revenues and thousands of miles of sophisticated track systems. These large carriers own and maintain almost all of the nation's main railroad track. Class I railroads prefer long-distance hauling as opposed to labor intensive local shipments and the switching and reconfiguring that accompany them. The Class I freight carriers currently serving Ohio are Consolidated Rail Corporation, CSX Transportation, Grand Trunk Railroad Company, and Norfolk Southern Railroad.

Class II railroads have gross annual revenues between $10 million and $50 million and generally operate on a regional basis. They concentrate on switching and branch line operations, often filling the gap between local rail operators and Class I carriers. Often these carriers own and operate thousands of miles of track, serving a specific region or economy such as the steel industry in northeast Ohio. Class II carriers currently serving the state are the Bessemer and Lake Erie Railroad Company, the Pittsburgh and Lake Erie Railroad, and the Youngstown and Southern Railroad Company. Since railroad deregulation in 1980, the number of Class II and III carriers has dramatically increased across the country and in Ohio.

Class III railroads are commonly referred to as short lines and usually operate on no more than 100 miles of track. These carriers often take over lines abandoned by larger Class I and II rail companies to serve a specific local market. They are generally more labor intensive than their larger regional and national counterparts but have lower wage and benefit standards. Thirteen Class III carriers are scattered across Ohio. All of these railroads except one, the Indiana and Ohio Railroad, exist within state lines.

While the future of rail-based transportation is uncertain—with reduced traffic and rail abandonment occurring across both nation and state—the shipment of goods via train continues to be important in Ohio (Figure 14.5). Amtrak, the nation's passenger rail system, uses Class I carrier track to serve several areas, but the vast majority of Ohio is not provided with passenger rail service at all. While these passenger trains contribute to movement, they do not play as important a role as they might. Unlike that of several surrounding states, the Ohio legislature has failed to support a comprehensive passenger rail system. The Ohio High Speed Rail Authority was formed in 1986 to investigate the potential of and possibly even operate an intercity railroad service, but the efforts of this group have amounted to very little. Although the authority is responsible for service "throughout the state" and is charged with connecting all large urban areas, early mention of a Cleveland-Columbus-Cincinnati route eroded statewide support for the project.

Public Transit

Every county and urban area in Ohio is served by some form of publicly funded transportation designated as public transit. Ohio has 51 public transit systems with 2,784 transit vehicles, 93 percent of which are buses, vans, or trolley buses, altogether comprising the fourth largest fleet in the nation. Public transit is a $500 million a year business in Ohio, with over 6,500 employees. Buses carry approximately 84.2 percent of Ohio's transit passengers, somewhat below

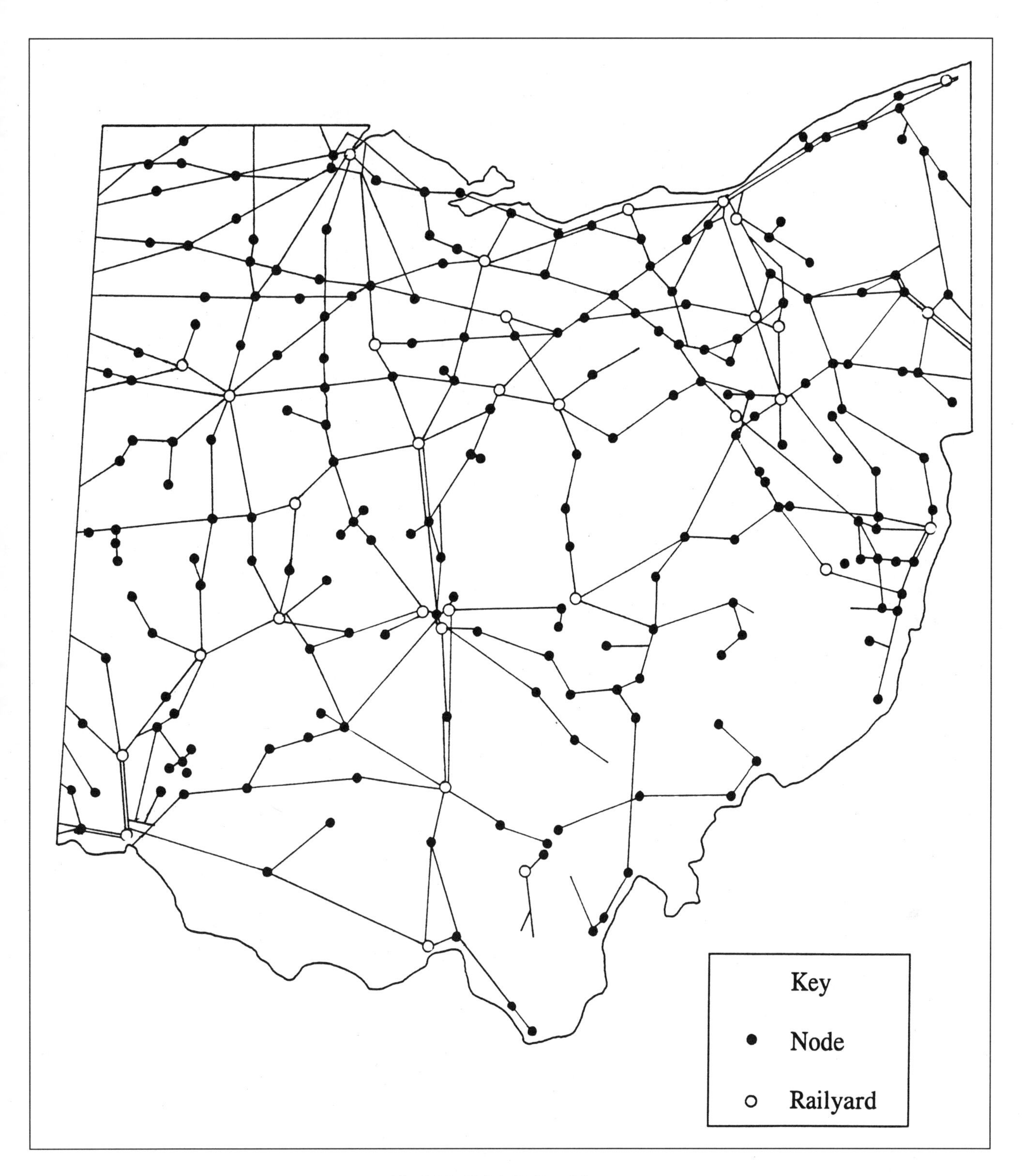

Fig. 14.3. Ohio's Railroad System

the national level of 93 percent (Piper and Huey 1986). Heavy rail is responsible for 13.5 percent of Ohio's public transit traffic but only 2.6 percent of such traffic nationally. Light rail, which accounts for 3.1 percent of the nation's transit traffic, is less utilized in Ohio, representing only 1.4 percent of the state's public transit traffic. The remaining passengers are carried by taxi, automobile, or trolley. There are

Fig. 14.4. Rail Hump Yard, Bellevue (Photo by Leonard Peacefull)

three basic types of public transit service currently available in Ohio: fixed route, commuter rail, and demand responsive.

A fixed-route system is one that features regularly offered specific passenger pickup and delivery along scheduled routes throughout a city or urban area. Each of the state's large urban areas provides a fixed-route public transit service with buses (Figure 14.6). In addition, Dayton offers a fixed-route electric trolley service. Of the state's fifty-one public transit systems, thirty-one now offer fixed-route service. Bus fleet size ranges from the over 800 operated by the Greater Cleveland Regional Transit Authority to the handfuls of vehicles operated by many small-town systems. Northeast Ohio features the state's largest public transit system and the only one operating heavy and light rail commuter service. The Cleveland authority—which manages this sophisticated, multimodal system—ranks among the nation's largest operators. This regional system features 32 miles of serviced track and over 200 rail cars with a combined seating capacity in excess of 15,000. In the past decade the lines have been upgraded and new light rail vehicles introduced. This network operates in conjunction with several fixed-route and demand-responsive systems. Total ridership on the entire system reached 130 million in 1980 but has continually fallen since. It was approximately 86 million in 1986 (Piper and Huey 1986).

Demand-responsive service exists at the individual passenger level and functions as a call-ahead, door-to-door pickup and delivery system. It is used in most of Ohio's rural communities and by urban systems to provide access to the transportation disadvantaged—namely, the handicapped and the elderly. Forty-one of the state's transit systems offer this specialized transit service. Most operate on a very limited basis, using several vans, automobiles, or taxis, frequently gearing lift-equipped vans to wheelchair-bound passengers. Fares are minimal and often nonexistent.

While all of Ohio's counties receive public transit funds, only a few offer what might be recognized as viable public transit. Systems are small-scale for the most part and limited to urban areas. Public transit in the state does not occur in the form of a logical and comprehensive system but as a set of scattered autonomous units. While most local/regional transit authorities are financially sound at this time, they offer a vastly divergent and frequently limited menu of ridership alternatives.

Waterborne Transportation

The Ohio Division of Water Transportation was formed within ODOT in 1984 to coordinate and pro-

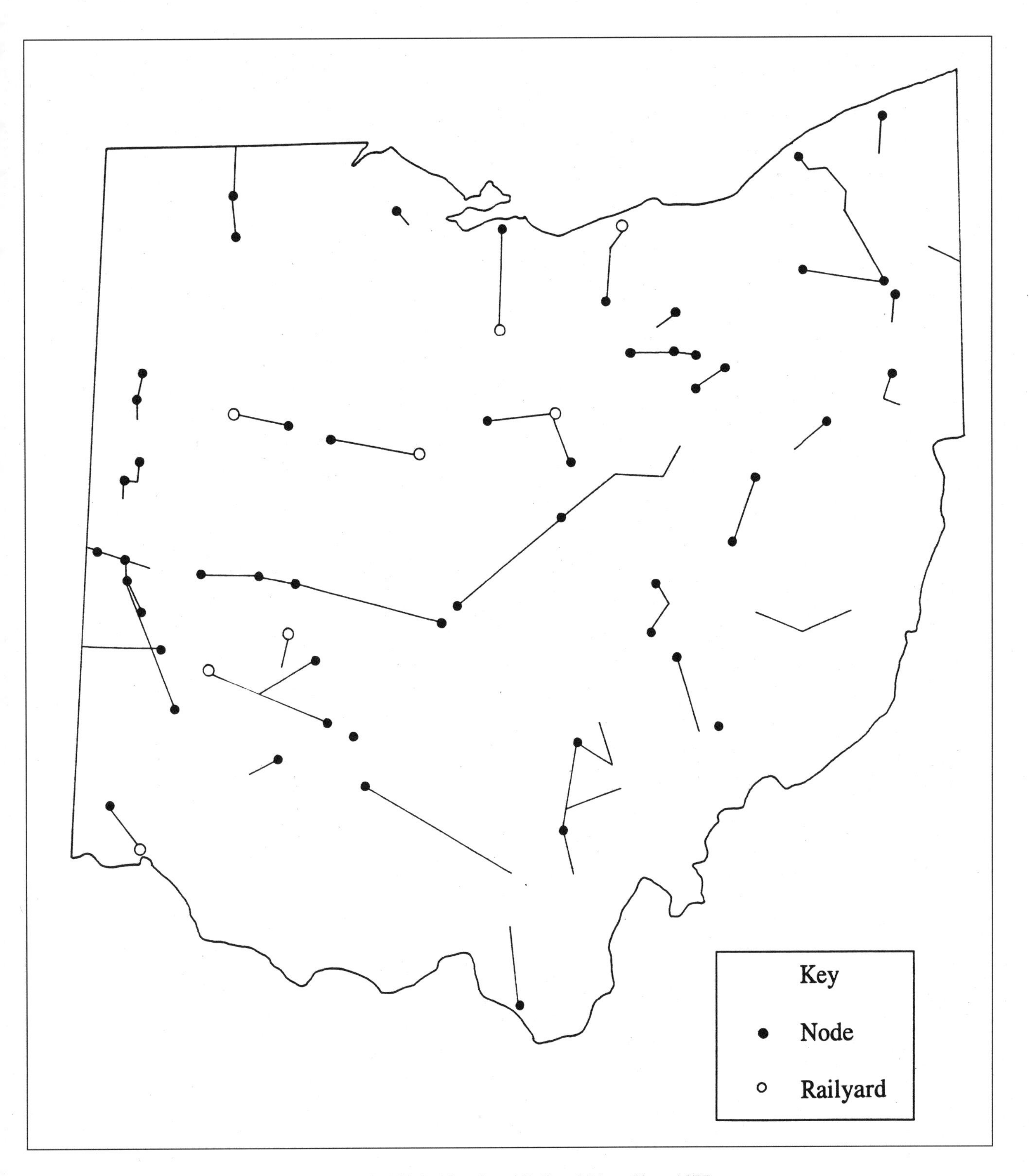

Fig. 14.5. Ohio's Abandoned Railroad Lines Since 1977

mote the state's waterborne transportation efforts (ODOT 1988). Prior to that time, each city, port, and terminal was independent in its effort to compete and to promote itself as a cargo originator or destination. Nine autonomous port authorities—seven along Lake Erie and two on the Ohio River—exist to actually

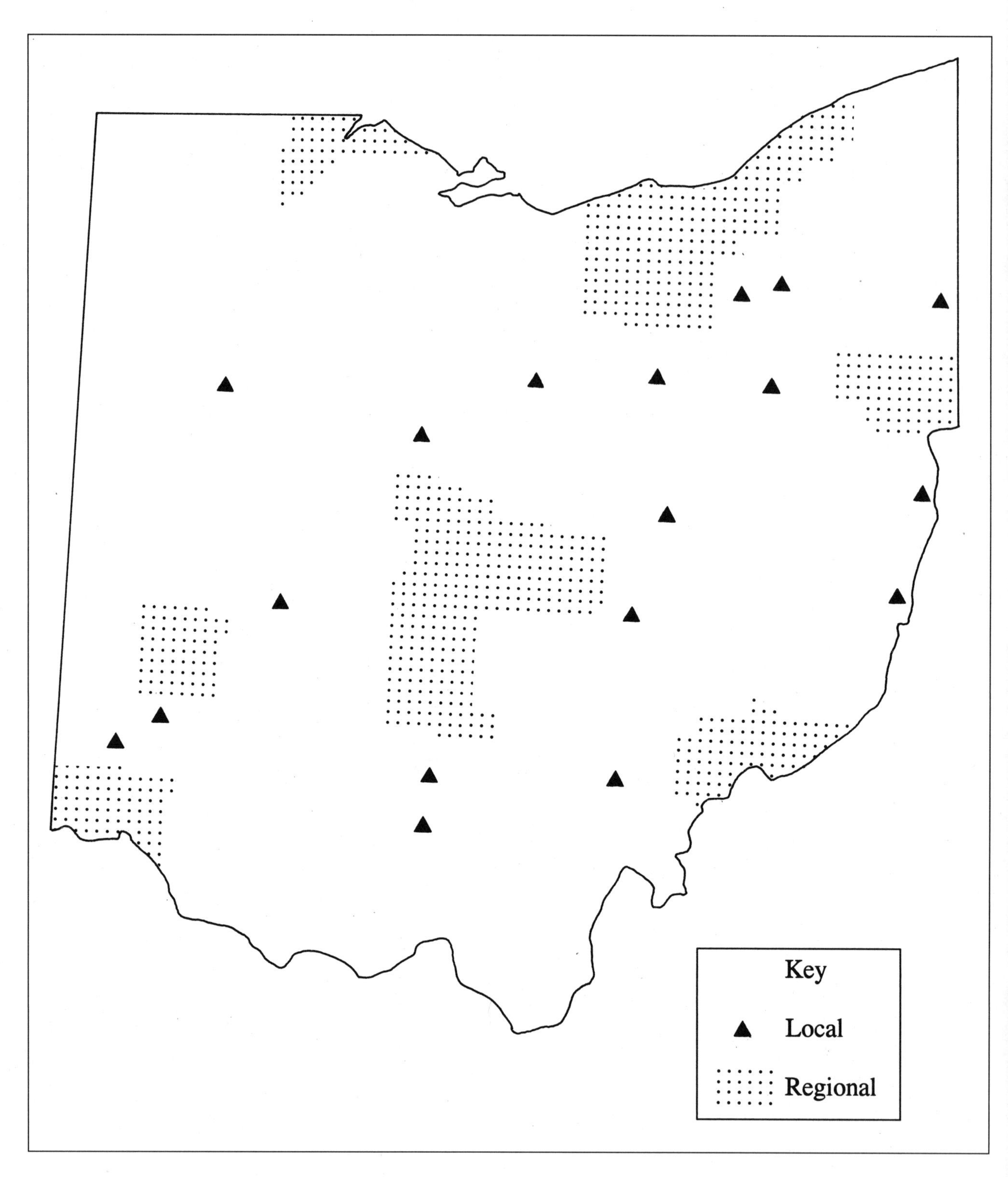

Fig. 14.6. Ohio's Fixed Route Transit Systems

operate the various facilities. Recent efforts have focused on issuing bonds and spearheading local and regional economic development efforts.

Lake Erie

Ohio's relationship with Lake Erie has played an integral role in the state's economic history. The 265-mile shoreline provides local manufacturers and markets with national and even international opportunities. Transportation across the Great Lakes and the St. Lawrence Seaway continues to feature low cost but relatively slow movement of commodities, raw materials, and in some cases finished products. Eight ports operate along Ohio's north coast, with most activity occurring in the Cleveland and Toledo areas (Figure 14.7). Toledo's 135-acre port is the busiest coal-handling facility on the Great Lakes, with over 12 milion tons handled in 1985 (ODOT 1988). This is the largest dockside foreign trade zone complex and the most active international cargo-handling operation on the Great Lakes. Toledo ranks twenty-second among United States ports in terms of cargo volume—ahead of Detroit, Boston, and Seattle. The other operating ports are Ashtabula, Conneaut, Fairport Harbor, Huron, Lorain, and Sandusky. A ship leaving northern Ohio can unload in Rotterdam, Netherlands, the world's busiest port, eleven days later.

Unfortunately, a number of global, regional, and local factors have reduced the viability of large-scale Lake Erie shipments. First and foremost, the depth of the St. Lawrence Seaway prohibits the passage of large, modern seafaring vessels. Many of the ships that operate on the Great Lakes cannot exit through the seaway. Lake ships range in length from 500 to 1,000 feet, with capacities of from 6,000 to 65,000 tons. The majority of these ships are self-loading, which means that they do not require onshore loading and unloading facilities. Second, regional markets for iron ore and coal are much different than they once were, now limited by decreasing steel production and heavy manufacturing in general. Local port operations in Cleveland and Toledo move fewer automobile parts, agricultural products, and food items than in the past (Table 14.4).

Of the eleven commodities that are tracked through Ohio's ports, eight are less active than in 1976 (ODOT 1989). The most significant decreases occurred in the shipment of metallic ores and coal. The three categories that show an increase involve chemicals and allied products (clay, concrete, glass, stone, and certain primary metals). One "commodity" that has received an inordinate amount of recent attention is garbage imported by barge from the northeast, but the movement of waste is still relatively minor compared to others. The simple equation that defines transportation costs has recently shifted its emphasis away from fiscal costs toward temporal costs, and that shift has hurt Ohio port activity. Transportation time has become much more important than it was just a few years ago. Just-in-time manufacturing and a new emphasis on suppliers, two trends in the automobile industry, are perhaps the most glaring examples of the shifts that have reduced the role of all slow forms of transportation. While the Canadian free trade agreement might heavily impact the amount of port activity along the Great Lakes and signify a positive trend in the shipment of some commodities, the degree of that influence remains uncertain.

The Ohio River

The Ohio River forms the southern boundary of the state, winding between East Liverpool and Cincinnati for roughly 450 miles. A northern component of the inland waterway network that operates between the Midwest and the Gulf of Mexico, the Ohio flows from Pennsylvania into the Mississippi near Paducah, Kentucky, separating Ohio from West Virginia and Kentucky. The United States Army Corps of Engineers works to maintain the river as a viable transportation route. It is channelized to maintain a minimum depth of 9 feet and a minimum width of 300 feet (ODOT 1988). Since 1954 all of the state's lock and dam operations have been modernized except that at Gallipolis, which is being replaced with a new facility.

The Ohio section of the Ohio River features nine primary city ports, nine lock and dam operations, eighty privately owned terminals, and twenty publicly owned terminals (Figure 14.8) (ODOT 1988). The average river barge is 35 × 195 feet and can transport 1,500 tons. A barge carries roughly the same quantity as fifteen large rail cars or fifty-eight large truck trailers. Tows consisting of fifteen to twenty barges are common along the river. Of the 53,043,000 tons of freight shipped via the river in 1985, over 40,248,000 tons (75.9 percent) were energy-related products—coal, crude oil, or petroleum (Table 14.5) (ODOT

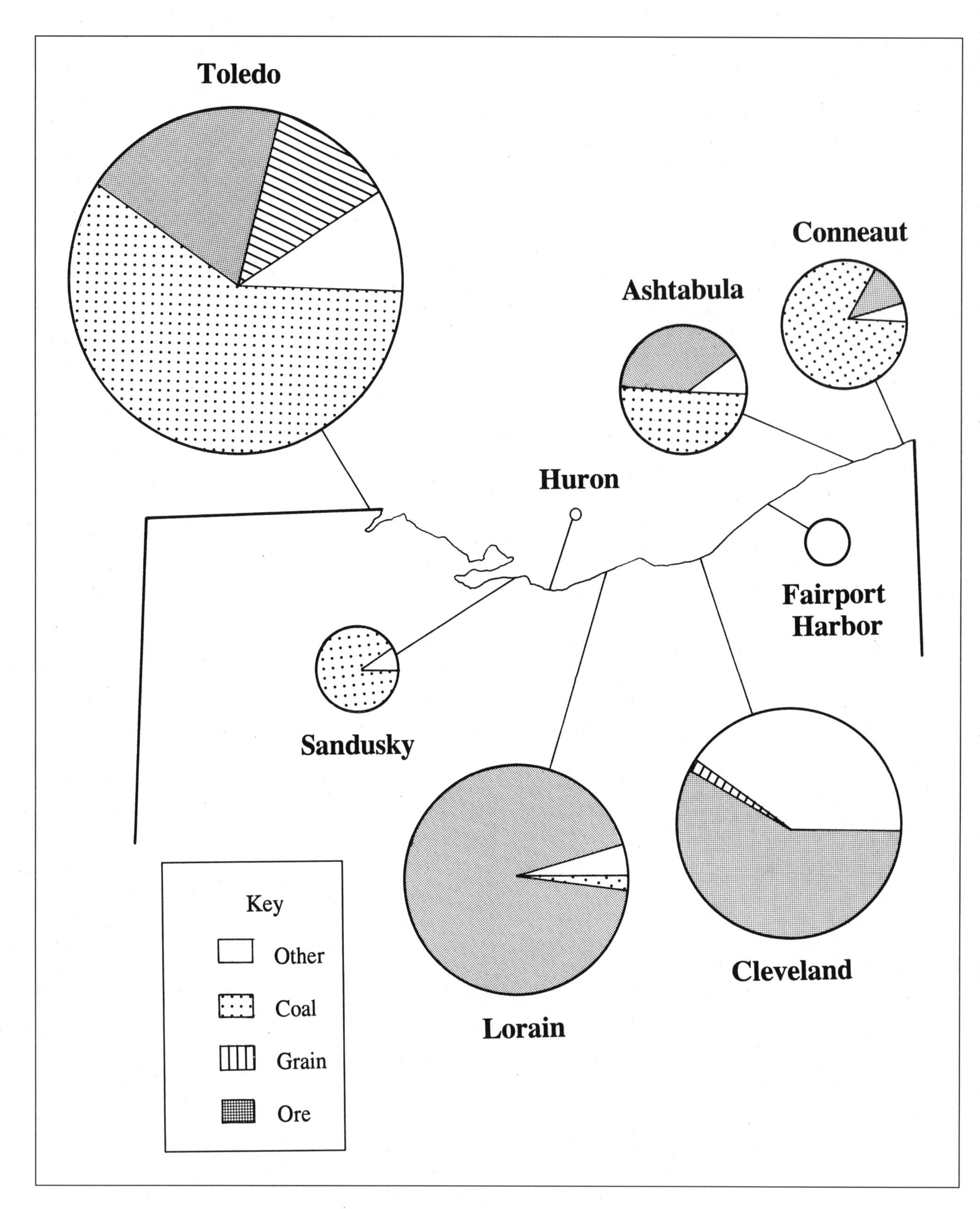

Fig. 14.7. Lake Erie Commodity Flows Through Ohio Ports

Table 14.4

Commodity Trends Through Ohio Ports on Lake Erie, 1980–85, by Tons

COMMODITY	1980	1981	1982	1983	1984	1985	SHORT TERM TREND	LONG TERM TREND
Farm Products	5,320	4,459	3,694	3,376	2,828	2,993	–	+
Metallic Ores	34,054	39,302	16,512	21,103	29,313	25,547	–	–
Coal	32,301	31,219	28,958	20,253	32,641	25,806	–	–
Nonmetallic Minerals	9,094	8,917	5,740	6,794	7,080	9,001	–	–
Food	74	43	11	61	6	10	–	+
Paper	48	42	1	0	25	18	–	–
Chemicals	482	853	795	651	788	1,014	+	+
Petroleum/ Coal Prod.	1,188	655	569	708	751	642	–	–
Clay, Concrete, Etc.	435	302	302	329	500	497	+	+
Primary Metals	474	810	563	1,973	696	1,518	+	–
Waste/Scrap	111	58	224	119	180	77	+	–
Total	83,581	86,660	57,367	55,367	74,808	67,122	–	–

Source: ODOT, 1989

Table 14.5

Commodity Trends Through Ohio River Ports, 1980–85, by Tons

COMMODITY	1980	1981	1982	1983	1984	1985	SHORT TERM TREND	LONG TERM TREND
Farm Products	1,080	1,141	2,044	1,985	1,326	2,546	+	+
Metallic Ores	527	787	379	647	903	916	+	–
Coal	42,472	32,701	27,648	25,208	31,121	35,860	–	–
Crude, Etc.	237	507	504	553	591	512	+	–
Nonmetallic Minerals	4,878	4,531	3,778	3,656	4,473	3,898	–	–
Food	354	399	367	492	597	511	–	–
Chemicals	2,754	2,608	2,270	2,906	3,589	3,458	+	+
Petroleum/ Coal Prod.	4,173	4,368	4,331	4,131	4,291	3,876	–	–
Clay, Concrete, Etc.	435	448	372	349	387	314	–	–
Primary Metal	1,125	1,397	721	911	1,240	1,004	–	–
Waste/Scrap	101	90	65	95	123	147	+	–
Total	58,136	48,977	42,479	40,933	48,641	53,043	–	–

Source: ODOT, 1989

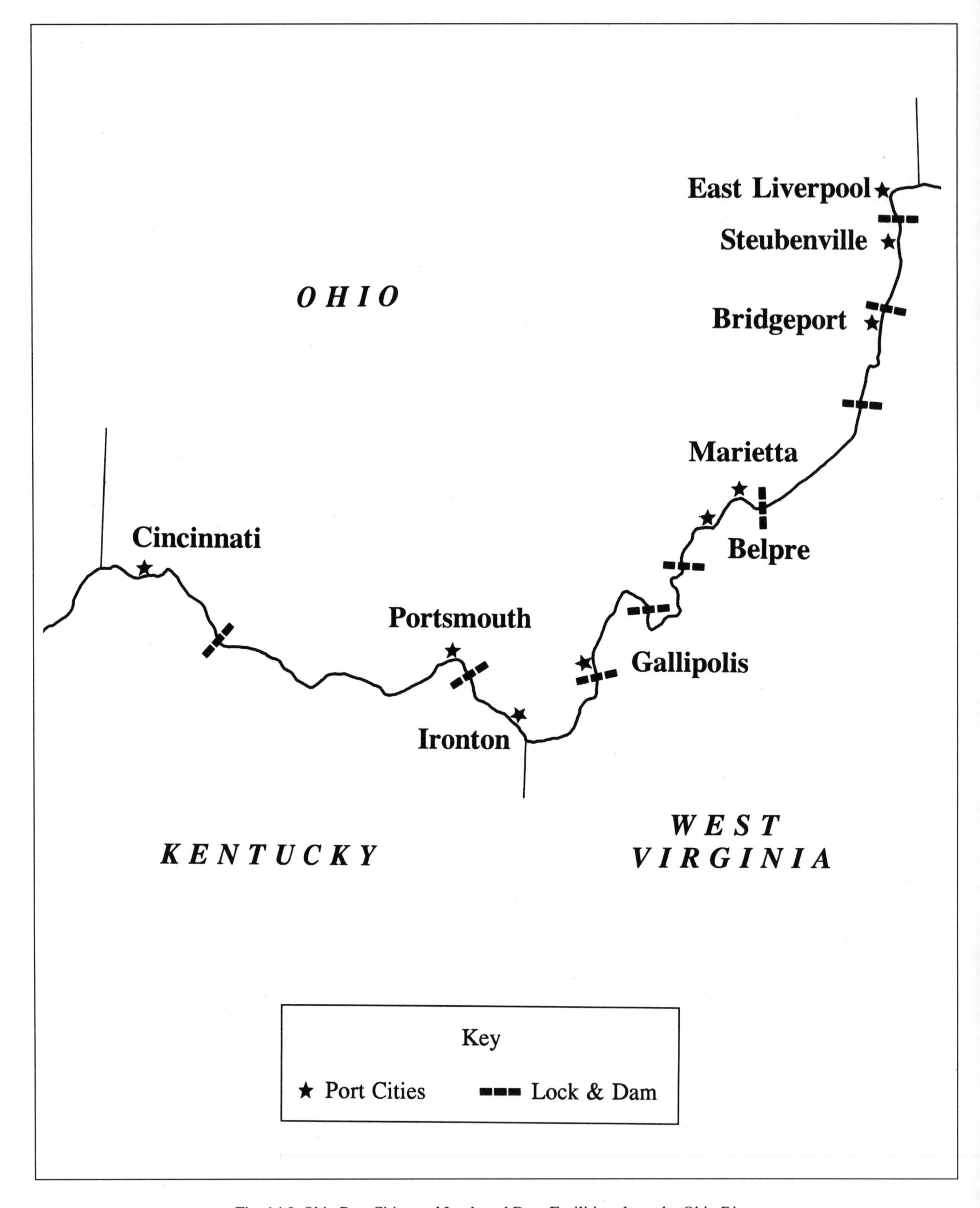

Fig. 14.8. Ohio Port Cities and Lock and Dam Facilities along the Ohio River

1989). Coal was by far the primary commodity, measuring 35,860,000 tons (67.6 percent). The Cincinnati area is the most active, with about 30 percent of the state's river traffic occurring around its forty terminals. Other busy port communities are East Liverpool, Steubenville, Marietta, Ironton, and Portsmouth. It takes approximately eighteen days for a shipment to reach Europe from Cincinnati through the river network.

The future of goods movement along both Lake Erie and the Ohio River is uncertain. Unless the St. Lawrence Seaway is deepened, the entire Great Lakes transportation system will remain extremely limited. While emerging foreign trade agreements and restrictions will alter the face of waterborne transportation as we now know it, the future is cloudy at best. While the shipment of some commodities is increasing, the current trend of looking toward our waterways more as arenas of recreation and less as avenues of trade will probably continue.

Airborne Transportation

As with other modes of transportation, air travel (passenger or cargo) is a derived demand, with use depending largely on the size and structure of national and international economies. Air transportation is vital to the service sectors of the economy, including the service or management aspects of both traditional and modern manufacturing. In 1985 the airline industry generated $46.7 billion in revenues and $1.4 billion in operating profits carrying 380 million passengers (Apogee Research, Inc. 1987). While the size and growth of air transport have been driven in part by technological efficiency, the ability to make more effective use of economic and physical resources is one key to economic development.

With its high cost, rapid speed, and relatively reliable service, the use of air freight has traditionally been dominated by high-value goods. Given the evolution of temporally shrinking societies and just-in-time manufacturing, however, the role of air transport in future goods movement will most certainly change. The rapidly growing overnight small-package business is perhaps the best example of an industry that would not be possible without advancing aviation technology, a global system of airports, and the direct encouragement provided by economic deregulation in the United States. The small-package industry now totals $6 billion a year, while more conventional air freight adds up to $2 billion a year (Apogee Research, Inc. 1987). Because of its high speed and reliability, aviation plays a role in helping to reduce inventories and control the costs incurred from delays in receiving components. The shipment of heavier air freight is expected to contribute a higher proportion to total air shipments in the future. Changing patterns and processes of air transport have international and national repercussions on economic development (Kuhlman, Smolen, and Moon 1989).

In addition to this impact on transportation, airports also provide employment and stimulate new business activity. Economic benefits provide the main justification for public support of airports. These benefits can be defined as direct, indirect, and induced impacts. Direct impacts represent economic activities that would not have occurred without the airport. They are consequences of economic activities carried out at the airport by airport management, airlines, and other tenants directly involved in aviation. Employment, the purchase of locally produced goods and services, and contracts for construction and capital improvements are examples. By contrast, most indirect impacts are typically located off-site or along the margins and are attributable to the airport. Examples are hotels, car rental agencies, travel agencies, restaurants, and storage rental firms. Like those firms or agencies providing more direct impacts, these businesses also provide employment and purchase goods and services; they invest in the community through their own efforts to expand and improve. None of these business activities would occur without the airport. Induced impacts are the multiplier effects that the direct and indirect impacts have on other sectors of the economy. These are created by successive rounds of spending and result in increases in employment and incomes over and above the combined direct and indirect impacts.

Airports

Ohio has 853 airports categorized according to size and use (ODOT 1989). The mean distance between these airports is approximately 20 miles. Seventeen of these airports, the largest, are considered Class I, including six that operate as national trunk carriers (Figure 14.9). Trunk carriers are major airlines flying national and in some cases international flights. Cities

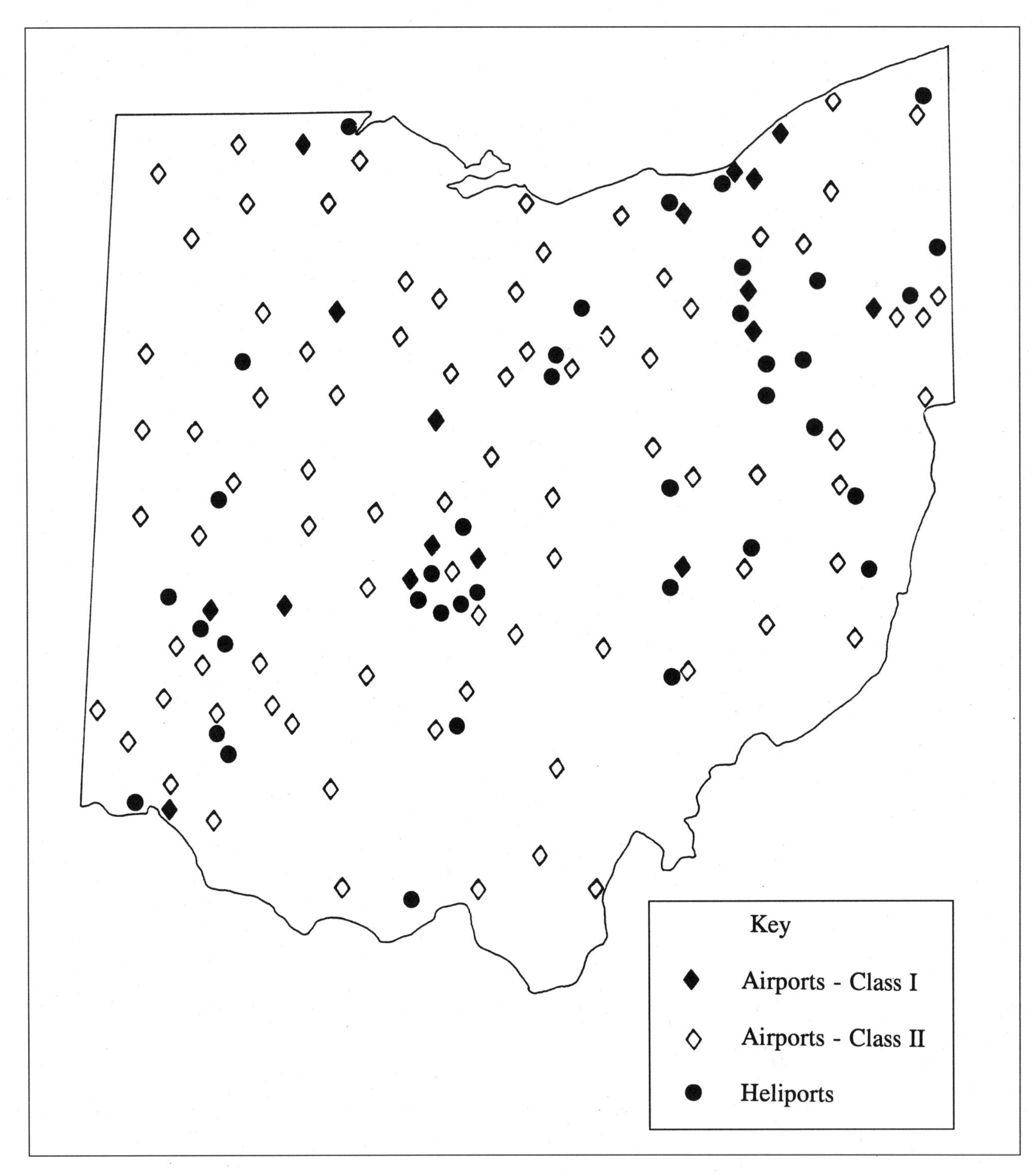

Fig. 14.9. Primary Aviation Facilities in Ohio

with these large airport operations include Cleveland, Columbus, Dayton, Toledo, Akron-Canton, and Youngstown (Table 14.6). The Cincinnati area is served by the Greater Cincinnati International Airport in northern Kentucky. Ohio ranks first in the United States in the proportion of paved and lighted runways to land area (1:207 square miles). Approximately 6,220 airplanes are currently registered in Ohio.

Table 14.6
Ohio's Primary and Other Commercial Airports

CITY/ AIRPORT	SERVICE LEVEL	ROLE	BASED AIRCRAFT	ENPLANED PASSENGERS	TOTAL OPERATIONS
Akron/ CAK I	Primary	Medium	134	166,000	108,000
Cleveland/ CLE II	Primary	Long	169	2,932,000	241,000
Cleveland/ BKL NP	Primary	Short	62	13,000	69,000
Columbus/ CMH I	Primary	Medium	184	1,621,000	233,000
Dayton/ DAY III	Primary	Long	107	1,503,000	165,000
Mansfield/ MFD I	Other	Short	87	6,000	55,000
Port Clinton/ 1G7 NP	Other	Short	33	<1,000	43,000
Toledo/ TOL I	Primary	Medium	111	290,000	82,000
Youngstown/ YNG I	Primary	Short	89	61,000	100,000

Source: FAA, 1987

The state contains 83 Class II airports that meet local and regional demand and act as relievers to larger Class I operations. Eighty of the state's airports are classified as Class III, or general aviation facilities. These smaller facilities are generally local in nature and often feature only one runway. Many support local commerce, but more frequently they meet the growing demand for recreational flying opportunities. Class III airports include publicly held facilities and privately owned operations that are open to the public. Ohio's Division of Aviation, also under the auspices of ODOT, declares that the primary function of the state's Class II and III airports is "the retention and expansion of business enterprises within their service areas. Many of the Class II facilities in Ohio serve as county airports in rural areas, and by enabling businesses to move executives and high priority freight by air they have added greatly to Ohio's attractiveness as a potential industrial location" (ODOT 1989). Every Ohio county contains a Class I or II airport except Hocking, Lawrence, Meigs, Paulding, Preble, Sandusky, Stark, Trumbull, and Washington (ODOT 1989).

Ohio's abundance of airports and its central location amid North America's population have made the state a center of air freight shipment. A number of national air freight hubs operate in Ohio, including the Emery-Purolator facility in Dayton and the Airborne Express facility at a former United States Air Force base at Wilmington. Burlington Air Express (BAX), an international air freight company that specializes in heavier packages, began a hub operation at Toledo Express Airport in 1990 (Kuhlman, Smolen, and Moon 1989). Currently, the air freight industry generates over 5,100 jobs across the state, with growth expected (ODOT 1989).

Heliports

Transportation via helicopter is a rapidly expanding sector of the aviation industry. Both passengers and

Table 14.7

Transportation Characteristics by County

COUNTY	INTERSTATE MILEAGE	U.S. ROUTE MILEAGE	STATE ROUTE MILEAGE	RAILROAD LINES	AIRPORTS	
					CLASS I	CLASS II
Adams	0	29	187	1	0	1
Allen	23	25	136	9	0	1
Ashland	16	71	171	3	0	1
Ashtabula	29	86	250	11	0	1
Athens	0	65	147	2	0	1
Auglaize	13	29	170	5	0	1
Belmont	34	32	215	10	0	1
Brown	0	89	119	2	0	1
Butler	11	48	173	10	0	3
Carroll	0	0	153	9	0	1
Champaign	0	45	165	5	0	1
Clark	31	58	100	7	1	0
Clermont	14	45	209	2	0	1
Clinton	15	50	142	2	0	2
Columbiana	0	50	253	7	0	1
Coshocton	0	34	181	2	0	1
Crawford	0	21	178	6	0	2
Cuyahoga	105	112	228	16	3	0
Darke	0	50	213	6	0	1
Defiance	0	32	132	2	0	1
Delaware	17	68	123	5	0	1
Erie	0	42	109	8	0	1
Fairfield	2	50	143	3	0	1
Fayette	15	69	86	4	0	1
Franklin	112	119	120	18	3	2
Fulton	0	57	85	5	0	1
Gallia	0	18	171	2	0	1
Geauga	0	57	137	2	0	1
Greene	22	73	66	3	0	1
Guernsey	52	40	154	4	0	1
Hamilton	96	118	86	30	1	1
Hancock	25	61	158	5	1	1
Hardin	0	22	154	4	0	1
Harrison	0	59	111	11	0	1
Henry	0	43	134	5	0	1
Highland	0	56	204	3	0	1
Hocking	0	19	149	1	0	0
Holmes	0	37	137	2	0	1
Huron	0	70	158	10	0	2
Jackson	0	28	141	6	0	1
Jefferson	0	19	150	13	0	1
Knox	0	59	140	1	0	1
Lake	31	41	135	6	1	1
Lawrence	0	23	167	3	0	0
Licking	29	51	185	6	0	1
Logan	0	50	176	4	0	1
Lorain	16	46	230	11	0	1

Table 14.7, continued

Transportation Characteristics by County

COUNTY	INTERSTATE MILEAGE	U.S. ROUTE MILEAGE	STATE ROUTE MILEAGE	RAILROAD LINES	AIRPORTS CLASS I	CLASS II
Lucas	34	65	113	27	1	0
Madison	27	46	124	5	0	1
Mahoning	31	59	174	16	1	3
Marion	0	20	181	4	1	1
Medina	45	39	164	8	0	2
Meigs	0	15	175	2	0	0
Mercer	0	45	166	1	0	1
Miami	20	28	154	3	0	1
Monroe	0	0	214	1	0	1
Montgomery	55	42	122	12	1	2
Morgan	0	0	189	0	0	1
Morrow	20	26	131	3	0	1
Muskingum	27	40	200	10	1	0
Noble	19	0	197	1	0	1
Ottawa	0	0	139	5	0	1
Paulding	0	34	134	2	0	0
Perry	0	14	172	5	0	1
Pickaway	3	55	136	4	0	1
Pike	0	16	129	5	0	1
Portage	23	23	204	5	0	1
Preble	18	61	113	2	0	0
Putnam	0	31	178	5	0	1
Richland	21	37	205	5	1	2
Ross	0	94	122	6	0	1
Sandusky	0	63	112	9	0	0
Scioto	0	61	139	5	0	1
Seneca	0	46	177	8	0	2
Shelby	21	0	144	4	0	1
Stark	19	72	235	16	0	0
Summit	77	6	186	14	2	1
Trumble	12	32	308	13	0	0
Tuscarawas	35	39	140	11	0	1
Union	0	51	144	1	0	1
Van Wert	0	71	97	5	0	1
Vinton	0	30	128	3	0	1
Warren	34	45	136	7	0	1
Washington	18	10	245	4	0	0
Wayne	7	58	187	8	0	1
Williams	0	81	108	6	0	1
Wood	43	61	207	13	0	2
Wyandot	0	49	151	4	0	1
Total	1,317	3,872	13,900	560	17	83
Mean	15.0	44.6	159.2	6.4	.2	.9

Source: ODUC, 1988; ODOT, 1989

goods are moved around urban areas on helicopters, and more injury victims are transported to hospitals by helicopter each year. Ohio contains 308 heliports, 269 of which are noncommercial (ODOT 1989). Roughly 180 helicopters are currently registered in the state. Each of Ohio's major urban areas features a downtown heliport. The majority of Ohio's heliports fall into the so-called power corridor between Cleveland, Columbus, and Cincinnati.

Conclusion

The key issues facing Ohio today and tomorrow center around transportation. Economic development, for example, has become the pet phrase of Ohioans concerned with our declining industrial base, urban decentralization, rural out-migration, the state's role in political and economic internationalization, and a plethora of other pressing topics. Foreign trade, global competition, and urban revitalization head the list of issues of which citizens must be cognizant. Among the common threads running through all of these uncertainties is transportation. How has the interstate highway system contributed to out-migration and decentralization? Should transport-based taxation be used to balance budgets and fund social programs? How can rural communities balance the positive and negative aspects of larger, urban centers by using interstate highways and interchanges? Should the Ohio Turnpike Commission continue? Who will pay to replace or repair Ohio's 100,000 miles of pavement, 43,000 bridges, dozens of locks and dams, and nearly 900 airports? From a historical perspective, transportation has defined Ohio. Today it shows no signs of becoming less important.

Ohio's natural environment rests alongside economic development in the forefront of state news. Both topics center around transportation. The shipment of hazardous wastes on our highways and rail lines, the transfer of radiative material to and from plants and reactors, and record levels of auto emissions demand attention from citizens and lawmakers alike. Highway construction and airport expansion threaten wetlands and further reduce the state's most sensitive ecosystems. As people and goods gain access to more remote places, those places are redefined. Economic development cannot occur without negatively impacting the natural environment. This direct conflict often pushes transportation to the forefront of some very complex and controversial decisions. How can planners balance the construction and expansion of transport systems with a fragile natural environment? What weight should environmental impact statements represent in planning and decision making? Have the local impacts of cargo plane takeoffs and landings been accurately measured? Can the U.S. Army Corps of Engineers fairly and responsibly balance the needs of north shore ecosystems, residents, and transport networks?

Many questions involving the relationship between the movement of people, goods, and services and the by-products of this movement remain unanswered but begging for clarification. Ohioans face the uneasy task of balancing transportation demands and a sensitive physical environment, a task that will keep both issues at the forefront of local, regional, and state agendas well into the foreseeable future. The level of accessibility afforded the citizens of Ohio is highly irregular and varies across the state. Some urban areas offer passengers and shippers a variety of transportation alternatives while others do not. Rural communities in general lag far behind their urban counterparts in terms of internal and interregional access. Counties that feature an interstate highway, a railroad line, a Class I or II airport, and an above-average number of national and state highway miles have advantages over those less endowed (Table 14.7 and Figure 14.10). In general, these "advantaged" counties lie between Cleveland and Cincinnati. Economic development, employment, health care, education, and, in a more holistic sense, equity all fluctuate with accessibility across Ohio. Few issues can be separated from the process of movement, especially when that process is constantly changing in its location, scale, and mode.

Fig. 14.10. Ohio Counties Given a Transportation Advantage

REFERENCES

Apogee Research, Inc. 1987. *Airports and Airways.* Washington, D.C.: National Council on Public Works.

Federal Aviation Administration. 1987. *National Plan of Integrated Airport Systems (NPIAS) 1986–1995.* Washington, D.C.: Dept. of Transportation.

High Speed Rail Association. 1986. *High Speed Rail Association Yearbook 1986.* Washington, D.C.: High Speed Rail Association.

Kulhman, Bruce R., Gerald E. Smolen, and Henry E. Moon. 1989. *Direct, Indirect, and Induced Impacts of the Toledo–Lucas County Port Authority.* Toledo, Ohio: The University of Toledo Business Research Center.

Moon, Henry. 1988. *Toward Industrializing Toledo Express Airport.* Toledo, Ohio: Toledo–Lucas County Port Authority.

———. 1992a. *The Interstate Highway System, Geographical Snapshots of North America.* New York: Guilford Publications.

———. 1992b. Vehicular Transportation (Chapter 7). *Region in Transition: An Economic and Social Atlas of Northeast Ohio.* Akron, Ohio: The University of Akron.

———. 1994a. Solid Waste Management in Ohio. *The Professional Geographer.* 46, no. 2: 191–98.

———. 1994b. *The Interstate Highway System.* Washington, D.C.: The Association of American Geographers Resource Publications in Geography Series.

Mould, David H. 1994. *Dividing Lines.* Dayton, Ohio: Wright State University Press.

Noble, Allen G., and Albert J. Korsok. 1975. *Ohio: An American Heartland.* Columbus: Ohio Dept. of Natural Resources, Division of Geological Survey.

Ohio Data Users Center. 1988. *1988 Ohio County Profiles.* Columbus: Ohio Data Users Center, Dept. of Development.

Ohio Department of Transportation. 1986. *Classification by Surface Type of Existing Mileage in Each County on State Highway, County, and Township Systems.* Columbus: Ohio Dept. of Transportation, Division of Planning and Design.

———. 1988. *Water Transportation Map of Ohio.* Columbus: Ohio Dept. of Transportation, Division of Water Transportation.

———. 1989. *Ohio Transportation Facts.* Columbus: Ohio Dept. of Transportation, Bureau of Transportation Technical Services.

Piper, Barbara A. and Paula J. Huey. 1986. *Ohio Transit—1986.* Washington, D.C.: U.S. Dept. of Transportation Urban Mass Transit Administration.

Rennie, Henry G. 1987. Ohio's MSAs and Economic Specialization. *The Ohio Economy.* (University of Toledo Business Research Center) 3, no. 3.

Scheiber, Harry N. 1987. *Ohio Canal Era.* Athens, Ohio: Ohio University Press.

Spychalski, John C. 1986. Railroads: Progress, Challenges, and Opportunities for Ohio and the Nation. *The Ohio Economy.* (University of Toledo Business Research Center) 2, no. 4.

U.S. Department of Commerce, Bureau of the Census. 1989. *Statistical Abstract of the United States 1988.* Washington, D.C.: U.S. Dept. of Commerce.

———. 1995. *Statistical Abstract of the United States 1994.* Washington, D.C.: U.S. Dept. of Commerce.

CHAPTER FIFTEEN

Outdoor Recreation and Tourism

Richard V. Smith

This chapter deals with the two most studied leisure activities in Ohio: outdoor recreation and tourism. A comprehensive inventory and analysis of all leisure activities that occur in the state has not yet been conducted. Some information is available from standard statistical sources or in atlases of popular culture on various leisure activities, such as attendance at spectator sporting events or museums. Increasingly it is apparent that leisure time resources and activities are of central importance in any area since they consume large amounts of time, make numerous and extensive impacts on the landscape, and have a significant economic role. Thus leisure opportunities and resources are a major component of the amenity base of a community or region and play a substantial role in helping to create an environment that is favorable for future economic development.

Outdoor Recreation

Statewide Recreation Planning

Ohio recently completed and published its 1993 Recreation Plan (Ohio Department of Natural Resources 1993). That plan is the sixth the state has completed, the fourth that is comprehensive in nature, and the second that is designed to lay the base for a continuing outdoor recreation planning process. The new plan is also the second that the state has undertaken in an era of markedly decreased federal funding for outdoor recreation, and it establishes a clear set of priorities for the further development of Ohio's leisure resources. A sampling of this plan's thirteen priorities includes:

- establishing a permanent, dedicated source of funding for recreation resource development statewide;
- expanding trail-related activities;
- increasing opportunities for water-based activities;
- developing educational programs to foster an outdoor ethic in Ohio's citizens;
- improving planners' abilities to identify recreation resource needs;
- improving recreational opportunities for special populations; and
- establishing mutually desirable partnerships with public, private, and independent recreation providers.

Previous plans usually had as their dominant theme a clear effort to achieve the best possible review by the Bureau of Outdoor Recreation (later the Heritage Conservation and Recreation Services and currently the National Park Service). Obtaining the maximum eligibility for Land and Water Conservation Fund allocations was the major goal. Secondary themes were those that state and federal officials agreed had a particular relevance at the time the study was structured. The 1975–1980 Statewide Comprehensive Outdoor Recreation Plan (SCORP) was launched in 1972 and completed in 1975 (Ohio Department of Natural Resources 1976). The special themes examined included detailed metropolitan area studies; impacts of energy costs and shortages (added in 1973); recreational needs of the handicapped; role of the private sector; the public's attitudes about Ohio's outdoor recreational opportunities; participation and needs analysis; and an exhaustive inventory of recreational resources. A large (5,542 household) citizen user survey was conducted by mail. Analyses of the data were prepared for state, regional, and county units. An effort was made to make this data available to local agency planners, even though the primary thrust was at the state level. The anticipated uneasiness with the accuracy of county-level data furthered an awareness that data generated for state-level use must be disaggregated by area with the utmost discretion. As a result, in part, the two most recent plans do not try to be all things to all people.

The 1980–1985 Statewide Comprehensive Outdoor Recreation Plan was launched in 1977 and completed in 1980 (Ohio Department of Natural Resources 1981). This planning study was established in a similar manner to the previous plan. A major inventory, another massive citizen user survey (this time conducted by telephone), attitude analysis, and a participation and needs analysis were once again dominant components of the study. Detailed special studies were also conducted on the changing roles of the several types of political units and on the recreational needs of the elderly.

The 1986 plan differed in significant ways from its predecessors. Major emphases included a focus on state-provided activities, a concern with trends in outdoor recreation (as opposed to forecasts of future participation and needs), and a response to the pragmatic issues then confronting providing agencies. Improved marketing of recreation and the linkages between outdoor recreation and the state's tourism industry were also addressed.

Participation, Trends, and Desires

The findings on Ohio's recent recreation patterns and problems are quite similar to those identified in the three previous SCORPs. The activities, enablers, barriers, and needs of the population vary only in degree from the earlier studies. Though Ohioans have a generally satisfactory hierarchy of recreational facilities available to them—ranging from the state to the most local levels—a significant part of the population continues to confront barriers to participation. As in earlier studies, these groups were identified as the elderly, the disadvantaged, the inner-city resident, and the increasing number of people living outside of incorporated communities. Nearly all parts of the population desire some recreational involvement, however infrequent. In addition, an increasing desire for lifelong, higher risk, and individual activities has further developed. As a result of these trends, the range of desires for recreation facilities and programming continues to expand.

Relatively little change in Ohio recreation participation and needs occurred during the 1980s. Ohioans do want an increase in the availability of facilities close to their points of residence. The "parks to people" movement is still important. Although the too-limited funds available to both the state park system and local agencies for operation and programming must be primarily devoted to maintaining quality in existing facilities, the acquisition of lands for future parks close to metropolitan areas deserves renewed consideration. Land banking may be the only way to assure the availability of lands that will be needed for park expansion in the next decade.

The factors affecting recreation locally and nationally are in a state of marked change. Demographic and economic changes, environmental concerns, affluence or poverty, and major life-style and value reorientations are all likely to affect recreation in Ohio. The aging of the population and the changing nature of household composition present the greatest significance to recreation planners. Since Ohio tends to adopt changes somewhat later than other states, a careful review of what is occurring in such pacesetter states as California, Washington, Connecticut, Florida, and Texas can be valuable.

Surveys of the population consistently demonstrate that Ohioans are pleased with their recreational opportunities. Though inevitably a part of the outdoor recreational experience sought by Ohioans is achieved outside of the state, the facilities Ohio provides satisfy most present in-state demands. Ohio has a state park system consisting of seventy-two units that provide many needed open-space facilities. In 1986, Ohio's state parks and recreation areas ranked second among the fifty states in popularity, trailing only California in the reported number of visitors (Bureau of Census 1987).

Many of Ohio's cities provide park and recreation program opportunities for their citizens; quite a few are highly regarded. Nearly one-half of the state's counties have a metropolitan park system providing additional open-space areas and related facilities. Such systems in Ohio have a mandate to focus on preservationist functions. An increasing number of townships provide the equivalent of community parks, and the state contains one sizable national forest and one national recreation area. Ohio does have a mature, generally efficient, and popularly supported recreation system. The immediate need is to further improve the quality of the experiences of the participants, to make opportunities available to those currently unable to participate for whatever reason, and to identify improved management techniques or recommend additional recreational facilities in order to better meet anticipated demand.

Participation in outdoor recreation activities can only be estimated from currently available data. Few activities attract a majority of the population—even as occasional participants. The bulk of the recreationally active engage in a small number of activities, and it is likely that a large part of the population engages at least occasionally in some activity. Unfortunately, consistent data sets are not available from the most recent Ohio household surveys, and thus precise patterns and trends cannot yet be documented.

Table 15.1, which gives an idea of the relative popularity of selected activities, provides some notion of change between 1983 and 1990. Though the two household surveys are not perfectly comparable, two features stand out: 1) every outdoor recreational activity listed has some notable participation; and 2) from 1983 to 1990, a number of activities increased in the proportion of Ohio households that participate. That suggestion of a trend toward greater participation is consistent with the evidence provided by national surveys like the periodic Nielsen participation polls and by numerous state and local studies. Allowing for the special advantages and disadvantages of Ohio's recreational resources and environment, there is a real correspondence between the recreational behavior of Ohioans and of citizens of other states.

Ohioans continue to want their recreational facilities relatively uncrowded, clean, orderly, and accessible to population centers. In response to the latter desire, which has been repeatedly documented in field survey efforts, present efforts are geared toward encouraging the creation of additional facilities near major cities and making these facilities accessible to as much of the population as possible. Growth in the number of local facilities is likely to be great in the years ahead. Since facility needs are divided between the several different providers, facilities commonly provided in state parks are just as desired as those provided by local governments and the private sector. Unhappily, budget constraints for both state and city systems have been severe, having a negative effect on facility operation, programming, and maintenance in

Table 15.1

Percentage of Households Participating in Selected Outdoor Recreational Activities

ACTIVITY	1983	1990
Powerboating	28.2	35.7
Sailing	7.9	4.4
Waterskiing	17.6	18.0
Other Boating	16.7	22.1
Canoeing	23.7	20.9
Fishing from Shore/Wading	NA	54.0
Fishing from Boat	NA	41.8
Backpack & Tent Camping	23.1	25.1
Group Camping	13.8	15.6
Motorized Camping	20.5	22.5
State Lodges and Cabins	17.1	15.5
Deer/Turkey Hunting	NA	25.0
Small Game Hunting	20.5	29.2
Waterfowl Hunting	4.4	6.5
Other Hunting	9.7	14.1
Field Sports	22.1	NA
Baseball	NA	15.9
Softball	NA	22.3
Outdoor Soccer	NA	6.6
Court Sports	20.0	NA
Outdoor Volleyball	NA	18.1
Outdoor Basketball	NA	18.1
Outdoor Tennis	NA	10.1
Winter Sports	20.1	NA
Snowmobiling	NA	6.8
Ice Skating	NA	9.5
Snow Skiing	NA	7.3
Picnicking	67.1	59.2
Trail Activities	40.0	NA
Walking for Pleasure	NA	64.9
Day Hiking	NA	31.8
Jogging/Running	NA	20.6
Bicycling	28.7	30.9
ORV Riding	7.9	11.7
Outdoor Pool Swimming	41.9	39.5
Beach Activities	48.1	52.1
Gold	26.5	23.4
Wildlife/Nature Observation	NA	48.9

Source: 1986 and 1993 Ohio Statewide Comprehensive Outdoor Recreation Plans

recent years. Budget constraints on staffing are seen by recreation experts to be of particular concern, especially since better management, operation, and programming might mean that existing facilities could meet the growing demand without adding new facilities. By the same token, cooperation between two or more political units and governmental assistance of private sector efforts also have the potential of providing for future increased demand.

Ohio has been in the forefront of the country's economic and political life for many decades. It has not, however, been a leader in social and cultural change, and in fact the characteristics of Ohio's recreation were quite similar in 1990 to earlier years.

Changes in participation by activity

Boating: Changes in boating preferences are occurring, as powerboating has had significant growth while canoeing and sailing have both declined in popularity during the past seven years. No clear explanation for these trends is evident although reasons may include congestion on inland lakes and streams as well as changes in individual preferences for outdoor recreation. If better access to Lake Erie is achieved, all forms of boating on that body of water may grow significantly. Interior lakes are close to capacity at peak periods, and only the newest lakes are likely to have major increases.

Fishing: Fishing has retained its popularity and relative place among leading outdoor recreation activities, and there is no evidence to suggest a change, provided the state's water bodies can avoid a further major decline in their quality. If access to major water bodies is improved, modest growth may be anticipated.

Camping: Camping has increased slightly in popularity during the 1983–90 period and is likely to remain at its present level in the future. An increased interest in backpack camping is apparent and will require additional attention in the years ahead. Group camping is also a potential growth area.

Trail activities: Hiking and walking for pleasure will continue to hold their own and may demonstrate growth, as walking continues to grow in popularity as both a fitness means and a favored outdoor activity. As with picnicking, the participation data are incomplete, because of the diverse settings the activities allow. Demands on park managers for increased walking facilities will not be great, though if environmental awareness continues to develop, the need for nature trails will grow.

Bicycling: Because of transportation funding programs and the increased desire for long distance rides, a gradual growth in bike paths will occur. Safety issues are also contributing to a greater interest in developing a better bicycling situation. The bulk of the participation will continue to parallel the character of walking and picnicking and, like these two activities, will be difficult to measure because of the lack of an absolute need for specialized facilities.

Field sports: Interest in field sports will continue to place programming and facility demands on local agencies. Shifts in the particular sports demanded will occur due to previously noted changes in American life. Baseball and softball for the younger age groups will decline because of changing demographics and the declining number of young people; more adult softball leagues will be in demand for some years to come; growth sports such as soccer will require major additional facilities. To keep pace with the changing nature of the demand and with adult organizations concerned with the availability of sport opportunities, local agencies will need to maintain close ties to local school systems.

Court sports: Tennis is growing in demand, as are a number of other nearly lifelong recreational activities. Space for informal court sports is important in a variety of recreational landscapes.

Golf: Golf is a growth area in which the assurance of facilities open to the public for a greens fee is important. Good courses are near capacity on peak days.

Beach activities: Popular in Ohio, major beaches are filled to capacity on peak days, when the weather is reasonable. Few additional facilities are likely to be acquired, so total participation will remain constant.

Pool swimming: Pool use is currently on a plateau, with an increasing part of the population served by private or club facilities. Swimming and a number of other water-based activities will continue to peak annually in June and July and then decline in August as the season lengthens and household vacations are more common.

Winter sports: Though winter sports have high potential, Ohio's undependable winters are a major constraint. Both cross-country and downhill skiing will draw increasing numbers if the conditions are right and as the nature of skiing fits into the lifestyles of more people.

There is nothing in the data developed during recent surveys suggesting major changes on the horizon. Those changes that appear possible will be balanced by shifts in the composition of Ohio's population. Since the number of questionnaire respondents who indicated that they will participate more is similar to the number who indicated that they will participate less, a balancing off seems likely. However, it should be noted that each year sees the desire for better recreational experiences grow. The demands placed on the managers of delivery agencies are likely to be qualitative in nature, replacing quantitative demands as the primary area of concern.

Factors Affecting Recreation

In the mid 1990s there are a number of factors impacting the provision of recreation in the state:

1. State parks will have leveled or reduced budgets. Planned facilities at newer parks will be completed, but the primary concern will be to maintain the quality of existing parks. Major emphasis will also be placed on parks near urban centers.
2. The limitations on budget and staffing changes will stimulate additional innovative approaches to problems. Ohio has had a number of model operations; old leaders and new leaders will become visible as copers with the constraints facing agency operation.
3. Local park systems will hold their own. The older central cities and suburbs will have a continuing series of problems with aging facilities, vandalism, and littering. Security concerns will continue to grow.
4. Ohio's population will continue to grow in rural nonfarm areas. The roles of townships and counties will continue to demand reexamination. People will want recreational services. The legislature may have to modify the conditions under which township government functions.
5. The implications of the major socioeconomic change now taking place merit continuing study. The changing age structure ("a graying of the population") and the increasing diversity in the nature of households will result in different demands being placed on delivery agencies.
6. The population will have a continued growth of discrete groups who will seek special services and opportunities. These groups may be age groups or interest groups. Recognizing the complex composition of the population will be essential for the effective operation of a parks and recreation department.
7. As governments are perceived to be less successful in delivering desired recreational services and as concern grows over the quality of the recreational experience, the private sector is likely to be increasingly important. Cooperative ventures will be more common between public and private organizations and agencies.

There do not appear to be major changes in store for recreation in Ohio in the near future. Clearly, Ohio has problems and needs with most aspects of its recreational facilities. Those facilities are not likely to increase in number, the primary concern of participants on the one hand and the managers of the delivery agencies on the other will be to maintain the facilities at an acceptable level along with a continued search for means to make available the best possible recreational experiences. The concerns of the Ohio Department of Natural Resources (ODNR) and local officials should be primarily with what is now present in the several systems. Expansion should be limited and very selective.

Tourism

Tourism has been an identifiable economic activity in Ohio for many decades. However, much of the industry's economic impact has been invisible to the casual observer. In the 1980s and 1990s, tourism emerged as a growth activity increasingly encouraged by state government and pursued with equal vigor by communities large and small within Ohio. Its promise is widely recognized.

Today's widespread interest in recognizing and developing tourism as an economic sector has two evident sources. First is the realization that there is a market for what the state has to offer and that Ohio can indeed attract tourists if its resources are enhanced and made known. Second, the past two decades have seen a dramatic change in the nature of the state's economy, and tourism is one of several activities that can contribute to the continuing redevelopment of a different but still-sturdy economy. Agriculture is no longer a principal source of employment, and manufacturing, long the dominant mainstay, has experienced a major restructuring. Ohio needs to diversify, and it seeks to do so partly through the expansion of central office functions, professional services, high-tech industries, and tourism.

The hope is that tourism will do, at a much higher level, what it has always done—that is, be the primary economic activity in certain communities and function as a major economic sector in major metropolitan areas. Though it is often difficult to see clear direct and indirect economic impacts from tourism in complex urban economies, it is now well-appreciated that these impacts are real and that they work with other forces to maintain both jobs and an area's overall amenity base. The U.S. Travel Data Center's "Impact of Travel on State Economies, 1990" indicated that travel expenditures in Ohio in 1990 totaled about $8 billion, created 141,000 jobs, and yielded substantial taxes to both state and local governments. Their data places Ohio tenth among the fifty states in travel expenditures, and present development programs are designed to improve on that performance.

Ohio's expanding tourism industry recognizes changes in the leisure time travel patterns of Americans. The most important of those changes is the increasing trend toward multiple short (2–5-day) trips. That trend is partly a consequence of changes in the American household such as the increasing complexities in family schedules due to employment of two or more household members, increasing household affluence, and a desire for more frequent breaks from daily routine. Ohio has objectively considered its tourism potential and, at least for the near future, believes that the short-trip market is consistent with the tourism resources presently available in the state. Ohio's development program is firmly based on that premise.

Ohio's Tourism Assets

Numerous opportunities for outdoor recreational activities are presented by Ohio's natural resources. Environmentally, the state has four distinct seasons. Whereas a flat plains area dominates the northwest and west, the remainder of the state is gently rolling, including the dissected Appalachian Plateau. Natural water bodies consist of Lake Erie and an elaborate river system, highlighted by the Ohio River. There is general agreement that Lake Erie provides the greatest single resource for the industry. All interior lakes are man-made. Second- and third-growth forests are abundant and maintain their original maple-beech and hickory-oak associations.

The state's 10 million people are overwhelmingly urban, with nearly half of the total population living in the three major metropolitan areas: Cincinnati, Cleveland, and Columbus. Many of the tourist resources reflect the historical, cultural, economic, and social characteristics of these urban centers. Numerous smaller cities, towns, and villages also have tourist resources that encourage substantial visitor flows.

Ohio's tourism resources may be classified in three major groups:

1) Resources based on the environment, which include the western, central, and eastern basins of Lake Erie; Appalachia; the Ohio River; and outliers to these areas.
2) Major metropolitan areas—primarily Cincinnati, Cleveland, and Columbus—which offer historical and cultural attractions; amusement and theme parks; spectator sports; hotels, restaurants, and shopping; convention facilities; people (visiting family and friends is a prime tourism motive); and special events.
3) Secondary centers and attractions, which include locations with all-year attractions (comparable at a different scale to the resources outlined for major metropolitan centers); locations with seasonal attractions; locations with short term attractions.

A number of the primary attractions are based on environmental resources. Though Lake Erie is the single most important resource, its value and potential have varied in recent decades in line with the quality of its waters. In the 1960s Lake Erie was a seriously ailing body of water, but a number of recent pollution control measures have helped reverse the decline in water quality. Today the waters are much improved, although potentially serious problems, such as heavy metal accumulations, remain. The lake's western basin is the section of greatest value; this is due both to its array of peninsulas, embayments, and islands and the presence of the finest walleye fishing waters on the North American continent.

Ohio's Appalachia contains both the state's greatest concentration of state parks and all three parts of the Wayne National Forest (Figure 15.1). The wooded hills of this eastern and southern region have much outdoor recreation appeal. Man-made lakes and streams dot the area, and the Ohio River forms its eastern and southern boundary. It has long been thought that this region possesses a great potential for outdoor recreation and tourism. Although both activities are important, the area has not yet achieved its potential. Additional developments that link resources and/or a decline in the ability of Ohioans and residents of neighboring states to travel to distant outdoor recreation sites may be needed before the region's potential is approached.

The major cities each possess a complex of accommodations, cultural and historical resources, and other attributes that are appealing to visitors, and each is making a major effort to attract large conventions, a critical component of urban tourism. Each city also has developed special events that symbolize its special qualities. These events serve both to boost the total number of visitors and to reduce the marked imbalance in tourist flows between the summer and holiday weekend peaks and the rest of the year. Cincinnati's Riverfest is one annual event that draws immense crowds, and 1992's Tall Stacks weekend, also in Cincinnati, drew well over a million visitors to a festival celebrating the steamboats that once ruled the inland rivers of the central United States. Cincinnati has made tentative plans to repeat the Tall Stacks event every 3–4 years. Columbus's Ohio State Fair is one of the most popular single events held in the state.

Numerous other centers and areas in the state have resources or events that pull large crowds of tourists. These range from a major air show in Dayton to innumerable local events centering on narrow themes. Both local government or development groups and the state's tourism agency are increasing their energetic promotion of all tourism elements.

Tourist Destinations and Tourism Regions

Although little good data is available on the distribution of visitors to tourism destinations across the state, it is clear that four major tourist regions exist (Figure 15.2): the three larger metropolitan areas (Cincinnati, Cleveland, and Columbus) and the so-called Lake Erie Vacationland clearly draw the largest proportion of Ohio's visitors.

The three metropolitan areas have a large and varied set of tourism resources. As large cities, they draw many visitors desiring to take advantage of their shopping, medical services, spectator sports offerings, museums, musical organizations, theaters, and universities. A vigorous program is in place in all three cities to boost convention traffic, with related efforts designed to keep conventioneers in town for an extra day or two as a visitor. Each has some special resource, such as the major theme parks found in the Cincinnati and Cleveland areas or the state's largest university in Columbus. Additional facilities and events are planned in all three metropolitan areas to further stimulate the flow of tourists.

The Lake Erie Vacationland in the northwest part of the state is the only major tourist region based on environmental resources. Lake Erie is directly and indirectly responsible for a tourist region that began in the years following the Civil War as a summer home region and continues today as a complex vacation area with a high rate of growth. The tourism industry present in this region today is based on a variety of factors. These include an enormous array of cottages, summer homes, condominiums, camping areas, trailer parks, and other forms of weekend, weekly, monthly, or summer-long accommodations; the variety of water-based forms of outdoor recreation; the opportunity to escape from such nearby cities as Cleveland, Detroit, and Toledo; and the pleasures of the region's many scenic drives and vantage points as well as its historic sites. The presence at Cedar Point, near Sandusky (Figure 15.3), of one of the country's most popular theme parks is another important factor. Finally, sport fishing in the area—especially for walleye but also for several other desirable species—has resulted in economic activity now valued at well over half a billion dollars a year.

Ottawa County, just northwest of Sandusky, epitomizes lake-related development. The center of the walleye fishing region, this county contains several thousand summer homes and 70 percent of Ohio-registered boats located on Lake Erie. Ottawa County's official population numbers slightly over 40,000, but that figure swells to an estimated 120,000 during summer weekdays and 250,000 during summer weekends, according to the Ottawa County

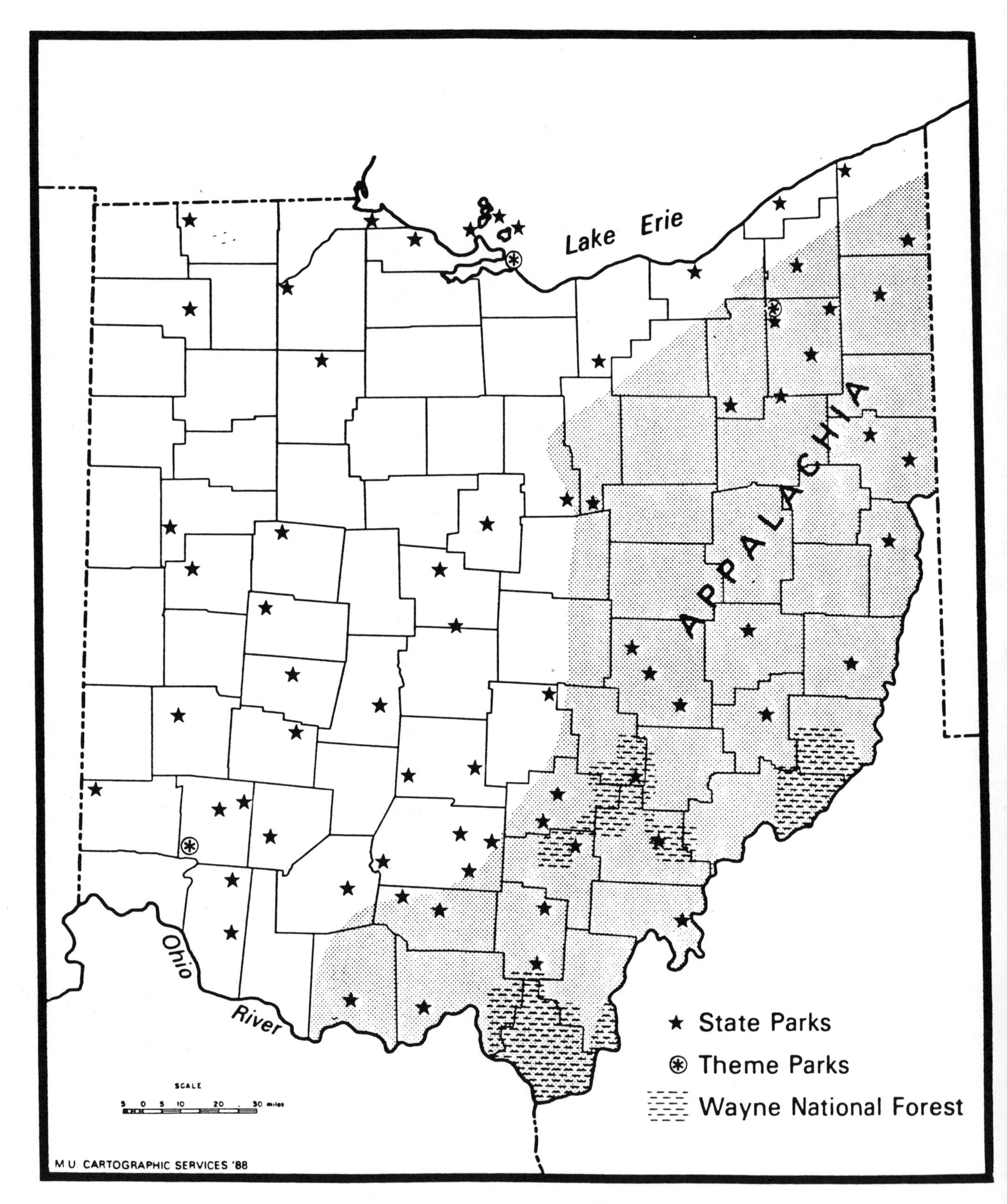

Fig. 15.1. Selected Ohio Tourism Resources

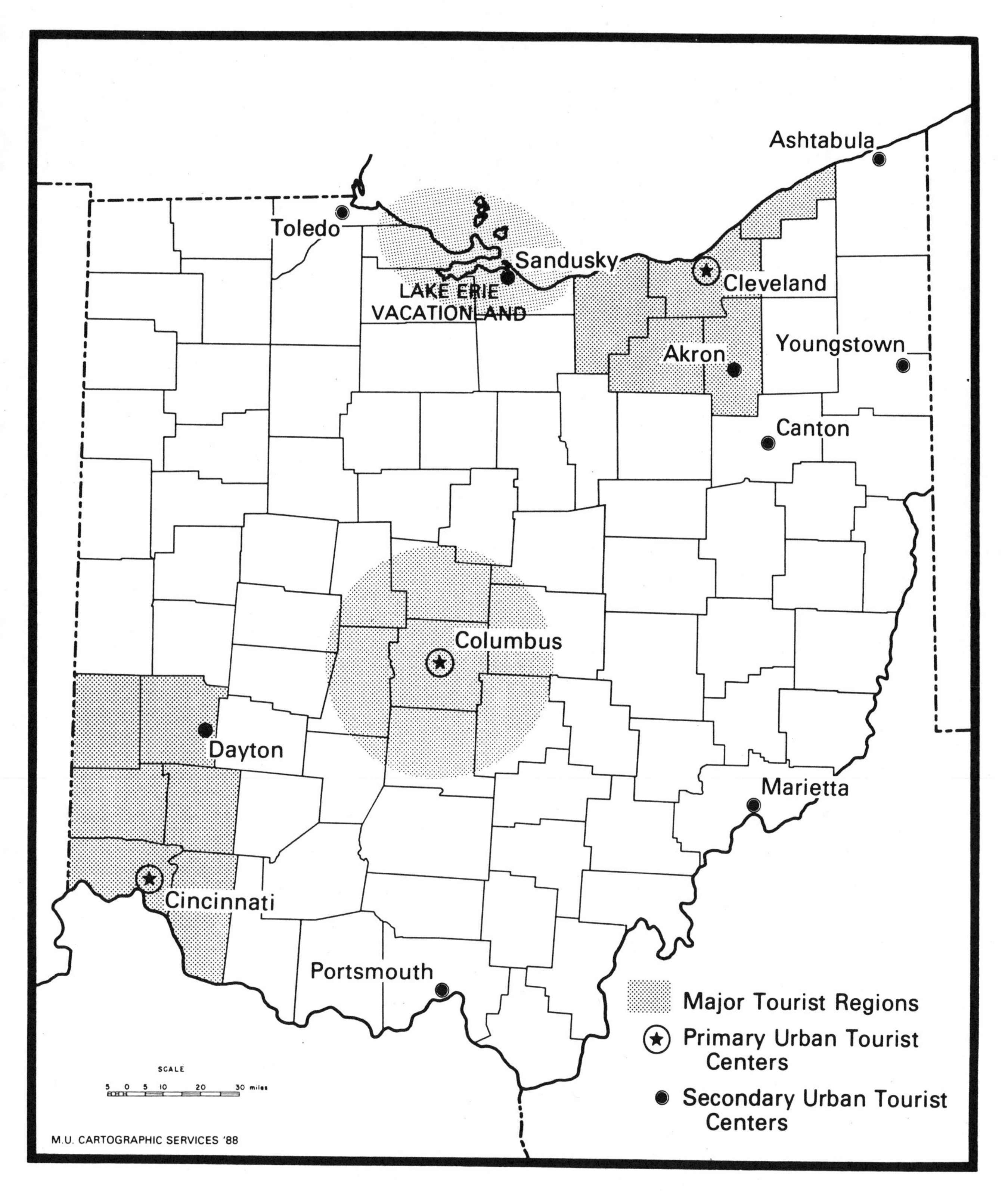

Fig. 15.2. Major Tourist Regions and Urban Tourist Centers

Fig. 15.3. Cedar Point, One of Ohio's Main Tourist Attractions (Photo by Dan Feicht; Courtesy of Cedar Point)

Visitors Bureau. Such figures are compelling evidence for describing this as a major tourist region. The enormous summer weekend volume of visitors is intensifying pressures on the lake, the road system, and other parts of the county's infrastructure, as Ottawa County experiences both the positive and the negative impacts of a tourism boom.

There is no precise information on how many communities in the state share in the tourist industry. Seventy-two state parks scattered throughout the state receive tourists. Over two thousand special events are offered by at least three hundred communities during the year. Mapping and analysis of the relationships among these places and the events they offer have begun. What is evident now is that many communities are actively encouraging the flow of visitors.

The state's four major tourism regions may soon be joined by another resource-based region: Ohio's Appalachia. Other distinct regions are not likely to develop, but it may be possible to map linear alignments of frequently visited places to complement the formal regions in the near future. The concept of tourist regions perhaps should be joined with a set of lesser but still important descriptive terms for alignments or clusters of tourism places. Mapping and structural studies now under way should permit testing a model of tourism places and regions that would better describe the spatial nature of tourism in large regions, states, or countries.

Tourism Planning

A significant level of planning for the further development of tourism is a phenomenon of the 1980s. The incoming administration of Governor Richard Celeste in 1983 included in its economic development effort a concerted effort to boost tourism as an economic sector with the hope of measurable benefits in jobs and taxes. By any reasonable measure, that effort has been a success and is being continued in the 1990s under Governor George Voinovich's administration. It is estimated that over 140,000 jobs have been created and that the volume of tourist spending has increased by $2–3 billion since 1987. It is difficult to identify exactly how much of this increase can be attributed to the state program, but there is universal agreement that it has been considerable.

The Office (now Division) of Travel and Tourism in 1983 started a marketing research program to identify the most promising means to enhance Ohio's tourism industry. With the help of a consulting firm, a distinctive plan evolved. One major element was its recognition that few people knew about Ohio's touristic possibilities. Therefore an information campaign was launched that included the distribution of attractive informational materials and the establishment of a toll-free telephone information service that provided useful suggestions and up-to-date information on events. Advertisements were placed in journals as well as on radio and television, and a discount coupon booklet was developed in cooperation with the operators of many tourist attractions.

The research conducted by state officials and their consultants also suggested that a market segment be identified. Given Ohio's tourism resources, what kinds of tourists should be the focus of the information campaign? Although the long-term development effort would continue to include all forms of tourism, the short-run focus was determined to be best aimed at short-term visitors. Thus a "getaway" theme was instituted that focused on the virtues of spending a two to four day holiday in Ohio. Ohioans themselves were part of the target for this approach, as were the residents of urban areas within 250 miles (approximately a 4-hour automobile trip) of Ohio.

In more recent years, especially 1987 and 1988, the focus of this program shifted, and the 250-mile target area was broadened to 500 miles (Figure 15.4), a shift that adds many tens of millions of Americans and Canadians to the targeted population. Additional efforts with new incentives are being aimed at group tour operators and travel industry executives.

In a similar vein, local areas are also increasing their developmental efforts. Every town and city has some agency or person responsible for encouraging an increased flow of visitors. Promotional materials flow in an effort to attract the family, the convention planner, and the group tour operator. Cleveland's Convention and Visitors Bureau notes that visitors spent $1.8 billion in 1986, creating 33,278 jobs in Cuyahoga County. Ottawa County believes that at least $600 million is spent in the county by visitors each year. Akron reports that overnight and day tour groups increased by 50 percent in 1988 over 1987. Such results are encouraging other communities to increase their promotional efforts.

The development effort is taking special forms in certain areas. For example, Cincinnati has become the first city in Ohio to have direct nonstop flights to Europe, and daily flights to London, Frankfurt, and Paris have become very successful. Both city officials and Delta Airlines (the carrier involved) are energetically encouraging European travel agents to make use of this connection, and positive results have been reported.

Potentials for the Future

Development officials are confident that recent growth trends in Ohio tourism will continue, due to several factors:

1. Survey results indicate that 44 percent of Ohio residents took a day or overnight trip for recreational purposes. Such results suggest that additional Ohio families and residents of neighboring states might also be encouraged to pursue in-state tourist trips and also that multiple trips by households to various Ohio locations are possible, further expanding the potential group of consumers. As it is believed that a considerable part of the market has not yet been touched, Ohio's tourism development program seeks both to encourage Ohioans to make more of their leisure time expenditures within the state and to attract a flow of money from outside state boundaries.
2. Ohio's resource base for tourism is still developing. A wide variety of current projects—for instance, new Lake Erie resort communities, new marinas, expanded and more attractive urban amenities, and enlarged convention facilities—points in this direction.
3. Foreign visitors, realizing that Ohio has much to offer, are including short stays in their itineraries. The already important Canadian tourist flow has been increased through the targeting of the whole of Ontario for media advertisements.

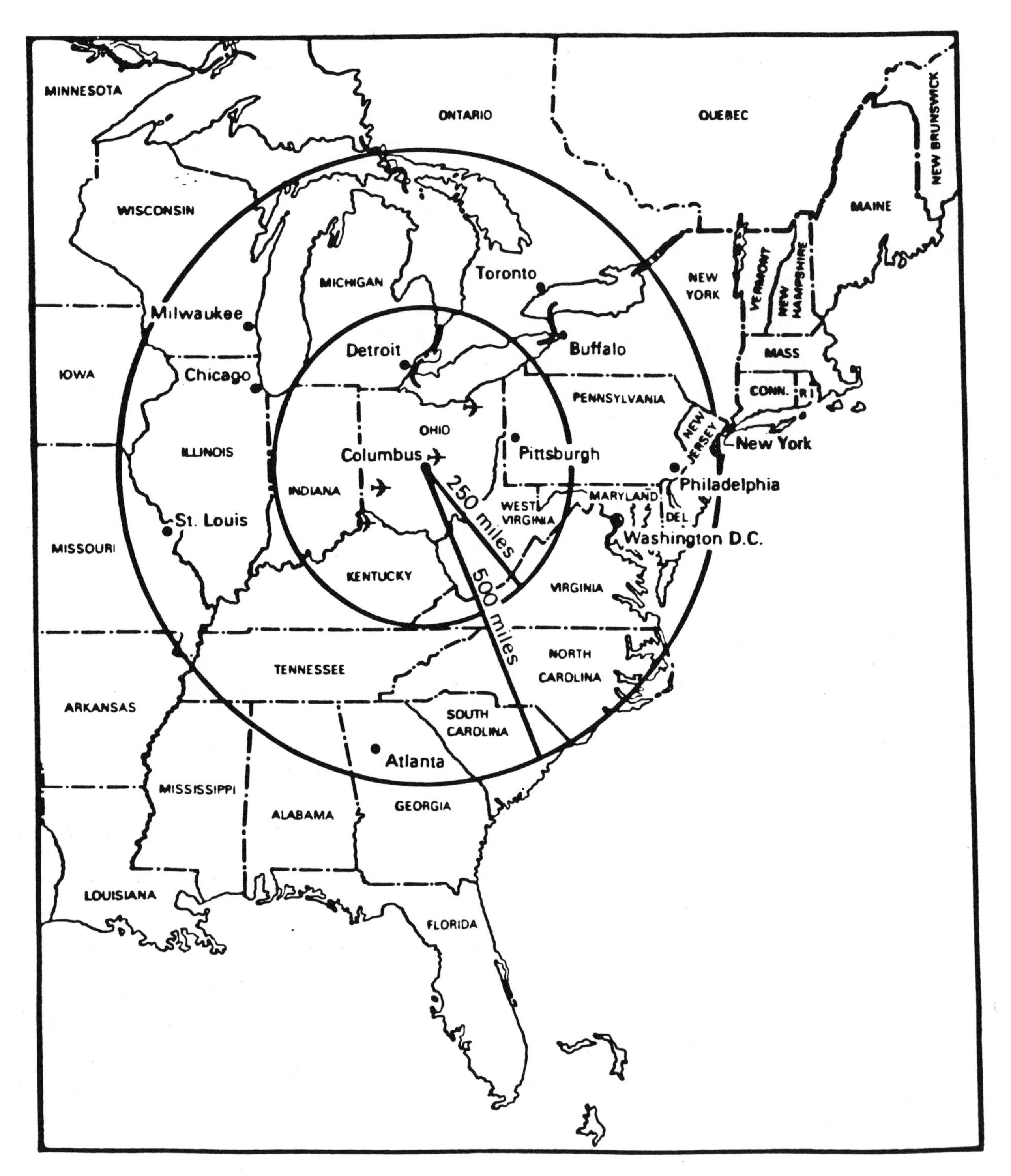

Fig. 15.4. Ohio's Targeted Market Areas

4. Ohio and the other Great Lakes states have joined together to push the Great Lakes region as an attractive place to visit. This effort is particularly important at worldwide travel fairs. The cooperative approach has much to offer and is likely to increase in its effectiveness in the years ahead.
5. Cities and areas are making use of special days and special events to draw visitors. It is believed that visitors who get a taste of what the area has to offer are likely to make repeat visits.
6. The greater reach and intensity of the state's promotional campaign, in conjunction with increased

local efforts, are making Ohio much more visible as a holiday alternative.

The tourism development efforts of the 1980s and 1990s have met with considerable success, and the potential for continued expansion in this industry is great. However, careful planning and increased cooperation between officials of neighboring localities and between states and local officials will be needed. Until the 1980s, few Americans would have included Ohio in their list of important tourist states. Recent events have changed that image, and tourism is commencing to have measurable impacts on the economic and physical landscape of the state.

REFERENCES

Division of Travel and Tourism. 1984. *Ohio Travel and Tourism: General Public Survey Results.* Columbus: Ohio Department of Development.

———. 1986. *Tourism in Ohio: A Prime Economic Incentive.* Columbus: Ohio Department of Development.

———. 1987. *Ohio: The Heart of It All!* Columbus: Ohio Department of Development.

———. 1988a. *Calendar of Events: May–August, 1988.* Columbus: Ohio Department of Development.

———. 1988b. *Trends in the Recreational Travel Market in Ohio, 1988.* Columbus: Ohio Department of Development.

Miami University Research Group. 1984. Fifteen research reports submitted to the Office of Outdoor Recreation Services, Ohio Department of Natural Resources, as follows: Outdoor Recreation in Ohio (1984 and 1990); Community Recreation; Backpack Camping; Beach Activities; Boating; Fishing; Group Camping; Organized Nature Programs; Picnicking; Golf; Social Camping; Trail Activities; Winter Activities; A General Summary Report; and Preliminary Study on Motivators, Enablers, and Barriers.

National Park Service. 1984. *Preliminary Findings of 1982–83 Nationwide Recreation Survey.* Washington, D.C.: Department of the Interior News Release.

Office of Outdoor Recreation Services. 1988. *1986 Ohio Statewide Comprehensive Outdoor Recreation Plan.* Columbus: Ohio Department of Natural Resources.

Ohio Department of Natural Resources. 1976. *1975–1980 Statewide Comprehensive Outdoor Recreation Plan.* Columbus: Ohio Department of Natural Resources.

———. 1981. *1980–1985 Statewide Comprehensive Outdoor Recreation Plan.* Columbus: Ohio Department of Natural Resources.

———. 1988. *1986 Ohio Statewide Comprehensive Outdoor Recreation Plan.* Columbus: Ohio Department of Natural Resources.

———. 1993. *1993 Ohio Statewide Comprehensive Outdoor Recreation Plan.* Columbus: Ohio Department of Natural Resources.

Smith, R. V. 1986. The Potential of Tourism as a Factor in Ohio's Economic Future. *Papers and Proceedings of Applied Geography Conference,* vol. 9.

———. 1990. Tourism and Tourism Planning in Ohio. *Tourism Recreation Research* 2.

U.S. Bureau of Census. 1987. *Statistical Abstract of the United States.* Washington, D.C.: U.S. Government Printing Office.

U.S. Travel Data Center. 1992. *Impact of Travel on State Economies, 1990.* Washington, D.C.: U.S. Travel Data Center.

U.S. Travel and Tourism Administration. 1985. *Measures of the Economic Impact of Foreign Visitor Spending in the U.S. by State Visited, 1983.* Washington, D.C.: U.S. Department of Commerce.

Young, Cyrus W.. 1985. *Facility Based State Park Participation Models.* Report prepared for the Office of Outdoor Recreation Services, Ohio Department of Natural Resources.

Editor's Note: This research was financed in part through a planning grant from the National Park Service, Department of the Interior, under the provisions of the Land and Water Conservation Fund Act of 1965 (P.L. 88–578); and in part through a grant from the Office of Outdoor Recreation Services, Ohio Department of Natural Resources. The outdoor recreation section was originally published in the "Focus on Ohio" issue of *Ohio Geographers* (vol. 16, 1988), and the tourism section in Miami University's Department of Geography working papers was released in January 1990 (Miami Geographical Research Paper, no. 2). Both sections have been revised and updated for this book. The maps included in the tourism section have also appeared elsewhere.

CHAPTER SIXTEEN

Change on the Edge of Ohio

Nancy R. Bain

Removing the economic bulk of Ohio—its industrial and agricultural areas—leaves an edge known as Appalachian Ohio. This area has been on the periphery since early statehood; only during initial settlement did the area boom, as early settlers crossed through the region or traveled up the Ohio River. This period was brief, and after a short time the region was relegated to relative backwater status. Figure 16.1 shows the movement of settlement through early nineteenth-century Ohio. At this time, Ohio's Appalachian edge had a settlement background—predominantly Middle Atlantic in origin with an overlay of Southern influences. By 1850 the region had generally less ethnic diversity than the remainder of the state.

To define a region as an edge implies transition, and it is this characteristic that has worked to create the southeast. Although it could be seen as the western fringe of the Appalachian physiographic region or the resource-rich boundary of Ohio's industrial heartland, there is yet another perspective, for the region is increasingly becoming the recreation playground of postindustrial Ohioans.

Physical and Human Backdrop

The physical geography of the southeast sets it apart from the glaciated and flatter sections of the state. Its divergent nature is apparent on many state maps: on a map of glacial deposits (see Figure 1.6) it is the unglaciated section; on a map of physiographic sections it is the eastern region (see Figure 1.1). Oil, gas, and coal deposits are predominant in this area. These resources and other environmental traits provide both opportunities and liabilities for the human occupants of this unglaciated section of Ohio.

Quite early in the European settlement phase, variations in land use came from the different choices made in facing a harsh physical environment. Originally the first state capital was in Ross County, but this shifted northward along with the major population movement that took place. This early period has been well considered by Hubert Wilhelm, who, in his work on the 1850 census, shows the area between the Ohio River corridor and the developing portion of the state as a stagnating land (1982). Similarly the area retained a strong ethnic link with the British Isles, while the rest of the state experienced the successive waves of immigration that led to a blossoming diversity of cultural groups.

In recent times these rural southeastern counties remain sparsely settled and different from the rest of the state in a number of ways. Mapping the traditional indicators of social and economic status, one finds the block of edge counties together in the same group. Figure 16.2 shows that the lowest group in median family income concentrates in the region. Likewise, an indicator of educational attainment, the lowest percentage of high school graduates in persons over 25 years of age, appears in the edge area (Figure 16.3). These and other indicators suggest an area with more concentrated disadvantaged population, less spatially concentrated than in the rest of the state. These conditions have attracted special attention and development programs for the area. One long-term initiative has been the Appalachian Regional Commission (ARC) that has worked to develop the area under a number of banners, from growth areas to enterprise zones. These planning districts are shown on Figure 16.4.

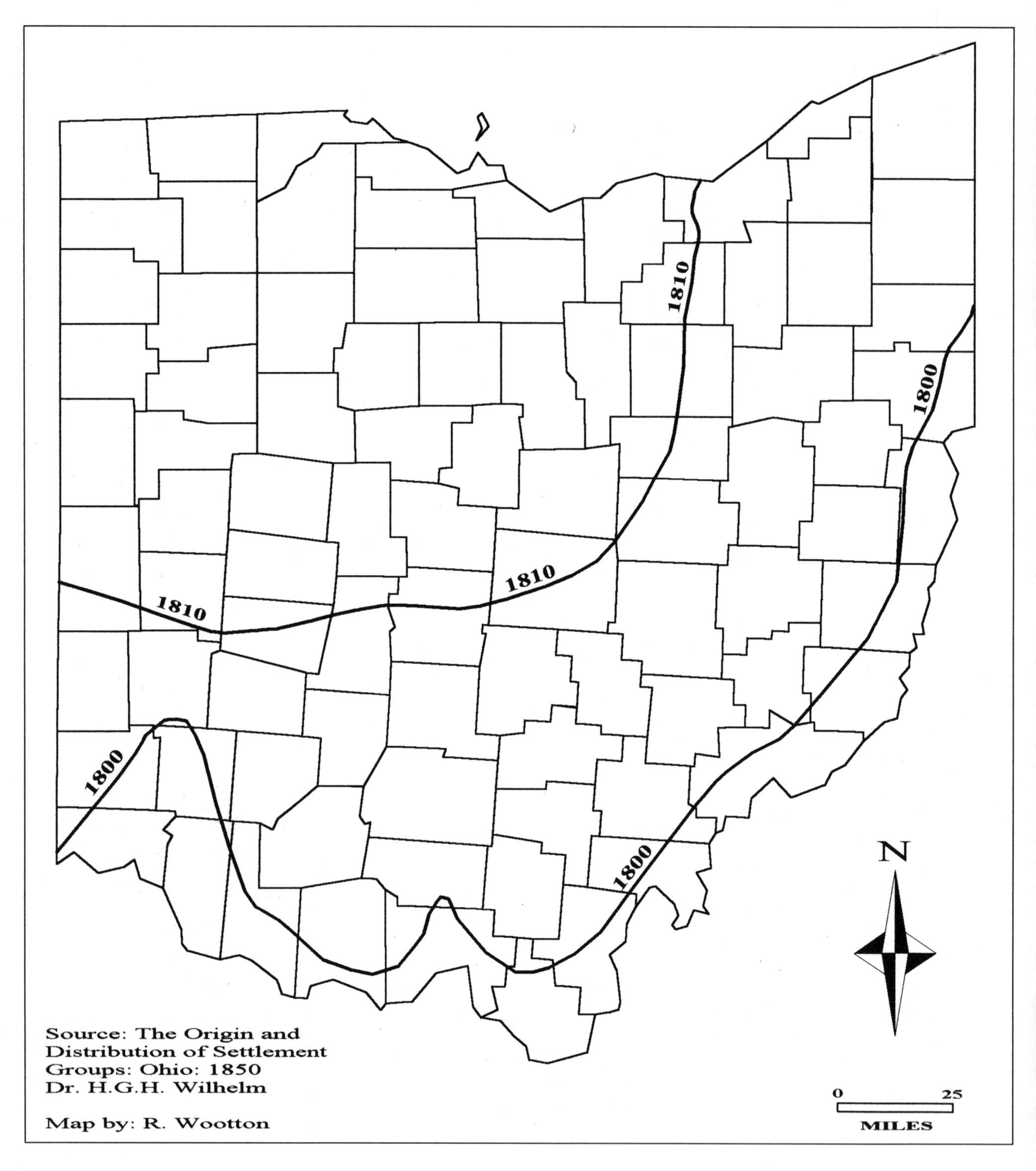

Fig. 16.1. Frontier Settlement Advance in Ohio

Resource Boom and Primary Activities

The edge has maintained a persistent emphasis on the primary activities of farming, mining, and logging. These began as local subsistence activities but grew to provide materials for the industrial development of the state. Initially, subsistence farming dominated the economy, but this was overtaken by mining and

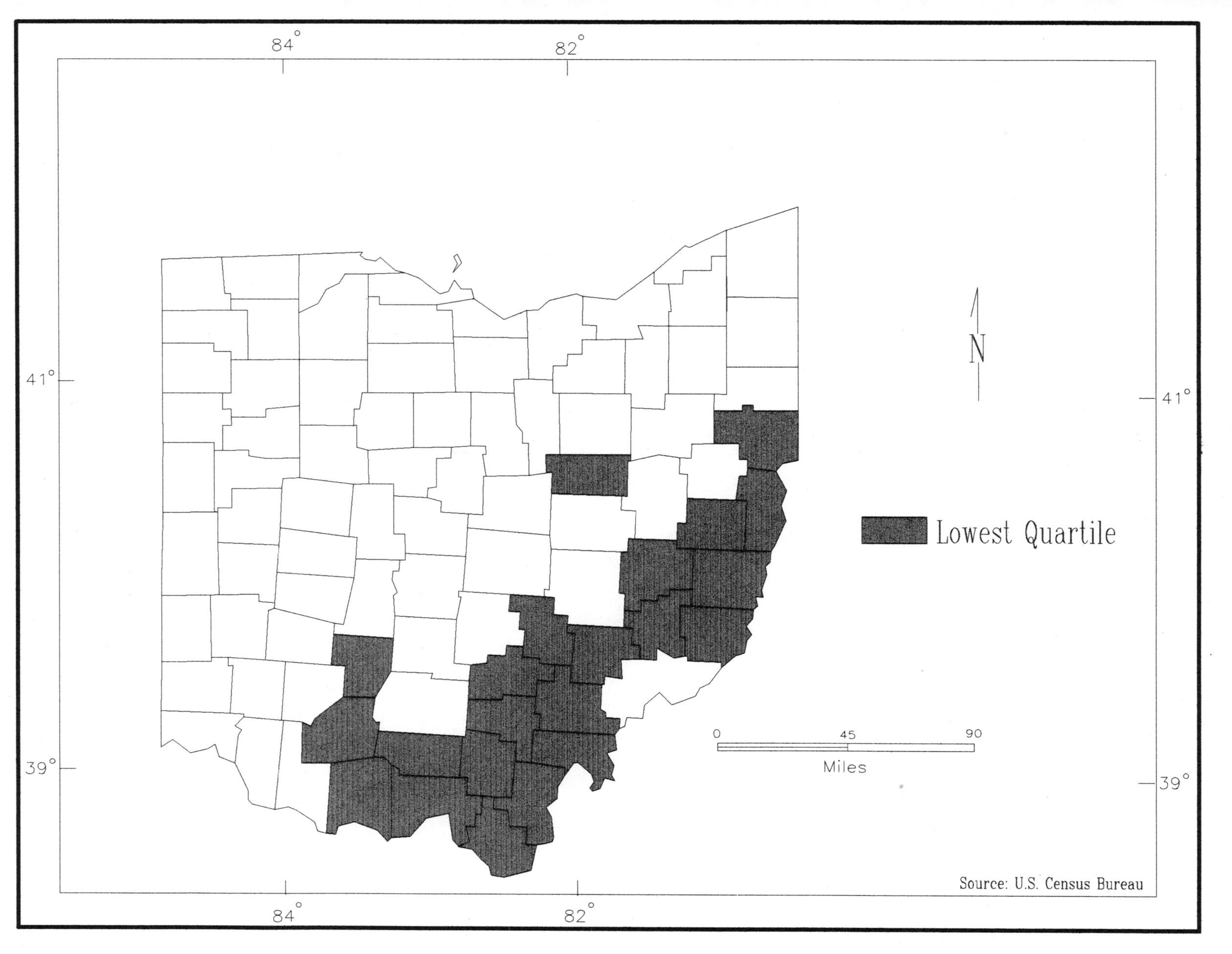

Fig. 16.2. Median Family Income in Ohio, 1989

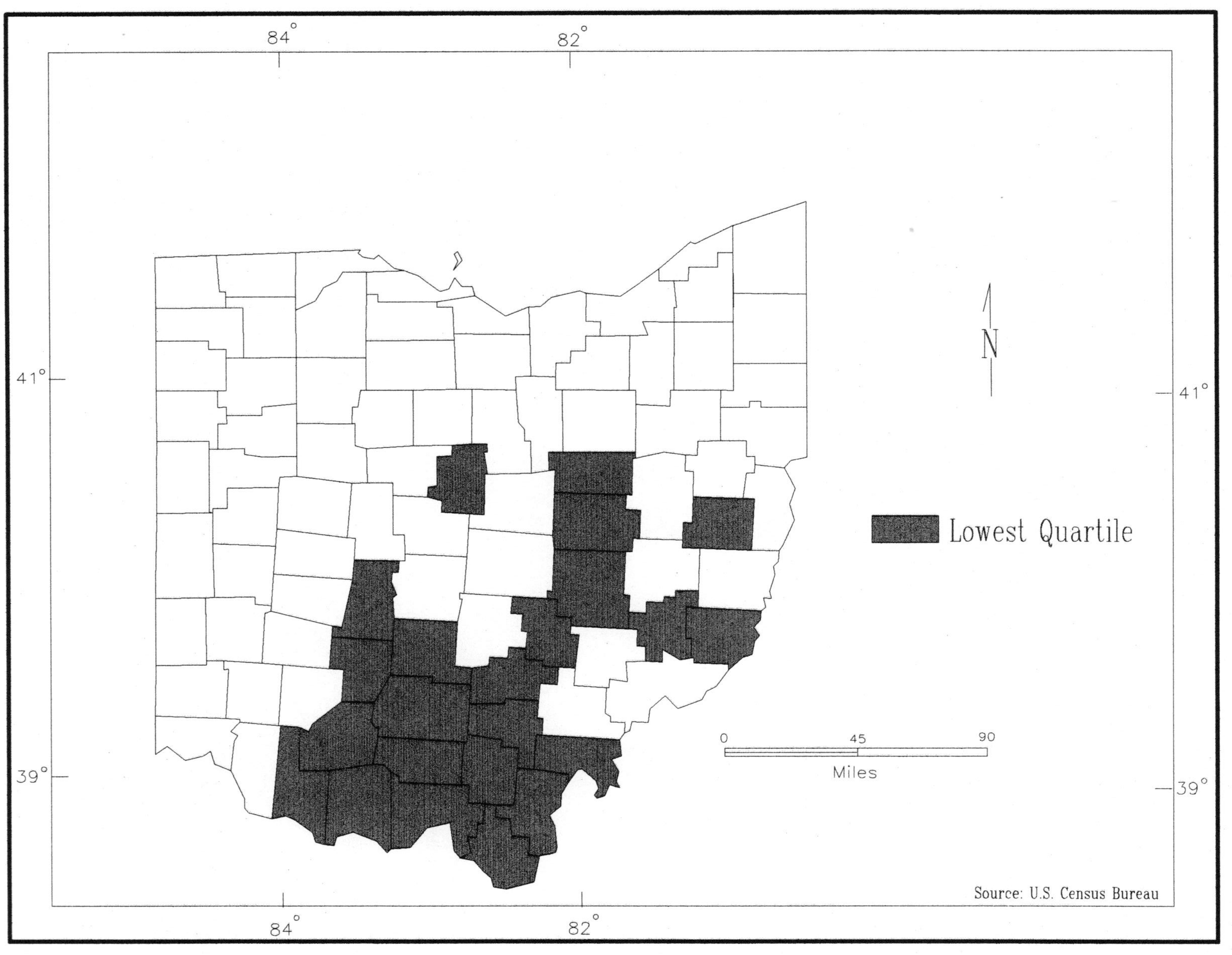

Fig. 16.3. Educational Attainment, 1990 (Percentage of People Over 25 Years of Age with a High School Diploma or More)

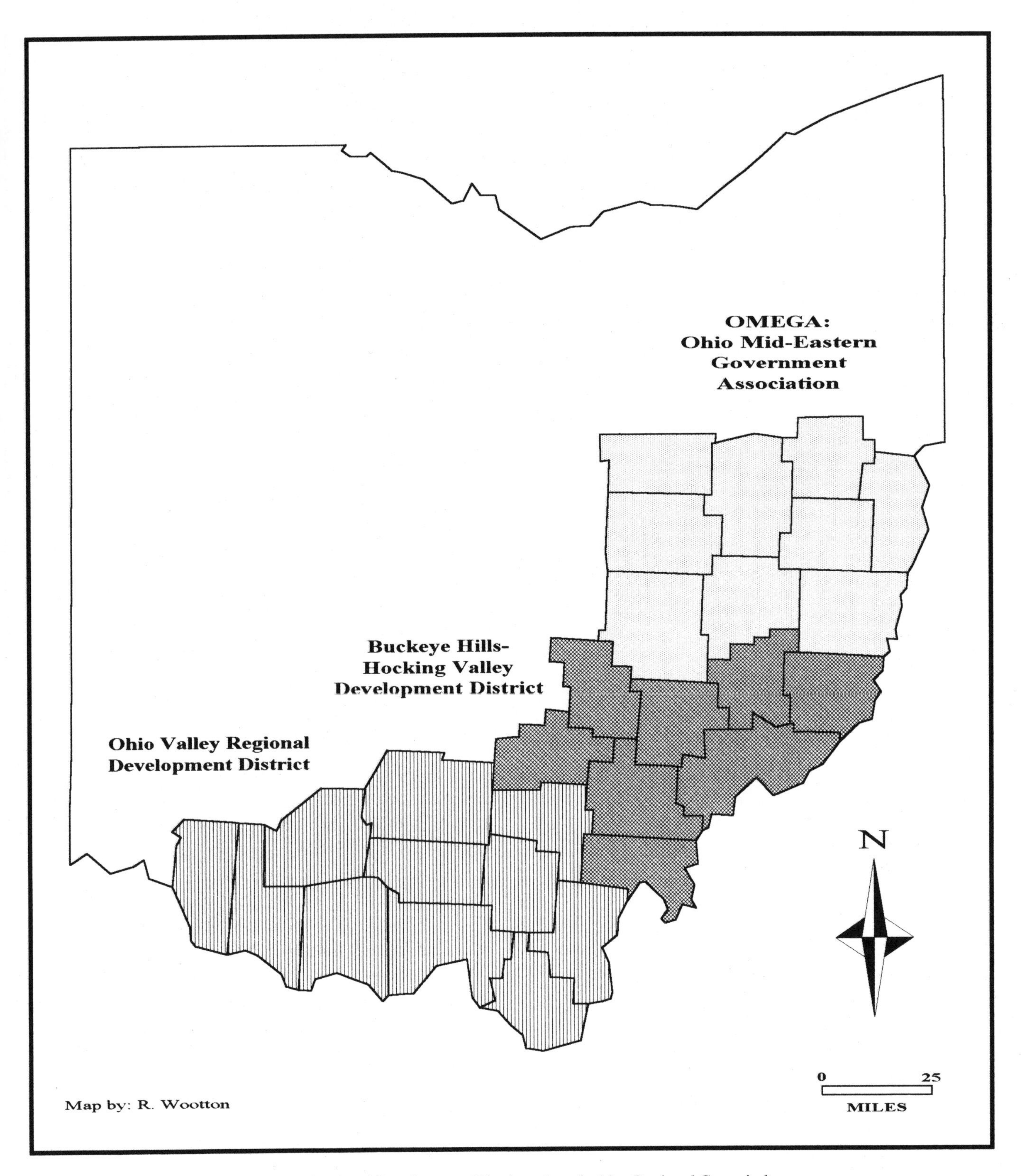

Fig. 16.4. Regional Development Districts; Appalachian Regional Commission

Table 16.1
Population in 1970 as a Percentage of 1900

	STATE	SUB 9	SUB 10
Number of Farms (%)	40.2	30.2	26.0
Area in Farms (%)	69.8	51.3	43.8
Total Population	256.3	124.0	117.0

Source: Sitterley 1976

logging as the state industrialized. With the statewide move to commercial farming, farming units disappeared more quickly here than across the rest of the state (Sitterley 1976). Consequently, farming declined as the primary economic activity. Subareas nine and ten (Sitterley's designation for Appalachian areas) include much of the physical region described above.

As industrialization developed, wood, iron, and coal emerged as the primary economic resources, especially in large tracts of privately held land owned by entrepreneurs who recognized their potential (LaVelle 1984). Initially the emphasis was on wood, or wood with iron; then it shifted to coal with iron. One important iron area was the Hanging Rock district along State Route 93. The early iron furnaces, which were based on the proximity of either wood or coal plus iron ore, served a more local market than the emergent blast furnaces of the industrial era.

With the industrial age came the expansion of the railroad into the area, and that development had a major impact on the region. The railroads provided both a market for coal and the means to export surplus coal to the Great Lakes transportation system. For a brief moment this region's coal deposits and deep mines were the most important in the nation, with a regional output that fueled the industrialization of the state.

During its early development in this region, the mining industry shifted from a craft activity to an industrial one. The craft basis of early coal mining came from a system in which each miner had individual job control and was paid by the piece. These miners' tasks included undercutting, drilling, and shooting, followed by the laborious handloading of the coal. In the early years of the boom, the Jeffrey Machine Company eliminated the tedious job of using human muscle power to undercut in cramped quarters, and replaced it with a machine that used compressed air. The Hocking Valley (Athens, Hocking, and Perry Counties) operators began using this technology in the 1890s. However, the rest of the mining procedures remained in the hands of the individual miner, who followed the undercutting machine with the tasks noted above. Area mines varied in the extent of the coal, the depth of the seams, and the presence of gas and excessive water within them. The craft basis of mining made the miners primarily responsible for personal judgments on safety. And all the while they were paid on a piecework basis.

During this early industrial period, mine owners formed groups called *syndicates* to increase marketing efficiency and to deal with blossoming labor problems. The introduction of machinery increased safety concerns from the pick days, by creating noise, dust, and variable roof conditions. These conditions, combined with the rapid expansion of the workforce without apprenticeship training, led to some terrible mine disasters (see Figure 16.5). The resulting conflicts between labor and mine owners spurred the formation of the trade unions that, at a meeting in Columbus in 1890, would amalgamate to form the United Mine Workers of America.

Urban Field

Coal mining in the region declined as new coalfields opened up in Southern and Western states. Responding to this decline, numerous residents moved out of the region. Instabilities in the population were common until the later years of this century, when the new reality became the urban field. Ohio's Appalachian edge became the opportunity area for expansion of urbanite residences, industrial developments, recreation, and service industry facilities.

New work on the evolving form of cities often emphasizes the urban fringe, the area between the built-up portion of the city and its surrounding rural area. Definition and treatment of this fringe in the literature tends to stress its hodgepodge nature, which evolved quite predictably out of its transitional status. More recently, Joel Garreau coined the term "edge city" (Garreau 1991). From this term one envisions a circular swath of low-density land uses intermingled with a number of smaller cities. This development requires, of course, a large city as its centerpiece. The southeast region lies on the peripheries of Ohio's three largest and most influential cities: Cleveland, Columbus, and Cincinnati—each of which creates potential edge developments.

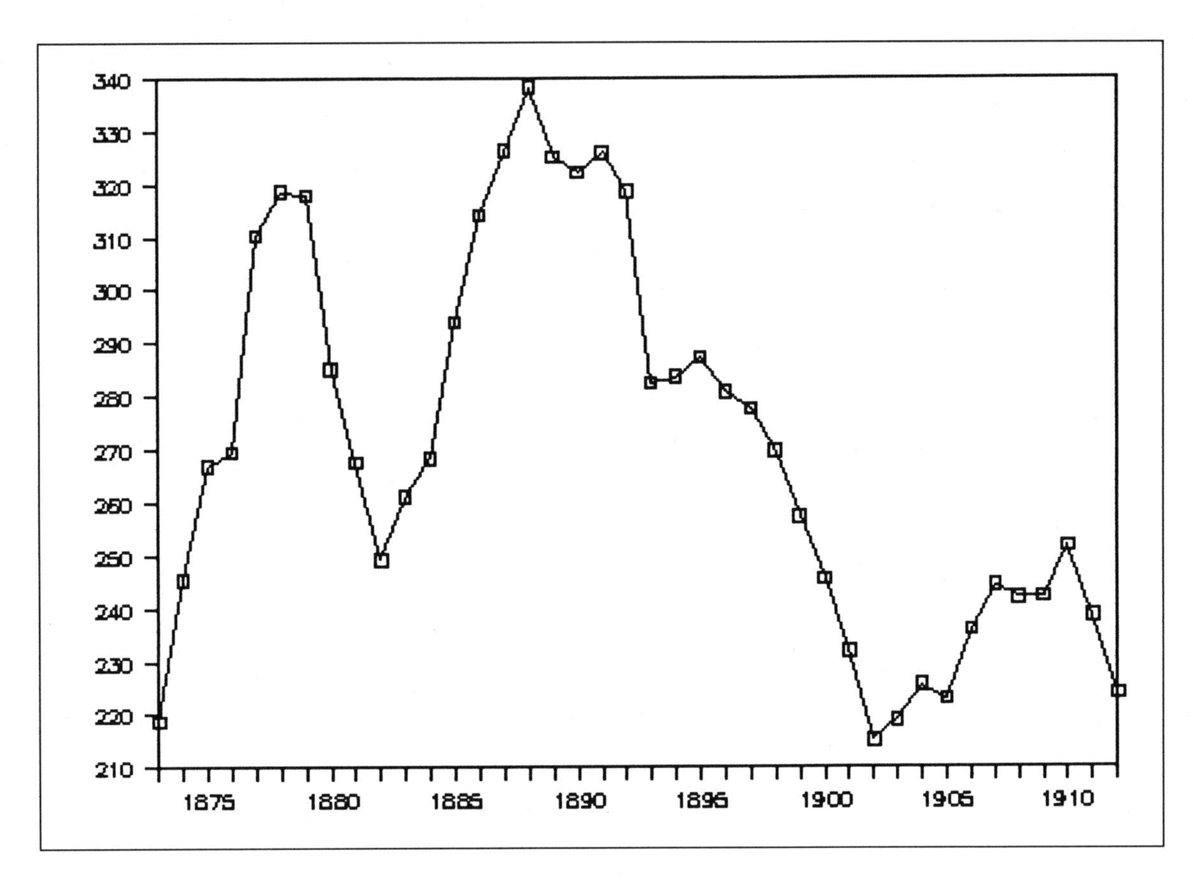

Fig. 16.5. Tons Per Mining Fatality

Rural Choice in Residential Location

Recent increases in rural population, termed by some the "Rural Renaissance," come from the desire for lower taxes, relief from urban threats, and a more comfortable environment. These reasons stand in contrast to the usual explanation for internal migration that stresses job-related moves. Although the shift of population to nonmetropolitan counties has concentrated in the small town and fringe areas, a portion of the new growth includes some movement into rural communities. This movement is transforming the rural landscape from an agrarian one to a nonfarm country landscape. It combines the needs of the modern urbanite with the residues of the past. The term *nonfarm* comes from demography and census work. It developed as the designation—with all its implications no longer fully appreciated—of the settlement and occupational traits of those residing in the country but working and living an urban life. This concept, from a census viewpoint, allows a clearer picture of the variations in rural life.

Public opinion surveys reveal that respondents prefer a rural setting. A typical response would be that three out of four people state a preference for the small city, village, or countryside. The numbers vary from survey to survey, but the stated preference, if not the actual behavior, would be for the rural nonfarm setting. Fringe areas with open land—no longer farmed and outside other primary pursuits—provide opportunities for recreation, lower-density residential development, and hobby farms. One summary of six separate county studies of rural land holdings in western Appalachian counties across several states showed substantial areas held by urbanites in small parcels for primarily recreational purposes (Bain 1984). This same study revealed that the combination of nonresidents and public ownership controlled at least 15–30 percent of Ohio's total land area.

Industrial Expansion into the Edge

Industry has also moved into rural areas with greenfield sites growing more in importance, beginning

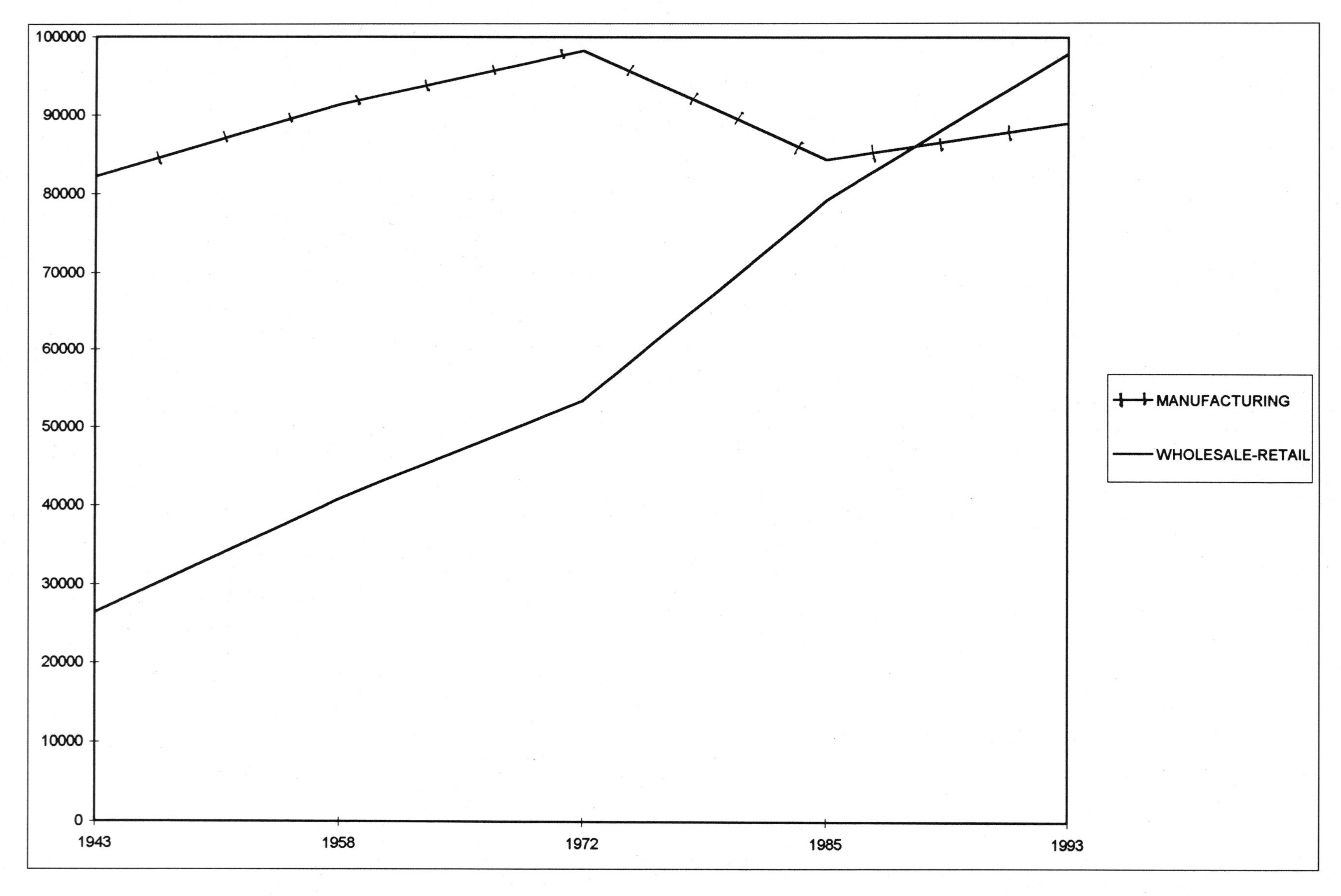

Fig. 16.6. Employment Totals (County Business Patterns)

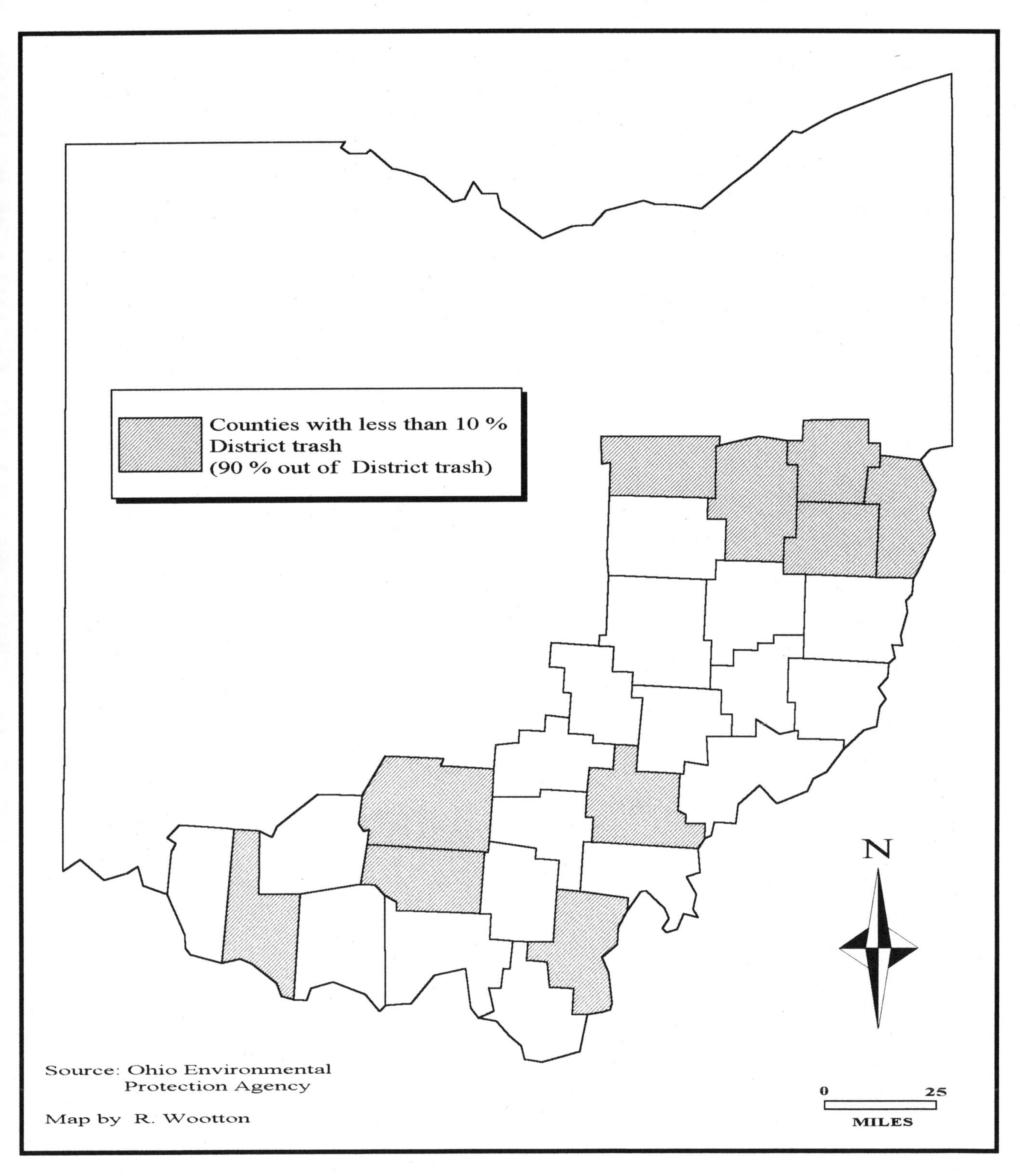

Fig. 16.7. Landfill Demand, 1990

with general industrial expansion, and later with environmental regulations, especially air pollution zoning and toxic cleanup liability. Tracing the overall pattern, Figure 16.6 displays the aggregated workforce in two major industrial categories, manufacturing and wholesale-retail, showing expansion in both groups. Although this is typical of many sectors, it is most apparent in the automobile industry, a result of the interplay of several additional factors.

Although the automobile industry has remained Ohio's largest manufacturing sector in terms of employment and value of production, since the 1960s the industry has declined in overall employment. During the 1970s and 1980s, nearly 40 percent of auto industry jobs were lost; however, these losses were not distributed evenly across the state (Hartzell 1991): specific locales suffered a complete loss of such employment, while other communities experienced job growth in the industry. Hartzell found that the more complex production has remained in the traditional centers of the industry, while less complex production has shifted southward and to rural locations.

These shifts in population and industrial production themselves create new problems for the southeast. In addition to new populations, land uses, manufacturing relationships, and service arrangements, the rest of Ohio is having yet another impact on the area. Despite air quality regulations and other environmental laws, the edge has become the disposal zone for the state. Such a conclusion is evidenced by the location of new power plants—both those constructed and proposed—since 1970. The same pattern appears in the solid waste disposal industry, which has established numerous sites in edge counties to house the new hazardous disposal and solid waste facilities. Depicting this postmodern phenomenon, Figure 16.7 shows the counties that had solid waste disposal facilities accepting 90 percent from outside of the county or the solid waste district.

Conclusion

The edge of the state is changing along with the rest of Ohio. After its initial importance, the area stagnated except where coal and other resources created localized, short-term booms. People left the land which made available reasonably priced space on the edge. An emerging growth spurt comes from expansion in the service sector, from industrial locations into greenfield locations, from home sites and recreation potential, and from low-cost open land available for solid waste and other disposal facilities. These seem to be conflicting uses, and along with these is residential distant suburbanization, made possible by telecommuting. These various facets of change suggest a different role for edge, or Appalachian, Ohio in the next century.

REFERENCES

Bain, Nancy R. 1984. Nonresident Land Ownership; A Potential Factor in Rural Change. *Land Use Policy* 1:64–71.

Garreau, Joel. 1991. *Edge City: Life on the New Frontier.* New York: Doubleday.

Hartzell, Eric S. 1991. Japanese Investment and the Automotive Parts Industry in Ohio. Honor's thesis, Honor's Tutorial College, Ohio University.

LaVelle, Abby A. 1984. Landscape Perceptions of the Wayne National Forest: A Comparative Analysis. Master's thesis, Geography Department, Ohio University.

Ohio Cooperative Extension. Community and Natural Resources. 1970. Rural Zoning 1970. Map.

———. 1988. Rural Zoning 1988. Map.

Ohio Department of Natural Resources. 1970. Oil and Gas Fields Map of Ohio. Map.

———. 1976. *Coal Production in Ohio 1800–1974.* Information Circular No. 44. Comp. Horace R. Collins. Columbus, Ohio.

———. N.d. Glacial Deposits of Ohio. Map.

———. N.d. Original Vegetation at the Time of the Earliest Surveys. Map.

———. N.d. Physiographic Sections of Ohio. Map.

Ohio Environmental Protection Agency. Division of Solid and Hazardous Waste Management. 1991. Ohio Solid Waste Facility Data Report—1990.

Ohio State Inspector of Mines. 1875–1914. *Annual Report.*

Rubenstein, J. 1989. The Impact of Japanese Investment in the United States. In *Motor of Change.* Ed. C. M. Law. London: Routledge.

Sitterley, John H. 1976. *Land Use in Ohio 1900–1970: How and Why It Has Changed.* Research Bulletin 1084. Ohio Agricultural Research and Development Center.

Sternlieb, George, and James W. Hughes, eds. 1988. "Prologue." In *America's New Market Geography: Nation, Region, and Metropolis.* Rutgers, N.J.: Center for Urban Policy Research.

Wilhelm, Hubert G. H. 1982. *The Origin and Distribution of Settlement Groups: Ohio 1850.* Athens, Ohio: Ohio University Geography Department.

CONCLUSION

A State of Change

Leonard Peacefull

As Ohio prepares to leave the twentieth century and enter a new millennium, it is poignant to reflect on the health of a state that has experienced a great deal of change in the present century and to shed some light on the prospects for the coming century. Once the preeminent manufacturing-industrial heart of America, Ohio has undergone more far-reaching changes in the 1900s than probably any other state in the nation. Some states have seen rapid socioeconomic progress in the past fifty years; others have seen their economies relentlessly decline. In still other states, new resources have been developed that have led to an economic boom, though in some cases this has been followed by a rapid downturn or bust. Few have endured a more deep-seated diversity of development in their economy as the Buckeye State.

At the beginning of the century, Ohio had a predominantly agriculturally led economy, producing both food and the machinery to carry out food production. This was soon surpassed by nonagriculture-related manufacturing. Companies that for decades had based their income on farming were forced to either change or go out of business. Witness the Hoover Company in New Berlin (now North Canton): At the turn of the century this company was manufacturing harness equipment but before long diversified into vacuum cleaning machines so as to use the surplus leather it had accumulated. Steel became the material of choice for manufacturing processes and the assembly line the procedure for production. Steel mills and large factories soon dominated the landscape.

As the century wore on, Ohio became the engine room of the nation's economy, building machines that produced the goods that in turn created the wealth of the country. From basic means within the new economy, companies emerged that grew to be not only national leaders but world leaders in their particular markets; some of these world leaders—NCR, Procter and Gamble, Timken—grew into multinational organizations. Other companies developed into worldwide organizations, only to be taken over by larger, foreign firms, as happened with Hoover and General Tire. In later years the state has seen an even larger shift in its economy, as its manufacturing base has been replaced by service industries.

The move from agriculture to manufacturing was an internally driven progression. On the other hand, the switch from manufacturing to service industries has mostly been influenced by events outside of the state. One such influential change was the oil crisis of the late 1970s, which led to the recession of the early 1980s. A general rise in prices caused by the large hikes in gasoline prices added to the need, already pressing, for a transformation in the structure of Ohio's economic base. A contributor to the demise of industry in many Ohio towns, this inevitable diversification away from manufacturing was forced on the state by global economic pressures. Competition from developing and newly industrialized countries, where cheap nonunionized labor could keep production costs down, forced prices so low that Ohio's industries could not compete. The availability of this cheap, overseas labor enticed manufacturers to vacate their Ohio real estate for greenfield sites in other lands, where they would be welcomed with open arms and lucrative tax breaks.

In the 1980s two further factors combined to influence changes in manufacturing in Ohio: the loss of jobs to the Sun Belt states and the decreasing power of labor unions. The former situation, which has mainly run its course, has been overtaken by other factors. And while the loss of bargaining power has weakened the position of unions to negotiate high wages for their members, this weakening of unions may have strengthened manufacturing's competitive position, as wages across the nation are leveling out, making Ohio's jobs as competitive as other parts of the United States. Although union workers may not receive wages as high as before, they will have work, which is good for the state's economy.

Now we move toward the end of the millennium; in ten years Ohio celebrates its bicentennial. What will the geography of the state be in the early part of the new millennium? In the mid 1990s two major international trade agreements are coming into force that could seriously influence the economy of the state in different ways. First, the North American Free Trade Agreement (NAFTA) will seriously affect the way goods and services are traded among the three signatory countries of Canada, the United States, and Mexico. Skeptics foresee that this agreement could end major manufacturing in the state. They predict that companies will move to Mexico where labor is cheap, in order to cut production costs. Goods manufactured will then be imported back into the country with little or no import tariffs but sold at the same price as if they had been manufactured in the United States. This could only happen if the cost of raw materials was kept at a low level and the inherent rise in transportation cost offset the other gains made. The agreement also means that there will be no tariffs on Ohio-made goods exported to our neighbors—a fact that could make Ohio goods less expensive, especially to customers north of the border. In the first year of NAFTA, 1994, America's trade with Canada and Mexico expanded at twice the rate of trade with non-NAFTA countries in 1993 (Economist 1994a); and Ohio shared proportionately in this prosperity.

In reality, NAFTA had been preceded by an even more controversial agreement between the United States and Canada. In 1988, the United States–Canada Free Trade Agreement (FTA) was effected. For Canada its aim was to secure greater access to U.S. markets; for the United States it resolved nagging problems in trade and investment in the auto and energy sectors. Many believed that this trade agreement would take jobs away from states such as Ohio and transfer manufacturing to Canada. Some predicted that the automobile industry would suffer greatly. Yet, as forecast by promoters of the agreement, this industry is the one manufacturing sector that has seen a large upturn in its fortunes, mostly because companies have been able to regionalize their production on either side of the border. Ohio lost few jobs to Canada because companies were aware that the costs of production, especially labor, north of the border would not be less. In fact when the socioeconomic costs were taken into account, some companies felt they were better off staying put. (For further information on the FTA, see Schott and Smith 1988 and McKee 1988.)

The General Agreement on Tariffs and Trade (GATT), ratified by Congress in December 1994, is the second of the two international agreements that factor into the future development of Ohio. It is early in the implementation of this long-fought trade agreement, but economists estimate that by 2005 the world will be $500 billion better off (Economist 1994b). Ohio should see a gain of at least 2 percent of its gross domestic profit, bringing further changes to the economic and social geography of the state.

These two trade agreements, NAFTA and GATT, will certainly have an impact on the economic structure of the state, affecting where business is carried out, the numbers of jobs created or lost, and, consequently, where people live. Changes in population will further impact local economies and so influence further changes in the geography. We see this clearly in the new structures of the industrial economy that are appearing today. Although most of the smokestack industries are gone, their headquarters remain to service the needs of multinational companies, such as the Goodyear Corporation, that underpin Ohio's economy.

Tertiary industries are now the major employers in Ohio, and one of the largest of these employers today is the health care industry. Often considered as merely a service to the community, this industry employs the largest number of people in the state, thus contributing greatly to the economy. The Cleveland Clinic, for example, is the largest employer in the city of Cleveland, yet it is not the only medical facility in the city. In the future, greater reliance will be placed on this industry. The state's inhabitants are living longer and thus will require greater medical attention in their later years. Ohioans along with the rest of the nation are becoming more health conscious, visiting their health professionals more regularly, and these professionals are putting greater emphasis on preventative medicine. In the next century this industry could be more important to Ohio than any other industry in the state's history.

Another industry of growing importance in the state is banking. Bank One, a Columbus-based commercial banking organization, has been a leader in the move to interstate banking. In 1993, Key Corp, a New York–based multistate banking group, joined forces with Cleveland's Society Bank to create one of the largest banking organizations in the nation. The biggest benefit to the state from this merger is the move of the new corporation's headquarters from Albany to Cleveland.

With the 1994 elections, change came to Ohio's government. The new state government, now under Republican rule, is set to take on a major downsizing of state government. This could dramatically impact many aspects of life that Ohioans have taken for granted over the years. Programs likely to be affected include Medicare, welfare, prisons, schools, and higher education. Reductions in funding in all of these programs will impact the state, but it is likely that cuts in education, particularly higher education, will have serious economic consequences in the future. Without an educated workforce, Ohio will find it difficult to attract jobs requiring skilled labor.

Changes in the industrial base of the state have led to the demise of the downtown in many of our large cities. Once thriving and bustling, city centers now lie vacant, like decaying, lifeless shells. As one observer of Toledo's Riverfront development wryly noted, you can tell the state of the economy of a downtown area when one can park all day for "a buck fifty." The larger cities of Cleveland, Cincinnati, and Columbus have managed to survive and flourish, maintaining high expectations for their downtown areas, only because they are major centers in the nation. Meanwhile, the CBD's of second tier cities have had mixed results with redevelopment plans. Toledo, for example, had no national importance to support the city center once its economic base started to decline. The Riverfront development, seen as a desperate attempt to revive the city's economy, has folded. A similar fate befell Akron, Youngstown, and Dayton in the 1970s, but late-1980s attempts have brought some prosperity to the downtowns of Akron and Dayton. All the while these cities continue to grow in size and population away from the downtown.

Such is the growth of the suburban fringe that we now consider metropolitan sprawl as part of the greater city. The city of Columbus has engulfed much of Franklin County. Greater Cleveland has spread into Medina and Summit Counties, and the city's influence is felt in Lorain, Lake, and Geauga Counties. Similarly, Greater Cincinnati has spread its influence across the Ohio River into Kentucky.

In the rural areas change is also occurring, perhaps at a more rapid pace than some would care for. The cities continue to spread out, as urban development soaks up farmland. Even the conservative Amish are leaving their enclaves in Geauga and Stark Counties for new homes in Kentucky. For the future it is clear that the continued movement of Ohioans from the dense urban areas of the northeast and southwest will continue. The consequence of this exodus is that counties that were once predominantly rural will be swallowed into the urban expansion now taking place.

On the transport front there are several developments under way that could impact the future of transportation in the state. The federal Intermodal Surface Transportation and Efficiency Act (ISTEA) will radically change transportation, both in the state and throughout the country. This act, which requires cities and states to incorporate nontransportation considerations into their transport planning (Plous 1993), attempts to redirect the focus of transportation policy and planning away from the private automobile and toward more public transport, bikeways, and pedestrian opportunities. This goal has been encapsulated in Governor Voinovich's Access Ohio effort, which reexamines every aspect of the state's transportation needs.

Moving away from road-building programs, there are also plans to build a high-speed rail system from Cleveland to Cincinnati via Columbus. These plans have been voiced, discussed, rejected, and resurrected for more than a decade, but something may come of this idea in the next decade. New emphasis has been put on this project, especially since the ISTEA gave a boost to high-speed rail when it authorized $30 million for grade-crossing improvements across the nation. This had led the railroad's supporters to look for more assistance from the federal government and for monies from the state government in Columbus. There are many instances around the world of successful high-speed rail projects, both in Europe and Japan. In France, for example, the journey between

Paris and Lyon is quicker by rail than by air. The competition has forced airfares down and, with the extension of the line to Marseilles, has strengthened ties between France's major cities. A similar linkage could be established in Ohio. If there is action on this project soon, the rail link could be up and running by the middle of the next decade. As a consequence, this linkage will further the development of a power corridor through the three cities that will have serious negative repercussions in other parts of the state.

The proposed building of a new railroad that would use electricity rather than diesel fuel is one of the environmental benefits brought about by changes to Ohio's economic structure. The demise of the steelworks caused much hardship to communities in which they were situated, but their pollution of the atmosphere will not be missed. The growing tertiary industries have far less impact on the environment than manufacturing. But as the population increases, so too will the need for landfill sites for domestic refuse. There remain a number of sites on the federal Super Fund list that await cleanup and probably will still be awaiting cleanup a century hence. What are the prospects for the new toxic waste incinerator in East Liverpool? While it may bring jobs and finance to the state, it certainly will not improve the public health. Ohioans are becoming more environmentally conscious and have made great strides to clean up their state. The water system is less polluted, most notably Lake Erie—once considered the lake that died. But what of our air? With more cars and trucks on the road each year, our urban areas are becoming more smog laden than ever.

It is hoped that some of the initiatives discussed earlier will help clean up the atmosphere. The best form of environmental control is awareness on the part of every Ohioan of the dangers that change can make to the environment. Both farmers who use large amounts of phosphates on their soils and homeowners who spray chemicals on their lawns are adding harmful substances to the soils and the water table. Is it really necessary to go to such extremes for an extra bushel of corn or a greener front lawn? Fortunately, in the greener years of the late-twentieth century, farmers are cutting down on fertilizers and homeowners are using less toxic methods of lawn manicure. The environment is the big issue and will continue to be so in the next century. Awareness of our part in the environment is an important element of our education.

This book recognizes the process of change that has enveloped the state throughout the twentieth century. In seeking to illustrate, highlight, and comment on the events that are shaping the future of the state, we have charted a course through Ohio's beginnings up to the present day and have glanced into the new millennium. Change is taking place so rapidly that one cannot keep up with the process. It can be said that a society evolves at the speed of its fastest means of communication. In the days of the Conestoga wagon our nation progressed, but at a slower pace than today. We are living today amid the communications superhighway—the internet. Ohio must take an active part in that great thrust forward and not hang back, waiting for something to happen.

REFERENCES

Economist. 1994a. Happy Ever NAFTA? *The Economist.* 10 December.

Economist. 1994b. Dancing round GATT. *The Economist.* 26 November.

McKee, David. 1988. *Canadian-American Economic Relations.* New York: Praeger.

Plous, F. K. 1993. Refreshing ISTEA. *Planning.* February.

Schott, Jeffrey, and Murray Smith. 1988. *The Canada-United States Free Trade Agreement: The Global Impact.* Washington, D.C.: Institute for International Economics.

Author's Note: Thanks are due to Thomas Maraffa, Henry Moon, Allen Noble, and David Stevens for their input on this chapter.

Contributors

Nancy R. Bain is professor in the Department of Geography at Ohio University. She earned her Ph.D. in Geography from the University of Minnesota. Her interests are land use planning, environmental planning, impact analysis, and rural development. Her current project is a study of the 1880s impacts of an early coal and iron mining company.

Warren Colligan is a Planning and Market Research Consultant. He received his M.A. from the University of Akron. His research interests include the history and development of Akron.

Frank Costa is professor of Urban Studies and Geography at the University of Akron. He received his Ph.D. from the University of Wisconsin-Madison. His research interests include land use planning and land use regulation, comparative urban planning, and urbanization. He has written *Public Planning in the Netherlands* and *Urbanization in Asia.*

Timothy D. Gerber is Soil Survey Coordinator for the Ohio Department of Natural Resources, Division of Soil and Water Conservation. He received his M.S. from The Ohio State University. His interest is soil inventory. He coauthored *Soil Survey of Carrol County, Ohio,* USDA, 1983.

Jeffrey J. Gordon is associate professor of Geography at Bowling Green State University. He received his Ph.D. from Syracuse University. His research interests are landscape perception, geography of popular culture, and geography of Native Americans. He has written extensively on pedagogy, cartography, and the geography of flea markets, and is a pioneer in the field of the geography of music. His latest Native American research is a book chapter on the Battle of Fallen Timbers.

Henry L. Hunker is professor of Geography and Public Policy and Management at The Ohio State University. He received his Ph.D. from The Ohio State University. His research interests include the geography of Ohio, Ohio's demographic trends, Ohio's economy, Columbus, urban geography, environmental issues, and resources.

Richard W. Janson served as a two-term president (1993–95) of the Ohio Academy of Science. He holds the office of chairman of Ohio's Thomas Edison Program and councillor of the American Geographical Society and is an adjunct professor of Geography at Kent State University. He and his brother Ray have owned and operated The Janson Industries, a manufacturing company in Canton, for 35 years. Dr. Janson received his undergraduate education at Denison University, followed by graduate work at Duke University, University of Chicago, and Kent State University (where he earned his Ph.D. in Geography). His principal areas of research are energy, economic development, and manufacturing.

Richard Lewis is assistant professor of Geography and undergraduate coordinator at Kent State University. His research interests include historical geography, urbanization, and the geography of the United States.

Thomas A. Maraffa is associate professor and chair of the Geography Department at Youngstown State University. He received his Ph.D. from The Ohio State University. His research interests include transportation, economic geography, and location analysis. He has published on a variety of topics including air transport, urban commuting, and manufacturing and store location.

Harry L. Margulis is associate professor at First College, Cleveland State University. He received his Ph.D. from Rutgers University. His research interests are housing policy, urban revitalization, multiculturalism, and ethnicity. He has recently published articles in *Housing Studies,* the *Journal of Architecture Education, Urban Geography,* and *Urban Affairs Quarterly.*

Mary-Ellen Mazey is chair of the Department of Urban Affairs and Director at the Center for Urban and Public Affairs at Wright State University. She received her Ph.D. from the University of Cincinnati. Her research interest is primarily urban and regional planning. She created the Center for Urban and Public Affairs at Wright State to focus upon applied research, training, and data base development. The majority of the center's work is oriented toward the Dayton-Springfield Metropolitan Area.

Henry Moon is professor of Geography and Planning and Dean of University College at the University of Toledo. He received his Ph.D. from the University of Kentucky. His research interests include transportation and land use. He is chair of the East Lakes Division of the Association of American Geographers. He has published in *Land Use Policy, Rural Development Perspectives, Transportation,* and *The Professional Geographer.* He has a recent book entitled *The Interstate Highway System,* published by the Association of American Geographers.

William A. Muraco is professor and chair for the Department of Geography and Planning at the University of Toledo. He received his Ph.D. from The Ohio State University. His interests are economic geography, health service location, and urban geography. His publications include a variety of research articles and monographs associated with economic geography, technology diffusion, health service delivery, and public facility location issues.

Thomas L. Nash is a professor emeritus of Geography at the University of Akron, and National Park Ranger, United States Department of the Interior, Cuyahoga National Recreation Area. He received his Ph.D. from Kent State University. His research interests include cartography, geographic information systems, and planning and economic development. He is working on GPS and GIS in the CUNRA and other parks in the National Parks Service. He has authored many works particularly on soil and remote sensing.

Allen G. Noble is professor and former head of the Department of Geography at the University of Akron. A specialist in cultural geography, he has published widely and coauthored *Ohio: An American Heartland.* He has been a visiting professor at Laurentian University, the University of Idaho, the University of Peredeniya (Sri Lanka), and the University of Tel Aviv (Israel). He is a former Fulbright Scholar and in 1988 received the Association of American Geographers Honors Award.

Leonard Peacefull is an independent consultant and lives in Cheshire, England, where he formerly served as subject leader of Geography at Manchester Metropolitan University. He received his Ph.D. from Kent State University and taught for nine years in Ohio. He is a Fellow of the Royal Geographic Society. His interests are in regional economics, resources, environmental impacts, and geographic systems. His publications include work on the proposed Nova Scotian Tidal Power Facility, dynamic systems in geographic research, and quantitative analysis. In 1990 he edited *The Changing Heartland: A Geography of Ohio.*

Thomas W. Schmidlin is associate professor and chair in the Department of Geography at Kent State University. He received his Ph.D. from Cornell University. His research interests are climates of cold regions, permafrost, and tornado hazards. He has published in *The Journal of Applied Meteorology, Cold Region Science and Technology, Physical Geography, The Ohio Journal of Science,* and *The Climatological Bulletin.* His book *Thunder in the Heartland: A Chronicle of Outstanding Weather Events in Ohio* is published by the Kent State University Press (1996).

Richard V. Smith is professor of Geography at Miami University. He received his Ph.D. from Northwestern University. His interests include recreational user and resource issues and problems, delimiting and structure of tourism regions, tourism and recreation planning, community information, and population/resource relationships. His professional contributions include numerous articles, papers, and book chapters, and he has served as a consultant to the Ohio Department of Natural Resources. He is a recipient of the Roy I. Wolfe Award from the Association of American Geographers' Recreation, Tourism, and Sport Specialty Group.

Howard A. Stafford is professor, Department of Geography, at the University of Cincinnati. He received his Ph.D. from the University of Iowa. His interests include industrial location and regional and economic development. He has published *Principles of Industrial Facility Location* (1980) and *Manufacturing Unit of the High School Geography Project* (1966), as well as numerous articles in professional journals.

David T. Stephens is professor of Geography at Youngstown State University. He received his Ph.D. from the University of Nebraska. He has taught a course in Ohio geography for the last twenty years. His research interests include cultural and historical geography.

Hubert G. H. Wilhelm is professor of Geography at Ohio University. He received his Ph.D. from Louisiana State University. His interests are cultural geography, rural settlement, migration, folk architecture, and landscape interpretation. He has recently produced three documentary videos dealing with Ohio's settlement landscape.

Index